Advances in Dam-Break Modeling for Flood Hazard Mitigation: Theory, Numerical Models, and Applications in Hydraulic Engineering

Advances in Dam-Break Modeling for Flood Hazard Mitigation: Theory, Numerical Models, and Applications in Hydraulic Engineering

Editors

Francesca Aureli
Andrea Maranzoni
Gabriella Petaccia

MDPI

Basel • Beijing • Wuhan • Barcelona • Belgrade • Novi Sad • Cluj • Manchester

Editors

Francesca Aureli
Department of Engineering and Architecture
University of Parma
Parma
Italy

Andrea Maranzoni
Department of Engineering and Architecture
University of Parma
Parma
Italy

Gabriella Petaccia
Department of Civil Engineering and Architecture
University of Pavia
Pavia
Italy

Editorial Office
MDPI
St. Alban-Anlage 66
4052 Basel, Switzerland

This is a reprint of articles from the Special Issue published online in the open access journal *Water* (ISSN 2073-4441) (available at: www.mdpi.com/journal/water/special_issues/dam_model_flood_hydraulic).

For citation purposes, cite each article independently as indicated on the article page online and as indicated below:

Lastname, A.A.; Lastname, B.B. Article Title. *Journal Name* **Year**, *Volume Number*, Page Range.

ISBN 978-3-7258-1152-6 (Hbk)
ISBN 978-3-7258-1151-9 (PDF)
doi.org/10.3390/books978-3-7258-1151-9

Cover image courtesy of Andrea Maranzoni
Remains of the Gleno dam (Scalve Valley in the province of Bergamo, Italy) which collapsed on 1 December 1923.

Contents

About the Editors . . . vii

Preface . . . ix

Francesca Aureli, Andrea Maranzoni and Gabriella Petaccia
Advances in Dam-Break Modeling for Flood Hazard Mitigation: Theory, Numerical Models, and Applications in Hydraulic Engineering
Reprinted from: *Water* **2024**, *16*, 1093, doi:10.3390/w16081093 . . . 1

Francesca Aureli, Andrea Maranzoni and Gabriella Petaccia
Review of Historical Dam-Break Events and Laboratory Tests on Real Topography for the Validation of Numerical Models
Reprinted from: *Water* **2021**, *13*, 1968, doi:10.3390/w13141968 . . . 18

Francesca Aureli, Andrea Maranzoni, Gabriella Petaccia and Sandra Soares-Frazão
Review of Experimental Investigations of Dam-Break Flows over Fixed Bottom
Reprinted from: *Water* **2023**, *15*, 1229, doi:10.3390/w15061229 . . . 39

Andrea Maranzoni and Massimo Tomirotti
Three-Dimensional Numerical Modelling of Real-Field Dam-Break Flows: Review and Recent Advances
Reprinted from: *Water* **2023**, *15*, 3130, doi:10.3390/w15173130 . . . 90

Jaromír Říha, Stanislav Kotaška and Lubomír Petrula
Dam Break Modeling in a Cascade of Small Earthen Dams: Case Study of the Čižina River in the Czech Republic
Reprinted from: *Water* **2020**, *12*, 2309, doi:10.3390/w12082309 . . . 116

Francisco N. Cantero-Chinchilla, Rafael J. Bergillos, Pedro Gamero, Oscar Castro-Orgaz, Luis Cea and Willi H. Hager
Vertically Averaged and Moment Equations for Dam-Break Wave Modeling: Shallow Water Hypotheses
Reprinted from: *Water* **2020**, *12*, 3232, doi:10.3390/w12113232 . . . 137

Sayed Masihullah Ahmadi and Yoshimichi Yamamoto
A New Dam-Break Outflow-Rate Concept and Its Installation to a Hydro-Morphodynamics Simulation Model Based on FDM (An Example on Amagase Dam of Japan)
Reprinted from: *Water* **2021**, *13*, 1759, doi:10.3390/w13131759 . . . 158

Aissam Gaagai, Hani Amir Aouissi, Andrey E. Krauklis, Juris Burlakovs, Ali Athamena and Ivar Zekker et al.
Modeling and Risk Analysis of Dam-Break Flooding in a Semi-Arid Montane Watershed: A Case Study of the Yabous Dam, Northeastern Algeria
Reprinted from: *Water* **2022**, *14*, 767, doi:10.3390/w14050767 . . . 180

Diego Bello, Hernán Alcayaga, Diego Caamaño and Alonso Pizarro
Influence of Dam Breach Parameter Statistical Definition on Resulting Rupture Maximum Discharge
Reprinted from: *Water* **2022**, *14*, 1776, doi:10.3390/w14111776 . . . 210

Malena Melo and Julian Eleutério
Probabilistic Analysis of Floods from Tailings Dam Failures: A Method to Analyze the Impact of Rheological Parameters on the HEC-RAS Bingham and Herschel-Bulkley Models
Reprinted from: *Water* **2023**, *15*, 2866, doi:10.3390/w15162866 . . . 223

Zijian Xue, Ling Zhou and Deyou Liu
Accurate Numerical Modeling for 1D Open-Channel Flow with Varying Topography
Reprinted from: *Water* **2023**, *15*, 2893, doi:10.3390/w15162893 . . . . **255**

Sergio Martínez-Aranda, Javier Fernández-Pato and Pilar García-Navarro
Non-Equilibrium Bedload Transport Model Applied to Erosive Overtopping Dambreach
Reprinted from: *Water* **2023**, *15*, 3094, doi:10.3390/w15173094 . . . . **271**

André Felipe Rocha Silva and Julian Cardoso Eleutério
Effectiveness of a Dam-Breach Flood Alert in Mitigating Life Losses: A Spatiotemporal Sectorisation Analysis in a High-Density Urban Area in Brazil
Reprinted from: *Water* **2023**, *15*, 3433, doi:10.3390/w15193433 . . . . **290**

Cagri Akgun, Salim Serkan Nas and Akin Uslu
2D and 3D Numerical Simulation of Dam-Break Flooding: A Case Study of the Tuzluca Dam, Turkey
Reprinted from: *Water* **2023**, *15*, 3622, doi:10.3390/w15203622 . . . . **308**

About the Editors

Francesca Aureli

Francesca Aureli received her Ph.D. in Hydraulic Engineering at Politecnico di Milano, Italy in January 2000. Her thesis focused on the physical and mathematical modeling of dam-break flows. She then continued her research activity at the University of Parma, holding the role of assistant professor from 2005 to 2018. Since 2019, she has been an associate professor of Hydraulic and Maritime Constructions and Hydrology at the Department of Engineering and Architecture of the University of Parma. Her main research interests are the physical and numerical modeling of unsteady free surface flows, the assessment of flood hazards due to dam or levee breaching, the application of imaging techniques for the acquisition of data related to free-surface flow phenomena, the definition of Synthetic Design Hydrographs for river flood applications, and the numerical and physical modelling of small hydrokinetic rotors. She is a co-author of several scientific works regarding the physical and numerical modeling of dam-break phenomena.

Andrea Maranzoni

Andrea Maranzoni received his Ph.D. degree from the University of Parma, Italy in March 2004. His Ph.D. dissertation focused on mathematical and physical modelling of dam-break flows. He has worked as an assistant professor at the University of Brescia, Italy (from 2005 to 2008) and the University of Parma (from 2009 to 2018). He is currently an associate professor of Hydraulics at the Department of Engineering and Architecture of the University of Parma. His main research interests include numerical and physical modelling of unsteady free surface flows; assessment of flood hazard due to levee or dam breaching; imaging techniques for free-surface measurement; mathematical modelling of mixed flows; seismic-induced flow motions; the performance of side weirs; and spatially varied flows. He has authored numerous journal and conference papers on dam-break modelling.

Gabriella Petaccia

Gabriella Petaccia received her Ph.D. in Hydraulic Engineering from the University of Pavia, Italy in February 2003. Her thesis focused on shock waves propagation in natural valleys due to barrage breaking. She continued her research activity at the University of Pavia, holding the role of assistant professor from 2007 to 2019. Since 2019, she has been an associate professor of Hydraulic and Maritime Constructions and Hydrology at the Department of Civil Engineering and Architecture of the University of Pavia. Her main research interests are the development of flood propagation models and their applications to real-world cases; mapping of floodable areas, development and the application of methodologies for flood simulation through hydraulic structures (bridges); development of a coupled 1D and 2D flood model; development of GPU techniques to perform real-time flood propagation simulations; experimental modeling of dam spillways; development of a 2D model for wood debris transport during floods; and lag time estimation through overland flow models. She is a co-author of several scientific works regarding the physical and numerical modeling of dam-break phenomena.

Preface

Dam-break modeling is an important research topic in hydraulic engineering, both in theory and practice. The collapse of a dam can cause a flood with catastrophic consequences for downstream areas, resulting in both human and economic losses. Moreover, the vulnerability of older dams to hydrological extreme events is increasing due to structural deterioration or inadequate spillway capacity. The exposure of floodable areas is also increasing due to urban development. Flood hazard assessment, identifying the elements at risk, and evaluating their vulnerability are fundamental steps of dam-break flood risk analysis for planning mitigation measures. Accurate simulation of dam-break flows on real-world topographies and mapping the related flood hazard require robust and efficient numerical models. To validate and improve these models, the availability of information about documented historical dam-break events and physical model data collected during experimental investigations are essential.

This reprint collects the articles published in the Special Issue on "Advances in Dam-Break Modeling for Flood Hazard Mitigation: Theory, Numerical Models, and Applications in Hydraulic Engineering". It includes original contributions, review papers, and discussions, which offer recent advancements and new insights into the research topic of dam-break modelling. The main objectives of the Special Issue were to (a) share recent advancements in the theoretical analysis of dam-break flow; (b) present new numerical schemes and techniques addressing the numerical complexities involved in the simulation of dam-break flows on real-world topographies; (c) introduce new laboratory experiments and test cases useful for the validation of numerical models; and (d) showcase practical applications of dam-break numerical models to real-field case studies for flood hazard and risk assessments. The considerable number of contributions received demonstrates the vitality of the academic community actively engaged in the field of dam-break modelling.

We hope this reprint can be a valuable resource for dam-break modelers, practitioners, and researchers and that it can stimulate further research on this classic and fascinating research topic, which holds great relevance in hydraulic engineering. We express our gratitude to the publisher for allowing us to serve as guest editors of this Special Issue. We appreciate the publisher's belief in the success of this editorial initiative and we are grateful for the constant assistance and support provided to us throughout this three-year experience.

Finally, we would like to thank all the authors contributing to the Special Issue, as well as all the reviewers, academic editors, and editorial assistants involved in the review processes, for their valuable work, which significantly contributed to ensuring the quality of this collection.

Francesca Aureli, Andrea Maranzoni, and Gabriella Petaccia
Editors

water

MDPI

Editorial

Advances in Dam-Break Modeling for Flood Hazard Mitigation: Theory, Numerical Models, and Applications in Hydraulic Engineering

Francesca Aureli [1], Andrea Maranzoni [1,*] and Gabriella Petaccia [2]

[1] Department of Engineering and Architecture, University of Parma, Parco Area delle Scienze 181/A, 43124 Parma, Italy; francesca.aureli@unipr.it

[2] Department of Civil Engineering and Architecture, University of Pavia, Via Ferrata 3, 27100 Pavia, Italy; petaccia@unipv.it

* Correspondence: andrea.maranzoni@unipr.it

1. Introduction

Despite significant advancements being made in recent decades (e.g., [1–3]), dam-break modeling is still an active field of theoretical and applied research, which is of great interest to hydraulic engineers. Flooding induced by the collapse of a dam has potentially catastrophic consequences in downstream areas, both in terms of human and economic losses [4]. In addition, the vulnerability of older dams to extreme hydrological events is increasing due to structural deterioration or inadequate spillway capacity [5]. Similarly, the exposure of urban areas to flooding is globally increasing as a consequence of urban development [6].

In this context, robust and efficient numerical models which enable the accurate simulation of dam-break flows on real-world topography are effective and useful tools for the assessment of flood hazard and risk associated with dam-break events (e.g., [7]). In fact, flood hazard assessment is a fundamental prerequisite to designing prevention and mitigation measures to reduce the number of people affected by water-related disasters and improve living conditions in urban areas. This objective fits perfectly with Goal 11 of the United Nations' 2030 Agenda, which aims to make cities and communities safe, resilient, and sustainable [8].

Experience and knowledge based on historical dam-break events [4], along with the availability of physical model data [3], allow modeling tools to be validated, constantly improved, and used with confidence in dam-break flood hazard and risk mapping activities. Robust and reliable climate and hydrological models also represent a strategic predictive resource in the definition of future hydrological scenarios under climate change [9].

In light of these premises, the Special Issue entitled "Advances in Dam-Break Modeling for Flood Hazard Mitigation: Theory, Numerical Models, and Applications in Hydraulic Engineering" aims to collect contributions regarding (a) recent advancements in the theoretical analysis of dam-break flows; (b) recent advancements in numerical schemes and techniques used to model dam-break flows on real-world topography; (c) new laboratory experiments and test cases for the validation of numerical models; and (d) practical applications of dam-break numerical models to real-field cases for flood hazard and risk assessments.

Recent contributions in the literature on these themes are analyzed and discussed in this editorial, highlighting the topics on which scientific research on dam-break modeling is currently most focused. Unfortunately, some of these topics were not addressed by the contributions presented in the Special Issue. A final comment is made on this fact.

Citation: Aureli, F.; Maranzoni, A.; Petaccia, G. Advances in Dam-Break Modeling for Flood Hazard Mitigation: Theory, Numerical Models, and Applications in Hydraulic Engineering. *Water* **2024**, *16*, 1093. https://doi.org/10.3390/w16081093

Received: 18 March 2024
Accepted: 20 March 2024
Published: 11 April 2024

2. Overview of the Papers Included in the Special Issue

The contributions to the Special Issue are listed and summarized in Table 1. They are also briefly presented in the following four subsections, each of which covers one main theme addressed in the Special Issue.

Table 1. Summary of the contributions to the Special Issue (https://www.mdpi.com/journal/water/special_issues/dam_model_flood_hydraulic, accessed on 16 February 2024).

(1) Contribution Number	(2) Contribution Reference	(3) Contribution Type	(4) Research Topic	(5) Research Focus	(6) Research Results
1	Aureli et al. (2021)	Review	Real-world dam-break modeling. Theme (d)	An overview of documents in the literature concerning historical dam-break events with real-field datasets available for the validation of numerical models; an overview of documents concerning investigations into dam-break flows on physical models reproducing real-world topographies	The analysis of the type and quality of the available data; the division of the retrieved cases into well-documented cases and cases with partial or inaccurate datasets; a help in identifying the test cases most suited to the modelers' needs; the achievement of improvements in data access; the identification of test cases worthy of future research
2	AlQasimi and Mahdi (2022)	Discussion	Real-world dam-break modeling. Theme (d)	A comment on Contribution No. 1 (Aureli et al., 2021) about the classification of the Lake Ha! Ha! dam test case	Due to its serious shortcomings potentially limiting its applicability, the Lake Ha! Ha! dam test case should be classified as a test case with a partial or inaccurate dataset
3	Aureli et al. (2022)	Reply	Real-world dam-break modeling. Theme (d)	A reply to the comment in Contribution No. 2 (AlQasimi and Mahdi, 2022) about the classification of the Lake Ha! Ha! dam test case	Due to the large and complete dataset available in digital format, the Lake Ha! Ha! dam test case could be considered as well documented
4	Aureli et al. (2023)	Review	Experimental dam-break modeling. Theme (c)	An overview of experimental investigations into dam-break flows over a non-erodible bottom for the validation of numerical models with a focus on fundamental dam-break wave physical characteristics; dam-break waves through geometric singularities; impacts against obstacles; dam-break propagation in idealized urban areas; tsunami bores; green water events; dam-break waves of non-Newtonian fluids; dam-breaks in cascade reservoirs; dike-break-induced flows on a lateral floodplain; the collapse of storage tanks and bunds or dike overtopping	The analysis of test conditions and measuring techniques; a help in identifying the most appropriate laboratory test for validation purposes; the achievement of improvements in experimental data access; the identification of physical aspects worthy of future experimental research
5	Maranzoni and Tomirotti (2023)	Review	Numerical dam-break modeling. Theme (b)	An overview of studies in the literature that use 3D hydrodynamic models for the simulation of large-scale dam-break flooding on irregular real-world topography	The analysis of governing equations, numerical methods, and the status of codes; the identification of recent advancements in numerical techniques, modeling accuracy, and computational efficiency; the identification of advantages and limitations of 3D dam-break models compared to 2D shallow-water ones; a help in choosing the most suitable numerical method for the application of interest; the identification of future numerical research directions

Table 1. *Cont.*

(1) Contribution Number	(2) Contribution Reference	(3) Contribution Type	(4) Research Topic	(5) Research Focus	(6) Research Results
6	Říha et al. (2020)	Original research	Numerical dam-break modeling. Themes (b) and (d)	A study focused on dam-breaks of small embankment dams in a cascade (overtopping and piping); the comparison of empirical formulae and 2D SWE simulations through HEC-RAS 2D; the attenuation of the peak discharge; the analysis of the case study of the Tetřeví Stream and Čižina River (Czech Republic)	Peak discharge attenuation is influenced by the flood volume, slope, and morphology of the floodplain and increases with the distance from the breached dam. Empirical formulae derived for a single dam break may underestimate the peak outflow by up to 10% in the case of a dam cascade
7	Cantero-Chinchilla et al. (2020)	Original research	Theoretical and numerical dam-break modeling. Themes (a) and (b)	A study focused on a VAM (vertically averaged and moment equation) model for the simulation of dam-break waves; the analysis of non-hydrostatic pressure effects; validation against available experimental data	Discussion of the role of the perturbation parameters in overcoming the limitations of the classic shallow water model; discussion of the complex and necessary interplay between the dynamic component of fluid pressure and the modeling of the velocity profile in producing accurate solutions for dam break flows. Acceptable agreement between numerical predictions and experimental data is shown
8	Ahmadi and Yamamoto (2021)	Original research	Theoretical and experimental dam-break modeling. Themes (a), (c), and (d)	A study focused on the outflow discharge equation for a sudden partial dam-break; verification based on the historical Malpasset dam-break event; application to the hypothetical collapse of the Amagase dam (Japan)	Discussion of the effect of the breach shape; the evaluation of a breach-shape coefficient via laboratory experiments; the integration of the outflow discharge equation in a 2D model. A simple and effective equation is proposed
9	Gaagai et al. (2022)	Original research	Numerical dam-break modeling. Theme (d)	A simulation of the floods resulting from the Yabous dam breach (Algeria) through HEC-RAS 1D; a sensitivity analysis on breach parameters (breach slope, width, and formation time)	The provision of flood inundation maps; the discussion of the effect of breach parameters on flooding variables and the effectiveness of the modeling tool for the simulation of the dam-break wave propagation
10	Bello et al. (2022)	Original research	Numerical dam-break modeling. Theme (d)	A statistical definition of the most influential dam breach parameters and assessment of their influence on the released maximum discharge	The most influential breach parameter on the maximum discharge is the breach formation time, followed by the final breach width and height. The statistical definition of breach parameters strongly affects the maximum discharge magnitude uncertainty, therefore influencing the estimation of the flood risk associated with the breaching, the delimitation of flooded areas, and the preparation of emergency action plans
11	Melo and Eleutério (2023)	Original research	Numerical dam-break modeling. Theme (d)	A study of the impact of the rheological variability and the non-Newtonian behavior of stored materials on the results of the modeling of flooding caused by mining tailings dam collapses	The variability of rheological parameters significantly affects flooded areas, maximum depths, and arrival times. The Bingham model may be applied as a first approximation as it leads to conservative results. The Herschel–Bulkley model proved to be more efficient for probabilistic analysis, with a smaller range of uncertainties and better probability distribution, improving the evaluation for risk management purposes

Table 1. *Cont.*

(1) Contribution Number	(2) Contribution Reference	(3) Contribution Type	(4) Research Topic	(5) Research Focus	(6) Research Results
12	Xue et al. (2023)	Original research	Numerical dam-break modeling. Themes (a) and (b)	A novel 1D model for the simulation of open-channel flows with variable width and topography; its application to rectangular channels with variable widths; verification against steady and unsteady flows with a reference solution; validation against experimental data	The implementation of new conservation and non-conservation terms allowed the authors to achieve satisfactory results in the simulation of selected test cases concerning dam-break and steady flows
13	Martínez-Aranda et al. (2023)	Original research	Numerical dam-break modeling. Theme (b)	A 2D finite-volume model of bedload sediment transport in non-equilibrium conditions during dike erosion or dam breach; its validation against experimental data of dike overtopping and dam-breach opening	A non-equilibrium bedload transport model is preferable compared to an equilibrium model in morphological dam-break modeling over a mobile bed
14	Silva and Eleutério (2023)	Original research	Dam-break flood risk assessment. Theme (d)	A study on the effect of the integration of early warning and evacuation systems in flood risk assessment in terms of loss of life; the analysis of the case study of the hypothetical Pampulha Lake dam breach (Brazil) using 2D hydrodynamic modeling coupled with an agent-based life loss model	The determination of an optimal time for warning issuance; the discussion of the crucial role of preparedness and the perception of mobilization for the minimization of the potential for loss of life; the discussion of a possible optimization strategy based on differencing alerts and actions in different sectors of the downstream area
15	Akgun et al. (2023)	Original research	Dam-break flood risk assessment. Theme (d)	A comparison of the 2D and 3D modeling of dam-breach flow with Flow3D software; validation against available experimental data; application to the case study of the hypothetical failure of Tuzluca dam (Turkey); the assessment of loss of life for different scenarios (overtopping and piping)	Discussion of the impact of breach geometry and failure time on expected casualties; 2D modeling is more practical and less time-consuming than 3D modeling, which, on the other hand, is more accurate and hence suitable for small-scale simulations

2.1. *Advancements in Theoretical Analysis of Dam-Break Flows*

Three papers address theme (a), providing valuable contributions to the theoretical analysis of dam-break flows.

Cantero-Chinchilla et al. (2020) [Contribution No. 7] proposed a new vertically averaged and moment equation (VAM) model to overcome the limitations of the classic shallow water equation (SWE) model in the simulation of dam-break waves over a wet bed. The importance of including the effects of non-hydrostatic pressure and the vertical flow velocity profile in the dam-break model was investigated and assessed by comparing numerical results of the VAM and SWE models with experimental data available in the literature.

Ahmadi and Yamamoto (2021) [Contribution No. 8] proposed a new, simple equation to predict dam-break outflow discharge in the case of a partial dam-break event. This equation can be used as an inflow boundary condition in the hydrodynamic modeling of dam-break wave propagation in downstream areas. Among the strengths of the proposed equation, there are few input parameters regarding the dam and the modeling of a partial breach through a percentage failure of the structure. The authors validated the equation against experimental data, investigated the effect of the breach shape, and, finally, showed the results obtained for selected real-field case studies.

Xue et al. (2023) [Contribution No. 12] proposed a one-dimensional (1D) model for the simulation of open-channel flows in rectangular channels with variable widths and non-flat bottoms. The shallow water momentum equation was rewritten in a convenient conservative form by introducing a pressure term in the numerical flux and a non-prismaticity term in the source term. Satisfactory results were obtained in validation and verification tests concerning dam-breaks and steady flows.

2.2. Advancements in Numerical Modeling of Dam-Break Flows on Real-World Topography

Regarding theme (b), Říha et al. (2020) [Contribution No. 6] used a simplified parametric model for piping and overtopping embankment dam erosion with parameter optimization based on a Monte-Carlo sampling procedure in order to predict the outflow discharge hydrograph. Standard 2D shallow-water modeling was adopted to simulate the dam-break wave propagation in a cascade dam-break flood.

Martínez-Aranda et al. (2023) [Contribution No. 13] applied a recently developed 2D finite-volume bedload transport model in equilibrium and non-equilibrium conditions to study dam breach and dyke erosion processes. The benchmarking of the non-equilibrium model based on dyke erosion and dam breach opening laboratory tests reproducing highly unsteady flows showed improved results compared with the classical equilibrium approach, despite the careful calibration of the exchange parameters required, especially in the regions where complex unsteady processes occur.

The papers by Cantero-Chinchilla et al. (2020) [Contribution No. 7] and Xue et al. (2023) [Contribution No. 12] also deal with relevant numerical issues in dam-break flow models. In particular, Cantero-Chinchilla et al. (2020) [Contribution No. 7] used a three-step semi-implicit hybrid finite-volume–finite-difference scheme for the numerical solution of the VAM equations. Xue et al. (2023) [Contribution No. 12] solved the 1D SWE model via a two-step upwind scheme based on the random choice method and the Harten, Lax, and van Leer (HLL) approximate Riemann solver.

Maranzoni and Tomirotti (2023) [Contribution No. 5] performed a comprehensive literature review inherent to theme (b) since it is focused on three-dimensional (3D) hydrodynamic models for the simulation of large-scale dam-break flows on irregular real-world topography. The authors discussed the advantages and limitations of the 3D models (in terms of complexity, accuracy, and computational efficiency) compared to conventional two-dimensional (2D) ones.

2.3. New Laboratory Experiments and Test Cases for the Validation of Numerical Models

Considering theme (c), Ahmadi and Yamamoto (2021) [Contribution No. 8] performed a new laboratory investigation to collect experimental data for the calibration of the shape parameter appearing in the partial dam-break outflow rate equation they proposed. This contribution is the only one published in the Special Issue that presents new experimental dam-break flow data.

An extensive review of the state-of-the-art experimental investigations into dam-break flows was performed by Aureli et al. (2023) [Contribution No. 4], limited to schematic test cases with fixed, non-erodible bottoms. This survey demonstrated the impressive number of experimental studies conducted in the last century which aimed to provide insight into the physical features of dam-break waves and collect experimental data for the validation of numerical models.

2.4. Applications of Dam-Break Numerical Models to Flood Hazard and Risk Assessment

Most contributions to the Special Issue fall under theme (d).

Říha et al. (2020) [Contribution No. 6] considered a case study concerning a cascade of three small reservoirs along the Tetřeví Stream and the Čižina River in the Moravian–Silesian Region, Czech Republic.

The dam-break outflow rate equation proposed by Ahmadi and Yamamoto (2021) [Contribution No. 8] was coupled with a two-dimensional flood simulation model applied to the Amagase dam (Kyoto Prefecture, Japan) dam-break case study. Consequently, this article also contributes to theme (d), supporting dam-break flood hazard assessment in highly populated areas.

Real-field dam-break case studies were also analyzed by Gaagai et al. (2022) [Contribution No. 9], Bello et al. (2022) [Contribution No. 10], Melo and Eleutério (2023) [Contribution No. 11], Silva and Eleutério (2023) [Contribution No. 14], and Akgun et al. (2023) [Contribution No. 15], who considered hypothetical dam-break events for selected dams in emerging countries. The commonality between all of these studies is that different dam-break scenarios were considered, and the effect of uncertainties in the model parameters was examined through sensitivity analyses to effectively delimitate the floodable areas and assess the risk that threatens the population exposed to dam-break flooding.

In detail, Gaagai et al. (2022) [Contribution No. 9] studied the flooding wave resulting from the hypothetical failure of the Yabous dam (northeastern Algeria). Bello et al. (2022) [Contribution No. 10] considered the dam-break of the Chacrillas dam (central Chile). Melo and Eleutério (2023) [Contribution No. 11] conducted a probabilistic analysis to investigate the sensitivity of tailings dam-break floods to relevant rheological parameters of non-Newtonian hydrodynamic models in a fictitious ICOLD (International Commission on Large Dams) dam-break case study. Based on the case study of the hypothetical failure of the Pampulha Lake dam (Minas Gerais State, Brazil), Silva and Eleutério (2023) [Contribution No. 14] investigated the effectiveness of specific alert systems in the mitigation of potential loss of life in densely populated urban areas. Finally, Akgun et al. (2023) [Contribution No. 15] analyzed different dam-break scenarios for the planned Tuzluca dam (Iğdır Province, Turkey), obtaining results in terms of dam-break flood risk and potential loss of life that are expected to guide the preparation of emergency action plans for the downstream region.

The review performed by Aureli et al. (2021) [Contribution No. 1] fits into theme (d) because it concerns historical dam-break events for which enough reliable information is available to set up real-field dam-break test cases, as well as datasets obtained by experimental dam-break studies on physical models reproducing real-world topography. A comment on this review (AlQasimi & Mahdi 2022; Contribution No. 2) and the subsequent reply (Aureli et al. 2022; Contribution No. 3) discuss the completeness of the available dataset for a specific historical dam-break event (the 1996 Lake Ha! Ha! breakout, Québec, Canada).

3. Recent Developments in Dam-Break Modeling: Gaps and Advancements

3.1. Recent Advancements in Theoretical Analysis of Dam-Break Flows

Dam-break modeling is usually performed through 1D (cross-sectional averaged) or 2D (depth-averaged) shallow water models. In particular, 2D depth-averaged models based on 2D shallow water equations are widely used to simulate the dynamics of a large variety of geophysical surface flows (e.g., [10,11]). However, in many hydraulic applications, such as the propagation of waves over obstacles, flows over spillways, and the motion of undular bores, the hydrostatic pressure assumption can no longer be considered valid [12–14]. Similarly, more complex non-hydrostatic models are required to accurately simulate coastal wave propagation phenomena, such as wave run-up and wave shoaling, due to the need for the reproduction of nearshore dynamics, preserving both nonlinear and dispersive effects [15,16]. In these kinds of flows, vertical acceleration plays a significant role, and a dynamic component affects the pressure distribution. Given the high computational cost of three-dimensional modeling, the development of enhanced non-hydrostatic depth-averaged models overcoming the limitations of the classical shallow water (hydrostatic) models is an active research topic in environmental and hydraulic engineering. For the sake of computational costs, several approaches based on the vertical averaging of the fully three-dimensional flow equations have mainly been investigated, such as the non-hydrostatic extension of shallow water models [17,18] and Boussinesq-type models [19,20], among which are the weighted residuals techniques proposed by Green and Naghdi [21] and widely used in the literature (e.g., [22]), which are precursors of the vertically averaged and moment equations model (VAM) [23]. Given their promising peculiarities, VAM models are currently being studied with the aim of improving numerical techniques and computational efficiency to effectively deal with non-hydrostatic simulations of free-surface flows [24], as well as in the presence of sediment transport [25]. Recent applications of these models also concern dam-break problems [26], as confirmed by Contribution No. 7 to this Special Issue.

Despite being computationally expensive, 3D models (based on the Navier–Stokes equations or the Reynolds-averaged Navier–Stokes equations coupled with a suitable free-surface-tracking algorithm) are increasingly used in dam-break modeling. Indeed, they overcome the limitations of depth-averaged models and offer a detailed and accurate prediction of flooding dynamics, inherently including the vertical acceleration of fluid and 3D effects due to the flow curvature ([27,28], and Maranzoni & Tomirotti 2023 [Contribution No. 5 to this Special Issue]). The current availability of powerful computational resources allows the efficient use of Computational Fluid Dynamics (CFD) in many hydrodynamic problems, including dam-break flow analysis, even when concerning large-scale applications on real-world topography (e.g., [29–32]). Despite this recent possibility, only one contribution to this Special Issue [Contribution No. 15] reports a 3D simulation (using commercial software) of real-field dam-break flooding resulting from the hypothetical failure of a real dam.

As observed by Maranzoni and Tomirotti (2023) [Contribution No. 5], coupled 2D–3D models can be a valid compromise between simulation accuracy and computational efficiency in dam-break analyses and offer a viable and modern alternative to conventional 2D and 3D hydrodynamic models. Such hybrid models are becoming popular in the hydrodynamic modeling of free-surface flows. According to this approach, hydrostatic shallow water equations are used in regions where vertical accelerations are negligible, while Boussinesq-type or Navier–Stokes equations are adopted in zones where the non-hydrostatic effects are significant [26]. Currently, some CFD software codes allow models to simulate large-scale shallow flows via the 2D depth-averaged model and near-field flows (near structures, obstacles, or topographic irregularities) via a 3D model (e.g., [33]). In general, the portions of the computational domain where each of the two models are applied must be preselected by the user; the interface between the two regions (with the related exchange of hydraulic information) is critical and requires efficient numerical treatment. Research into the development of coupling strategies and algorithms to implement hybrid models is currently active. For example, Sarkhosh and Jin [34] recently proposed a hybrid Lagrangian solver of one-dimensional (1D) shallow water flow and a 2D large-eddy simulation with moving coupling to simulate dam-break flows over frictional beds. However, the application of hybrid models to real-world case studies seems prohibitive.

New, innovative dam-break mathematical models are continually being proposed in the literature, confirming that research in this field is still lively. Surrogate models, which mimic the input–output dynamics of a complex system by simple approximated models, can be adopted as an alternative to physically-based models. Such a modeling approach has recently gained substantial popularity, mainly due to the increasing development of machine learning and artificial intelligence. Surrogate models are based on different possible approaches based on data and intrinsic physics, such as deep neural networks (characterized by a strong ability to catch complex nonlinearity in high-dimensional data), stochastic approaches based on Gaussian processes, reduced-order models or low-complexity models, and hybrid physics models, which combine physics and data-driven methods [35]. Moreover, surrogate models generally possess characteristics that make their use attractive in inverse modeling and in the determination of model parameters through differentiation for optimization purposes. In any case, one of the main motivations behind the development of such models is the reduction in the computational cost compared to conventional physically-based approaches. However, long model training and data generation times, as well as the scarce ability to identify correct solutions outside of the situations for which training was carried out, make these models mainly suitable in scenarios where repetitive simulations are necessary, without including the physics of the process. Physics-Informed Neural Networks (PINNs) have been proposed in the literature to overcome these limitations, embedding physical laws into models' training processes. Promising results have been obtained in dam-break modeling, albeit only concerning geometrically simple situations [36].

A crucial step in dam-break modeling and flood risk assessment is the estimation of the outflow discharge hydrograph induced by the dam failure. This information can be imposed as an inflow boundary condition in numerical simulations of dam-break wave propagation [37–39]. In fact, sometimes, dam-break hydrodynamic models do not include the reservoir behind the dam (for example, when the reservoir bathymetric data are unavailable) and rarely incorporate a breach formation model [40]. Dam breach modeling actually requires several parameters to be estimated, mainly related to breach geometry and development. This task can be accomplished through empirical methods (based on the regression analysis of historical dam-break data [41,42]) or physically-based methods (which reproduce the breach formation, e.g., [43]) and is particularly challenging in embankment dams, for which different breach mechanisms are possible (namely, overtopping, piping, and internal erosion), involving several geometric, structural, and hydrological factors [44,45]. Additionally, dam breach parameters are affected by high uncertainty, which propagates to the prediction of breach peak outflow and, consequently, of the inundation extent, flood hazard level, dam-break wave arrival time, and, in general, flood risk [46–49]. The accurate estimate of dam breach parameters is addressed by Contributions No. 9 and 10 in the Special Issue and by recent papers published in the literature which propose innovative approaches [44], thereby confirming that this topic is still relevant and attractive and that there is room for further research in this area [50].

Sometimes, dam-break flow develops over a mobile bed, incorporating large quantities of sediment from the bottom and inducing significant morphological changes. In these cases, geomorphic models which can account for the motion of water and sediment in non-equilibrium conditions and predict contextual changes in bottom morphology are required [51]. Scientific research on this topic is still active (as confirmed by Contribution No. 13 of this Special Issue), although different modeling approaches have been proposed in the recent past in the literature, including variable-density shallow water models (where the mixture of water and sediment is idealized as an equivalent homogeneous continuum [52,53]), two-layer models [54,55], and two-phase models [56]. Such complex models include closure equations with several model parameters, and, hence, they must be extensively validated against experimental data to accurately assess their predictive capabilities [51,57,58].

3.2. Recent Advancements in Numerical Modeling of Dam-Break Flows on Real-World Topography

The hydrodynamic simulation of dam-break flow over irregular, real-world topography requires accurate, robust, and efficient numerical models. Generally, these qualities can be found at an acceptable level in 2D shallow water models, provided that the numerical scheme used to solve the basic equations can cope with the difficulties induced by complex flow phenomena, such as critical transitions, the formation of shocks, and the motion of wetting and drying fronts (e.g., [59–62]). An appropriate treatment of the source terms, allowing the numerical model to satisfy the *C*-property and to be well balanced (i.e., preserving steady and stationary conditions) [63,64], is desirable, if not essential, especially in dam-break analyses over irregular topography, as quiescent water in the reservoir must be preserved until the arrival of the depression wave. Many researchers have focused on the robustness of the discretization of the source term; in this framework, research on this topic is still ongoing [65], and today, upwind schemes are preferred to centered ones [63]. Well-balanced schemes were proved to preserve a steady state even in the presence of geometric discontinuities [66]. For example, in this field, an energy-balanced version of the augmented Roe schemes was introduced [67], where the source term was accounted for by increasing the number of elementary waves.

The numerical treatment of the bottom and friction source terms influences the accuracy and stability of the numerical scheme, especially in the presence of wetting and drying fronts or overland flows due to the very shallow water depths involved (e.g., [68,69]).

Various methodologies have been proposed in the literature to enforce the balance between numerical fluxes and the bottom source term, based on a classic hydrostatic reconstruction strategy [70,71], a surface reconstruction technique [72], the correction of fluxes [73,74], or the incorporation of the bottom source in the numerical fluxes [60,75]. Also, the positivity-preserving property (which avoids the calculation of unphysical negative water depths) and a mass-conservation treatment of the motion of wetting and drying fronts are advantageous in the obtainment of accurate and stable solutions [76–78]. For this reason, the implementation of new formulations of the source terms capable of ensuring the greater accuracy and stability of the numerical resolution schemes is a constant subject of investigation, especially in the context of the development of Godunov-type finite-volume hydrodynamic models. Finally, particular attention is paid to providing new schemes with which to reconstruct the values of flow variables at the boundaries of grid cells and to improve and simplify the discretization of the stiff friction term [79,80].

Conventional finite-element methods (e.g., [81,82]) and discontinuous Galerkin finite-element schemes (e.g., [83]) were also adopted to solve shallow water equations with source terms and constitute a significant research area, especially in dam-break modeling, given their ability to cope with complex geometries and reach high-order accuracies.

A popular modern strategy to solve hydrodynamic governing equations in dam-break modeling, both in 2D and 3D frameworks, is based on meshless particle-based methods, such as Smoothed Particle Hydrodynamics (SPH) [84] and the Moving Particle Semi-Implicit Method (MPS) [85,86]. Such methods offer the advantage of avoiding the construction of a computational mesh and the adoption of complex free-surface-tracking algorithms (e.g., [87–90]). Additionally, Lattice Boltzmann methods (which are based on a discrete form of the Boltzmann equation and simulate flow through collision models of fictitious particles moving on a discrete lattice grid) were applied to shallow water equations [91] but are rarely used in dam-break flooding simulations (e.g., [92,93]), despite their efficiency and flexibility. Therefore, they deserve further research, especially in 3D dam-break flow applications.

3.3. *New Laboratory Experiments and Test Cases for the Validation of Numerical Models*

The comprehensive review of experimental studies on dam-break flow reported by Aureli et al. (2023) [Contribution No. 4] shows that in the past, an impressive number of experimental investigations were performed on this topic in a variety of test conditions, and many extensive and accurate datasets are available for the validation of dam-break numerical models. However, as observed by the same authors [Contribution No. 4], most datasets are not available in digital format. Hence, the creation of a public repository would be advisable to facilitate access to these data and their use.

Despite the numerous experimental investigations carried out in the past, the experimental approach remains very popular in dam-break flow analysis for the provision of insights into specific physical aspects that are not yet completely clear and for the collection of further experimental data concerning new geometries and test conditions to validate numerical models. In the Special Issue, only one original experimental investigation (on dam-break outflow) appears; however, experimental data available in the literature have been used in several contributions to this collection for validation purposes (Contributions No. 7, 12, 13, and 15). Additionally, some new contributions to experimental dam-break modeling have been published in the literature in the last year (e.g., [86,94,95]). In some of these, dam-break waves are used as incident waves to investigate the features of specific phenomena, such as green water events [96], the propagation of tsunami waves in the swash zone [97], or interaction with floating structures [98]. Non-intrusive measurement techniques are preferably adopted to collect experimental data. Special attention is devoted to imaging techniques that can provide data distributed in a relatively large area with a frequency suitable for rapidly varying flows at the cost of complex calibration procedures and post-processing actions (distortion correction, denoising, etc.) [94].

In dam-break experiments, dam-break waves are typically generated by suddenly removing a lift gate (which can be moved upwards or, more rarely, downwards) or a swing gate. The gate-opening modality can significantly influence the resulting dam-break flow [99] and must be carefully considered during the setup design. In any case, the opening time should be short enough to reproduce a nearly instantaneous dam collapse [100].

It is worth noting that model validation based on small-scale laboratory tests is affected by errors induced by scale effects (see the discussion reported in Contribution No. 4 of this Special Issue). This fact must be considered when applying numerical models to practical problems in large-scale, real-world situations. The impact of scale effects, especially in physical models, deserves further analysis. Similarly, dam-break flows generated by a partial dam-break obtained through partial gate opening [101] or partial gate removal (in both horizontal and vertical directions) [102] are topics worthy of further experimental and numerical investigation.

Finally, further efforts to conduct new experiments into partial dam-break (see Contribution No. 8 of this Special Issue) or breach formation [103], as well as dam-break flows involving sediment transport and morphological changes in an erodible bed or interactions with erodible structures [104], are advisable to collect additional experimental data for validation purposes. Modern non-intrusive, imaging-based measurement techniques are available to this end, allowing the accurate acquisition of water levels, flow velocities, and bottom levels with high temporal and spatial resolution [105]. The contextual measurement of both the free surface and the erodible bottom surface in dam-break flows with distinct 3D characters is an ongoing challenge.

3.4. *Applications of Dam-Break Numerical Models to Flood Hazard and Risk Assessment*

Most of the contributions in this Special Issue address theme (d), confirming the current considerable interest in the assessment of the hydraulic hazard and risk associated with dam-break flooding in scientific research and practical applications to provide accurate and exhaustive information for use in flood risk management and emergency planning. To this end, the identification of meaningful flood hazard indicators and associated flood hazard classifications is crucial to the correct quantification of flood hazards in areas that would be potentially floodable in the event of a dam failure [106]. Flood damage and life loss models are thus essential to predict the potential impact of a catastrophic dam accident, especially when different flood alert systems or non-structural mitigation measures must be tested or compared (e.g., [107,108]).

However, several uncertainty sources potentially affect the inundation model results and, consequently, decision-making [109]. In dam-break analysis, such uncertainty sources, which include the breaching mechanism, the breach width, the breaching duration, and the reservoir level at the breach time, must be considered in the definition of dam-break scenarios [47,110]. Probabilistic maps, commonly used in the analysis of riverine- or levee-breach-induced flooding (e.g., [111–113]), were recently extended to dam-break flood hazard and risk assessment and were shown to facilitate the informative and comprehensive quantification of the uncertainty associated with flood hazard estimates [47,48].

The need for effective flood risk management is even more urgent in developing countries (it is worth noting that three contributions to the Special Issue consider dam-break case studies concerning dams located in Algeria, Turkey, and Brazil). Climate change may increase the likelihood of dam failures, and the severity of the dam-break consequences may be exacerbated by the increasing exposure and lack of preparedness of the population at risk [108].

The paper by Říha et al. (2020) [Contribution No. 6] highlights the recent interest in cascading dam-break floods and the impact of scenario parameters on the evolution of the inundation process and the related consequences (e.g., [114,115]). Similarly, the article by Melo and Eleutério (2023) [Contribution No. 11] demonstrates the increasing interest in the assessment of flood hazards and risks associated with the potential failure of tailings dams used to store waste materials [116]. As stated by the same authors in the introduction to their contribution, "failures of tailings dams occur at a significantly higher frequency compared to those of water storage dams", and "recent incidents have highlighted the immense destructive potential associated with tailings inundations" or waste soil landslides (e.g., [117]). This is particularly true in emerging countries rich in raw materials, where mining activities are carried out on a large scale and often in the absence of adequate safety protocols. Usually, a mudflow originates from a tailings dam failure, and hydrodynamic simulations must take into account the non-Newtonian behavior of the released material [118]. The influence of the rheological model and the related parameters on the inundation dynamics and the extent of the flood inundation area is a crucial problem on which Contribution No. 11 of the Special Issue focuses. Similarly, non-Newtonian behavior is typically assumed in the simulation of dam-break debris flows based on the single-phase hypothesis [93].

4. What Was Missing

The contributions submitted to the Special Issue highlight the importance of some relevant lines of research in dam-break modeling. However, the absence of contributions concerning new modeling approaches and research currently at the forefront highlights, to some extent, the unpreparedness of the related state of the art towards practical applications.

In this Special Issue, it would have been interesting to host contributions regarding surrogate models applied to dam-breaks. However, the current research does not seem capable of addressing applications of practical interest yet. Due to their high complexity, these models do not currently represent (and probably will not be able to represent even in the short term) a credible alternative to conventional methods such as finite-difference, finite-element, or finite-volume methods. The simulation of naturally complex flows over irregular real-world topography in real-field case studies is not on the horizon yet [36], despite attempts to integrate fundamental physical laws and domain knowledge into the models, thereby trying to ensure generalization capabilities and credible predictions.

No contributions to the Special Issue focused on the theme of dam-break wave propagation in urban areas. However, this remains a very significant topic [119], especially, for example, in light of the application of hydrodynamic numerical models to the study of levee-breach-induced flooding [120], frequently involving densely urbanized lowland areas. An adequate description of built-up areas incompatible with a low spatial resolution (which is suitable for limiting calculation times, especially from the perspective of real-time applications) could be obtained by enhancing the capabilities of simplified approaches (e.g., porosity, 1D-2D coupling, etc.) [121], which therefore prove to be strategic and deserve further research.

A viable option to overcome the limitation related to the computational inefficiency of physically-based models is to accelerate model calculations via high-performance computing (HPC) and graphics processing unit (GPU) technology to reduce model running times. In particular, parallel computing is a valid and widely used method to improve computational efficiency, possibly exploiting GPU computing power and processing capabilities. Reducing the computational time of model executions also allows larger domains to be considered with high spatial resolution and the simulation of flooding dynamics for a longer physical time. Moreover, a large number of model runs for different dam-break scenarios can be performed in sensitivity analyses. Even in the presence of promising runtimes, the scalability in recently proposed multi-GPU codes is still unsatisfactory for a large number of GPUs due to thread divergence for dry cells, poor load balancing between MPI ranks with static decomposition, and, above all, the time consumed by I/O and MPI communication for large-scale problems with respect to the computation time. These times (in the order of minutes each) are difficult to minimize, even in the face of constant effort in this direction. In fact, they do not depend on the number of GPUs and represent the highest percentage of the total runtime when trying to achieve operational purposes. Future perspectives should therefore aim to design optimized I/O parallel algorithms and explore new communication techniques [122]. This line of research is cutting-edge. However, given the high stakes and the very specific topic, researchers interested in this field are likely to turn to more sector-specific journals rather than generalist journals, such as MDPI's *Water*.

5. Conclusions

This Special Issue concerning "Advances in Dam-Break Modeling for Flood Hazard Mitigation: Theory, Numerical Models, and Applications in Hydraulic Engineering" consists of a collection of 15 papers. The topics addressed range from recent advancements in the theoretical analysis and numerical modeling of dam-break flows, both in schematic configurations and on irregular real-world topography, to new laboratory experiments and test cases suitable for the validation of numerical models. Special attention is devoted to case studies in which dam-break modeling is used for flood hazard and risk assessment.

During the three years in which manuscript submission was open, research on dam-break modeling has continued, also through the contributions included in this collection, thereby demonstrating great liveliness and fidelity in its ultimate goal of protection from flood hazards. In fact, under the aegis of the general topic of the total or partial failure of a dam, developments of fundamental importance have originated in the modeling of flooding phenomena, which allow populations to be protected in an increasingly targeted and effective manner from flood risks.

Many foods for thought came from the analysis of the papers included in the Special Issue, and numerous new questions arose from the fact that some relevant aspects of the many topics related to dam-break modeling have been overlooked. As Guest Editors, we hope that the articles included in the Special Issue and the reflections and proposals advanced in this editorial will be useful and stimulating for researchers involved in this interesting research field, which is so rich in history and has profoundly influenced the evolution of hydrodynamic modeling for over a century.

Author Contributions: F.A., A.M. and G.P. contributed equally to conceiving, designing, and writing the editorial. All authors have read and agreed to the published version of the manuscript.

Funding: This work was supported by the Italian Ministry of University and Research through the PRIN 2017 Project RELAID (REnaissance of LArge Italian Dams), project number 2017T4JC5K. The support from Italian Ministry of University and Research and University of Pavia (Italy) within the program "Dipartimenti di Eccellenza 2023–2027" is acknowledged.

Acknowledgments: All the authors who contributed to the Special Issue are acknowledged for their valuable contributions, as well as the editorial managers and anonymous reviewers for the time they dedicated.

Conflicts of Interest: The authors declare no conflicts of interest.

List of Contributions:

1. Aureli, F.; Maranzoni, A.; Petaccia, G. Review of Historical Dam-Break Events and Laboratory Tests on Real Topography for the Validation of Numerical Models. *Water* **2021**, *13*, 1968. https://doi.org/10.3390/w13141968
2. AlQasimi, E.; Mahdi, T.-F. Comment on Aureli et al. Review of Historical Dam-Break Events and Laboratory Tests on Real Topography for the Validation of Numerical Models. *Water* **2021**, *13*, 1968. *Water* **2022**, *14*, 264. https://doi.org/10.3390/w14020264
3. Aureli, F.; Maranzoni, A.; Petaccia, G. Reply to AlQasimi, E.; Mahdi, T.-F. Comment on "Aureli et al. Review of Historical Dam-Break Events and Laboratory Tests on Real Topography for the Validation of Nu Models. *Water* **2021**, *13*, 1968". *Water* **2022**, *14*, 267. https://doi.org/10.3390/w14020267
4. Aureli, F.; Maranzoni, A.; Petaccia, G.; Soares-Frazão, S. Review of Experimental Investigations of Dam-Break Flows over Fixed Bottom. *Water* **2023**, *15*, 1229. https://doi.org/10.3390/w15061229
5. Maranzoni, A.; Tomirotti, M. Three-Dimensional Numerical Modelling of Real-Field Dam-Break Flows: Review and Recent Advances. *Water* **2023**, *15*, 3130. https://doi.org/10.3390/w15173130

6. Říha, J.; Kotaška, S.; Petrula, L. Dam Break Modeling in a Cascade of Small Earthen Dams: Case Study of the Čižina River in the Czech Republic. *Water* **2020**, *12*, 2309. https://doi.org/10.3390/w12082309
7. Cantero-Chinchilla, F.N.; Bergillos, R.J.; Gamero, P.; Castro-Orgaz, O.; Cea, L.; Hager, W.H. Vertically Averaged and Moment Equations for Dam-Break Wave Modeling: Shallow Water Hypotheses. *Water* **2020**, *12*, 3232. https://doi.org/10.3390/w12113232
8. Ahmadi, S.M.; Yamamoto, Y. A New Dam-Break Outflow-Rate Concept and Its Installation to a Hydro-Morphodynamics Simulation Model Based on FDM (An Example on Amagase Dam of Japan). *Water* **2021**, *13*, 1759. https://doi.org/10.3390/w13131759
9. Gaagai, A.; Aouissi, H.A.; Krauklis, A.E.; Burlakovs, J.; Athamena, A.; Zekker, I.; Boudoukha, A.; Benaabidate, L.; Chenchouni, H. Modeling and Risk Analysis of Dam-Break Flooding in a Semi-Arid Montane Watershed: A Case Study of the Yabous Dam, Northeastern Algeria. *Water* **2022**, *14*, 767. https://doi.org/10.3390/w14050767
10. Bello, D.; Alcayaga, H.; Caamaño, D.; Pizarro, A. Influence of Dam Breach Parameter Statistical Definition on Resulting Rupture Maximum Discharge. *Water* **2022**, *14*, 1776. https://doi.org/10.3390/w14111776
11. Melo, M.; Eleutério, J. Probabilistic Analysis of Floods from Tailings Dam Failures: A Method to Analyze the Impact of Rheological Parameters on the HEC-RAS Bingham and Herschel-Bulkley Models. *Water* **2023**, *15*, 2866. https://doi.org/10.3390/w15162866
12. Xue, Z.; Zhou, L.; Liu, D. Accurate Numerical Modeling for 1D Open-Channel Flow with Varying Topography. *Water* **2023**, *15*, 2893. https://doi.org/10.3390/w15162893
13. Martínez-Aranda, S.; Fernández-Pato, J.; García-Navarro, P. Non-Equilibrium Bedload Transport Model Applied to Erosive Overtopping Dambreach. *Water* **2023**, *15*, 3094. https://doi.org/10.3390/w15173094
14. Silva, A.F.R.; Eleutério, J.C. Effectiveness of a Dam-Breach Flood Alert in Mitigating Life Losses: A Spatiotemporal Sectorisation Analysis in a High-Density Urban Area in Brazil. *Water* **2023**, *15*, 3433. https://doi.org/10.3390/w15193433
15. Akgun, C.; Nas, S.S.; Uslu, A. 2D and 3D Numerical Simulation of Dam-Break Flooding: A Case Study of the Tuzluca Dam, Turkey. *Water* **2023**, *15*, 3622. https://doi.org/10.3390/w15203622

References

1. Galland, J.-C.; Morris, M. CADAM European Concerted Action on Dam-Break Modelling. In Proceedings of the 28th IAHR World Congress, Graz, Austria, 22–27 August 1999; Paper ID 13675. Available online: https://www.iahr.org/library/infor?pid=13675 (accessed on 5 March 2024).
2. Maijala, T. *RESCDAM Development of Rescue Actions Based on Dam-Break Flood Analysis*; Final Report; Finnish Environment Institute: Helsinki, Finland, 2001; Available online: https://civil-protection-humanitarian-aid.ec.europa.eu/system/files/2014-11/rescdam_rapportfin.pdf (accessed on 5 March 2024).
3. Zech, Y.; Soares-Frazão, S. Dam-Break Flow Experiments and Real-Case Data. A Database from the European IMPACT Research. *J. Hydraul. Res.* **2007**, *45* (Suppl. S1), 5–7. [CrossRef]
4. Zhang, L.; Peng, M.; Chang, D.; Xu, Y. *Dam Failure Mechanisms and Risk Assessment*; John Wiley & Sons: Singapore, 2016.
5. Perera, D.; Smakhtin, V.; Williams, S.; North, T.; Curry, A. *Ageing Water Storage Infrastructure: An Emerging Global Risk*; UNU-INWEH Report Series, Issue 11; United Nations University Institute for Water, Environment and Health: Hamilton, ON, Canada, 2021.
6. Cao, W.; Zhou, Y.; Güneralp, B.; Li, X.; Zhao, K.; Zhang, H. Increasing Global Urban Exposure to Flooding: An Analysis of Long-Term Annual Dynamics. *Sci. Total Environ.* **2022**, *817*, 153012. [CrossRef] [PubMed]
7. Federal Emergency Management Agency. *Federal Guidelines for Inundation Mapping of Flood Risks Associated with Dam Incidents and Failures*; FEMA P-946; FEMA, US Department of Homeland Security: Washington, DC, USA, 2013. Available online: https://www.fema.gov/sites/default/files/2020-08/fema_dam-safety_inundation-mapping-flood-risks.pdf (accessed on 5 March 2024).
8. Transforming Our World: The 2030 Agenda for Sustainable Development. Available online: https://sdgs.un.org/2030agenda (accessed on 5 March 2024).
9. Fluixá-Sanmartín, J.; Altarejos-García, L.; Morales-Torres, A.; Escuder-Bueno, I. Review article: Climate Change Impacts on Dam Safety. *Nat. Hazards Earth. Syst. Sci.* **2018**, *18*, 2471–2488. [CrossRef]
10. Toro, E.F. *Shock-Capturing Methods for Free-Surface Shallow Flows*; John Wiley & Sons: Chichester, UK, 2001.
11. Castro-Orgaz, O.; Hager, W.H. *Shallow Water Hydraulics*; Springer: Cham, Switzerland, 2019.
12. Castro-Orgaz, O.; Hager, W.H. *Non-Hydrostatic Free Surface Flows*; Springer: Cham, Switzerland, 2017.
13. Escalante, C.; Morales de Luna, T.; Cantero-Chinchilla, F.; Castro-Orgaz, O. Vertically Averaged and Moment Equations: New Derivation, Efficient Numerical Solution and Comparison with Other Physical Approximations for Modeling Non-Hydrostatic Free Surface Flows. *J. Comput. Phys.* **2024**, *504*, 112882. [CrossRef]

14. Echeverribar, I.; Brufau, P.; García-Navarro, P. Extension of a Roe-type Riemann Solver Scheme to Model Non-Hydrostatic Pressure Shallow Flows. *Appl. Math. Comput.* **2023**, *440*, 127642. [CrossRef]
15. Escalante, C.; de Luna, T.M. A General Non-Hydrostatic Hyperbolic Formulation for Boussinesq Dispersive Shallow Flows and Its Numerical Approximation. *J. Sci. Comput.* **2020**, *83*, 62. [CrossRef]
16. Nadiga, B.T.; Margolin, L.G.; Smolarkiewicz, P.K. Different Approximations of Shallow Fluid Flow over an Obstacle. *Phys. Fluids* **1996**, *8*, 2066–2077. [CrossRef]
17. Stelling, G.S.; Zijlema, M. An Accurate and Efficient Finite-Difference Algorithm for Non-Hydrostatic Free-Surface Flow with Application to Wave Propagation. *Int. J. Numer. Methods Fluids* **2003**, *43*, 1–23. [CrossRef]
18. Cantero-Chinchilla, F.N.; Castro-Orgaz, O.; Dey, S.; Ayuso, J.L. Nonhydrostatic Dam Break Flows. I: Physical Equations and Numerical Schemes. *J. Hydraul. Eng.* **2016**, *142*, 04016068. [CrossRef]
19. Special Collection on Boussinesq-Type Modeling of Free-Surface Flows: Necessity or Luxury? *J. Hydraul. Eng.* Guest Editors: Castro-Orgaz, O.; Hager, W.H.; Katopodes, N.D. Available online: https://ascelibrary.org/jhend8/boussinesq_type_free_surface_necessity_luxury (accessed on 11 March 2024).
20. Mohapatra, P.K.; Chaudhry, M.H. Numerical Solution of Boussinesq Equations to Simulate Dam-Break Flows. *J. Hydraul. Eng.* **2004**, *130*, 156–159. [CrossRef]
21. Green, A.E.; Naghdi, P.M. A Derivation of Equations for Wave Propagation in Water of Variable Depth. *J. Fluid Mech.* **1976**, *78*, 237–246. [CrossRef]
22. Cantero-Chinchilla, F.N.; Castro-Orgaz, O.; Khan, A.A. Depth-Integrated Nonhydrostatic Free-Surface Flow Modeling Using Weighted-Averaged Equations. *Int. J. Numer. Methods Fluids* **2018**, *87*, 27–50. [CrossRef]
23. Khan, A.A.; Steffler, P.M. Vertically Averaged and Moment Equations Model for Flow over Curved Beds. *J. Hydraul. Eng.* **1996**, *122*, 3–9. [CrossRef]
24. Gamero, P.; Bergillos, R.J.; Cantero-Chinchilla, F.N.; Castro-Orgaz, O. A MATLAB Software Platform for Modelling Vertically-Integrated Non-Hydrostatic Flows with Moment Equations. *Environ. Model. Softw.* **2020**, *127*, 104674. [CrossRef]
25. Cantero-Chinchilla, F.N.; Castro-Orgaz, O.; Khan, A.A. Vertically-Averaged and Moment Equations for Flow and Sediment Transport. *Adv. Water Resour.* **2019**, *132*, 103387. [CrossRef]
26. Castro-Orgaz, O.; Chanson, H. Undular and Broken Surges in Dam-Break Flows: A Review of Wave Breaking Strategies in a Boussinesq-type Framework. *Environ. Fluid Mech.* **2020**, *20*, 1383–1416. [CrossRef]
27. Biscarini, C.; Di Francesco, S.; Manciola, P. CFD Modelling Approach for Dam Break Flow Studies. *Hydrol. Earth Syst. Sci.* **2010**, *14*, 705–718. [CrossRef]
28. Munoz, D.H.; Constantinescu, G. 3-D Dam Break Flow Simulations in Simplified and Complex Domains. *Adv. Water Resour.* **2020**, *137*, 103510. [CrossRef]
29. TELEMAC-3D—3D Hydrodynamics. Available online: http://www.opentelemac.org/index.php/presentation?id=18 (accessed on 25 January 2024).
30. FLOW-3D. Available online: https://www.flow3d.com/products/flow-3d/ (accessed on 25 January 2024).
31. DualSPHysics: From Fluid Dynamics to Multiphysics Problems. Available online: https://dual.sphysics.org/ (accessed on 25 January 2024).
32. Biscarini, C.; Di Francesco, S.; Ridolfi, E.; Manciola, P. On the Simulation of Floods in a Narrow Bending Valley: The Malpasset Dam Break Case Study. *Water* **2016**, *8*, 545. [CrossRef]
33. FLOW-3D Modeling Capabilities. Hybrid Shallow Water/3D Flow. Available online: https://www.flow3d.com/modeling-capabilities/hybrid-shallow-water-3d-flow/ (accessed on 25 January 2024).
34. Sarkhosh, P.; Jin, Y.C. A Hybrid 1D-2D Lagrangian Solver with Moving Coupling to Simulate Dam-Break Flow. *Adv. Water Resour.* **2023**, *178*, 104487. [CrossRef]
35. Song, Y.; Shen, C.; Liu, X. A Surrogate Model for Shallow Water Equations Solvers with Deep Learning. *J. Hydraul. Eng.* **2023**, *149*, 04023045. [CrossRef]
36. Li, C.; Han, Z.; Li, Y.; Li, M.; Wang, W.; Dou, J.; Xu, L.; Chen, G. Physical Information-Fused Deep Learning Model Ensembled with a Subregion-Specific Sampling Method for Predicting Flood Dynamics. *J. Hydrol.* **2023**, *620*, 129465. [CrossRef]
37. Pilotti, M.; Tomirotti, M.; Valerio, G.; Bacchi, B. Simplified Method for the Characterization of the Hydrograph Following a Sudden Partial Dam Break. *J. Hydraul. Eng.* **2010**, *136*, 693–704. [CrossRef]
38. Aureli, F.; Maranzoni, A.; Mignosa, P. A Semi-Analytical Method for Predicting the Outflow Hydrograph due to Dam-Break in Natural Valleys. *Adv. Water Resour.* **2014**, *63*, 38–44. [CrossRef]
39. Pilotti, M.; Milanesi, L.; Bacchi, V.; Tomirotti, M.; Maranzoni, A. Dam-Break Wave Propagation in Alpine Valley with HEC-RAS 2D: Experimental Cancano Test Case. *J. Hydraul. Eng.* **2020**, *146*, 05020003. [CrossRef]
40. Dazzi, S.; Vacondio, R.; Mignosa, P. Integration of a Levee Breach Erosion Model in a GPU-Accelerated 2D Shallow Water Equations Code. *Water Resour. Res.* **2019**, *55*, 682–702. [CrossRef]
41. Wahl, T.L. Uncertainty of Predictions of Embankment Dam Breach Parameters. *J. Hydraul. Eng.* **2004**, *130*, 389–397. [CrossRef]
42. Froehlich, D.C. Embankment Dam Breach Parameters and Their Uncertainties. *J. Hydraul. Eng.* **2008**, *134*, 1708–1721. [CrossRef]
43. Macchione, F. Model for Predicting Floods due to Earthen Dam Breaching. I: Formulation and Evaluation. *J. Hydraul. Eng.* **2008**, *134*, 1688–1696. [CrossRef]

44. Sammen, S.S.; Mohamed, T.A.; Ghazali, A.H.; Sidek, L.M.; El-Shafie, A. An Evaluation of Existent Methods for Estimation of Embankment Dam Breach Parameters. *Nat. Hazards* **2017**, *87*, 545–566. [CrossRef]
45. da Silva, A.Â.C.L.; Eleutério, J.C. Identifying and Testing the Probability Distribution of Earthfill Dam Breach Parameters for Probabilistic Dam Breach Modeling. *J. Flood Risk Manag.* **2023**, *16*, e12900. [CrossRef]
46. Bellos, V.; Tsakiris, V.K.; Kopsiaftis, G.; Tsakiris, G. Propagating Dam Breach Parametric Uncertainty in a River Reach Using the HEC-RAS Software. *Hydrology* **2020**, *7*, 72. [CrossRef]
47. Rizzo, C.; Maranzoni, A.; D'Oria, M. Probabilistic Mapping and Sensitivity Assessment of Dam-Break Flood Hazard. *Hydrol. Sci. J.* **2023**, *68*, 700–718. [CrossRef]
48. Maranzoni, A.; D'Oria, M.; Rizzo, C. Probabilistic Mapping of Life Loss due to Dam-Break Flooding. *Nat. Hazards* **2024**, *120*, 2433–2460. [CrossRef]
49. Ahmadisharaf, E.; Kalyanapu, A.J.; Thames, B.A.; Lillywhite, J. A Probabilistic Framework for Comparison of Dam Breach Parameters and Outflow Hydrograph Generated by Different Empirical Prediction Methods. *Environ. Model. Softw.* **2016**, *86*, 248–263. [CrossRef]
50. Azmi, M.; Thomson, K. Dam Breach Parameters: From Data-Driven-Based Estimates to 2-Dimensional Modeling. *Nat. Hazards* **2024**, *120*, 4423–4461. [CrossRef]
51. Capart, H.; Spinewine, B.; Young, D.L.; Zech, Y.; Brooks, G.R.; Leclerc, M.; Secretan, Y. The 1996 Lake Ha! Ha! Breakout Flood, Québec: Test Data for Geomorphic Flood Routing Methods. *J. Hydraul. Res.* **2007**, *45* (Suppl. S1), 97–109. [CrossRef]
52. Canelas, R.; Murillo, J.; Ferreira, R.M. Two-Dimensional Depth-Averaged Modelling of Dam-Break Flows over Mobile Beds. *J. Hydraul. Res.* **2013**, *51*, 392–407. [CrossRef]
53. Duan, J.G.; Yu, C.; Ding, Y. Numerical Simulation of Sediment Transport in Unsteady Open Channel Flow. *Water* **2023**, *15*, 2576. [CrossRef]
54. Swartenbroekx, C.; Zech, Y.; Soares-Frazão, S. Two-Dimensional Two-Layer Shallow Water Model for Dam Break Flows with Significant Bed Load Transport. *Int. J. Numer. Methods Fluids* **2013**, *73*, 477–508. [CrossRef]
55. Li, J.; Cao, Z.; Pender, G.; Liu, Q. A Double Layer-Averaged Model for Dam-Break Flows over Mobile Bed. *J. Hydraul. Res.* **2013**, *51*, 518–534. [CrossRef]
56. Di Cristo, C.; Greco, M.; Iervolino, M.; Leopardi, A.; Vacca, A. Two-Dimensional Two-Phase Depth-Integrated Model for Transients over Mobile Bed. *J. Hydraul. Eng.* **2016**, *142*, 04015043. [CrossRef]
57. Spinewine, B.; Zech, Y. Small-Scale Laboratory Dam-Break Waves on Movable Beds. *J. Hydraul. Res.* **2007**, *45* (Suppl. 1), 73–86. [CrossRef]
58. Soares-Frazão, S.; Canelas, R.; Cao, Z.; Cea, L.; Chaudhry, H.M.; Moran, A.D.; El Kadi, K.; Ferreira, R.; Cadórniga, I.F.; Gonzalez-Ramirez, N.; et al. Dam-Break Flows over Mobile Beds: Experiments and Benchmark Tests for Numerical Models. *J. Hydraul. Res.* **2012**, *50*, 364–375. [CrossRef]
59. Brufau, P.; Vázquez-Cendón, M.E.; García-Navarro, P. A Numerical Model for the Flooding and Drying of Irregular Domains. *Int. J. Numer. Methods Fluids* **2002**, *39*, 247–275. [CrossRef]
60. Liang, Q.; Marche, F. Numerical Resolution of Well-Balanced Shallow Water Equations with Complex Source Terms. *Adv. Water Resour.* **2009**, *32*, 873–884. [CrossRef]
61. Song, L.; Zhou, J.; Guo, J.; Zou, Q.; Liu, Y. A Robust Well-Balanced Finite Volume Model for Shallow Water Flows with Wetting and Drying over Irregular Terrain. *Adv. Water Resour.* **2011**, *34*, 915–932. [CrossRef]
62. Hou, J.; Simons, F.; Mahgoub, M.; Hinkelmann, R. A Robust Well-Balanced Model on Unstructured Grids for Shallow Water Flows with Wetting and Drying over Complex Topography. *Comput. Methods Appl. Mech. Engrg.* **2013**, *257*, 126–149. [CrossRef]
63. Bermudez, A.; Vazquez, M.E. Upwind Methods for Hyperbolic Conservation Laws with Source Terms. *Comput. Fluids* **1994**, *23*, 1049–1071. [CrossRef]
64. Greenberg, J.M.; Leroux, A.Y. A Well-Balanced Scheme for the Numerical Processing of Source Terms in Hyperbolic Equations. *SIAM J. Numer. Anal.* **1996**, *33*, 1–16. [CrossRef]
65. Gómez-Bueno, I.; Boscarino, S.; Castro, M.J.; Parés, C.; Russo, G. Implicit and Semi-Implicit Well-Balanced Finite-Volume Methods for Systems of Balance Laws. *Appl. Numer. Math.* **2023**, *184*, 18–48. [CrossRef]
66. Del Grosso, A.; Castro, M.J.; Chan, A.; Gallice, G.; Loubère, R.; Maire, P.-H. A Well-Balanced, Positive, Entropy-Stable, and Multi-Dimensional-Aware Finite Volume Scheme for 2D Shallow-Water Equations with Unstructured Grids. *J. Comput. Phys.* **2024**, *503*, 112829. [CrossRef]
67. Mairal, J.; Murillo, J.; García-Navarro, P. The Entropy Fix in Augmented Riemann Solvers in Presence of Source Terms: Application to the Shallow Water Equations. *Comput. Methods Appl. Mech. Eng.* **2023**, *417*, 116411. [CrossRef]
68. Cea, L.; Bladé, E. A Simple and Efficient Unstructured Finite Volume Scheme for Solving the Shallow Water Equations in Overland Flow Applications. *Water Resour. Res.* **2015**, *51*, 5464–5486. [CrossRef]
69. Aureli, F.; Prost, F.; Vacondio, R.; Dazzi, S.; Ferrari, A. A GPU-Accelerated Shallow-Water Scheme for Surface Runoff Simulations. *Water* **2020**, *12*, 637. [CrossRef]
70. Audusse, E.; Bristeau, M.-O. A Well-Balanced Positivity Preserving "Second-Order" Scheme for Shallow Water Flows on Unstructured Meshes. *J. Comput. Phys.* **2005**, *206*, 311–333. [CrossRef]
71. Chen, G.; Noelle, S. A New Hydrostatic Reconstruction Scheme Based on Subcell Reconstructions. *SIAM J. Numer. Anal.* **2017**, *55*, 758–784. [CrossRef]

72. Xia, X.; Liang, Q.; Ming, X.; Hou, J. An Efficient and Stable Hydrodynamic Model with Novel Source Term Discretization Schemes for Overland Flow and Flood Simulations. *Water Resour. Res.* **2017**, *53*, 3730–3759. [CrossRef]
73. Begnudelli, L.; Sanders, B.F. Unstructured Grid Finite-Volume Algorithm for Shallow-Water Flow and Scalar Transport with Wetting and Drying. *J. Hydraul. Eng.* **2006**, *132*, 371–384. [CrossRef]
74. Song, L.; Zhou, J.; Li, Q.; Yang, X.; Zhang, Y. An Unstructured Finite Volume Model for Dam-Break Floods with Wet/Dry Fronts over Complex Topography. *Int. J. Numer. Methods Fluids* **2011**, *67*, 960–980. [CrossRef]
75. Valiani, A.; Begnudelli, L. Divergence Form for Bed Slope Source Term in Shallow Water Equations. *J. Hydraul. Eng.* **2006**, *132*, 652–665. [CrossRef]
76. Brufau, P.; García-Navarro, P.; Vázquez-Cendón, M.E. Zero Mass Error Using Unsteady Wetting-Drying Conditions in Shallow Flows over Dry Irregular Topography. *Int. J. Numer. Methods Fluids* **2004**, *45*, 1047–1082. [CrossRef]
77. Aureli, F.; Maranzoni, A.; Mignosa, P.; Ziveri, C. A Weighted Surface-Depth Gradient Method for the Numerical Integration of the 2D Shallow Water Equations with Topography. *Adv. Water Resour.* **2008**, *31*, 962–974. [CrossRef]
78. Yan, R.; Tong, W.; Chen, G. A Mass Conservative, Well Balanced and Positivity-Preserving Central Scheme for Shallow Water Equations. *Appl. Math. Comput.* **2023**, *443*, 127778. [CrossRef]
79. Zhao, J.; Liang, Q. Novel Variable Reconstruction and Friction Term Discretisation Schemes for Hydrodynamic Modelling of Overland Flow and Surface Water Flooding. *Adv. Water Resour.* **2022**, *163*, 104187. [CrossRef]
80. Hou, J.; Wang, T.; Li, P.; Li, Z.; Zhang, X.; Zhao, J.; Hinkelmann, R. An Implicit Friction Source Term Treatment for Overland Flow Simulation Using Shallow Water Flow Model. *J. Hydrol.* **2018**, *564*, 357–366. [CrossRef]
81. Zhang, T.; Peng, L.; Feng, P. Evaluation of a 3D Unstructured-Mesh Finite Element Model for Dam-Break Floods. *Comput. Fluids* **2018**, *160*, 64–77. [CrossRef]
82. Savant, G.; Berger, C.; McAlpin, T.O.; Tate, J.N. Efficient Implicit Finite-Element Hydrodynamic Model for Dam and Levee Breach. *J. Hydraul. Eng.* **2011**, *137*, 1005–1018. [CrossRef]
83. Kesserwani, G.; Ghostine, R.; Vazquez, J.; Ghenaim, A.; Mosé, R. Application of a Second-Order Runge-Kutta Discontinuous Galerkin Scheme for the Shallow Water Equations with Source Terms. *Int. J. Numer. Methods Fluids* **2008**, *56*, 805–821. [CrossRef]
84. Violeau, D. *Fluid Mechanics and the SPH Method: Theory and Applications;* Oxford University Press: Oxford, UK, 2012.
85. Geng, C.; Wang, W.-H.; Zhou, Z.-Q. Application of an Improved MPS Method in Simulation of Dam Break. *Ocean Eng.* **2023**, *287*, 115779. [CrossRef]
86. Sanchez-Mondragon, J.; Fontes, J.V.H.; Espinoza-Haro, M.P.; Esperança, P.T.T. Interaction of Wet Dam-Break Flows with a Structure: On the Effects of Surface Tension and Gate Opening Velocity in Water Surface Evolution Using MPS Method. *Comp. Part. Mech.* **2023**, *10*, 1221–1244. [CrossRef]
87. Prakash, M.; Rothauge, K.; Cleary, P.W. Modelling the Impact of Dam Failure Scenarios on Flood Inundation Using SPH. *Appl. Math. Model.* **2014**, *38*, 5515–5534. [CrossRef]
88. Xu, X. An Improved SPH Approach for Simulating 3D Dam-Break Flows with Breaking Waves. *Comput. Methods Appl. Mech. Eng.* **2016**, *311*, 723–742. [CrossRef]
89. Wu, Y.; Tian, L.; Rubinato, M.; Gu, S.; Yu, T.; Xu, Z.; Cao, P.; Wang, X.; Zhao, Q. A New Parallel Framework of SPH-SWE for Dam Break Simulation Based on OpenMP. *Water* **2020**, *12*, 1395. [CrossRef]
90. Xu, X.; Jiang, Y.-L.; Yu, P. SPH Simulations of 3D Dam-Break Flow Against Various Forms of the Obstacle: Toward an Optimal Design. *Ocean Eng.* **2021**, *229*, 108978. [CrossRef]
91. Zhou, J.G. A Lattice Boltzmann Model for the Shallow Water Equations. *Comput. Methods Appl. Mech. Eng.* **2002**, *191*, 3527–3539. [CrossRef]
92. Zhou, J.G. *Lattice Boltzmann Methods for Shallow Water Flows;* Springer: Berlin, Germany, 2004.
93. Kefayati, G.; Tolooiyan, A.; Dyson, A.P. Finite Difference Lattice Boltzmann Method for Modeling Dam Break Debris Flows. *Phys. Fluids* **2023**, *35*, 013102. [CrossRef]
94. Zhang, L.; Zhang, F.; Xu, W.; Bo, H.; Zhang, X. An Innovative Method for Measuring the Three-Dimensional Water Surface Morphology of Unsteady Flow Using Light Detection and Ranging Technology. *Ocean Eng.* **2023**, *276*, 114079. [CrossRef]
95. Tan, T.; Ma, Y.; Zhang, J.; Niu, X.; Chang, K.A. Experimental Study on Flow Kinematics of Dam-Break Induced Surge Impacting onto a Vertical Wall. *Phys. Fluids* **2023**, *35*, 025127. [CrossRef]
96. Rodríguez-Ocampo, P.E.; Fontes, J.V.H.; Ring, M.; Mendoza, E.; Esperança, P.T.T.; Silva, R. Evaluation of the Influence of Deck Slope on Green Water Loads Using the Wet Dam-Break Approach. *J. Braz. Soc. Mech. Sci. Eng.* **2024**, *46*, 42. [CrossRef]
97. Yang, W.; Dou, Y.; Yang, S.; Geng, X.; Hou, H.; Zhang, Y. Experimental Study on Failure Mode and Failure Mechanism of Multi-Box Girder Bridge Under Attack of Tsunami. *Ocean Eng.* **2023**, *280*, 114717. [CrossRef]
98. Wang, D.; Dong, S. A Numerical Model for Simulating the Interaction of Dam-Break Waves and Floating Structures. *Ocean Eng.* **2023**, *285*, 115469. [CrossRef]
99. von Häfen, H.; Goseberg, N.; Stolle, J.; Nistor, I. Gate-Opening Criteria for Generating Dam-Break Waves. *J. Hydraul. Eng.* **2019**, *145*, 04019002. [CrossRef]
100. Lauber, G.; Hager, W.H. Experiments to Dambreak Wave: Horizontal Channel. *J. Hydraul. Res.* **1998**, *36*, 291–307. [CrossRef]
101. Lazzarin, T.; Viero, D.P.; Defina, A.; Cozzolino, L. Flow Under Vertical Sluice Gates: Flow Stability at Large Gate Opening and Disambiguation of Partial Dam-Break Multiple Solutions. *Phys. Fluids* **2023**, *35*, 024114. [CrossRef]

102. De Marchi, G. Sull'Onda di Piena che Seguirebbe al Crollo della Diga di Cancano. Prove su Modello per il Tronco Fluviale dalla Diga a Ponte Cepina–Calcolo per il Tronco da Ponte Cepina a Tirano [On the Dam-Break Wave Resulting from the Collapse of the Cancano Dam. Physical Model for the River Reach from the Dam to Ponte Cepina–Numerical Model for the River Reach from Ponte Cepina to Tirano]. *L'Energia Elettr.* **1945**, *22*, 319–340. (In Italian)
103. Morris, M.W.; Hassan, M.A.A.M.; Vaskinn, K.A. Breach Formation: Field Test and Laboratory Experiments. *J. Hydraul. Res.* **2007**, *45* (Suppl. S1), 9–17. [CrossRef]
104. Di Cristo, C.; Evangelista, S.; Greco, M.; Iervolino, M.; Leopardi, A.; Vacca, A. Dam-Break Waves over an Erodible Embankment: Experiments and Simulations. *J. Hydraul. Res.* **2018**, *56*, 196–210. [CrossRef]
105. Soares-Frazão, S. Review of Imaging-Based Measurement Techniques for Free Surface Flows Involving Sediment Transport and Morphological Changes. *J. Hydroinform.* **2020**, *22*, 958–971. [CrossRef]
106. Maranzoni, A.; D'Oria, M.; Rizzo, C. Quantitative Flood Hazard Assessment Methods: A Review. *J. Flood Risk Manag.* **2023**, *16*, e12855. [CrossRef]
107. Silva, A.F.R.; Eleutério, J.C. Analysis of Flood Warning and Evacuation Efficiency by Comparing Damage and Life-Loss Estimates with Real Consequences Related to the São Francisco Tailings Dam Failure in Brazil. *Nat. Hazards Earth Syst. Sci.* **2023**, *23*, 3095–3110. [CrossRef]
108. Wang, Y.; Fu, Z.; Cheng, Z.; Xiang, Y.; Chen, J.; Zhang, P.; Yang, X. Uncertainty Analysis of Dam-Break Flood Risk Consequences Under the Influence of Non-Structural Measures. *Int. J. Disaster Risk Reduct.* **2024**, *102*, 104265. [CrossRef]
109. Balbi, M.; Lallemant, D.C.B. The Cost of Imperfect Knowledge: How Epistemic Uncertainties Influence Flood Hazard Assessments. *Water Resour. Res.* **2023**, *59*, e2023WR035685. [CrossRef]
110. Sarchani, S.; Koutroulis, A.G. Probabilistic Dam Breach Flood Modeling: The Case of Valsamiotis Dam in Crete. *Nat. Hazards* **2022**, *114*, 1763–1814. [CrossRef]
111. Smemoe, C.M.; Nelson, E.J.; Zundel, A.K.; Miller, A.W. Demonstrating Floodplain Uncertainty Using Flood Probability Maps. *J. Am. Water Resour. Assoc.* **2007**, *43*, 359–371. [CrossRef]
112. Vorogushyn, S.; Merz, B.; Lindenschmidt, K.-E.; Apel, H. A New Methodology for Flood Hazard Assessment Considering Dike Breaches. *Water Resour. Res.* **2010**, *46*, W08541. [CrossRef]
113. D'Oria, M.; Maranzoni, A.; Mazzoleni, M. Probabilistic Assessment of Flood Hazard due to Levee Breaches Using Fragility Functions. *Water Resour. Res.* **2019**, *55*, 8740–8764. [CrossRef]
114. Wang, T.; Li, Z.; Ge, W.; Zhang, Y.; Jiao, Y.; Zhang, H.; Sun, H.; van Gelder, P. Risk Assessment Methods of Cascade Reservoir Dams: A Review and Reflection. *Nat. Hazards* **2023**, *115*, 1601–1622. [CrossRef]
115. Hu, L.; Yang, X.; Li, Q.; Li, S. Numerical Simulation and Risk Assessment of Cascade Reservoir Dam-Break. *Water* **2020**, *12*, 1730. [CrossRef]
116. Lumbroso, D.; Davison, M.; Body, R.; Petkovšek, G. Modelling the Brumadinho Tailings Dam Failure, the Subsequent Loss of Life and How It Could Have Been Reduced. *Nat. Hazards Earth Syst. Sci.* **2021**, *21*, 21–37. [CrossRef]
117. Gold Mine Disaster in Turkey Poses International Public Health Risk. Available online: https://www.wsws.org/en/articles/2024/02/14/izwr-f14.html (accessed on 27 February 2024).
118. Sreekumar, U.; Nistor, I.; Rennie, C.; Mohammadian, A. Numerical Modeling of Downstream Morphological Evolution during Mount Polley Tailings Dam Failure. *J. Hydraul. Eng.* **2024**, *150*, 05023003. [CrossRef]
119. Cea, L.; Costabile, P. Flood Risk in Urban Areas: Modelling, Management and Adaptation to Climate Change. A Review. *Hydrology* **2022**, *9*, 50. [CrossRef]
120. Ferrari, A.; Dazzi, S.; Vacondio, R.; Mignosa, P. Enhancing the Resilience to Flooding Induced by Levee Breaches in Lowland Areas: A Methodology Based on Numerical Modelling. *Nat. Hazards Earth Syst. Sci.* **2020**, *20*, 59–72. [CrossRef]
121. Guinot, V.; Delenne, C.; Soares-Frazão, S. Self-Similar Solutions of Shallow Water Equations with Porosity. *J. Hydraul. Res.* **2023**, *61*, 109–119. [CrossRef]
122. Morales-Hernández, M.B.; Sharif, M.B.; Kalyanapu, A.; Ghafoor, S.K.; Dullo, T.T.; Gangrade, S.; Kao, S.-C.; Norman, M.R.; Evans, K.J. TRITON: A Multi-GPU Open Source 2D Hydrodynamic Flood Model. *Environ. Model. Softw.* **2021**, *141*, 105034. [CrossRef]

water

MDPI

Review

Review of Historical Dam-Break Events and Laboratory Tests on Real Topography for the Validation of Numerical Models

Francesca Aureli [1,*], Andrea Maranzoni [1] and Gabriella Petaccia [2]

1 Department of Engineering and Architecture, University of Parma, Parco Area delle Scienze 181/A, 43124 Parma, Italy; andrea.maranzoni@unipr.it
2 Department of Civil Engineering and Architecture, University of Pavia, via Ferrata 3, 27100 Pavia, Italy; petaccia@unipv.it
* Correspondence: francesca.aureli@unipr.it

Abstract: Dam break inundation mapping is essential for risk management and mitigation, emergency action planning, and potential consequences assessment. To quantify flood hazard associated with dam failures, flooding variables must be predicted by efficient and robust numerical models capable to effectively cope with the computational difficulties posed by complex flows on real topographies. Validation against real-field data of historical dam-breaks is extremely useful to verify models' capabilities and accuracy. However, such catastrophic events are rather infrequent, and available data on the breaching mechanism and downstream flooding are usually inaccurate and incomplete. Nevertheless, in some cases, real-field data collected after the event (mainly breach size, maximum water depths and flood wave arrival times at selected locations, water marks, and extent of flooded areas) are adequate to set up valuable and significant test cases, provided that all other data required to perform numerical simulations are available (mainly topographic data of the floodable area and input parameters defining the dam-break scenario). This paper provides a review of the historical dam-break events for which real-field datasets useful for validation purposes can be retrieved in the literature. The resulting real-field test cases are divided into well-documented test cases, for which extensive and complete data are already available, and cases with partial or inaccurate datasets. Type and quality of the available data are specified for each case. Finally, validation data provided by dam-break studies on physical models reproducing real topographies are presented and discussed. This review aims at helping dam-break modelers: (a) to select the most suitable real-field test cases for validating their numerical models, (b) to facilitate data access by indicating relevant bibliographic references, and (c) to identify test cases of potential interest worthy of further research.

Keywords: dam-break modelling; real-field test cases; validation of numerical models; historical dam-break events; laboratory tests; real topography; review

Citation: Aureli, F.; Maranzoni, A.; Petaccia, G. Review of Historical Dam-Break Events and Laboratory Tests on Real Topography for the Validation of Numerical Models. *Water* **2021**, *13*, 1968. https://doi.org/10.3390/w13141968

Academic Editor: Karl-Erich Lindenschmidt

Received: 30 April 2021
Accepted: 13 July 2021
Published: 17 July 2021

1. Introduction

Artificial dams are hydraulic works of fundamental importance for human development since ancient times due to the great benefits provided by the storage of large water volumes, such as a water supply for drinking and irrigation purposes. The first dams known date back to 3000 BC in Mesopotamia. In recent times, as early as 1800, technological progress has allowed construction of large dams. The most impressive ones have been built within the last century, when, moreover, the number of dams has significantly increased according to water demand. In addition to supplying water and controlling floods, modern dams are often designed for power production.

However, despite its great benefits, the storage of large volumes of water poses serious risks for the downstream areas. Unfortunately, over the centuries, the history of retaining dams has been studded with disasters of various types, sometimes of great magnitude, with loss of human lives and destruction of downstream properties, agricultural lands, historical sites, industrial and productive settlements, and urban areas. Even though a time-related analysis shows that the frequency of failure of large dams has been reduced by a factor of four or more over the last forty years worldwide [1], dam incidents still occur at present with a non-negligible frequency [2–4]. In the current year, the failure of the Rishiganga dam (India), induced by a glacier avalanche, caused a catastrophic flood in the downstream valley with many casualties and huge damages to four hydropower plants. Moreover, the state of emergency was declared over toxic wastewater leaks in Florida due to the dreaded breach of the Piney Point reservoir. In this case a potential disaster was ultimately fortunately averted.

Wider dissemination of awareness on risk factors affecting dam safety is constantly necessary due to the presence of old and new hazardous conditions that can have an adverse effect on stability and efficiency of retention barrages. Some examples of such adverse factors are dam aging [5] and insufficient spillway capacity due to long-term alteration of weather patterns and exacerbated climatic extremes (e.g., [6,7]).

All these elements can increase dam-break flood risk in downstream areas, which is further amplified by growing exposure of human settlements and potential high vulnerability to flooding [8]. For this reason, research related to the assessment of the hydraulic risk resulting from the failure of retaining structures is constantly ongoing and involves scientists from all over the world. Moreover, the lessons learnt by dam accidents and catastrophes in the past are still very relevant [9,10]. Although there are countless works on the general subject of dam safety (e.g., [11–17]), only the ones that significantly helped in drafting the review are mentioned in this paper.

Among the possible aspects linked to the topic of dam safety [18,19], the assessment of flood hazard associated with a potential total or partial dam collapse is of key relevance and is strictly required by national technical guidelines worldwide [20–23]. To this end, relevant flooding variables must be predicted by numerical models, which have seen a growing and unstoppable development for decades, especially thanks to the constant advances in computational techniques (e.g., [24–26]). For use in practical applications, dam-break numerical models must be efficient and robust, and capable to accurately track wetting and drying fronts and handle all the complexities that characterize unsteady free surface flows on real topographies. The verification and validation of these models are then necessary before the application to real case studies. Model verification can be performed by comparing numerical results with available analytical solutions of dam-break problems (which usually hold under the shallow water approximation [27]). However, such analytical solutions, which in some countries have even influenced the drafting of legislation on dams, deal with schematic geometries and cannot adequately represent the complex phenomena occurring on real topographies. On the other hand, numerical models can be validated against field or experimental data. Accordingly, in recent decades, great attention has been devoted to set up suitable databases from laboratory dam-break investigations [28–30].

Field data can be derived from the analysis of historical dam-break events. However, such catastrophic events are fortunately rather rare, and available data on flood dynamics are typically inaccurate and incomplete. Indeed, there is no targeted preparation or general attitude for collecting of a rich amount of information at the dam site and in the downstream areas during a calamitous event of this kind. Nevertheless, in some cases, data collected after or occasionally during the event (mainly maximum water depths and flood wave arrival times at selected locations, and extent of the flooded areas) are sufficient to set up interesting validation test cases, provided that all other essential data required to perform numerical simulations are available (i.e., the digital elevation model of the floodable area and, possibly, of the reservoir, and the input parameters defining the dam-break scenario).

This paper reviews and provides an exhaustive list of the historical dam-break events for which datasets are available for validation purposes. The cases identified are organized in two groups: the former includes well-documented cases with extensive and complete data; the latter refers to cases with incomplete or inaccurate data, which could actually be completed without excessive effort with further research [31–39]. Laboratory investigations on dam-break flow in physical models reproducing real topographies are also considered.

The main aim of this review is to provide modelers with a set of real-field test cases for validation purposes, facilitating the retrieval of the corresponding data. An additional aim is to contribute to the studies related to dam safety, according to the philosophy of the 2019 ICOLD World declaration on Dam Safety: *The profession of dam engineering has a profound ethical responsibility to carry out its professional duties so that dams and reservoirs are designed, constructed and operated in the most effective and sustainable way, while also ensuring that both new and existing dams are safe during their entire lifespan, from construction to decommissioning* [40].

The paper is organized as follows. Well-documented dam-break real events are presented and analyzed in Section 2. Historical dam-break events open to further study to complete the validation datasets are presented in Section 3. In Section 4, dam-break investigations on physical models reproducing real topographies are discussed. Type and quality of the available data for each test case are highlighted in specific tables for the three categories. Finally, some conclusions are drawn in Section 5.

2. Well-Documented Dam-Break Test Cases

Table 1 reports twenty well-documented historical dam-breaks found in the literature for which the writers have deemed the available data adequate to validate numerical models. These events are indeed documented with exhaustive information and real-field data concerning breach formation and the water volume stored in the upstream reservoir before breaching. Data on the propagation of the dam-break flood wave downstream are also available in most of these cases. Some events listed in Table 1 have already been inserted in other databases, especially concerning the peak outflow resulting from an earthen dam failure [41–46]. The same applies to some dam-break cases presented in the next section.

Table 1 is organized in 23 columns. Columns from 2 to 4 report dam name, location and type, respectively. The table is divided into six different sections according to the dam type. Columns 5 and 6 specify the cause and year of the disaster. Among the twenty dam-break events considered, 60% refers to earthfill dams, 25% to concrete dams, 10% to rockfill dams, and 5% to tailing dams. The most frequent causes of failure for this set of historical events are piping (PI, 35%) and overtopping (OT, 30%), followed by quality problems and design errors (QP, 15%), erosion (ER, 15%), and poor management (PM, 5%). Most of the disasters included in this set occurred between 1950 and 2000 (55%), while only 5% occurred before 1900, 15% between 1900 and 1950, and 25% after 2000 (15% between 2000 and 2010, and 10% between 2010 and 2020). The fact that 25% of the well-documented disasters examined here occurred in the last two decades might seem surprising, but it has to be considered that in recent years input and validation data, such as a digital terrain model (DTM), rainfall data, and aerial photographs before and after the flooding, are available more frequently than in the past. Therefore, in the writers' opinion, the presence in Table 1 of a high number of dam-break cases occurred in the last twenty years is probably due to greater availability of information rather than to a real increase in the incidence of this calamitous event.

Table 1. Well-documented dam-break test cases.

(1) N.	(2) Dam Name	(3) Country	(4) Type [1]	(5) Cause [2]	(6) Year	(7) References	Available Information [3] (8) Dam Char.	(9) Reserv. Char.	(10) Reserv. Level	(11) Phot. Docum.	(12) Breach Char.	(13) Dam Mater.	(14) Breach Devel.	(15) DTM	(16) Storm	(17) Peak Flow	(18) Breach Outfl.	(19) Water Marks	(20) Hydrogr.	(21) Flood Timing	(22) Flooded Areas	(23) Sim. Flood [4]
1	Baldwin Hills	Calif. (USA)	EF	PI	1963	[47–52]	•	•	•	•	•	•	•	•		•	○		•	•	•	2D-FV
2	Xe Pian – Xe Namnoy	Laos	EF	ER	2018	[53]	○	•	•	•		•		•	•			•		○	•	1D-FD; 2D-FV
3	Gangneung	South Korea	EF	OT	2002	[54,55]	•	•	•	•				•	•			•	•	○	•	2D-FV
4	Kelly Barnes	Geo. (USA)	EF	PI	1977	[56–59]	•	•	•	•	•	•		○	•	•		•			•	
5	Lake Ha! Ha!	Canada	EF (DK)	OT	1996	[60–68]	•	•	•	•	•	•		•	•	•	•				•*	1D-FD; 2D-FV
6	Lawn Lake & Cascade Lake	Col. (USA)	EF	PI	1982	[69–72]	•	•	•	•	•	•	•	○		•		•		•	○	1D-FV
7	Quail Creek	Utah (USA)	EF	PI	1989	[73–75]	•	•	•		•		•	○		•		•		•		1D-FD
8	Teton	Idaho (USA)	EF	PI	1976	[76–81]	•	•	•	•	•	•	•	○		•		•	•		•	1D-FD
9	Um Al-Khair	S. Arabia	EF	PI	2011	[82]	•	•	•	•	•	•		•	•	•		•				1D-FD
10	Zeyzoun	Syria	EF	PM	2002	[83,84]	•	•						•		•		•	•	•		1D-FD
11	Gleno	Italy	CO	QP	1923	[85]	•	•	•	•	•		•	•				○		○	•	1D-FV
12	Malpasset	France	CO	QP	1959	[86–88]	•	•	•	•	•		•	•				•		•	○	2D-FD, FV; 3D
13	Palagnedra	Switzerland	CO	OT	1978	[89]	•	•	•	•				•	•	•		•	•	•		2D-FD
14	Sella Zerbino	Italy	CO	OT	1935	[90]	•	•	•	•				•	•	•		•		•		2D-FV
15	St. Francis	Calif. (USA)	CO	PI	1928	[91–93]	•	•	•					•				•		•	•	2D-FV; 3D
16	Taum Sauk Up.	Miss. (USA)	RF	QP	2005	[94–98]	•	•	•	•				•		•		•			•	1D-FD
17	Tous	Spain	RF	OT	1982	[99–101]	•	•	•	•				•	•	•		•		•	•	2D-FV
18	South Fork	Penns. (USA)	CF	OT	1889	[73,102–105]	•	•		•	•			•	•	•		•		•	•	
19	La Josephina	Ecuador	LD	ER	1993	[106,107]	•	•	•	•		•	•	○		•	○	•				1D-FD
20	Stava	Italy	TA m	PM-QP	1985	[108–111]	•			•		•		○				•		•		1D-FV; 2D-FV

[1] Type: CF = composite fill; CO = concrete; EF = earthfill; DK = dike; LD = landslide; RF = rockfill; TA = tailing. [2] Cause: ER = erosion; OT = overtopping; PI = piping; PM = poor management; QP = quality problems and design errors. [3] • = complete; ○ = uncertain, incomplete or not readily available; * = geomorphic effects; m = mudflow. [4] Numerical model: approach-method; FD = finite difference; FV = finite volume.

Column 7 of Table 1 provides relevant references in which the selected dam-break events are described. The remaining columns detail the type of available information for validation purposes for each entry. The quality or completeness of this information is indicated through different symbols. The full dot indicates that the available information is rather detailed, exhaustive, and accurate. The empty dot denotes incomplete or vague information (i.e., not accurate or not immediately available in quantitative terms). The availability of dam and reservoir characteristics, as well as of data concerning the water level in the reservoir before the failure, is indicated in Columns 8–10. Column 11 informs if photographic documents on the dam remnants (from which important data on the final breach shape and dimensions can be derived) and on the consequences of the dam-break flood in the downstream area are available. Data availability on breach formation, breach geometry, and dam materials is specified in Columns 12 to 14, while the availability of topographic data and rainfall data of the storm event (which sometimes is related to the dam-break) is indicated in Columns 15 and 16. Finally, the availability of other real-field data useful to validate numerical models is highlighted in Columns 17 to 22 (peak outflow and breach outflow hydrograph, water marks, water elevation or discharge or velocity data at selected locations, flood timing, extent of the flooded areas, respectively).

The peak flow at the dam was recorded or estimated from observations in 70% of cases considered in Table 1. Historical photographic documentation is available in 85% of cases. Rainfall data were recorded only in 45% of cases. A terrain model or topographic cross-sections are available for all the cases. Focusing on data concerning the dam-break flooding, the most available information is related to water marks (90% of cases), while time series of flooding variables are provided only in 25% of cases. Flood timing and maps of flooded areas are available in 70% and 65% of cases, respectively.

As regards the ten dam-break events involving earthfill (EF) dams, breach outflow is rarely known (only in 20% of cases), while breach characteristics and breach development data are available in 70% and 40% of cases, respectively.

Among the five concrete (CO) dam failures listed in Table 1, the Malpasset and St. Francis ones (entry no. 12 and 15, respectively) are probably the most famous. However, water marks and flood timing are available for all these five dam-break cases. Flooded areas are present in 60% of cases, while hydrographs of relevant flooding variables are available only in 20% of cases. An estimate of peak outflow was provided in 40% of cases.

Only two historical disasters involving rockfill (RF) dams were considered as well-documented. Water marks and flooded areas are known for both cases. The Tous dam failure is probably the most famous disaster involving a rockfill dam.

Column 23 indicates whether numerical simulations of the dam-break events were already carried out and which types of models were used. The literature review has shown that 90% of the well-documented cases listed in Table 1 have already been numerically simulated and that only two events have not yet been modelled (i.e., Kelly Barnes and South Fork, entry no. 4 and no. 18 in Table 1, respectively), although enough data for numerical modelling seem to be available in both cases.

Finally, it is worth noting that the Lake Ha! Ha! breakout flood (entry no. 5) was characterized by significant geomorphic effects along the flooded valley, while a highly destructive mudflow was generated by the collapse of two tailing dams in the Stava disaster (entry no. 20). The La Josefina dam-break (entry no. 19) was caused by the erosion of a natural dam produced by a landslide, which obstructed a river confluence forming two ephemeral lakes.

3. Dam-Break Test Cases Open to Further Study

Numerical modelers are usually very interested in lesser known historical dam-break events. Indeed, the simulation of new challenging cases supports a robust assessment of models' potentialities and effectiveness. In addition, real events always pose new challenges to numerical models, which encourage the development of more sophisticated and performant computational codes. Finally, the ability of the numerical models to adequately reproduce different real phenomena is one of the fundamental prerequisites for their use for hydraulic risk assessment.

Table 2 lists thirty seven historical dam-breaks for which the data retrieved in the literature are inaccurate or incomplete. This is confirmed by the fact that only a few of these cases were simulated and used for validation purposes. Often these events are documented with relevant, albeit scarce, data concerning breach formation and geometry, water level in the upstream reservoir before breaching, and propagation of the dam-break wave downstream. However, in the writers' opinion, an in-depth research is possible to gather further information and complete the validation datasets for most of these scarcely documented real cases.

Table 2 is organized with the same structure as Table 1. Almost 60% of the dam-break events included in this table refer to homogeneous (EF) or composite earthfill (CF) dams, 11% to rockfill (RF) dams, 11% to tailing (TA) dams, 6% to concrete (CO) dams; coal waste (CW), gravity (GR), masonry (MA), tephra (TE) or natural (NA) dam categories amount to approximately 3% each. In the cases involving tailing dams, the dam failure resulted in a destructive mudflow. Some events concerning earthfill dams were already considered in the literature, especially in the framework of breach formation and evolution modeling [4,6,28,29,33]. Nevertheless, for these cases, further data are often available, allowing the setting up of interesting benchmark tests for hydrodynamic dam-break models, too. Considering the list of Table 2, the most frequent cause of failure is overtopping (OT, 24%), followed by erosion (ER, 16%), quality problems and design error (QP, 14%), piping (PI, 11%), poor management and liquefaction (PM and LI, 8% each). Only occasionally dam failure is due to sliding (SL), undercut erosion (UE), lahars (LA), earthquakes (EQ), or a combination of different processes. As regards the year of failure, most of the disasters included in this set occurred between 1950 and 2000 (51%), while only 8% occurred before 1900, 14% between 1900 and 1950, and 27% after 2000. Also for this category of dam-break cases, it can be stressed that the presence of a high number of cases occurred in the last twenty years is probably due to greater availability of information rather than to a real increase in the incidence of such calamitous events.

The geometric characteristics of the dam and reservoir are known in 95% and 73% of cases, respectively. The characterization of the dam material is available in 27% of cases. The water level in the reservoir at the time of the collapse was reported only in 62% of selected events. The peak discharge at the dam was estimated in 41% of cases. Historical photographic documentation is available in 73% of cases. Data of the rainfall event, which potentially triggered the dam collapse, were recorded only in 35% of cases. A terrain model or topographical cross-sections are available only for 49% of cases.

Time series of breach outflow are very rare (available only in 5% of cases), while breach characteristics and breach development data are available for 68% and 24% of cases, respectively. Focusing on data concerning dam-break flooding, water marks are documented in 57% of cases. As for the previous set of well-documented cases, hydrographs of flooding variables are rather sporadic (only 24%), as well as data on flood timing (30%). The boundaries of flooded areas are available in 46% of cases.

Column 23 of Table 2 shows that only seven cases (19%) were already simulated by 1D or 2D numerical models. No numerical simulations of the other cases were found in the literature, even though a fair amount of information is present for some of these cases, both in terms of dam and reservoir characteristics and of the hydro-meteorological event that triggered the disaster.

It is worth noting that topographic data concerning the reservoir and downstream floodable area are currently available only for a limited number of events listed in Table 2. In some cases, especially for the more recent events, this gap could be filled considering terrain models freely downloadable from the web and other available resources. Freely downloadable world DTMs (from satellite images) are increasingly popular and are experiencing a constant and rapid development in terms of spatial resolution and accuracy (e.g., [112,113]). It can be predicted that DTMs with a spatial resolution in the order of 10 m will be probably freely available (or available at a low cost) soon. As an example, consider the Ivex dam (entry no. 8 in Table 2), located on the Chagrin River immediately upstream of the village of Chagrin Falls, northeast OH, USA. Thanks to the historical images from Google Earth, it is possible to geolocalize the two cascade reservoirs and reconstruct with satisfactory accuracy the main characteristics of the territory and the built-up areas at the time of the event. The availability of the ALOS Global Digital Surface Model "ALOS World 3D–30m" (AW3D30), freely downloadable from the JAXA website [114], could provide the topographic information (with acceptable spatial resolution) required to perform numerical simulations. Also, the MERIT Hydro dataset and other global digital elevation models (GDEMs) could be considered to set up the test case [115,116]. The roughness parameter involved in shallow water models can be set on the basis of the local land cover, which can be obtained from suitable databases (e.g., [117]). Moreover, in the absence of first-hand data concerning the breaching and flooding processes, useful validation data can be obtained through the analysis of remotely sensed images, which can occasionally be available. As an example, some authors used multi-spectral images from the Sentinel-2A satellite, commonly exploited for forest and land cover change monitoring, to integrate the observations collected in post-event survey campaigns [118].

Approximately half of the earthfill (EF) dam failures present a significant amount of test and validation data (at least 50% of the fifteen data categories considered under the heading of "Available Information" in Table 2). Among these, entry no. 4, referring to the Edenville and Sanford dams (failed on 19 May 2020), shows data in eleven of the fifteen categories considered. For this event, a large number of extraordinarily interesting videos and aerial photos is available online concerning dam breaching and the propagation of the flood wave downstream (e.g., [119,120]). As for the group of earthfill (EF) dams, in only one case the available information covers less than 25% of the data categories considered. Information in more than 50% data categories exists for four of the six composite fill (CF)-type dams and for all the rock fill (RF)-type dams. Only one of the four tailing (TA) dam-break cases is rather well documented. It is worth noting that a considerable amount of data, albeit rather inaccurate, is available for the flood induced on 26 February 1972 by the failure of a coal slurry impoundment at the Buffalo Creek coal mine, in West Virginia, USA (entry no. 33 in Table 2).

Other dam-break events, not included in Table 2 due to the current substantial lack of publicly available data, could be however very interesting. For example, an exceptionally rich photographic documentation is available for the Frenchman dam-break event [121], which occurred on 15 April 1952. This catastrophic dam failure was already mentioned in well-known literature studies on breach formation and evolution modeling [122].

Table 2. Dam-break test cases open to further study.

(1) N.	(2) Dam Name	(3) Country	(4) Type [1]	(5) Cause [2]	(6) Year	(7) References	Available Information [3]															(23) Sim. Flood [4]
							(8) Dam Char.	(9) Reserv. Char.	(10) Reserv. Level	(11) Phot. Docum.	(12) Breach Char.	(13) Dam Mater.	(14) Breach Devel.	(15) DTM	(16) Storm	(17) Peak Flow	(18) Breach Outfl.	(19) Water Marks	(20) Hydrogr.	(21) Flood Timing	(22) Flooded Areas	
1	Apishapa	Col. (USA)	EF	PI	1923	[13,123–127]	●	●	●		●	●				●						
2	Black Hills	S.Dak. (USA)	EF	OT	1972	[128–131]	○			○					●	○		●	●	○	●	
3	Centennial Narrows	Ariz. (USA)	EF	PI	1997	[132–135]	○			○	○				●	●			●			
4	Edenville + Sanford	Mich. (USA)	EF	ER-OT	2020	[136–143]	○	○	○	●	○		●	○	○			○	○		●	2D
5	Fujinuma	Japan	EF	SL-OT	2011	[144,145]	●		●	●	○	○		○							○	
6	Hatchtown	Utah (USA)	EF	QP	1914	[146,147]	○			●	○			○				○		○		
7	Ivanovo	Bulgaria	EF	PM	2012	[148,149]		○		●				●				○			●	
8	Ivex	Ohio (USA)	EF	PI	1994	[150]	●	●	●	●	●		●		●	●						
9	Lake Lee	Mass. (USA)	EF	ER	1968	[151,152]	●	●	●	●				○				●		●	●	
10	Little Deer Creek	Utah (USA)	EF	QP	1963	[153,154]	●	●	●	○	●		●			●		○			○	
11	Nahal Oz	Israel	EF	QP	2001	[155]	●	●	●	●	●			○		●					●*	
12	Niedow	Poland	EF	OT	2010	[156,157]	●	●	●	●	●		●	●		●		●				1D-FD
13	Opuha	New Zealand	EF	PM	1997	[158]	●	●	●	●	●			○		○				●		
14	Otto Run	Penn. (USA)	EF	OT	1977	[48,73,102,159]	●	●			●				●	●						
15	Panshet	India	EF	OT	1961	[160]	●	●							●							1D-FD
16	Zhugou	China	EF	OT	1975	[161]	●	●	●						●	●						
17	Banquiao	China	CF	OT	1975	[161–165]	●	○	○	○	○		●		●	●			○			
18	Belci	Romania	CF	OT	1991	[33,166–168]	●	○	●	○	○				○		○	●	○		○	
19	Big Bay Lake	Mississippi	CF	PI	2004	[73,169,170]	●	○	○	○	○	○	○					●		○	●	1D-FD
20	Bila Desna	Czech. Rep.	CF	ER	1916	[171]	○	○		○	○			○				○			○	
21	Dale Dike	England	CF	ER	1864	[172–176]	○			○	○	○						○			○	
22	Lake Delhi	Iowa (USA)	CF	OT-ER	2010	[177–181]	●		○	●	●		●	○	●				○			
23	Overholser	Okla.(USA)	CO	OT	1923	[182]	●	●			●			●								1D-FD
24	Vega de Tera	Spain	CO	QP	1959	[183]	●			●				●				●				1D-FD
25	Aznalcollar	Spain	RF	LI	1998	[184–187]	●	●		●	●	●		○				○		○	○	
26	Castlewood	Col. (USA)	RF	OT	1933	[122,188–191]	●	●	●	●	●	●			○	●				○	○	
27	Gouhou Dam	China	RF	ER	1993	[163,192–195]	●	●	●	○	●	●				●			○			
28	Hell Hole	Cal. (USA)	RF	PM	1964	[196]	●	●	●	●	○			○	●	●	●	○*	●			
29	Brumadinho	Brazil	TA [m]	LI	2019	[197–202]	○	○		○	○	○	○	●						○	●	2D-FV
30	El Cobre Old	Chile	TA [m]	EQ	1965	[203–205]	●	●										●				
31	Fundão	Brazil	TA [m]	LI	2015	[206,207]	○		○	●				○				○			●	
32	Merriespruit	S. Africa	TA [m]	SL-OT	1994	[208–211]	●	●	●		●			○							●	
33	Buffalo Creek	W.Virg. (USA)	CW	ER	1972	[212–216]	○	○	○	○		○			○			○	○	○		
34	Pantano de Puentes	Spain	GR	QP	1802	[217,218]	●	●										●				
35	Austin	Tex. (USA)	MA	UE	1900	[36,219,220]	●		○	○	○		○					○		○		
36	Tangiwai	New Zealand	TE	LA	1953	[221–223]		●	●							●		●		●		
37	Huohua Lake Natural Dam	China	NA	EQ	2017	[118]	●	○	●	●	●	●		●				○*			○	2D-FD

[1] Type: CF = composite fill embankment; CO = concrete; CW = coal waste; EF = earthfill; GR = gravity; MA = masonry; NA = natural dam; RF = rockfill; TA = tailing; TE = tephra. [2] Cause: EQ = earthquake; ER = erosion; LA = lahar; LI = liquefaction; OT = overtopping; PI = piping; PM = poor management; QP = quality problems and design errors; SL = sliding; UE = undercut erosion. [3] ● = complete; ○ = uncertain or incomplete or not readily available; * = geomorphic effects; [m] = mudflow. [4] Numerical model: approach-method; FD = finite difference; FV = finite volume.

4. Dam-Break Test Cases Based on Physical Modelling

Physical model investigations of dam-break floods over real topographies are rather rare in the literature. Table 3 shows 4 cases found by the authors with extensive experimental datasets. Experimental dam-break test cases concerning schematic idealized geometries are much more numerous and are based on dam-break experiments aimed at analyzing specific issues or hydrodynamic effects of selected topographic singularities on both fixed [224–228] or movable bed (e.g., [229]). However, the complexities occurring, all together, in physical models where a real topography is reproduced cannot be found in these idealized cases.

Among the test cases listed in Table 3, the Toce River test (entry no. 3) concerns a flash flood not induced by a dam-break but produced by an inflow hydrograph supplied through a pump. In this test, the rapidity of the flood propagation is comparable to that typical of dam-break floods. Therefore, the Toce river test case is usually included among the physical model tests for dam-break numerical models, although it is not a dam-break case in the strict sense. In all other cases listed in Table 3, the dam-break flood is generated by the sudden removal of a gate, miming an almost instantaneous collapse of a dam. Physical model test cases concerning flooding (on real topography) caused by levee breaches (e.g., [230]) are not considered in this review.

Probably, the best known dam-break dataset obtained from a physical model concerns the Malpasset dam-break event (entry no. 2 in Table 3). In 1964, an undistorted 1:400-scale physical model was built in the EDF laboratories to reproduce the historical catastrophic flooding induced by the total collapse of the dam, which occurred in 1959. Original data on the model topography and details about the construction technique are not available. However, Goutal [87] provides a dataset of 13,541 points (obtained from digitalization of historical topographic maps) describing the topography of the valley and the floodplain downstream of the dam. Hervouet and Petitjean [86] support the hypothesis that model bottom roughness was adjusted to reproduce correctly the field observations regarding the real event, and suggest that a roughness Manning coefficient ranging from 0.025 to 0.033 $m^{-1/3}s$ is used in the numerical computations based on shallow water equations. During the laboratory investigation, maximum water levels and dam-break wave arrival times were measured at nine points located along the valley downstream of the dam. These data are usually merged with the field data of the real event (water marks and arrival times at some locations) to form a complete dataset for validation purposes. This test case was used by many modelers (e.g., [88,231–233]) and was also considered within the frame of the CADAM European project [28].

A lesser known dam-break test case based on physical modeling concerns the collapse of the Cancano I dam (entry no. 1 in Table 3). The consequences of this hypothetical event were studied by De Marchi [234] on a 1:500-scale physical model reproducing a stretch (approximately 16 km long) of the very irregular alpine valley downstream of the dam. Unfortunately, elevation data of the model bottom (built by connecting several cross-sections of the valley through a concrete-lined surface) are not available. To fill this gap, Pilotti et al. [235] provide a 60-m DTM extracted from a current DTM of the valley. Moreover, they suggest a range of values for the Manning roughness coefficient at the real scale (0.04–0.055 $m^{-1/3}s$), based on dimensional analysis considerations. Moreover, the bathymetry of the Cancano I reservoir was not exactly reproduced in the physical model, and detailed information about this point was not reported in the original work by De Marchi [234]. Despite these shortcomings, this test case is exceptional in the availability of measured discharge hydrographs at three control cross-sections (namely the dam section and two significant cross-sections downstream) for two different scenarios of total and partial dam-break. In addition, the measured boundaries of the flooded areas along a stretch of the downstream valley are available for both the collapse scenarios. Pilotti et al. [236] and Pilotti et al. [235] simulated this test case at the real scale using 1D and 2D models, discussing the issue of potential scale effects that can affect the comparison between numerical results and experimental data.

Table 3. Physical models of dam-break on real topography.

(1) N.	(2) Dam Name	(3) Country	(4) Type [1]	(5) Scale	(6) Year [2]	(7) References	Available Information [3]														(22) Sim. Flood [4]
							(8) Dam Char.	(9) Reserv. Char.	(10) Reserv. Level	(11) Phot. Docum.	(12) Breach Char.	(13) Breach Develop.	(14) Peak Flow	(15) DTM	(16) Rough.	(17) Waterm./ Gauge	(18) Arrival Times	(19) Hydrogr.	(20) Flooded Areas	(21) Breach Outfl.	
1	Cancano I	Italy	CO	1:500	1943	[233–235]	•	○	•	•	•	•	•	○	•		•	•	•	•	1D-FV; 2D-FD, FV
2	Malpasset	France	CO	1:400	1964	[86–88,231–233]	•	•	•		•	•		•	•	•	•				2D-FD, FE, FV, LB; 3D
3	Toce River *	Italy	-	1:100	1999	[237–242]				•			•	•	•	•	•	•			2D-FV
4	Ürkmez	Turkey	EF	1:150 h–1:30 v	2013	[243,244]	•	•	•	•	•	•	•	○	○	•	•	•	○	•	2D-FD

[1] Type: CO = concrete; EF = earthfill. [2] Year: year of construction of the physical model. * = Flash flood (No dam-break). [3] • = complete; ○ = uncertain or incomplete or not readily available. [4] Numerical model: approach-method; FD = finite difference; FE = finite elements; FV = finite volume; LB = lattice Boltzmann.

Another interesting test case proposed in the framework of the European IMPACT research program [29,30] refers to a hypothetical flash flood in a 1:100-scale physical model reproducing a 5-km long stretch of the Toce valley (entry no. 3 in Table 3), in which an idealized urban district is placed [237]. Data concerning the model topography (9928 bottom elevation points) and the locations of the buildings (organized according to two different patterns) are provided (at the model scale) in supplemental files. Moreover, the authors suggest the optimal value of 0.0162 $m^{-1/3}s$ for the roughness coefficient of the concrete bottom of the physical model. Three inflow hydrographs with different peak discharges were considered as upstream boundary conditions. Water depth time series were measured at ten selected points, eight of them located within or near the urban area. This case was used by a large number of modelers to validate numerical models of urban flooding (e.g., [238–242]).

The last test case reported in Table 3 (entry no. 4) refers to the experimental investigation carried out by Guney et al. [243] on a distorted (1:150 horizontal scale and 1:30 vertical scale) physical model reproducing the reservoir of the Ürkmez dam and the downstream region (including an urban area) potentially floodable as a result of the hypothetical collapse of the dam. Only a topographic map of the floodable area is available and no indications are provided on reasonable roughness values for the concrete bottom of the model. A partial dam-break with trapezoidal breach was assumed. Flood depth time series were measured at eight selected positions in the floodable area, along with time series of vertically-averaged velocities and vertical profiles of flow velocity at selected times at other four points. Finally, some information about the timing of the dam-break wave propagation was obtained from videos of the experimental runs. Haltas et al. [244] simulated this test case at the real scale using a combined 1D-2D numerical model and compared predicted flow depth hydrographs with the (scaled) experimental ones. A 1-m DTM was used to create the computational grid and maps of land use were utilized to estimate a spatial distribution of Manning's roughness coefficient.

The physical model test cases considered in this review were chosen on the basis of the presence of experimental data on the propagation of the dam-break wave in downstream areas. However, it is worth noting that several laboratory experiments on breach formation and growth in earthfill dams are reported in the literature (e.g., [245,246]). Valuable and interesting results of experimental field investigations at the prototype scale are also available (e.g., [247]). All related experimental data can be useful for validating numerical models of breach development.

5. Conclusions

This paper outlines historical dam-break events currently and potentially available for the validation of dam-break models. This kind of data is indeed very useful to assess the predictive capabilities of numerical models on real topographies. Data available for each test case have been listed in two different tables concerning well-documented and scarcely documented cases, respectively.

These tables can be useful for modelers to select the most suitable real-field test cases for validation purposes, thereby supporting the development of robust and accurate computational models of dam-break flow. The extensive literature review helps the modelers to identify relevant references where available information on the selected dam-break events can be retrieved.

It is worth noting that two of the well-documented dam-break cases have not been simulated yet, despite the available dataset appears extensive and complete. The list of scarcely documented historical dam-break events (for which the dataset is rather incomplete or inaccurate) indicates potential test cases open to further research, which in addition are of potential scientific and historical interest. For this set, approximately half of the cases are characterized by a significant amount of data, covering at least half of the model and validation data categories considered. Gaps in the availability of topographic data and roughness information can be filled using freely downloadable world DEMs and land cover databases.

Dam-break test cases based on physical models reproducing real topographies have been listed in a separate table, in which the experimental data available for validation purposes are specified. Further physical model investigations concerning dam-break flows on real topographies would be welcome to widen available experimental datasets. Actually, extensive and accurate datasets of relevant hydraulic variables can easily be obtained from laboratory investigations on physical models under controlled conditions. However, in this case, particular care must be paid to the model scale to avoid scale effects.

Author Contributions: F.A., A.M. and G.P. contributed equally to the conceptualization and implementation of the research, to the analysis of the results and to the writing of the manuscript. All authors have read and agreed to the published version of the manuscript.

Funding: This work was supported by the Italian Ministry of University and Research through the PRIN 2017 Project RELAID "REnaissance of LArge Italian Dams", project number 2017T4JC5K.

Data Availability Statement: The data that support the findings of this study were derived from public domain resources.

Conflicts of Interest: The authors declare no conflict of interest.

References

1. ICOLD. Dams' Safety. Available online: https://www.icold-cigb.org/GB/dams/dams_safety.asp (accessed on 24 April 2021).
2. Douglas, K.; Spannagle, M.; Fell, R. Analysis of Concrete and Masonry Dam Incidents. *Int. J. Hydropower Dams* **1999**, *6*, 108–115.
3. Foster, M.; Fell, R.; Spannagle, M. The Statistics of Embankment Dam Failures and Accidents. *Can. Geotech. J.* **2000**, *37*, 1000–1024. [CrossRef]
4. ICOLD. *Dam Failures—Statistical Analysis*; Bulletin 188; ICOLD: Paris, France, 2021.
5. Perera, D.; Smakhtin, V.; Williams, S.; North, T.; Curry, A. *Water Storage Infrastructure: An Emerging Global Risk*; UNU-INWEH Report Series, Issue 11; United Nations University Institute for Water, Environment and Health: Hamilton, ON, Canada, 2021. Available online: https://inweh.unu.edu/wp-content/uploads/2021/01/Ageing-Water-Storage-Infrastructure-An-Emerging-Global-Risk_web-version.pdf (accessed on 7 July 2021).
6. Bocchiola, D.; Rosso, R. Safety of Italian Dams in the Face of Flood Hazard. *Adv. Water Resour.* **2014**, *71*, 23–31. [CrossRef]
7. ICOLD. *Global Climate Change, Dams, Reservoirs and Related Water Resources*; Bulletin 169; ICOLD: Paris, France, 2018.
8. ICOLD. *Dam Break Flood Analysis—Review and Recommendations*; Bulletin 111; ICOLD: Paris, France, 1998.
9. ICOLD. *Lessons from Dam Incidents*; ICOLD: Paris, France, 1973.
10. Charles, J.A.; Tedd, P.; Warren, A. *Lessons from Historical Dam Incidents*; Environment Agency: Bristol, UK, 2011. Available online: https://assets.publishing.service.gov.uk/government/uploads/system/uploads/attachment_data/file/290812/scho0811buba-e-e.pdf (accessed on 7 July 2021).
11. ICOLD. *Dam Failures—Statistical Analysis*; Bulletin 99; ICOLD: Paris, France, 1995.
12. U.S. Department of the Interior, Bureau of Reclamation. *RCEM–Dam Failure and Flood Event Case History Compilation*; U.S. Department of the Interior, Bureau of Reclamation, 2015. Available online: https://www.usbr.gov/ssle/damsafety/documents/RCEM-CaseHistories2015.pdf (accessed on 7 July 2021).
13. Wahl, T.L. *Prediction of Embankment Dam Breach Parameters—A Literature Review and Needs Assessment*; DSO-98-004, Dam Safety Research Report; U.S. Department of the Interior, Bureau of Reclamation, Dam Safety Office: Denver, CO, USA, 1998.
14. Wahl, T.L. *Evaluation of Erodibility-Based Embankment Dam Breach Equations*; Hydraulic Laboratory Report HL-2014-02; U.S. Department of the Interior, Bureau of Reclamation, Technical Service Center: Denver, CO, USA, 2014.
15. Froehlich, D.C. Embankment Dam Breach Parameters and Their Uncertainties. *J. Hydraul. Eng.* **2008**, *134*, 1708–1721. [CrossRef]
16. Rong, G.; Wang, X.; Xu, H.; Xiao, B. Multifactor Regression Analysis for Predicting Embankment Dam Breaching Parameters. *J. Hydraul. Eng.* **2020**, *146*, 04019051. [CrossRef]
17. USCOLD. *Lessons from Dam Incidents*; American Society of Civil Engineers: New York, NY, USA, 1975.
18. Adamo, N.; Al-Ansari, N.; Ali, S.H.; Laue, J.; Knutsson, S. Special Issue–Dam Safety Review: General Principles and Procedures-Part 1. *J. Earth Sci. Geotech. Eng.* **2020**, *10*, 1–348.
19. Adamo, N.; Al-Ansari, N.; Ali, S.H.; Laue, J.; Knutsson, S. Special Issue-Dam Safety Review: General Principles and Procedures-Part 2. *J. Earth Sci. Geotech. Eng.* **2021**, *11*, 1–438.
20. FEMA. Summary for Policymakers. In *Climate Change 2013—The Physical Science Basis*; Intergovernmental Panel on Climate Change, Ed.; Cambridge University Press: Cambridge, UK, 2013.
21. FEMA. *Federal Guidelines for Inundation Mapping of Flood Risks Associated with Dam Incidents and Failures*; FEMA P-946; 2013. Available online: https://www.fema.gov/sites/default/files/2020-08/fema_dam-safety_inundation-mapping-flood-risks.pdf (accessed on 7 July 2021).

22. European Council Directive 2007/60/EC of the European Parliament and of the Council of 23 October 2007 on the Assessment and Management of Flood Risks. 2007, pp. 27–34. Available online: https://eur-lex.europa.eu/legal-content/EN/ALL/?uri=CELEX%3A32007L0060 (accessed on 7 July 2021).
23. ICOLD European Club. Working Group on Dam Legislation. *Dam Legislation*. Final Report. 2018. Available online: https://britishdams.org/assets/documents/News%20Item%20Docs/2018/ICOLD%20EurClub%20-Dam%20Legislation%20Report%20-%20Dec%202017.pdf (accessed on 7 July 2021).
24. Almeida, A.B.; Franco, A.B. Modeling of Dam-Break Flow. In *Computer Modeling of Free-Surface and Pressurized Flows*; NATO ASI Series (Series E: Applied, Sciences); Chaudhry, M.H., Mays, L.W., Eds.; Springer: Dordrecht, The Netherlands, 1994; Volume 274, pp. 343–373.
25. Toro, E.F.; Garcia-Navarro, P. Godunov-type Methods for Free-Surface Shallow Flows: A Review. *J. Hydraul. Res.* **2007**, *45*, 736–751. [CrossRef]
26. Teng, J.; Jakeman, A.J.; Vaze, J.; Croke, B.F.W.; Dutta, D.; Kim, S. Flood Inundation Modelling: A Review of Methods, Recent Advances and Uncertainty Analysis. *Environ. Model. Softw.* **2017**, *90*, 201–216. [CrossRef]
27. Stoker, J.J. *Water Waves*; John Wiley & Sons: Hoboken, NJ, USA, 1992.
28. Morris, M. *CADAM: Concerted Action on Dambreak Modeling-Final Report*; Rep. No. SR 571; HR Wallingford: Wallingford, UK, 2000. Available online: http://www.ib-nujic.de/Dammbruch/CADAM_final.pdf (accessed on 7 July 2021).
29. Morris, M.; Hassan, M. IMPACT: Investigation of Extreme Flood Processes and Uncertainty—A European Research Project. In Proceedings of the Defra Flood and Coastal Management Conference, York, UK, 5–7 July 2005. Available online: https://eprints.hrwallingford.com/555/1/HRPP346_IMPACT_Investigation_of_extreme_flood_processes_and_uncertainty_-_a_European_research_project.pdf (accessed on 7 July 2021).
30. Zech, Y.; Soares-Frazão, S. Dam-Break Flow Experiments and Real-case Data. A Database from the European IMPACT Research. *J. Hydraul. Res.* **2007**, *45*, 5–7. [CrossRef]
31. Saxena, K.R.; Sharma, V.M. *Dams: Incidents and Accidents*; CRC Press: Boca Raton, FL, USA, 2004.
32. Singh, V.P.; Scarlatos, P.D. Analysis of Gradual Earth–Dam Failure. *J. Hydraul. Eng.* **1988**, *114*, 21–42. [CrossRef]
33. Singh, V. *Dam Breach Modeling Technology*; Springer Science & Business Media: Berlin/Heidelberg, Germany, 1996; Volume 17.
34. De Wrachien, D.; Mambretti, S. *Dam-Break Problems, Solutions and Case Studies*; de Wrachien, D., Mambretti, S., Eds.; WIT Press: Boston, MA, USA, 2009.
35. Zhang, L.M.; Xu, Y.; Jia, J.S. Analysis of Earth Dam Failures: A Database Approach. *Georisk* **2009**, *3*, 184–189. [CrossRef]
36. Bartholomew, C.L. Failure of Concrete Dams. In Proceedings of the 6th Association of State Dam Safety Officials Annual Conference, Albuquerque, NM, USA, 1–5 October 1989; pp. 428–445. Available online: https://damfailures.org/wp-content/uploads/2015/07/Failure-of-Concrete-Dams.pdf (accessed on 7 July 2021).
37. You, L.; Li, C.; Min, X.; Xiaolei, T. Review of Dam-Break Research of Earth-rock Dam Combining with Dam Safety Management. *Procedia Eng.* **2012**, *28*, 382–388. [CrossRef]
38. Zhang, L.; Peng, M.; Chang, D.; Xu, Y. *Dam Failure Mechanisms and Risk Assessment*; John Wiley & Sons Singapore Pte. Ltd.: Singapore, 2016.
39. Chanson, H. *Environmental Hydraulics for Open Channel Flows*; Elsevier Butterworth-Heinemann: Oxford, UK, 2004.
40. ICOLD. World Declaration on Dam Safety. 2019. Available online: https://www.ussdams.org/wp-content/uploads/2020/01/World-Declaration-on-Dam-Safety_final.pdf (accessed on 7 July 2021).
41. Hood, K.; Perez, R.A.; Cieplinski, H.E.; Hromadka, T.V.; Moglen, G.E.; McInvale, H.D. Development of an Earthen Dam Break Database. *JAWRA J. Am. Water Resour. Assoc.* **2019**, *55*, 89–101. [CrossRef]
42. Wang, B.; Liu, W.; Zhang, J.; Chen, Y.; Wu, C.; Peng, Y.; Wu, Z.; Liu, X.; Yang, S. Enhancement of Semi-Theoretical Models for Predicting Peak Discharges in Breached Embankment Dams. *Environ. Fluid Mech.* **2020**, *20*, 885–904. [CrossRef]
43. Wang, B.; Chen, Y.; Wu, C.; Peng, Y.; Song, J.; Liu, W.; Liu, X. Empirical and Semi-Analytical Models for Predicting Peak Outflows Caused by Embankment Dam Failures. *J. Hydrol.* **2018**, *562*, 692–702. [CrossRef]
44. Ashraf, M.; Soliman, A.H.; El-Ghorab, E.; El Zawahry, A. Assessment of Embankment Dams Breaching using Large Scale Physical Modeling and Statistical Methods. *Water Sci.* **2018**, *3*, 362–379. [CrossRef]
45. Sattar, A.M.A. Gene Expression Models for Prediction of Dam Breach Parameters. *J. Hydroinform.* **2014**, *16*, 550–571. [CrossRef]
46. Zhong, Q.; Chen, S.; Fu, Z.; Shan, Y. New Empirical Model for Breaching of Earth-Rock Dams. *Nat. Hazards Rev.* **2020**, *21*, 06020002. [CrossRef]
47. Gallegos, H.A.; Schubert, J.E.; Sanders, B.F. Two-dimensional, High-resolution Modeling of Urban Dam-Break Flooding: A Case Study of Baldwin Hills, California. *Adv. Water Resour.* **2009**, *32*, 1323–1335. [CrossRef]
48. MacDonald, T.C.; Langridge–Monopolis, J. Breaching Characteristics of Dam Failures. *J. Hydraul. Eng.* **1984**, *110*, 567–586. [CrossRef]
49. Leonards, G.A. Baldwin Hills Reservoir Failure. In Proceedings of the International Conference on Case Histories in Geotechnical Engineering, Arlington, VA, USA, 6–11 May 1984; Missouri University of Science and Technology, Ed.; University of Missouri-Rolla: St. Rolla, MO, USA, 1984; pp. 1581–1588.
50. Scott, R.F. Baldwin Hills Reservoir Failure in Review. *Eng. Geol.* **1987**, *24*, 103–125. [CrossRef]
51. Costa, J.E. *Floods from Dam Failures*; U.S. Department of the Interior Geological Survey; Open-File Report 85-560; U.S. Department of the Interior: Washington, DC, USA, 1985.

52. Hamilton, D.H.; Meehan, R.L. Ground Rupture in the Baldwin Hills. *Science* **1971**, *172*, 333–344. [CrossRef] [PubMed]
53. Latrubesse, E.M.; Park, E.; Sieh, K.; Dang, T.; Lin, Y.N.; Yun, S.-H. Dam Failure and a Catastrophic Flood in the Mekong Basin (Bolaven Plateau), Southern Laos, 2018. *Geomorphology* **2020**, *362*, 107221. [CrossRef]
54. Chikamori, H. Rainfall-Runoff Analysis of Flooding Caused by Typhoon RUSA in 2002 in the Gangneung Namdae River Basin, Korea. *J. Nat. Disaster Sci.* **2004**, *26*, 95–100.
55. Kim, B.; Sanders, B.F. Dam-Break Flood Model Uncertainty Assessment: Case Study of Extreme Flooding with Multiple Dam Failures in Gangneung, South Korea. *J. Hydraul. Eng.* **2016**, *142*, 05016002. [CrossRef]
56. Crisp, R.L.; Fox, W.E.; Robison, R.C.; Sauer, V.B.; Geological Survey Federal Investigative Board. *The 1977 Toccoa Flood, Report of Failure of Kelly Barnes Dam Flood and Findings*; USGS: Washington, DC, USA, 1977.
57. Sowers, G.F. *Reconnaissance Report on the Failure of Kelly Barnes Lake Dam, Toccoa Falls, Georgia*; Committee on Natural Disaster, Commission on Sociotechnical Systems, National Research Council; National Academies Press: Washington, DC, USA, 1978.
58. Land, L.F. Mathematical Simulations of the Toccoa Falls, Georgia, Dam-Break Flood. *J. Am. Water Resour. Assoc.* **1980**, *16*, 1041–1048. [CrossRef]
59. Reed, S.; Halgren, J. Validation of a New GIS Tool to Rapidly Develop Simplified Dam Break Models. In Proceedings of the Association of State Dam Safety Officials Annual Conference 2011, Washington, DC, USA, 25–29 September 2011; Volume 2, pp. 1860–1877.
60. Capart, H.; Spinewine, B.; Young, D.L.; Zech, Y.; Brooks, G.R.; Leclerc, M.; Secretan, Y. The 1996 Lake Ha! Ha! Breakout Flood, Québec: Test Data for Geomorphic Flood Routing Methods. *J. Hydraul. Res.* **2007**, *45*, 97–109. [CrossRef]
61. Brooks, G.R.; Lawrence, D.E. The Drainage of the Lake Ha! Ha! Reservoir and Downstream Geomorphic Impacts along Ha! Ha! River, Saguenay Area, Quebec, Canada. *Geomorphology* **1999**, *28*, 141–167. [CrossRef]
62. Lin, C.A. A Coupled Atmospheric-Hydrological Modeling Study of the 1996 Ha! Ha! River Basin Flash Flood in Québec, Canada. *Geophys. Res. Lett.* **2002**, *29*, 1026. [CrossRef]
63. Lapointe, M.F.; Secretan, Y.; Driscoll, S.N.; Bergeron, N.; Leclerc, M. Response of the Ha! Ha! River to the Flood of July 1996 in the Saguenay Region of Quebec: Large-scale Avulsion in a Glaciated Valley. *Water Resour. Res.* **1998**, *34*, 2383–2392. [CrossRef]
64. Ferreira, R.M.L.; Leal, J.G.B.; Cardoso, A.H. Mathematical Modeling of the Morphodynamic Aspects of the 1996 Flood in the Ha! HA! River. In Proceedings of the 31st IAHR World Congress 2005, Seoul, Korea, 11–16 September 2005; pp. 3434–3445.
65. Tremblay, M.; Guillaud, C. The 1996 Saguenay Flood Event and its Impacts. *Nat. Hazards* **2019**, *98*, 79–89. [CrossRef]
66. AlQasimi, E.; Mahdi, T.-F. Flooding of the Saguenay Region in 1996: Part 1—Modeling River Ha! Ha! Flooding. *Nat. Hazards* **2019**, *96*, 1–15. [CrossRef]
67. AlQasimi, E.; Pelletier, P.; Mahdi, T.-F. Flooding of the Saguenay Region in 1996. Part 2: Aux Sables River Flood Mitigation and Environmental Impact Assessment. *Nat. Hazards* **2019**, *96*, 17–32. [CrossRef]
68. Yu, W.; Lin, C.A.; Benoit, R. High Resolution Simulation of the Severe Precipitation Event over the Saguenay, Quebec Region in July 1996. *Geophys. Res. Lett.* **1997**, *24*, 1951–1954. [CrossRef]
69. Jarrett, R.D.; Costa, J.E. *Hydrology, Geomorphology, and Dam-Break Modeling of the 15 July 1982 Lawn Lake Dam and Cascade Lake Dam failures, Larimer County, Colorado (USA)*; U.S. Geological Survey Professional Paper 1369; Department of the Interior, U.S. Geological Survey: Washington, DC, USA, 1986.
70. Paquier, A. New Methods for Modelling Dam-Break Wave. In Proceedings of the Specialty Conference on Modelling of Flood Propagation Over Initially Dry Areas, Milan, Italy, 29 June–1 July 1994; pp. 229–240.
71. Paquier, A.; Robin, O. CASTOR: Simplified Dam-Break Wave Model. *J. Hydraul. Eng.* **1997**, *123*, 724–727. [CrossRef]
72. Baker, M.E.; McCormick, B. 30th Anniversary of the Lawn Lake Dam Failure: A Look Back at the State and Federal Response. In Proceedings of the Association of State Dam Safety Officials Annual Conference, Denver, CO, USA, 16–20 September 2012; Volume 2, pp. 1495–1525. Available online: https://damfailures.org/wp-content/uploads/2017/11/Baker_30thAnniver.pdf (accessed on 7 July 2021).
73. Pierce, M.W.; Thornton, C.I.; Abt, S.R. Predicting Peak Outflow from Breached Embankment Dams. *J. Hydrol. Eng.* **2010**, *15*, 338–349. [CrossRef]
74. Trieste, D.J. Evaluation of Supercritical/Subcritical Flows in High-Gradient Channel. *J. Hydraul. Eng.* **1992**, *118*, 1107–1118. [CrossRef]
75. Singh, A.K.; Kothyari, U.C.; Ranga Raju, K.G. Rapidly Varying Transient Flows in Alluvial Rivers. *J. Hydraul. Res.* **2004**, *42*, 473–486. [CrossRef]
76. Muhunthan, B.; Pillai, S. Teton dam, USA: Uncovering the Crucial Aspect of its Failure. *Proc. Inst. Civ. Eng. Civ. Eng.* **2008**, *161*, 35–40. [CrossRef]
77. Solava, S.; Delatte, N. Lessons from the Failure of the Teton Dam. In Proceedings of the 3rd Forensic Engineering Congress, San Diego, CA, USA, 19–21 October 2003; pp. 178–189.
78. Chen, T.-Y.K.; Capart, H. Kinematic Wave Solutions for Dam-Break Floods in Non-Uniform Valleys. *J. Hydrol.* **2020**, *582*, 124381. [CrossRef]
79. Lai, W.; Khan, A.A. Discontinuous Galerkin Method for 1D Shallow Water Flows in Natural Rivers. *Eng. Appl. Comput. Fluid Mech.* **2012**, *6*, 74–86. [CrossRef]
80. Land, L.F. *Evaluation of Selected Dam-Break Flood-Wave Models by Using Field Data*; Water-Resources Investigations 80-44; Department of the Interior, U.S. Geological Survey: Washington, DC, USA, 1980.

81. Balloffet, A.; Scheffler, M.L. Numerical analysis of the Teton Dam failure flood. *J. Hydraul. Res.* **1982**, *20*, 317–328. [CrossRef]
82. Azeez, O.; Elfeki, A.; Kamis, A.S.; Chaabani, A. Dam Break Analysis and Flood Disaster Simulation in Arid Urban Environment: The Um Al-Khair Dam Case Study, Jeddah, Saudi Arabia. *Nat. Hazards* **2020**, *100*, 995–1011. [CrossRef]
83. Irvem, A.; Ozbuldu, M. Evaluation of Flood Simulation for Zeyzoun Dam-Break in Syria using HEC-RAS Model. *Fresenius Environ. Bull.* **2020**, *29*, 1250–1255.
84. Saleh, H.; Allaert, G. Mitigating Urban Flood Disasters in Syria: A Case Study of the Massive Zeyzoun Dam Collapse. In Proceedings of the World Water Week 2009, Stockholm, Sweden, 16–22 August 2009; Ghent University, Stockholm International Water Institute: Stockholm, Sweden, 2009; pp. 192–193.
85. Pilotti, M.; Maranzoni, A.; Tomirotti, M.; Valerio, G. 1923 Gleno Dam Break: Case Study and Numerical Modeling. *J. Hydraul. Eng.* **2011**, *137*, 480–492. [CrossRef]
86. Hervouet, J.-M.; Petitjean, A. Malpasset Dam-Break Revisited with Two-Dimensional Computations. *J. Hydraul. Res.* **1999**, *37*, 777–788. [CrossRef]
87. Goutal, N. The Malpasset Dam Failure. An Overview and Test Case Definition. In Proceedings of the 4th CADAM Meeting, Zaragoza, Spain, 18–19 November 1999.
88. Valiani, A.; Caleffi, V.; Zanni, A. Case Study: Malpasset Dam—Break Simulation using a Two-Dimensional Finite Volume Method. *J. Hydraul. Eng.* **2002**, *128*, 460–472. [CrossRef]
89. Bruschin, J.; Bauer, S.; Delley, P.; Trucco, G. The Overtopping of the Palagnedra Dam. *Int. Water Power Dam Constr.* **1982**, *34*, 13–19.
90. Petaccia, G.; Natale, L. 1935 Sella Zerbino Dam-Break Case Revisited: A New Hydrologic and Hydraulic Analysis. *J. Hydraul. Eng.* **2020**, *146*, 05020005. [CrossRef]
91. Waltham, T. St Francis: The World's Worst Dam Site. *Geol. Today* **2018**, *34*, 100–108. [CrossRef]
92. Begnudelli, L.; Sanders, B.F. Simulation of the St. Francis Dam-Break Flood. *J. Eng. Mech.* **2007**, *133*, 1200–1212. [CrossRef]
93. Witcher, T.R. From Fame to Failure: The St. Francis Dam (Part 2). *Civ. Eng. Mag. Arch.* **2019**, *89*, 42–45.
94. Rogers, J.D.; Watkins, C.M.; Chung, J.W. The 2005 Upper Taum Sauk Dam failure: A case history. *Environ. Eng. Geosci.* **2010**, *16*, 257–289. [CrossRef]
95. Luna, R.; Wronkiewicz, D.; Rydlund, P.; Krizanich, G.; Shaver, D. Damage Evaluation of the Taum Sauk Reservoir Failure Using LiDAR. In Proceedings of the Geo-Denver 2007, Denver, CO, USA, 18–21 February 2007.
96. Watkins, C.M.; David Rogers, J. Overview of the Taum Sauk Pumped Storage Power Plant Upper Reservoir Failure, Reynolds County, MO. In Proceedings of the Association of State Dam Safety Officials Annual Conference 2010, Seattle, WA, USA, 19–23 September 2010; Volume 2, pp. 1859–1879.
97. Gehring, Q.; Luna, R. Evaluation of the Taum Sauk Upper Reservoir Failure. In Proceedings of the 6th International Conference on Case Histories in Geotechnical Engineering, Arlington, VA, USA, 11–16 August 2008; Missouri University of Science and Technology: St. Rolla, MO, USA, 2008.
98. Rydlund, P.H. *Peak Discharge, Flood Profile, Flood Inundation, and Debris Movement Accompanying the Failure of the Upper Reservoir at the Taum Sauk Pump Storage Facility near Lesterville, Missouri*; Scientific Investigations Report 2006-5284; U.S. Department of the Interior, U.S. Geological Survey: Washington, DC, USA, 2006. Available online: https://pubs.usgs.gov/sir/2006/5284/pdf/SIR06-5284_508.pdf (accessed on 7 July 2021).
99. Alcrudo, F.; Mulet, J. Description of the Tous Dam Break Case Study (Spain). *J. Hydraul. Res.* **2007**, *45*, 45–57. [CrossRef]
100. Serra-Llobet, A.; Tàbara, J.D.; Sauri, D. The Tous Dam Disaster of 1982 and the Origins of Integrated Flood Risk Management in Spain. *Nat. Hazards* **2013**, *65*, 1981–1998. [CrossRef]
101. Bellos, V.; Tsakiris, G. 2D Flood Modelling: The Case of Tous Dam Break. In Proceedings of the 36th IAHR World Congress, The Hague, The Netherlands, 28 June–3 July 2015.
102. Coleman, N.M.; Kaktins, U.; Wojno, S. Dam-Breach Hydrology of the Johnstown Flood of 1889—Challenging the Findings of the 1891 Investigation Report. *Heliyon* **2016**, *2*, e00120. [CrossRef]
103. Kaktins, U.; Todd, C.D.; Wojno, S.; Coleman, N. Revisiting the Timing and Events Leading to and Causing the Johnstown Flood of 1889. *Pa. Hist.* **2013**, *80*, 335–363. [CrossRef]
104. Ward, S.N. The 1889 Johnstown, Pennsylvania Flood: A Physics-Based Simulation. In *The Tsunami Threat—Research and Technology*; IntechOpen: London, UK, 2011. Available online: https://www.intechopen.com/books/the-tsunami-threat-research-and-technology/the-1889-johnstown-pennsylvania-flood-a-physics-based-simulation (accessed on 7 July 2021).
105. Brua, S.A. *Floods of 19–20 July 1977 in the Johnstown Area, Western Pennsylvania*; Open-File Report 78-963; U.S. Department of the Interior, U.S. Geological Survey: Harrisburg, PA, USA, 1978. Available online: https://pubs.er.usgs.gov/publication/ofr78963 (accessed on 7 July 2021).
106. United Nations, Department of Humanitarian Affairs. *Landslide La Josefina on the Paute River, Cuenca, Ecuador: Report on Disaster Management*; Open-File Report 78-963; Department of Humanitarian Affairs: Geneva, Switzerland, 1993. Available online: https://digitallibrary.un.org/record/229691?ln=en (accessed on 7 July 2021).
107. Cadier, E.; Zevallos, O.; Basabe, P. Le Glissement de Terrain et les Inondations Catastrophiques de La Josefina en Équateur. *Bull. Inst. Fr. Études Andines* **1996**, *25*, 421–441. Available online: https://horizon.documentation.ird.fr/exl-doc/pleins_textes/divers14-12/010011840.pdf (accessed on 7 July 2021).
108. Luino, F.; De Graff, J.V. The Stava Mudflow of 19 July 1985 (Northern Italy): A Disaster that Effective Regulation Hight Have Prevented. *Nat. Hazards Earth Syst. Sci.* **2012**, *12*, 1029–1044. [CrossRef]

109. Pirulli, M.; Barbero, M.; Marchelli, M.; Scavia, C. The Failure of the Stava Valley Tailings Dams (Northern Italy): Numerical Analysis of the Flow Dynamics and Rheological Properties. *Geoenviron. Disasters* **2017**, *4*, 3. [CrossRef]
110. Schippa, L.; Pavan, S. Numerical Modelling of Catastrophic Events Produced by Mud or Debris Flows. *Int. J. Saf. Secur. Eng.* **2011**, *1*, 403–422. [CrossRef]
111. Chandler, R.J.; Tosatti, G. The Stava Tailings Dams Failure, Italy, July 1985. *Proc. Inst. Civ. Eng. Geotech. Eng.* **1995**, *113*, 67–79. [CrossRef]
112. Musa, Z.N.; Popescu, I.; Mynett, A. A Review of Applications of Satellite SAR, Optical, Altimetry and DEM Data for Surface Water Modelling, Mapping and Parameter Estimation. *Hydrol. Earth Syst. Sci.* **2015**, *19*, 3755–3769. [CrossRef]
113. Yan, K.; Di Baldassarre, G.; Solomatine, D.P.; Schumann, G.J.P. A Review of Low-Cost Space-Borne Data for Flood Modelling: Topography, Flood Extent and Water Level. *Hydrol. Process.* **2015**, *29*, 3368–3387. [CrossRef]
114. Japan Aerospace Exploration Agency, E.O.R.C. ALOS Global Digital Surface Model "ALOS World 3D-30m (AW3D30)"—Product Description. Available online: https://www.eorc.jaxa.jp/ALOS/en/aw3d30/ (accessed on 7 July 2021).
115. McClean, F.; Dawson, R.; Kilsby, C. Implications of Using Global Digital Elevation Models for Flood Risk Analysis in Cities. *Water Resour. Res.* **2020**, *56*, e2020WR028241. [CrossRef]
116. Azizian, A.; Brocca, L. Determining the Best Remotely Sensed DEM for Flood Inundation Mapping in Data Sparse Regions. *Int. J. Remote Sens.* **2020**, *41*, 1884–1906. [CrossRef]
117. CORINE Land Cover—Copernicus Land Monitoring Service. Available online: https://land.copernicus.eu/pan-european/corine-land-cover (accessed on 7 July 2021).
118. Wang, D.; Zhou, Y.; Pei, X.; Ouyang, C.; Du, J.; Scaringi, G. Dam-Break Dynamics at Huohua Lake Following the 2017 Mw 6.5 Jiuzhaigou Earthquake in Sichuan, China. *Eng. Geol.* **2021**, *289*, 106145. [CrossRef]
119. Petley, D. Edenville Dam failure: The Astonishing Video of the Collapse Sequence; The Landslide Blog, AGU Blogosphere. Available online: https://blogs.agu.org/landslideblog/2020/05/21/edenville-dam-failure-2/ (accessed on 7 July 2021).
120. Wilkinson, M. Before and after Satellite Photos Show Damage Wrought by Mid-Michigan Floods. Bridge Michigan Website. Available online: https://www.bridgemi.com/michigan-environment-watch/and-after-satellite-photos-show-damage-wrought-mid-michigan-floods (accessed on 7 July 2021).
121. US National Weather Service Glasgow Montana. April 1952 Frenchman Dam Failure. Available online: https://www.facebook.com/media/set/?set=a.1001592366586578.1073741845.108063722606118&type=3 (accessed on 7 July 2021).
122. Froehlich, D.C. Peak Outflow from Breached Embankment Dam. *J. Water Resour. Plan. Manag.* **1995**, *121*, 90–97. [CrossRef]
123. Hakimzadeh, H.; Nourani, V.; Amini, A.B. Genetic Programming Simulation of Dam Breach Hydrograph and Peak Outflow Discharge. *J. Hydrol. Eng.* **2014**, *19*, 757–768. [CrossRef]
124. Wu, W. Simplified Physically Based Model of Earthen Embankment Breaching. *J. Hydraul. Eng.* **2013**, *139*, 837–851. [CrossRef]
125. Rozov, A.L. Modeling of Washout of Dams. *J. Hydraul. Res.* **2003**, *41*, 565–577. [CrossRef]
126. Zhong, Q.; Wu, W.; Chen, S.; Wang, M. Comparison of Simplified Physically Based Dam Breach Models. *Nat. Hazards* **2016**, *84*, 1385–1418. [CrossRef]
127. Xu, Y.; Zhang, L.M. Breaching Parameters for Earth and Rockfill Dams. *J. Geotech. Geoenviron. Eng.* **2009**, *135*, 1957–1970. [CrossRef]
128. Schwarz, F.K.; Huges, L.A.; Marshall Hansen, E. *The Black Hills—Rapid City Flood of June 9–10, 1972: A Description of the Storm and Flood*; Geological Survey Professional Paper 877; U.S. Geological Survey and National Oceanic and Atmospheric Administration: Washington, DC, USA, 1975. Available online: https://pubs.usgs.gov/pp/0877/report.pdf (accessed on 7 July 2021).
129. Graham, W.J.; Major, U.S. Dam Failures: Their Cause, Resultant Losses, and Impact on Dam Safety Programs and Engineering Practice. In Proceedings of the Great Rivers History Symposium at World Environmental and Water Resources Congress 2009, Kansas City, MO, USA, 17–21 May 2009; American Society of Civil Engineers: Reston, VA, USA, 2009; pp. 52–60.
130. Carter, J.M.; Williamson, J.E.; Teller, R.W. *The 1972 Black Hills-Rapid City Flood Revisited*; USGS Fact Sheet FS-037-02; U.S. Department of the Interior, U.S. Geological Survey: Washington, DC, USA, 2002. Available online: https://pubs.usgs.gov/fs/fs-037-02/pdf/fs-037.pdf (accessed on 7 July 2021).
131. Larimer, O.J. *Flood of June 9–10, 1972, at Rapid City, South Dakota*; Hydrologic Investigations Atlas HA-511; U.S. Department of the Interior, U.S. Geological Survey: Washington, DC, USA, 1973. Available online: https://pubs.usgs.gov/ha/511/plate-1.pdf (accessed on 7 July 2021).
132. Lukman, S.; Otun, J.A.; Adie, D.B.; Ismail, A.; Oke, I.A. A Brief Assessment of a Dam and Its Failure Prevention. *J. Fail. Anal. Prev.* **2011**, *11*, 97–109. [CrossRef]
133. Waters, S.D. Storm Report—Tropical Storm Nora—September 1997. Flood Control District of Maricopa County. Available online: http://alert.fcd.maricopa.gov/alert/nora/nora_rpt.html (accessed on 7 July 2021).
134. Newhouse, S. Averted Piping Failure—Earth Dam on Permeable Foundation. In Proceedings of the 6th International Conference on Case Histories in Geotechnical Engineering, Arlington, VA, USA, 11–16 August 2008; Missouri University of Science and Technology: St. Rolla, MO, USA, 2008.
135. Benoist, J.M.; Cox, G.D. Failure of Centennial Narrows Dam. In Proceedings of the Association of State Dam Safety Officials Annual Conference, Las Vegas, NV, USA, 11–14 October 1998.

136. Ayres Associates. *Probable Maximum Flood Determination—Tittabawasse River Hydroelectric Projects*; Ayres: Eau Claire, WI, USA, 2020. Available online: https://www.four-lakes-taskforce-mi.com/uploads/1/2/3/1/123199575/ttbw_pmf_final_report-may2020.pdf (accessed on 7 July 2021).
137. Petley, D. Edenville Dam Breach: Interpreting the Failure; The Landslide Blog, AGU Blogosphere. Available online: https://blogs.agu.org/landslideblog/2020/05/22/edenville-dam-breach/ (accessed on 7 July 2021).
138. Petley, D. Edenville Dam: A Major Dam Collapse in Michigan; The Landslide Blog, AGU Blogosphere. Available online: https://blogs.agu.org/landslideblog/2020/05/20/edenville-dam-1/ (accessed on 7 July 2021).
139. Yuan, Y.; Bazzett, D.; Padnani, R.; Wang, R. Unlocking Data from the Online Footage of the Edenville Dam Failure. Submitted to Geo-Extreme 2021. Available online: https://www.researchgate.net/publication/342639092_Unlocking_data_from_the_online_footage_of_the_Edenville_Dam_Failure (accessed on 7 July 2021).
140. NASA Science. Michigan Floods and Dam Failures May 2020. NASA Applied Science. Available online: https://appliedsciences.nasa.gov/what-we-do/disasters/disasters-activations/michigan-floods-and-dam-failures-may-2020 (accessed on 7 July 2021).
141. USGS, Earth Resources Observation and Science Center. Dam Breaks in Michigan. Available online: https://www.usgs.gov/centers/eros/dam-breaks-michigan?utm_source=twitter&utm_medium=social&utm_term=22f79331-0ea7-4adb-bbc3-7596b1432776&utm_content=&utm_campaign=usgs_eros&qt-science_support_page_related_con=0#qt-science_support_page_related_con (accessed on 7 July 2021).
142. USGS, National Water Information System. Current Conditions for USGS 04156000 Tittabawassee River at Midland, MI. Available online: https://nwis.waterdata.usgs.gov/usa/nwis/uv/?cb_00060=on&cb_00065=on&cb_70969=on&cb_70969=on&format=gif_stats&site_no=04156000&period=&begin_date=2020-05-14&end_date=2020-05-21 (accessed on 7 July 2021).
143. Hergott, M. Modeling Flooding from the Dam Failures in Michigan. AIR Worldwide. Available online: https://www.air-worldwide.com/blog/posts/2020/6/modeling-flooding-from-the-dam-failures-in-michigan/ (accessed on 7 July 2021).
144. Harder, L.F.; Kelson, K.I.; Kishida, T.; Kayen, R. *Preliminary Observations of the Fujinuma Dam Failure Following the March 11, 2011 Tohoku Offshore Earthquake, Japan*; Geotechnical Extreme Events Reconnaissance (GEER); Preliminary Association Report No. GEER-25e; Folsom, CA, USA, 2011. Available online: http://learningfromearthquakes.org/2011-03-11-tohoku-japan/images/2011_03_11_tohoku_japan/pdfs/QR5_Preliminary-Observations-of-Fujinuma-Dam-Failure_06-06-11.pdf (accessed on 7 July 2021).
145. Pradel, D.; Wartman, J.; Tiwari, B. Failure of the Fujinuma Dams during the 2011 Tohoku Earthquake. In Proceedings of the Geo-Congress 2013: Stability and Performance of Slopes and Embankments III, San Diego, CA, USA, 3–7 March 2013; pp. 1559–1573.
146. Clay, R.A. 100th Anniversary of Hatchtown Dam. In Proceedings of the Association of State Dam Safety Officials Annual Conference, San Diego, CA, USA, 21–25 September 2014.
147. Timmins, W.M. The Failure of the Hatchtown Dam, 1914. *Utah Hist. Q.* **1968**, *36*, 263–273.
148. German Aerospace Center. Bulgaria: Biser Flood as of February 10, 2012—Disaster Extent Map. Available online: https://reliefweb.int/sites/reliefweb.int/files/resources/map_1736.pdf (accessed on 7 July 2021).
149. Bocheva, L.; Pophristov, V. Seasonal Analysis of Large-Scale Heavy Precipitation Events in Bulgaria. *AIP Conf. Proc.* **2019**, *2075*, 200017.
150. Evans, J.E.; Mackey, S.D.; Gottgens, J.F.; Gill, W.M. Lessons from a Dam Failure. *Ohio J. Sci.* **2000**, *100*, 121–131.
151. Environmental Business Council of New England. Dam Management Webinar Series: Liability & Insurance Considerations for Dam Owners and Engineers—Part 1: Historical Dam Failures in New England. 2020. Available online: https://ebcne.org/wp-content/uploads/2020/10/Presentations-Dam-Management-Webinar-Series-Part-One.pdf (accessed on 7 July 2021).
152. Wooten, L. The Grande Old Dam of Foster's Pond Turns 162: An Engineer's Take. GEI Consultants, 2020. Available online: https://www.fosterspond.org/Lee_Wooten_presentation_%202020_Annual_Meeting.pdf (accessed on 7 July 2021).
153. Lindon, M.C. Forensics of a Dam Failure Fatality; Water and Whatever Blog. Available online: http://waterandwhatever.blogspot.com/2013/05/forensics-of-fatality.html (accessed on 7 July 2021).
154. Collins, W.E. *Report on the Cause of Failure of Little Deer Creek Dam, Provo River Project, Utah*; U.S. Department of the Interior, Bureau of Reclamation: Washington, DC, USA, 1964.
155. Bergman, N.; Sholker, O.; Roskin, J.; Greenbaum, N. The Nahal Oz Reservoir Dam-Break Flood: Geomorphic Impact on a Small Ephemeral Loess-Channel in the Semi-Arid Negev Desert, Israel. *Geomorphology* **2014**, *210*, 83–97. [CrossRef]
156. Kostecki, S.; Rędowicz, W. The Washout Mechanism of the Niedów Dam and its Impact on the Parameters of the Flood Wave. *Procedia Eng.* **2014**, *91*, 292–297. [CrossRef]
157. Kosierb, R. Process of Wave Transformation in the Riverbed on the Example of a Flood on the Nysa Łużycka River. *WasserWirtschaft* **2014**, *104*, 18–23. [CrossRef]
158. Lees, P.; Thomson, D. Emergency Management, Opuha Dam Collapse, Waitangi Day 1997. In Proceedings of the NZ Society on Large Dams (NZSOLD) 2003 Symposium "Dams–Consents and Current Practice", IPENZ, Wellington, New Zealand, 26 August 2003; pp. 84–89. Available online: https://www.epa.govt.nz/assets/FileAPI/proposal/NSP000028/Hearings-Week-06/dd571f3ec6/29-Colin-Riden-Emergency-management-Opuha-Dam-collapse-Waitangi-Day-1997.pdf (accessed on 7 July 2021).
159. Azimi, R.; Vatankhah, A.R.; Kouchakzadeh, S. Predicting Peak Discharge from Breached Embankment Dams. In Proceedings of the 36th IAHR World Congress, The Hague, The Netherlands, 28 June–3 July 2015.
160. Jansen, R.B. *Dams and Public Safety*; U.S. Department of the Interior, Bureau of Reclamation: Denver, CO, USA, 1983. Available online: https://www.usbr.gov/tsc/techreferences/mands/mands-pdfs/AZ1130.pdf (accessed on 7 July 2021).

161. Xu, Y.; Zhang, L.; Jia, J. Lessons from Catastrophic Dam Failures in August 1975 in Zhumadian, China. In Proceedings of the GeoCongress 2008, New Orleans, LA, USA, 9–12 March 2008; American Society of Civil Engineers: Reston, VA, USA, 2008; pp. 162–169.
162. Froehlich, D.C. Predicting Peak Discharge from Gradually Breached Embankment Dam. *J. Hydrol. Eng.* **2016**, *21*, 1–15. [CrossRef]
163. Zhong, Q.; Chen, S.; Deng, Z. A Simplified Physically-Based Breach Model for a High Concrete-Faced Rockfill Dam: A Case Study. *Water Sci. Eng.* **2018**, *11*, 46–52. [CrossRef]
164. Zhong, Q.; Chen, S.; Deng, Z. A Simplified Physically-Based Model for Core Dam Overtopping Breach. *Eng. Fail. Anal.* **2018**, *90*, 141–155. [CrossRef]
165. Graham, W.J. The Banqiao and Shimantan Dam Failures: Factors Affecting the Warning and Evacuation Success. In Proceedings of the Association of State Dam Safety Officials Annual Conference, Seattle, WA, USA, 19–23 September 2010; Volume 2, pp. 1055–1069.
166. Diacon, A.; Stematiu, D.; Mircea, N. An Analysis of the Belci Dam Failure. *Int. Water Power Dam Constr.* **1992**, *44*, 67–72.
167. Asman, I.; Bratianu, G. Project for Rehabilitation of Belci Dam after the 1991 Failure. In Proceedings of the 9th ICOLD European Club Symposium 2013, Venice, Italy, 10–12 April 2013. Available online: https://www.academia.edu/21861832/An_analysis_of_the_Belci_Dam_failure (accessed on 7 July 2021).
168. Sharma, R.P.; Kumar, A. Case Histories of Earthen Dam Failures. In Proceedings of the 7th International Conference on Case Histories in Geotechnical Engineering, Chicago, IL, USA, 29 April–4 May 2013.
169. Yochum, S.E.; Goertz, L.A.; Jones, P.H. Case Study of the Big Bay Dam Failure: Accuracy and Comparison of Breach Predictions. *J. Hydraul. Eng.* **2008**, *134*, 1285–1293. [CrossRef]
170. Ferguson, K.A.; Anderson, S.; Sossenkina, E. Reexamination of the 2004 Failure of Big Bay Dam, Mississippi. In Proceedings of the 34th Annual United States Society on Dams Conference, San Francisco, CA, USA, 7–11 April 2014. Available online: https://damfailures.org/wp-content/uploads/2017/12/62Ferguson_BigBayDam.pdf (accessed on 7 July 2021).
171. Raška, P.; Emmer, A. The 1916 Catastrophic Flood Following the Bílá Desná Dam Failure: The Role of Historical Data Sources in the Reconstruction of its Geomorphologic and Landscape Effects. *Geomorphology* **2014**, *226*, 135–147. [CrossRef]
172. Sheffield Local Studies Library. *Sources for the Study of the Sheffield Flood 1864*; Sheffield Libraries Archives and Information; Sheffield City Council: Sheffield, UK, 2015. Available online: https://nanopdf.com/download/flood-study-guide-v1-6-sheffield-city-council_pdf (accessed on 7 July 2021).
173. Burns, W. The Sheffield Flood: A Critical Study of Charles Reade's Fiction. *PMLA* **1948**, *63*, 686–695. Available online: www.jstor.org/stable/459435 (accessed on 7 July 2021). [CrossRef]
174. Binnie, G.M. The Collapse of the Dale Dyke Dam in Retrospect. *Q. J. Eng. Geol. Hydrogeol.* **1978**, *11*, 305–324. [CrossRef]
175. Binnie, G.M. Erratum to "The collapse of Dale Dyke Dam in Retrospect", Quarterly Journal of Engineering Geology and Hydrogeology, 11, 305–324 (1978). *Q. J. Eng. Geol. Hydrogeol.* **1979**, *12*, 61. [CrossRef]
176. Binnie, G.M. Postscript to 'The Collapse of the Dale Dyke dam in Retrospect' . *Q. J. Eng. Geol. Hydrogeol.* **1983**, *16*, 357–358. [CrossRef]
177. McDaniel, L.; Garton, J.; Fiedler, W.; King, W.; Schwanz, N. Lake Delhi Dam Breach—Two perspectives. In Proceedings of the Association of State Dam Safety Officials Annual Conference, Washington, DC, USA, 25–29 September 2011; p. 21.
178. Fiedler, B.W.; King, W.; Schwanz, N.; Garton, J.; McDaniel, L. Dam Safety: What Happened to Lake Delhi Dam? Available online: https://www.hydroreview.com/world-regions/dam-safety-what-happened-to-lake-delhi-dam/ (accessed on 7 July 2021).
179. Garton, J.; Welty, C. How Dams Fail and Proper Dam Maintenance. Iowa Department of Natural Resources, 2020. Available online: https://www.iowadnr.gov/Portals/idnr/uploads/water/dams/dams_damsfailures_powerpoint.pdf (accessed on 7 July 2021).
180. Stanley Consultants. Lake Delhi Dam—Design Alternatives Report–App. B: Hydrologic and Hydraulic Studies Report. 2011. Available online: https://www.scribd.com/document/311562178/2011DelhiStudy-Appendix-B-Hydrologic-and-Hydraulic-Studies-Report (accessed on 7 July 2021).
181. Eash, D.A. *Floods of July 23–26, 2010, in the Little Maquoketa River and Maquoketa River Basins, Northeast Iowa*; Open-File Report 2011-1301; U.S. Department of the Interior, U.S. Geological Survey: Washington, DC, USA, 2012; p. 45.
182. Shivers, M. Dam Breach Study and Inundation Mapping of Eleven Dams Owned by Oklahoma City, Oklahoma. Master's Thesis, Oklahoma State University, Oklahoma City, OK, USA, 2016.
183. Calderón, J.L.P. Reconstrucción Histórica, Estructural, Hidrológica, Hidráulica y Socioeconómica de la Catástrofe de Ribadelago (Rotura de la Presa de Vega de Tera). Master's Thesis, University of Vigo, Vigo, Spain, 2014.
184. Martí, J.; Riera, F.; Martínez, F. Interpretation of the Failure of the Aznalcóllar (Spain) Tailings Dam. *Mine Water Environ.* **2021**, *40*, 189–208. [CrossRef]
185. Gallart, F.; Benito, G.; Martín-Vide, J.P.; Benito, A.; Prió, J.M.; Regüés, D. Fluvial Geomorphology and Hydrology in the Dispersal and Fate of Pyrite Mud Particles Released by the Aznalcóllar Mine Tailings Spill. *Sci. Total Environ.* **1999**, *242*, 13–26. [CrossRef]
186. Lyu, Z.; Chai, J.; Xu, Z.; Qin, Y.; Cao, J. A Comprehensive Review on Reasons for Tailings Dam Failures Based on Case History. *Adv. Civ. Eng.* **2019**, *2019*, 4159306. [CrossRef]
187. McDermott, R.K.; Sibley, J.M. The Aznalcóllar tailings dam accident—A case study. *Miner. Resour. Eng.* **2000**, *9*, 101–118. [CrossRef]
188. Dollman, D.S. *Colorado's Deadliest Floods*; Arcadia Publishing: Charleston, SC, USA, 2017.

189. Groom, K.M.; Allen, C.D. Denver's Forgotten Flood: The Geomorphologic Impacts of the 1933 Castlewood Dam Failure. *J. West.* **2014**, *53*, 54–65.
190. Mauney, L. 80th Anniversary of the Castlewood Dam Failure. Decade Dam Failure Series—ASDSO. Available online: https://damfailures.org/wp-content/uploads/2018/09/ASDSO-Castlewood_Mauney.pdf (accessed on 7 July 2021).
191. Rudolph, K. August 3, 1933: Castlewood Dam Breaks, Floods Denver. Denver Public Library. Available online: https://history.denverlibrary.org/news/august-3-1933-castlewood-dam-breaks-floods-denver (accessed on 7 July 2021).
192. Zhang, L.M.; Chen, Q. Seepage Failure Mechanism of the Gouhou Rockfill Dam during Reservoir Water Infiltration. *Soils Found.* **2006**, *46*, 557–568. [CrossRef]
193. Chen, Q.; Zhang, L.M. Three-dimensional Analysis of Water Infiltration into the Gouhou Rockfill Dam Using Saturated-unsaturated Seepage Theory. *Can. Geotech. J.* **2006**, *43*, 449–461. [CrossRef]
194. Shen, G.; Lu, Y.; Zhang, S.; Xiang, Y.; Sheng, J.; Fu, J.; Fu, S.; Liu, M. Risk Dynamics Modeling of Reservoir Dam Break for Safety Control in the Emergency Response Process. *Water Supply* **2021**, *21*, 1356–1371. [CrossRef]
195. Zhong, Q.; Chen, S.; Fu, Z. Failure of Concrete-Face Sand-Gravel Dam Due to Water Flow Overtops. *J. Perform. Constr. Facil.* **2019**, *33*, 04019007. [CrossRef]
196. Scott, K.M.; Gravlee, G.C., Jr. *Flood Surge on the Rubicon River, California—Hydrology, Hydraulics, and Boulder Transport*; Geological Survey Professional Paper 422-M; USGS: Washington, DC, USA, 1968.
197. Palmer, J. Anatomy of a Tailings Dam Failure and a Caution for the Future. *Engineering* **2019**, *5*, 605–606. [CrossRef]
198. Pereira, L.F.; Cruz, G.d.B.; Guimarães, R.M.F. Impactos do Rompimento da Barragem de Rejeitos de Brumadinho, Brasil: Uma Análise Baseada nas Mudanças de Cobertura da Terra. *J. Environ. Anal. Prog.* **2019**, *4*, 122–129. [CrossRef]
199. Raman, A.; Liu, F. An Investigation of the Brumadinho Dam Break with HEC RAS Simulation. *arXiv* **2019**, arXiv:1911.05219.
200. Silva Rotta, L.H.; Alcântara, E.; Park, E.; Negri, R.G.; Lin, Y.N.; Bernardo, N.; Mendes, T.S.G.; Souza Filho, C.R. The 2019 Brumadinho Tailings Dam Collapse: Possible Cause and Impacts of the Worst Human and Environmental Disaster in Brazil. *Int. J. Appl. Earth Obs. Geoinf.* **2020**, *90*, 102119. [CrossRef]
201. de Lima, R.E.; de Lima Picanço, J.; da Silva, A.F.; Acordes, F.A. An Anthropogenic Flow Type Gravitational Mass Movement: The Córrego do Feijão Tailings Dam Disaster, Brumadinho, Brazil. *Landslides* **2020**, *17*, 2895–2906. [CrossRef]
202. Lumbroso, D.; Davison, M.; Body, R.; Petkovšek, G. Modelling the Brumadinho Tailings Dam Failure, the Subsequent Loss of Life and How It Could Have Been Reduced. *Nat. Hazards Earth Syst. Sci.* **2021**, *21*, 21–37. [CrossRef]
203. Rico, M.; Benito, G.; Díez-Herrero, A. Floods from Tailings Dam Failures. *J. Hazard. Mater.* **2008**, *154*, 79–87. [CrossRef]
204. Rico, M.; Benito, G.; Salgueiro, A.R.; Díez-Herrero, A.; Pereira, H.G. Reported Tailings Dam Failures. *J. Hazard. Mater.* **2008**, *152*, 846–852. [CrossRef]
205. Quelopana, H. Released Volume Estimation for Dam Break Analysis. In Proceedings of the 6th International Seminar on Tailings Management, Santiago, Chile, 10–12 July 2019.
206. do Carmo, F.F.; Kamino, L.H.Y.; Junior, R.T.; de Campos, I.C.; do Carmo, F.F.; Silvino, G.; do Castro, K.J.D.S.X.D.; Mauro, M.L.; Rodrigues, N.U.A.; de Souza Miranda, M.P.; et al. Fundão Tailings Dam Failures: The Environment Tragedy of the Largest Technological Disaster of Brazilian Mining in Global Context. *Perspect. Ecol. Conserv.* **2017**, *15*, 145–151. [CrossRef]
207. Fernandes, G.W.; Goulart, F.F.; Ranieri, B.D.; Coelho, M.S.; Dales, K.; Boesche, N.; Bustamante, M.; Carvalho, F.A.; Carvalho, D.C.; Dirzo, R.; et al. Deep into the Mud: Ecological and Socio-economic Impacts of the Dam Breach in Mariana, Brazil. *Nat. Conserv.* **2016**, *14*, 35–45. [CrossRef]
208. Van Niekerk, H.J.; Viljoen, M.J. Causes and Consequences of the Merriespruit and other Tailings-dam Failures. *Land Degrad. Dev.* **2005**, *16*, 201–212. [CrossRef]
209. Wagener, F.V.; Craig, H.J.; Blight, G.; McPhail, G.; Williams, A.A.B.; Strydom, J.H. The Merriespruit Tailings Dam Failure—A Review. In Proceedings of the 5th International Conference on Tailings and Mine Waste, Fort Collins, CO, USA, 26–28 January 1998; pp. 925–952.
210. Petkovšek, G.; Hassan, M.A.A.M.; Lumbroso, D.; Roca Collell, M. A Two-Fluid Simulation of Tailings Dam Breaching. *Mine Water Environ.* **2021**, *40*, 151–165. [CrossRef]
211. Visser, J.K. Safety Practices and Procedures Used When Three Major Accidents Occurred in South Africa. In Proceedings of the European Safety and Reliability Conference, ESREL 2005: Advances in Safety and Reliability, Tri City (Gdynia-Sopot-Gdansk), Poland, 27–30 June 2005; Kolowrocki, K., Ed.; CRC Press: Boca Raton, FL, USA, 2005; pp. 2001–2008.
212. Davies, E.W.; Bailey, F.J.; Kelly, D.B. *West Virginia's Buffalo Creek Flood: A Study of the Hydrology and Engineering Geology*; Geological Survey Circular 667; U.S. Department of the Interior, U.S. Geological Survey: Washington, DC, USA, 1972; p. 32. Available online: https://pubs.usgs.gov/circ/1972/0667/report.pdf (accessed on 7 July 2021).
213. Kelley, J.H.; Kealy, D.; Hylton, C.D.; Hallana, E.V.; Ashcraft, J.; Murrin, J.F.; Davies, W.E.; Erwin, R.B.; Latimer, I.S. The Buffalo Creek Flood and Disaster: Official Report from the Governor's Ad Hoc Commission of Inquiry. West Virginia Department of Arts, Culture and History, 1973. Available online: http://www.wvculture.org/history/disasters/buffcreekgovreport.html (accessed on 7 July 2021).
214. Gee, N. Case Study: Buffalo Creek Dam (West Virginia, 1972). Association of State Dam Safety Officials Lessons Learned from Dam Incidents and Failures. Available online: https://damfailures.org/case-study/buffalo-creek-dam-west-virginia-1972/ (accessed on 7 July 2021).

215. Waler, W.A. *Associates. Analysis of Coal Refuse Dam Failure Middle Fork Buffalo Creek, Saunders, West Virginia*; OFR 10-73; U.S. Department of the Interior, Bureau of Mines: Washington, DC, USA, 1973; Volume I, p. 217.
216. Waler, W.A. *Associates. Analysis of Coal Refuse Dam Failure Middle Fork Buffalo Creek, Saunders, West Virginia*; OFR 10-73; U.S. Department of the Interior, Bureau of Mines: Washington, DC, USA, 1973; Volume II, p. 178.
217. Chabal, J.P.; Bordes, J.L. Puentes, 1802: La Rupture du Plus Grand Barrage du Monde, ou le Double Echec d'Antonio de Robles. Le rapport Betancourt. *Quad. d'Història l'Enginyeria* **2009**, *X*, 151–167.
218. Úbeda Romero, E. La rotura del pantano de Puentes—Región de Murcia Digital. *Revista Murgetana* **1963**, *21*, 5–33.
219. Burnett, J. *Flash Floods in Texas*; Texas A&M University Press: College Station, TX, USA, 2008.
220. David Rogers, J. The Failure of the Lower Colorado River Dam at Austin, Texas. In Proceedings of the World Environmental and Water Resources Congress, Austin, TX, USA, 17–21 May 2015; Karvaz, K., Webster, V.L., Eds.; American Society of Civil Engineers: Reston, VA, USA, 2015; pp. 147–160.
221. Neall, V.E. Lahars as Major Geological Hazards. *Bull. Int. Assoc. Eng. Geol.* **1976**, *13*, 233–240. [CrossRef]
222. Manville, V.; Hodgson, K.A.; Nairn, I.A. A Review of Break-out Floods from Volcanogenic Lakes in New Zealand. *N. Z. J. Geol. Geophys.* **2007**, *50*, 131–150. [CrossRef]
223. *New Zealand Railways Board of Inquiry into Derailment of Wellington-Auckland Express at Whangaehu River Bridge between Tangiwai and Karioi Railway Stations on 24 December 1953. Tangiwai Railway Disaster/Report of Board of Inquiry*; Govt. Printer: Wellington, New Zealand, 1954.
224. Soares-Frazão, S. Experiments of Dam-break Wave over a Triangular Bottom Sill. *J. Hydraul. Res.* **2007**, *45*, 19–26. [CrossRef]
225. Soares-Frazão, S.; Zech, Y. Experimental Study of Dam-break Flow against an Isolated Obstacle. *J. Hydraul. Res.* **2007**, *45*, 27–36. [CrossRef]
226. Soares-Frazão, S.; Zech, Y. Dam-break Flow through an Idealised City. *J. Hydraul. Res.* **2008**, *46*, 648–658. [CrossRef]
227. Aureli, F.; Dazzi, S.; Maranzoni, A.; Mignosa, P.; Vacondio, R. Experimental and Numerical Evaluation of the Force Due to the Impact of a Dam-break Wave on a Structure. *Adv. Water Resour.* **2015**, *76*, 29–42. [CrossRef]
228. Elkholy, M.; LaRocque, L.A.; Chaudhry, M.H.; Imran, J. Experimental Investigations of Partial-Breach Dam-Break Flows. *J. Hydraul. Eng.* **2016**, *142*, 04016042. [CrossRef]
229. Spinewine, B.; Zech, Y. Small-scale Laboratory Dam-break Waves on Movable Beds. *J. Hydraul. Res.* **2007**, *45*, 73–86. [CrossRef]
230. LaRocque, L.A.; Elkholy, M.; Hanif Chaudhry, M.; Imran, J. Experiments on Urban Flooding Caused by a Levee Breach. *J. Hydraul. Eng.* **2013**, *139*, 960–973. [CrossRef]
231. Biscarini, C.; Di Francesco, S.; Ridolfi, E.; Manciola, P. On the Simulation of Floods in a Narrow Bending Valley: The Malpasset Dam Break Case Study. *Water* **2016**, *8*, 545. [CrossRef]
232. Venturi, S.; Di Francesco, S.; Geier, M.; Manciola, P. Modelling Flood Events with a Cumulant CO Lattice Boltzmann Shallow Water Model. *Nat. Hazards* **2021**, *105*, 1815–1834. [CrossRef]
233. Liang, D.; Lin, B.; Falconer, R.A. A Boundary-fitted Numerical Model for Flood Routing with Shock-capturing Capability. *J. Hydrol.* **2007**, *332*, 477–486. [CrossRef]
234. De Marchi, G. Sull'Onda di Piena che Seguirebbe al Crollo della Diga di Cancano. [On the Dam-break Wave Resulting from the Collapse of the Cancano Dam]. *L'Energia Elettr.* **1945**, *22*, 319–340.
235. Pilotti, M.; Milanesi, L.; Bacchi, V.; Tomirotti, M.; Maranzoni, A. Dam-Break Wave Propagation in Alpine Valley with HEC-RAS 2D: Experimental Cancano Test Case. *J. Hydraul. Eng.* **2020**, *146*, 05020003. [CrossRef]
236. Pilotti, M.; Maranzoni, A.; Milanesi, L.; Tomirotti, M.; Valerio, G. Dam-break Modeling in Alpine Valleys. *J. Mt. Sci.* **2014**, *11*, 1429–1441. [CrossRef]
237. Testa, G.; Zuccalà, D.; Alcrudo, F.; Mulet, J.; Soares-Frazão, S. Flash Flood Flow Experiment in a Simplified Urban District. *J. Hydraul. Res.* **2007**, *45*, 37–44. [CrossRef]
238. Soares-Frazão, S.; Lhomme, J.; Guinot, V.; Zech, Y. Two-dimensional Shallow-water Model with Porosity for Urban Flood Modelling. *J. Hydraul. Res.* **2008**, *46*, 45–64. [CrossRef]
239. El Kadi Abderrezzak, K.; Paquier, A.; Mignot, E. Modelling Flash Flood Propagation in Urban Areas Using a Two-dimensional Numerical Model. *Nat. Hazards* **2009**, *50*, 433–460. [CrossRef]
240. Guinot, V. Multiple Porosity Shallow Water Models for Macroscopic Modelling of Urban Floods. *Adv. Water Resour.* **2012**, *37*, 40–72. [CrossRef]
241. An, H.; Yu, S. Well-balanced Shallow Water Flow Simulation on Quadtree Cut Cell Grids. *Adv. Water Resour.* **2012**, *39*, 60–70. [CrossRef]
242. Miliani, S.; Montessori, A.; La Rocca, M.; Prestininzi, P. Dam-Break Modeling: LBM as the Way towards Fully 3D, Large-Scale Applications. *J. Hydraul. Eng.* **2021**, *147*, 04021017. [CrossRef]
243. Güney, M.S.; Tayfur, G.; Bombar, G.; Elci, S. Distorted Physical Model to Study Sudden Partial Dam Break Flows in an Urban Area. *J. Hydraul. Eng.* **2014**, *140*, 05014006. [CrossRef]
244. Haltas, I.; Tayfur, G.; Elci, S. Two-dimensional Numerical Modeling of Flood Wave Propagation in an Urban Area due to Ürkmez Dam-break, İzmir, Turkey. *Nat. Hazards* **2016**, *81*, 2103–2119. [CrossRef]
245. Morris, M.W.; Hassan, M.A.A.M.; Vaskinn, K.A. Breach Formation: Field Test and Laboratory Experiments. *J. Hydraul. Res.* **2007**, *45*, 9–17. [CrossRef]

246. Morris, M. Modelling Breach Initiation and Growth; Report Number T06-08-02; FLOODsite project Integrated Flood Risk Analysis and Management Methodologies; European Community's Sixth Framework Programme. 2008. Available online: http://www.floodsite.net/html/partner_area/project_docs/t06_08_01_breach__execsumm_v3_5_p01.pdf (accessed on 7 July 2021).
247. Shuibo, P.; Mingsen, Q.; Lianxiang, W.; Guoyi, X.; Yongqiang, S.; Longda, X.; Cuiyu, M.; Loukola, E.; Pyyny, J.; Reiter, P.; et al. *Chinese-Finnish Cooperative Research Work on Dam Break Hydrodynamics*; Shuibo, P., Loukola, E., Eds.; National Board of Waters and the Environment: Helsinki, Finland, 1993; pp. 1–92. Available online: https://helda.helsinki.fi/bitstream/handle/10138/29795/VYHA%20julkaisujaVYHA%20julkaisuja%20A%20167.pdf?sequence=1&isAllowed=y (accessed on 7 July 2021).

water

MDPI

Review

Review of Experimental Investigations of Dam-Break Flows over Fixed Bottom

Francesca Aureli [1], Andrea Maranzoni [1,*], Gabriella Petaccia [2] and Sandra Soares-Frazão [3]

1. Department of Engineering and Architecture, University of Parma, 43124 Parma, Italy
2. Department of Civil Engineering and Architecture, University of Pavia, 27100 Pavia, Italy
3. Institute of Mechanics, Materials and Civil Engineering, Université Catholique de Louvain, 1348 Louvain-la-Neuve, Belgium

* Correspondence: andrea.maranzoni@unipr.it

Abstract: Laboratory experiments of dam-break flows are extensively used in investigations of geophysical flows involving flood waves, to provide insight into relevant aspects of the physics of the process and collect experimental data for validating numerical models. A dam-break flow is a typical example of a highly unsteady free surface flow with high reproducibility. Indeed, dam-break experiments can be repeated several times under the same test conditions obtaining large amounts of different types of data (possibly using various measuring techniques) that can be combined in a single rich dataset. Moreover, laboratory tests on dam-break flows are widely considered a valuable benchmark for the validation of numerical models, since field data from historical events are scarce, sparse, and highly uncertain. However, no systematic review of laboratory investigations of dam-break flows and existing related datasets are available in the literature to provide a comprehensive overview of the test conditions considered, the measuring techniques used, and the experimental data collected. This review article aims to fill this gap, focusing on laboratory tests in schematic and idealized setups with a fixed, non-erodible bed. In particular, this review aims to help researchers and modelers to: (a) select the most appropriate laboratory tests for validating their numerical models; (b) facilitate access to databases by indicating relevant bibliographic references; (c) identify specific challenging aspects worthy of further experimental research; and (d) support the development of new or improved technologies for the mitigation of the impact of dam-break flood waves. The references reviewed are organized into tables according to the purposes of the laboratory investigation, and comprehensive information is provided on test conditions, datasets, and data accessibility. Finally, suggestions for future experimental research on dam-break flows are provided.

Keywords: dam-break flow; experimental tests; datasets; validation of numerical models; review

check for updates

Citation: Aureli, F.; Maranzoni, A.; Petaccia, G.; Soares-Frazão, S. Review of Experimental Investigations of Dam-Break Flows over Fixed Bottom. *Water* **2023**, *15*, 1229. https://doi.org/10.3390/w15061229

Academic Editor: Giuseppe Pezzinga

Received: 19 February 2023
Revised: 12 March 2023
Accepted: 14 March 2023
Published: 21 March 2023

1. Introduction

The technique of suddenly removing a gate placed between a reservoir storing a mass of water initially at rest and a downstream area is extensively used to generate unsteady free surface flows in experimental investigations of a variety of geophysical phenomena involving flood waves, such as dam-break floods and tsunamis. Despite active research (both theoretical and numerical) in this field in the last decades, physical modelling remains a widely used approach to provide insight into the features of the flow and collect valuable data for validating numerical models.

A dam-break flow is a typical (albeit extreme) example of an unsteady and rapidly varying flow. It is characterized by rapid and abrupt flow depth and velocity changes and by the presence of wetting and drying fronts. Hence, a dam-break flow is usually considered a stringent and probative validation test for numerical models. Indeed, it can be assumed that a numerical model able to cope with dam-break flows will also be able to simulate accurately less severe, slower floods.

Physical modelling in laboratory conditions on schematic, idealized geometries allows "for assessment within a controlled environment, enabling the isolation of individual processes and close study of their effect on the modelled system. The complexities of modelled systems are reduced to what is practical and feasible to model in a physical scale environment" [1]. Dam-break flow experiments are then relatively 'easy' to perform on the laboratory scale, since a small quiescent water volume must be released without the need to set up complex recirculation systems and regulation devices. In addition, dam-break flows are highly reproducible in controlled laboratory conditions, which allows experimental runs to be repeated several times under the same test conditions to collect a large amount of data of different types and merge them to form a complete database. This ease of implementation has fostered the investigation of various scenarios and situations characterized by different geometries and the presence of singularities and obstacles of various shapes. Therefore, even though laboratory setups do not reproduce, in general, the complexity of real situations in which various singularities and complex flow features simultaneously occur, conducting multiple idealized experiments focusing on specific singular features allows for an in-depth investigation of possible realistic flow conditions.

The physical quantities relevant to describe the process can be acquired with high accuracy in the laboratory via advanced and sophisticated measuring techniques. In particular, the advances in measuring techniques in the last two decades, especially the non-intrusive ones, have considerably enlarged the types of data that can be collected, and have improved their accuracy (e.g., [2–4]). Conversely, recovering reliable and accurate validation data from historical documents on real dam-breaks is unlikely because such catastrophic events are, fortunately, rare and seldom well documented [5]. Moreover, laboratory dam-break tests on scale physical models with real topography (which sometimes combine the real topography with an idealized situation [6]), are sporadic [5,7,8] although recent examples can be found in the literature [9].

Due to the advantages previously mentioned, a large number of laboratory tests on dam-break flows were performed in the past, and high-quality datasets are now available to the scientific community. However, a systematic review of laboratory investigations of dam-break flows, which provides a comprehensive overview of the test conditions, measuring techniques and available datasets, is missing in the literature. Only fragmentary or partial information is reported in some documents (e.g., [10–12]). Therefore, this review attempts to fill this gap, focusing on experiments conducted in schematic, idealized laboratory setups with a fixed, non-erodible bottom. It covers a period from the beginning of the 1900s (when the noteworthy early experiments on dam-break waves were performed) until the end of 2022. In particular, this review aims at helping researchers and dam-break modelers: (a) to select the most appropriate laboratory test cases for validating their numerical models; (b) to facilitate access to datasets and reference material through the indication of relevant bibliographic references; (c) to identify specific aspects regarding dam-break flows worthy of further insight and future research; and (d) to support the development of improved technologies to mitigate the impact of dam-break flood waves.

This review is limited to investigations with flood waves or bores generated by a typical dam-break mechanism, characterized by the total removal of a gate, and releasing the liquid mass stored behind. Investigations using different wave generation mechanisms (based on piston- or pumping-type wave makers, vertical release systems, and underflow gates) have not been considered. Furthermore, experiments on dam-break flows over an erodible bed with sediment transport, gravity currents, granular flows, and debris flows are not considered here in order not to overextend the scope of the review. Each of these topics would deserve a specific review (e.g., [3]) due to the relevance of the related applications and the amount of experimental research carried out.

2. State of the Art Experimental Investigations of Dam-Break Flows

Typical setups for dam-break flow studies are illustrated in Figure 1. Figure 1a shows an experimental facility for the simulation of a total dam-break, consisting of a rectangular flume equipped with a sluice gate, which can be suddenly removed to release a mass of quiescent water behind it. In the beautiful historical photo taken during Dressler's experiments in the 1950s [13], the wall of water released by the gate removal can be appreciated. The side walls of the flume are typically transparent to allow direct observation of the phenomenon and the use of image processing techniques. Figure 1b shows a typical laboratory setup for the study of partial dam-break phenomena. It consists of a tank in which a portion acting as a reservoir is separated from a floodable area through a partition wall, in which a sluice gate is located. In the case shown in the picture, the bottom of the tank (made of opalescent material) is backlit in order to apply a colorimetric technique based on light absorption to measure the free surface [14].

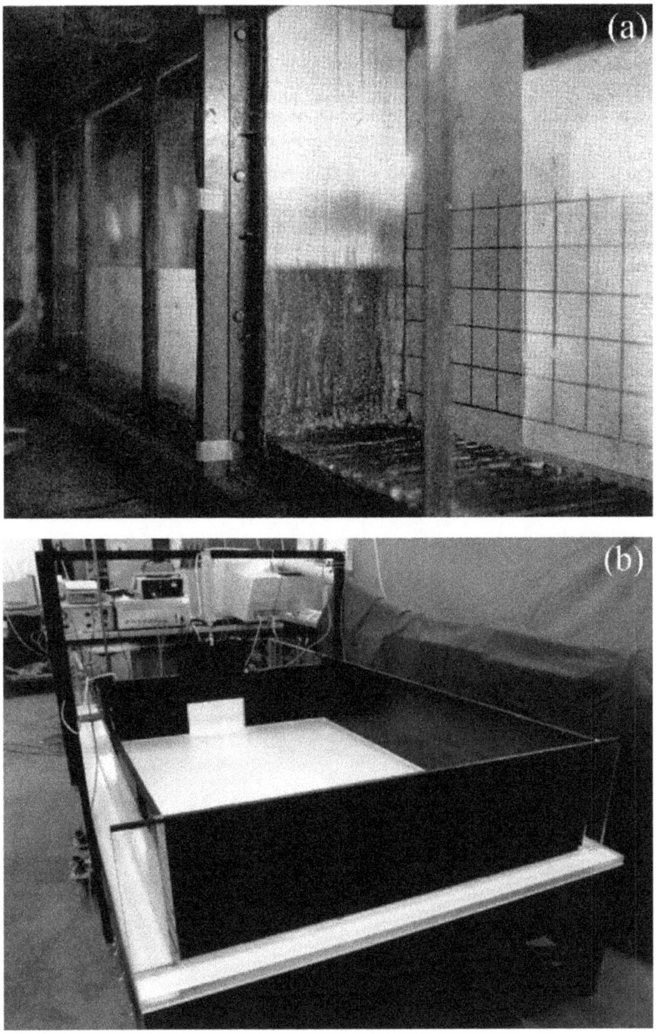

Figure 1. Pictures of typical experimental facilities for dam-break flow investigations. (**a**) Total dam-break in a rectangular channel (reprinted with permission from Ref. [13]; courtesy of International Association of Hydrological Sciences). (**b**) Prismatic tank for partial dam-break experiments (reprinted with permission from Ref. [14]).

References retrieved in the literature review are classified according to the objectives of the experimental investigation and organized into different tables.

Table 1 reports basic investigations of the physical characteristics of dam-break flows in straight (typically rectangular) channels or spreading on a plane. Such investigations mainly aim to explore the fundamental aspects and features of dam-break wave generation and propagation. Most reported cases concern smooth horizontal channels, but some studies also consider sloping channels or rough beds. Figure 1 shows typical laboratory setups for the study of total and partial dam-break flows.

Table 2 includes laboratory investigations of dam-break waves through geometric singularities (channel constrictions, bottom sills, curves or bends, etc.) to examine the effect of geometric elements and transition structures on the flow.

Table 3 lists experimental investigations of the dam-break wave impact against isolated obstacles, such as walls or vertical columns of various shapes. The disturbance induced on the flow by the presence of the obstacle is mainly analyzed in such experiments. Sometimes, the wave impact dynamics and the hydrodynamic load on the structure are also investigated.

Table 4 shows laboratory investigations of dam-break floods in idealized urban areas aiming to offer insights and an advanced understanding of urban flooding resulting from a dam-break event. In this field, the problem can be considered an extension of that presented in Table 3, since multiple obstacles are placed in the floodable area to reproduce a structured urban layout where more complex flow processes occur.

Table 5 reports experimental investigations concerning the propagation of tsunami bores (generated by a gate removal) in the swash zone. Such studies typically analyze the run-up over an adverse slope or the effect of coastal protective structures. Although a tsunami bore cannot strictly be considered a dam-break wave, these two wave types have many affinities, so tsunami bores are sometimes generated in the laboratory through the sudden removal of a gate. This review includes only investigations which use this technique to simulate a tsunami bore.

Table 6 lists experimental investigations of green water events in ships or offshore structures. In naval and maritime engineering, a 'green water' event is related to the presence of water on the deck of a ship or platform due to high waves exceeding the freeboard. Only the studies in which the wave overtopping onto the deck is produced by the sudden removal of a gate are considered in this review.

Table 7 includes experimental investigations and databases of dam-break waves of non-Newtonian liquids. Such phenomena are commonly observed in nature as well as in many industrial processes.

Table 8 shows laboratory investigations of dam-breaks in cascade reservoirs formed by multiple dams placed in sequence along a channel. In this case, a dam-break flood hazard assessment should consider that the collapse of the upstream dam could cause a flood wave involving the downstream dams, potentially inducing their overtopping or failure in a domino effect. Cascade reservoirs ensure flood hazard mitigation depending on their filling level and mutual distance.

Table 9 reports experimental investigations of dike-break-induced flows on a lateral floodplain. The break of a lateral structure produces significantly different effects compared to the collapse of a frontal one. Indeed, the flooding resulting from a dike-break is asymmetric, characterized by a long-term evolution, and is strongly influenced by the flow conditions in the main channel.

Finally, Table 10 contains details of experimental studies on the catastrophic failure of storage tanks with consequent potential overtopping of secondary containment systems (such as dikes or bunds). Such an application is of considerable interest in the industry when cylindrical tanks are used to store hazardous liquids whose sudden release could cause catastrophic effects.

Studies investigating multiple topics of those previously mentioned and potentially falling into many categories appear in all relevant tables for clarity.

Each table consists of 11 columns, which contain the information described below.

Column 1 provides the references in which the experimental investigations are presented and described.

Column 2 indicates the test conditions and the main characteristics of the dam-break flows investigated. In particular, this column specifies whether the dam-break is total or partial and whether the downstream channel is initially dry or wet. Moreover, it reports the types and dimensions of the singularities or obstacles interacting with the dam-break wave.

Column 3 describes the geometric configurations of the laboratory facilities and provides their main dimensions and roughness conditions.

Column 4 indicates the initial conditions of the experimental tests, namely the water depth behind the dam and the downstream water depth in wet bed conditions.

Column 5 reports the breach width, which can be different from the channel width in the event of a partial dam-break.

Columns 6 and 7 indicate the laboratory and the year in which the experiments were performed, respectively.

Column 8 gives the physical quantities measured, and Column 9 lists the measurement techniques and devices used.

Column 10 indicates whether experimental databases are freely available and downloadable in digital format.

Finally, Column 11 informs whether the experimental data were used to validate the dam-break numerical models in the original reference. In particular, this column specifies the types of numerical models used (among the many existing dam-break and flooding models reported, e.g., in [15–18]) and the value of the roughness coefficient set in the numerical simulations, if available. The knowledge of the roughness values proposed in the related references facilitates modelers, who can thus avoid laborious calibrations of this model parameter. In drafting Column 11, we have neglected the use of the data to develop and validate theoretical approaches or analytical solutions. Moreover, we have not investigated the subsequent use that other modelers may have made of the various databases in subsequent numerical studies.

Table 1. Basic experimental investigations of fundamental dam-break wave physical characteristics.

(1) Reference	(2) Dam-Break Type	(3) Setup Characteristics [1]	(4) Initial Conditions [2]	(5) Breach Width	(6) Laboratory	(7) Year	(8) Measured Data	(9) Measuring Technique [3]	(10) Data [4]	(11) Numerical Simulation [5]
Schoklitsch [19]	Total; dry bottom	Rectangular channel Exp. (a) $L = 26$ m, $W = 0.6$ m Exp. (b) $L = 150$ m, $W = 1.3$ m $Lr > 8$ m, $S = 0$; smooth	(a) $h_u < 0.25$ m (b) $h_u < 1$ m	(a) 0.6 m (b) 1.3 m	Technischen Hochschule, Graz, Austria	1917	Wave profiles; depth at the dam section as a function of h_u	Metal plates covered with washable colored stripes quickly dipped and lifted	✗	
Trifonov [20,21]	Total; dry bottom	Rectangular channel $L = 30$ m, $W = 0.4$ m $Lr =$ N.A., $S = 0.004$; smooth	$h_u = 0.3$, 0.4 m	0.4 m	Research Institute of Hydraulic Engineering, Leningrad, Russia	1933	Wave profiles	N.A.	✗	
Eguiazaroff [22]	Total (partial opening of the gate with different velocities)	Rectangular channel $L = 30$ m, $W =$ N.A. $Lr =$ N.A., $S = 0$; smooth and rough	$h_u = 0.3$ m	N.A.	Hydro-electric Laboratory, Leningrad, Russia	1935	Negative wave: free surface profiles at selected times; flow depth time series at six locations Positive wave: wave front celerity; free surface profiles at selected times	Electric chronograph; floating flow level recorder	✗	($\gamma = 0.056$ m$^{1/2}$, $\gamma = 0.4$ m$^{1/2}$)
Levin [7]	Total; dry and wet bottom	Rectangular, triangular, and trapezoidal channels $L =$ N.A., $W =$ N.A. $Lr =$ N.A., $S = 0$; smooth and rough	$h_d/h_u = 0$–0.75	N.A.	Belgrade Polytechnic, Serbia	1952	Flow depth at the dam site and at some representative sections of the wave profile	N.A.	✗	1D SWE (graphical method) ($n = 0.007$ s m$^{-1/3}$, $n = 0.026$ s m$^{-1/3}$)
Martin and Moyce [23]	Collapse of a liquid column; dry bottom	Tank $L > 3\,Lr$, $W = 0.057$ m, $Lr = 0.057$ m, $S = 0$; smooth	$h_u = 0.114$, 0.057 m	0.057 m	N.A.	1952	Wave front position; stage hydrographs	Video camera (300 fps)	✗	
Dressler [13]	Total; dry bottom	Rectangular channel $L = 65$ m, $W = 0.225$ m $Lr =$ N.A., $S = 0$; rough (3 roughness values)	$h_u = 0.22$, 0.11, 0.055 m	0.225 m	US Bureau Standard, USA	1954	Front positions, water depth profiles	Video cameras (1800 fps)	✗	–
WES [24]	Total; dry bottom	Rectangular channel $L = 121.92$ m, $W = 1.22$ m $Lr = 60.96$ m, $S = 0.005$; smooth	$h_u = 0.3048$ m	0.07–1.22 m	Vicksburg, Mississippi, USA	1960	Stage and discharge hydrographs	Video cameras (16 mm movies, 8–12 fps)	✗	($n = 0.009$ s ft$^{-1/3}$)
WES [25]	Total; dry bottom	Rectangular channel $L = 121.92$ m, $W = 1.22$ m $Lr = 60.96$ m, $S = 0.005$; rough	$h_u = 0.09$, 0.18, 0.30 m	0.18–1.22 m	Vicksburg, Mississippi, USA	1961	Stage and discharge hydrographs	Video cameras (16 mm movies, 8–12 fps)	✗	($0.04 < n < 0.12$ s ft$^{-1/3}$)
Faure and Nahas [26]	Total; dry bottom	Rectangular channel $L = 40.6$ m, $W = 0.25$ m $Lr = 20.3$ m, $S = 1.2 \cdot 10^{-4}$; rough	$h_u = 0.23$ m	0.25 m	Laboratoire National d'Hydraulique de Chatou, France	1961	Water depth time series; front propagation	Video cameras	✗	1D SWE MOC ($n = 0.016$ s m$^{-1/3}$, $n = 0.036$ s m$^{-1/3}$)
Estrade [27]	Total; dry bottom	Rectangular channel $L =$ N.A., $W = 0.25$, 0.5 m $Lr =$ N.A., $S = 0$; smooth	$h_u = 0.2$–0.3 m	0.25, 0.5 m	N.A.	1967	Wave profiles at different times	N.A.	✗	
Nakagawa et al. [28]	Total; dry and wet bottom	Rectangular channel $L = 30$ m, $W = 0.5$ m $Lr = 5$ m, $S = 0$; smooth	$h_u = 0.15$–0.4 m $h_d = 0$–0.35 m	0.5 m	Kyoto University, Japan	1969	Wave profiles; flow depth hydrographs at three positions; flow velocity at two locations; wave celerity; bore height	Video cameras (8–64 fps); pressure gauges	✗	
Chervet and Dallèves [29]	Total; dry bottom	Rectangular channel $L = 35$ m, $W = 0.3$ m $Lr = 5$, 7.5, 15 m, $S = -1$, 4, 10%; rough	$h_u = 0.3$ m	0.3 m	Laboratory of Hydraulics, Hydrology and Glaciology, Zurich, Switzerland	1970	Water depth and discharge hydrographs; front position and velocity	Video cameras	✗	1D SWE MOC ($n = 0.0077$–0.0167 s m$^{-1/3}$)
Cunge [30], Cavaillé [31]	Total; dry and wet bottom	Rectangular channel $L = 40$ m, $W = 0.25$ m $Lr = 20$ m, $S = 0$; rough	$h_u = 0.23$ m $h_d = 0$, 0.005, 0.01, 0.04 m	0.25 m	National Laboratory of Hydraulics, Chatou, France	1970	Water depth hydrographs; propagation path and discontinuity height	N.A.	✗	1D SWE FD ($n = 0.01$ s m$^{-1/3}$, $n = 0.0125$ s m$^{-1/3}$)

Table 1. *Cont.*

(1) Reference	(2) Dam-Break Type	(3) Setup Characteristics [1]	(4) Initial Conditions [2]	(5) Breach Width	(6) Laboratory	(7) Year	(8) Measured Data	(9) Measuring Technique [3]	(10) Data [4]	(11) Numerical Simulation [5]
Maxworthy [32]	Total; wet bottom; reflection against the closed end wall; interaction between solitary waves	Rectangular channel $L = 5$ m, $W = 0.2$ m, Lr = N.A., $S = 0$; smooth;	$h_d = 0.045$–0.067 m solitary waves with height of 0.31–0.5 h_d	0.2 m	University of Southern California, Los Angeles, USA	1976	Wave motion; maximum wave amplitude; qualitative wave profiles at selected times	Video camera (64 fps)	✗	
Xanthopoulos and Koutitas [33]	Total; dry bottom	Rectangular channel $L = 6$ m, $W = 0.25$ m $Lr = 1.2$ m, $S = 0$; rough	$h_u = 0.02$–0.15 m	0.25 m	Aristoteles University, Thessaloniki, Greece	1976	Water depth and discharge hydrographs; front propagation	Video cameras	✗	2D SWE FD ($n = 0.033$ s m$^{-1/3}$)
Barr and Das [34]	Total; dry bottom; reflections against the end wall	Rectangular channel (a) $L = 33.5$ m, $W = 1.5$ m, $Lr = 7.62$ m, $S = 0$; (b) $L = 4.4$ m, $W = 0.38$ m $Lr = 1.0$ m, $S = 0$; smooth and rough	(a) $h_u = 0.3048$ m (b) $h_u = 0.1676$–0.3048 m	(a) 1.5 m (b) 0.38 m	University of Strathclyde, Glasgow, UK	1980	Water depth hydrographs; water surface profiles; front trajectories	Video cameras	✗	1D SWE FD ($\varepsilon = 0.0134$–0.0387 m)
Barr and Das [35]	Total; wet bottom; reflections against the end wall	Rectangular channel $L = 33.5$ m, $W = 1.5$ m $Lr = 7.62$ m, $S = 0$; rough	$h_u = 0.3048$ m $h_d = 0.0762$ m	1.5 m	University of Strathclyde, Glasgow, UK	1981	Water depth hydrographs; water surface profiles; front trajectories	Video cameras	✗	1D SWE FD ($\varepsilon = 0.0134$ m, $\varepsilon = 0.0387$ m)
Memos et al. [36]	Total; dry bottom	Tank $L = 2.5$ m, $W = 1.5$ m Plane $W = -$, $S = 0$; rough	$h_u = 0.03$–0.105 m	0.05 m	National Technical University of Athens, Greece	1983	Front propagation, velocity of the front along the x axis, flow profile near the dam	Video camera (18 fps)	✗	($n = 0.01$ s m$^{-1/3}$)
Townson and Al-Salihi [37]	Total; dry and wet bottom	Rectangular channel $L = 4$ m, $W = 0.1$ m, $Lr \approx 1.9$ m, $S = 0$; smooth	$h_u = 0.10$ m $h_d/h_u = 0.176$	0.1 m	University of Strathclyde, Glasgow, UK	1989	Water depth hydrographs; water surface profiles at selected times	Video camera; resistance wave probes; pressure transducers	✗	1D SWE (radial) MOC
Menendez and Navarro [38]	Total; dry bottom (different gate removal times)	Rectangular channel $L = 30$ m, $W = 0.31$ m, $Lr \approx 15$ m, $S = 0$; smooth	$h_u = 0.38$ m (max)	0.31 m	University of Buenos Aires, Argentina	1990	Flow images; discharge and flow depth hydrographs at the gate site	Wire gages; video cameras	✗	
Iverson et al. [39] Logan et al. [40]	Total; dry bottom (steep bottom slope)	Rectangular channel $L = 95$ m, $W = 2$ m, $Lr = 12$ m, $S = 0.6$; smooth and rough	Water volume: 6 m^3	2 m	H.J. Andrews Experimental Forest, Oregon, USA	1992–2017	Flow depth time series at three locations; bottom pressure, bottom normal and shear loads at selected locations; propagation of the front wave	Ultrasonic distance meters; pressure and force transducers; video cameras	✓ (videos)	
Antunes Do Carmo et al. [41]	Total; wet bottom	Rectangular channel $L = 7.5$ m, $W = 0.3$ m, $Lr = 3.85$ m, $S = 0$; smooth	$h_u = 0.099$ m $h_d/h_u = 0.587, 0.515$	0.3 m	University of Coimbra, Portugal	1993	Water depth hydrographs at four positions	Water depth gauges	✗	2D SGN FD
Tingsanchali and Rattanapitikon [42]	Partial; dry bottom	Downstream plane $L = 4$ m, $W = 1.9$ m, $Lr = 2.8$ m (Reservoir, $W = 1.7$ m; bottom step at the plane inlet: 0.4 m) $S = 0$ and 1/200; smooth	$h_u = 0.1, 0.2, 0.25$ m	0.1 m	Asian Institute of Technology, Bangkok, Thailand	1993	Wave front propagation; water depth hydrographs at selected positions	Video camera; water depth gauges; mini-current meter	✗	2D SWE FD ($n = 0.001$–0.03 s m$^{-1/3}$)
Braschi et al. [43]	Partial; dry and wet bottom	Tank $L = 1.4$ m, $W = 0.5$ m, $Lr = 0.4$ m, $S = 0$; smooth	$h_u = 0.14$ m $h_d = 0, 0.005$ m	0.05 m	University of Pavia, Italy	1994	Contour maps of water depth at different times	Video camera (25 fps)	✗	2D SWE MOC-based ($n = 0.01$ s m$^{-1/3}$)
Manciola et al. [44]	Total; wet and dry bottom; open and closed downstream end (three different gate opening velocities)	Rectangular channel $L = 9$ m, $W = 0.49$ m, $Lr = 3.366$, 5.876 m, $S = 0$; smooth	$h_u = 0.2, 0.22, 0.3, 0.35$ m $h_d = 0, 0.021$ m	0.49 m	University of Pavia, Italy	1994	Discharge hydrograph at the gate section; front celerity hydrographs; water depth time series at the gate section; wave front propagation	Video cameras (25 fps)	✗	1D SWE FD ($n = 0.015$ s m$^{-1/3}$)
Aguirre-Pe et al. [45]	Total; dry bottom; highly viscous fluid	Rectangular channel $L = 7$ m, $W = 1$ m, $Lr = h_u/\sin\theta$, $S = 0.03, 0.05, 0.07, 0.1, 0.15$; smooth	$h_u = 0.05, 0.08, 0.1$ m	1 m	University of Los Andes, Mérida, Venezuela	1995	Wave front propagation; wave profile at selected times; flow depth time series at selected locations	Video camera (30 fps)	✗	1D SWE FD
Fraccarollo and Toro [46]	Partial; dry bottom	Plane $L = 3$ m, $W = 2$ m, $Lr = 1$ m, $S = 0$ and 7%; smooth	$h_u = 0.6$ m (0.64 m)	0.4 m	University of Trento, Italy	1995	Bottom pressure time series at 14 points; water depth time series at nine points; time series of flow velocity components at 14 locations	Pressure transducers; capacitance wave meters; electromagnetic velocity meters	✗	2D SWE FV ($n = 0$)

Table 1. *Cont.*

(1) Reference	(2) Dam-Break Type	(3) Setup Characteristics [1]	(4) Initial Conditions [2]	(5) Breach Width	(6) Laboratory	(7) Year	(8) Measured Data	(9) Measuring Technique [3]	(10) Data [4]	(11) Numerical Simulation [5]
Jovanović and Djordjević [47]	Total; dry bottom	Rectangular channel $L = 4.5$ m, $W = 015$ m, $Lr = 2.25$ m, $S = 0.1\%$; smooth	$h_u = 0.3$ m	0.15 m	University of Belgrade, Yugoslavia	1995	Water depth hydrographs, water depth profiles	Water depth capacity probes and video camera	✗	2D SWE FD ($n = 0.009$ s m$^{-1/3}$)
Jovanović and Djordjević [47]	Partial; dry bottom	Downstream plane $L = 1$ m, $W = 0.8$ m, $Lr = 1$ m (Reservoir, $W = 1$ m), $S = 0$; smooth	$h_u = 0.15$ m	0.1 m	University of Belgrade, Yugoslavia	1995	Water depth hydrographs, water depth profiles	Water depth capacity probes and video camera	✗	2D SWE FD ($n = 0.01$ s m$^{-1/3}$)
Koshizuka and Oka [48]; Koshizuka et al. [49]	Total, dry bottom; impact on a vertical wall	Rectangular channel $L = 0.584$ m, $W =$ N.A., $Lr = 0.146$ m, $S = 0$; smooth	$h_u = 0.292$ m	N.A.	University of Tokyo, Japan	1996	Water depth profiles, wave front evolution	Video camera (50 fps)	✗	2D NSE MPS
Lauber and Hager [50]	Total; dry bottom	Rectangular channel $L = 14$ m, $W = 0.5$ m, $Lr = 3.5$ m $S = 0$; smooth	$h_u = 0.3$ m	0.5 m	ETH Zurich, Switzerland	1998	Free surface profiles, velocity and discharge profiles, wave front position	Video camera (50 fps)	✗	($\varepsilon = 5 \times 10^{-6}$ m)
Lauber and Hager [51]	Total; dry bottom	Rectangular channel $L = 14$ m, $W = 0.5$ m, $Lr = 3.5$ m $S = 0.1$, 0.5; smooth	$h_u = 0.3$ m	0.5 m	ETH Zurich, Switzerland	1998	Surface profiles velocity distribution at fixed positions; discharge hydrographs	Video camera (50 fps)	✗	($\varepsilon = 5 \times 10^{-6}$ m)
Stansby et al. [52]	Total; dry and wet bottom	Rectangular channel $L = 15.24$ m, $W = 0.4$ m, $Lr = 9.6$ m, $S = 0$; smooth	$h_u = 0.1$, 0.36 m $h_d = 0$, $0.01h_u$, $0.45h_u$	0.4 m	University of Manchester, UK	1998	Water elevation profiles	Laser, video camera (25 fps)	✗	
Blaser and Hager [53]	Total; dry bottom	Rectangular channel $L = 14$ m, $W = 0.5$ m, $Lr =$ N.A. $S = 0$–0.5; rough	$h_u = 0.2$–0.6 m	0.5 m	ETH Zurich, Switzerland	1999	Wave front velocity and location	N.A.	✗	($\varepsilon = 2.5 \times 10^{-3}$ m)
Nsom et al. [54]	Total; dry bottom; Newtonian solution (glucose syrup-water)	Rectangular channel $L = 5$ m, $W = 0.3$ m, $Lr = h_u/S$, $S = 3$–$12°$; smooth	$h_u = 0.055$ m	0.3 m	Université de Savoie, Cedex, France	2000	Flow depth time series at a selected section; front wave propagation	Video camera (1000 fps); ultrasonic distance meters	✗	
Gallati and Braschi [55]	Total; dry and wet bottom	Tank $L = 1.2$ m, $W = 0.05$ m, $Lr = 0.3$ m; rough	$h_u = 0.1$ m $h_d = 0$–0.02 m	0.05 m	University of Pavia, Italy	2000	Water elevation profiles	Video camera (24 fps)	✗	2D EUL SPH
Liem et al. [56]	Total; dry bottom	Rectangular channel $L = 14$ m, $W = 0.5$ m, $Lr = 5$ m, $S = 0$; smooth	$h_u = 0.3$, 0.35, 0.4, 0.45 m	0.5 m	Aachen University of Technology, Germany	2001	Front position and velocity	Video camera (4500 fps)	✗	1D SWE FE, FV
Briechle and Köngeter [57]	Total; dry and wet bottom; inflow in the reservoir	Rectangular channel $L = 12.2$ m, $W = 0.5$ m, $Lr = 2.65$ m, $S = 0.002$; smooth	$h_u = 0.3$, 0.35, 0.4, 0.45 m; steady inflow: 0, 40, 80, 120 l s^{-1}	0.5 m	Aachen University of Technology, Germany	2002	Water depth hydrographs in six sections; front position and velocity	Video camera (4500 fps)	✗	
Soares-Frazão and Zech [58]	Total; wet bottom (undular bore)	Tank: $L = 10$ m, $W > 1$ m Channel: $L = 26.15$ m, $W = 1$ m $S = 0$; smooth	Different values of h_u–h_d	1.0 m	Université Catholique de Louvain, Belgium	2002	Water depth hydrographs at six positions	Water level gauges	✗	1D BOU Hybrid FV–FD ($n = 0$)
Shige-eda and Akiyama [59]	Partial (asymmetric); dry bottom	Tank $L = 4.8$ m, $Wr = 2.98$ m $Lr = 1.93$ m, $S = 0$; smooth	$h_u = 0.4$ m	0.5 m	Kyushu Institute of Technology, Kitakyushu, Japan	2003	Wave front position, flow depths and surface velocity hydrographs at six points	Digital video tape recorder; PTV	✗	2D SWE FV ($n < 0.07$ s m$^{-1/3}$)
Stelling and Duinmeijer [60]; Duinmeijer [61]	Partial; dry and wet bottom	Tank $L = 31$ m, $W = 7.56$ m, $Lr = 2.4$ m, $S = 0$; smooth	$h_u = 0.6$ m $h_d = 0$, 0.03–0.05 m	0.4 m	Delft University of Technology, The Netherlands	2003	Water depth hydrographs; front position and velocity	Water depth resistance probes; video camera (30 fps)	✗	2D SWE FD ($n = 0.012$ s m$^{-1/3}$)
Chegini et al. [62]	Total; dry bottom	Rectangular channel $L = 15.24$ m, $W = 0.4$ m, $Lr = 9.76$ m, $S = 0$; smooth	$h_u = 0.1$ m $h_d = 0.1$–$0.55\,h_u$	0.4 m	University of Manchester, UK	2004	Flow field and velocity	Particle tracking and streak velocimetry	✗	
Gallati and Sturla [63]	Partial; dry bottom	Tank $L = 1.4$ m, $W = 0.5$ m, $Lr = 0.4$ m, $S = 0$; smooth	$h_u = 0.08$ m	0.155 m	University of Pavia, Italy	2004	Images of the flow field in the flood plain at different time steps	Video camera (25 fps)	✗	2D SWE SPH ($n = 0.01$ s m$^{-1/3}$)
Jánosi et al. [64]	Total; dry and wet bottom	Tank $L = 9.93$ m, $W = 0.15$ m, $Lr = 0.38$ m, $S = 0$; smooth	$h_u = 0.11$–0.25 m $h_d = 0$, 0.018, 0.038 m	0.15 m	Eötvös University, Budapest, Hungary	2004	Water surface profiles; front position and velocity	Video cameras	✗	

Table 1. *Cont.*

(1) Reference	(2) Dam-Break Type	(3) Setup Characteristics [1]	(4) Initial Conditions [2]	(5) Breach Width	(6) Laboratory	(7) Year	(8) Measured Data	(9) Measuring Technique [3]	(10) Data [4]	(11) Numerical Simulation [5]
Bukreev and Gusev [65]	Total; dry and wet bottom	Rectangular channel $L >> 1.3$ m, $W = 0.2$ m, $Lr >> 0.3$ m, $S = 0$; rough	$h_u = 0.205$ m $h_d = 0.0, 0.02$ m	0.2 m	Russian Academy of Sciences, Novosibirsk, Russia	2005	Water level profiles	Wavemeters, video camera	✗	
Eaket et al. [66]	Partial; dry and wet bottom	Tank $L = 4.75$ m, $W = 2.31$ m, $Lr = 2.32$ m, $S = 0$; smooth	$h_u = 0.1, 0.2, 0.3$ m $h_d = 0.05, 0.1$ m	0.89 m	University of Alberta, Edmonton AB, Canada	2005	Water surface profiles and velocities	Video stereoscopy, Video cameras (30 fps)	✗	
Piau and Debiane [67]	Total; dry bottom; highly viscous Newtonian solution (12, 85, 130 Pa s)	Rectangular channel $L = 5$ m, $W = 0.3$ m, $Lr = 2, 4, 6, 8h_u$, $S = 0$; smooth	$h_u = 0.054, 0.055$ m	0.3 m	Université Joseph Fourier, Grenoble, France	2005	Wave front position with time; flow depth profiles at selected times	Video cameras (25, 1000 fps); ultrasonic distance meters	✗	
Barnes and Baldock [68]	Total; dry bottom	Rectangular channel $L = 4.0$ m, $W = 0.4$ m, $Lr = 2.25$ m, $S = 0$; rough	$h_u = 0.2$ m	0.4 m	University of Queensland, Brisbane, Australia	2006	Shear stress; free surface elevation; velocity	Shear plate, ADV, acoustic displacement sensors	✗	($\varepsilon = 0.1 \times 10^{-3}$ m)
Bateman et al. [69]	Total; dry bottom; end platform	Channel: $L = 9.0$ m, $W = 0.4$ m, $Lr = 2.0$ m, $S = 27^{\circ}$; rough; Platform: 4 m × 2.4 m	$h_u = 0.5$ m	0.4 m	Technical University of Catalonia, Barcelona, Spain	2006	Water surface profiles	Video cameras (30, 1000 fps)	✗	
Cruchaga et al. [70]	Total; dry bottom; impact on a vertical wall (two different fluids: shampoo and water)	Tank $L = 0.42$ m, $W = 0.228$ m, $Lr = 0.114$ m, $S = 0$; smooth	$h_u = 1Lr, 2Lr$	0.228 m	University of Santiago, Chile	2007	Water depth time series at selected sections; wave front position	Video cameras	✗	2D NSE, ETILT FE
Maranzoni et al. [71]	Total; dry bottom; horizontal and sloping channel	Tank $L = 11$ m, $W = 0.18$ m, $Lr = 0.114$ m, $S = 0, 6\%$; smooth	$h_u = 0.1$ m	0.18 m	University of Brescia, Italy	2007	Water surface profiles; Water depth hydrographs	Video camera (25 fps)	✗	1D SWE FV; 2D EUL, VOF FV
Aureli et al. [14,72]	Partial; dry and wet bottom	Tank $L = 2.6$ m, $W = 1.2$ m, $Lr = 0.8$ m, $S = 0$; smooth	$h_u = 0.15$ m $h_d = 0.01$ m	0.3 m	University of Parma, Italy	2008	Water surface at 10 times; water depth time series at a gauge point	Video camera (3 fps); ultrasonic distance meters	✓	2D SWE FV ($n = 0.007$ s m$^{-1/3}$)
Mohamed [73]	Total; dry and wet bottom	Rectangular channel $L = 12.2$ m, $W = 1.22$ m, $Lr = 3.60$ m, $S = 0$; concrete bottom and glass side walls, smooth	$h_u = 0.3, 0.45, 0.6$ m $h_d = 0, 0.025, 0.05$ m	1.22 m	University of Hawaii at Manoa	2008	Water surface profiles in time, bore height, shape and speed	Video camera (30 fps)	✗	
Ancey et al. [74]	Total; dry bottom; highly viscous Newtonian fluid (glucose solution)	Rectangular channel $L = 4$ m, $W = 0.3$ m $S = 0, 6, 12, 18, 24^{\circ}$; smooth	Mass in the reservoir: 50.8–57.6 kg	0.3 m	EPFL, Lausanne, Switzerland	2009	Free surface (imaging technique) and flow depth profiles at selected times; front position with time	Video camera	✗	
Yang et al. [75]	Partial; wet bottom	Rectangular channel $L = 28$ m, $W = 1.6$ m, $Lr = 10$ m, $S = 0$; concrete bottom and glass side walls; smooth	$h_u = 0.4$ m $h_d = 0.12$ m	0.2 m	Tsinghua University, Beijing, China	2010	Water depth hydrographs; velocity fields at fixed times	Pressure probes, PIV, video cameras	✗	3D RANS, VOF FV
Ozmen-Cagatay and Kocaman [76,77]	Total; dry and wet bottom	Rectangular channel $L = 9$ m, $W = 0.3$ m, $Lr = 4.65$ m, $S = 0$; smooth;	$h_u = 0.25$ m $h_d = 0, 0.025, 0.1$ m	0.3 m	Cukurova University, Adana, Turkey	2010	Water depth profiles at different time steps	Video camera (50 fps)	✗	2D RANS, VOF FV; 2D SWE FV
Duarte et al. [78]; Boillat et al. [79]; Ribeiro et al. [80]	Total; silted-up reservoir; dry bottom; multiphase flow	Rectangular channel $L = 5.5$ m, $W = 0.42$ m, $Lr = 1.5$ m, $S = 0$; smooth (2 mean grain size diameters)	$h_u = 0.4, 0.41, 0.42$ m (sediment depth: 0.22–0.39 m)	0.42 m	EPFL, Lausanne, Switzerland	2011	Video images; water and sediment surface profiles at selected times; sediment deposition; water front propagation; maximum wave depth profile	Video camera (15 fps)	✗	
Marra et al. [81]	Total; dry bottom	Rectangular channel $L = 3$ m, $W = 0.1$ m, $S = 1.5–24^{\circ}$; smooth and rough	Water volume in the reservoir = 3, 4, 5, 6, 7, 8 l	0.1 m	EPFL, Lausanne, Switzerland	2011	Wave front position and velocity; water surface profiles at selected times; water depth hydrographs at two positions	Video camera (500–800 fps)	✗	(two rough bottoms: $n = 0.0133$ s m$^{-1/3}$, $n = 0.0153$ s m$^{-1/3}$)

Table 1. *Cont.*

(1) Reference	(2) Dam-Break Type	(3) Setup Characteristics [1]	(4) Initial Conditions [2]	(5) Breach Width	(6) Laboratory	(7) Year	(8) Measured Data	(9) Measuring Technique [3]	(10) Data [4]	(11) Numerical Simulation [5]
Aleixo et al. [82–85]	Total; dry bottom; first stages (upward and downward moving gate)	Rectangular channel $L = 6$ m, $W = 0.25$ m, $Lr = 3$ m, $S = 0$; smooth	$h_u = 0.325, 0.4$ m	0.25	Université Catholique de Louvain, Belgium	2011	Flow images; velocity field and components at selected sections	Video camera (100 fps); PIV	✗	
Feizi Khankandi et al. [86]	Total; four different reservoir geometries; dry and wet bottom	1: $Lr = 0.89$ m, $W = 2$ m, 2: $Lr = 1.79$ m, $W = 1.5$ m, 3: $Lr = 1.5$–2.5 m, $W = 0.51$ m, 4: $Lr = 3.5$ m, $W = 0.51$ m, Channel: $L = 9.3$ m, $W = 0.51$ m, $S = 0$; smooth	$h_u = 0.35, 0.4, 0.45$ m $h_d = 0, 0.08$ m	0.51m	Amirkabir University of Technology, Tehran, Iran	2012	Water depth, velocity and discharge hydrographs at different positions; water surface profile at different times	Ultrasonic distance meters; ADV, video camera (110 fps)	✗	($n = 0.011$ s m$^{-1/3}$)
Oertel and Bung [87]	Total; dry bottom	Rectangular channel $L = 22$ m, $W = 0.3$ m, $Lr = 13$ m, $S = 0$; smooth	$h_u = 0.1, 0.2, 0.3, 0.4$ m	0.3 m	Bergische Universität Wuppertal, Germany	2012	Water depth in seven measuring points; water depth profiles at selected times; velocity field at selected times	Ultrasonic distance meters; video camera (1000 fps); PIV	✗	2D RANS, VOF FV ($\varepsilon = 0.0015 \times 10^{-3}$ m)
LaRocque et al. [88]	Total; dry bottom	Rectangular channel $L = 7.31$ m, $W = 0.18$ m, $Lr = 3.37$ m, $S = 0.93\%$; smooth	$h_u = 0.25, 0.3, 0.35$ m	0.18 m	University of South Carolina, USA	2013	Water surface profiles at selected times; velocity vertical profiles at eight locations	Ultrasonic distance meters; ultrasonic Doppler velocity profilers	✗	2D RANS, VOF FV ($\varepsilon = 0.01 \times 10^{-3}$ m)
Miani et al. [89]	Total; wet bottom	Rectangular channel $L = 10$ m, $W = 0.5$ m, $Lr = 1$ m, $S = 0$; smooth	$h_u = 0.4$ m $h_d = 0.2, 0.3$ m; $h_u = 0.4$ m $h_d = 0.1, 0.2, 0.4$ m	0.5 m	Joint Research Centre, Ispra, Italy	2013	Water depth hydrographs at 10 locations	Ultrasonic distance meters	✗	1D SWE FV
Hooshyaripor and Tahershamsi [90]	Total; dry bottom	Rectangular channel $L = 9.3$ m, $W = 0.51$ m, $Lr = 4.5$ m, $S = 0$; smooth	$h_u = 0.35$ m	0.51 m	Amirkabir University of Technology, Iran	2015	Water depth hydrographs at 11 points; velocity and discharge hydrographs at six locations	Ultrasonic distance meters, ADV	✗	3D RANS, VOF FV ($n = 0.011$ s m$^{-1/3}$)
Jiang and Baldock [91]	Total; dry bottom	Rectangular channel $L = 3$ m, $W = 0.4$ m, $Lr = 1.7$ m, $S = 0$; smooth	$h_u = 0.1, 0.15, 0.2$ m	0.4 m	University of Queensland, St. Lucia, Australia	2015	Flow depth and bottom shear stress time series	Acoustic displacement sensors; shear plate; PIV	✗	2D SWE FV ($n = 0.01, 0.011,$ 0.019 s m$^{-1/3}$)
Jiang and Baldock [91]	Total; dry bottom (fixed sand false bed, two grain sizes $d_{50} = 0.22, 2.85$ mm)	Rectangular channel $L = 3$ m, $W = 0.4$ m, $Lr = 1$ m, $S = 0, 1/10$; rough (fine and coarse)	$h_u = 0.08$–0.22 m	0.4 m	University of Queensland, St. Lucia, Australia	2015	Flow depth and bottom shear stress time series	Acoustic displacement sensors; shear plate; PIV	✗	2D SWE FV ($n = 0.01, 0.011,$ 0.019 s m$^{-1/3}$)
McMullin [92]	Total; dry and wet bottom (two gate removal mechanisms)	Rectangular channel $L = 0.5$ m, $W = 0.175$ m, $Lr = 0.2$ m, $S = 0$; smooth	$h_u = 0.06$–0.14 m $h_d = 0.005$–0.02 m	0.175 m	University of Nottingham, UK	2015	Wave front position in time; wave profiles at selected times; horizontal and vertical velocity at selected times and positions	Video cameras, PIV	✗	2D NSE, VOF FD
Mrokowska et al. [93]	Total; wet bottom; closed downstream end	Rectangular channel $L = 60$ m, $W = 0.6$ m $Lr = 5$ m, $S = 0.002$; smooth;	$h_u = 0.31, 0.36$ m $h_d = 0.04, 0.06, 0.08$ m	0.6 m	Polish Academy of Science, Warsaw, Poland	2015	Water depth hydrographs at seven locations; velocity fields	Water level sensors; video camera (520 fps); PIV	✗	
Aleixo et al. [94]	Total; silted-up reservoir (tailings dam-break); dry bottom; sudden enlargement	Plane $L = 7.66$ m, $W = 3.66$ m, $S = 0$; smooth Reservoir $Lr = 3.24$ m, $Wr = 0.5$ m	$h_u = 0.4$ m (sediment depth 0.2 m)	0.5 m	National Sedimentation Laboratory, Oxford, Mississippi, USA	2016	Velocity fields	Video cameras (400 fps); PIV-PTV	✗	
Elkholy et al. [95]	Partial; dry bottom	Tank $L = 11$ m, $W = 4.3$ m, $Lr = 3$ m, $S = 0$; smooth	$h_u = 0.25, 0.5, 0.75$ m	0.4 m	University of South Carolina, USA	2016	Pressure head at the bottom in nine points; water surface elevations and surface velocity; velocity profile at the center of the gate section	Pressure sensors; PTV (video cameras, 60 fps); ultrasonic velocity profiler	✗	
Javadian et al. [96]	Total; dry bottom closed downstream end	Rectangular channel $L = 2$ m, $W = 0.2$ m, $Lr = 1$ m, $S = 0$; smooth;	$h_u = 0.11, 0.12, 0.13$ m	0.2 m	Sharif University of Technology, Tehran, Iran	2016	Water surface profiles at selected times; wavefront position in time	Video camera (24 fps)	✗	

Table 1. *Cont.*

(1) Reference	(2) Dam-Break Type	(3) Setup Characteristics [1]	(4) Initial Conditions [2]	(5) Breach Width	(6) Laboratory	(7) Year	(8) Measured Data	(9) Measuring Technique [3]	(10) Data [4]	(11) Numerical Simulation [5]
Hooshyaripor et al. [97]	Total; dry bottom	Rectangular channel L = 9.3 m, W = 0.51 m, S = 0; smooth Reservoir: Lr = 4.5 m, W = 2.25 m (different side slopes and lengths)	h_u = 0.35 m	0.51 m	Amirkabir University of Technology, Tehran, Iran	2017	Water depth and flow velocity time series at selected locations	Ultrasonic distance meters; ADV	✗	
Liu and Liu [98,99]	Total; dry and wet bottom	Rectangular channel L = 6.5 m, W = 0.4 m, Lr = 1.5 m, S = 0; smooth	h_u = 0.16–0.36 m h_d = 0, 0.02, 0.04 m	0.4 m	Zhejiang University, Hangzhou, China	2017	Water surface profiles at selected times; water depth time series; flow velocity time series	Video camera (150 fps); capacitive wave gauges; ADV	✗	
Cordero et al. [100]	Patial; dry bottom	Reservoir Lr = 1 m; W = 1 m Floodable area L = 4 m, W, 3 m S = 0, 12°; smooth	h_u = 0.1, 0.15, 0.2 m	$2h_u$ (triang. 1H:1V slope)	Polytechnic University of Turin, Italy	2018	Water surface at selected times; water depth time series; water depth profiles	Video camera (100 fps)	✗	
Liu et al. [101]	Total; dry and wet bottom	Rectangular channel L = 18 m, W = 1 m, Lr = 8.37 m, S = 0; smooth	h_u = 0.6 m h_d = 0.06, 0.12, 0.18, 0.24 m	1 m	Sichuan University, Chengdu, China	2018	Water surface and average flow velocity profiles at selected times; wave front celerities	Video cameras (48 fps)	✗	1D SWE
Hamid et al. [102,103]	Total; dry bottom open and closed downstream end	Rectangular channel L = 6.7 m, W = 0.3048 m, Lr = 2.13 m, S = 0.002; smooth	h_u = 0.762 m	0.3048 m	University of Engineering and Technology, Peshawar, Pakistan	2018	Water depth and flood wave velocity time series at selected sections	Point gauges and velocity sensor	✗	2D SWE FV
Stolle et al. [104]; von Häfen et al. [105]	Total; wet bottom; swing gate (opening time influence)	Rectangular channel L = 30 m, W = 1.5 m, Lr = 21.55 m, S = 0; rough	h_u = 0.2, 0.3, 0.4, 0.5 m	1.4 m	University of Ottawa, Canada	2018	Water depth time series at four locations; flow velocity at a selected location; wave front arrival time	Capacitance wave gauges; propeller velocity flowmeter; video cameras (70, 120 fps)	✗	(ε = 0.001·10^{-3} m, λ = 0.014, 0.0293)
Liu et al. [106]	Total; wet bottom	Rectangular channel L = 18 m, W = 1 m, Lr = 8.37 m S = 0; smooth	h_u = 0.4 m h_d = 0.02, 0.04, 0.08, 0.12, 0.16 m	1 m	Sichuan University, Chengdu, China	2019	Video images; water surface profiles at selected times; water depth time series at selected locations	Video cameras (48 fps)	✗	2D RANS, VOF FV
Melis et al. [107]	Total; dry bottom; effect of vegetation (polymeric cylinders)	Rectangular channel L = 11.6 m, W = 0.5 m, Lr = N.A., S = 0, 1, 2, 3%; smooth, rough	h_u = 0.15, 0.2, 0.25, 0.3 m	0.5 m	Polytechnic University of Turin, Italy	2019	Water surface profiles	Video cameras (30 fps)	✓	1D SWE FD (n = 0.05 s m$^{-1/3}$)
Turhan et al. [108]; Turhan et al. [109]	Total; dry and wet bottom; closed downstream end; salt water	Rectangular channel L = 1.216 m, W = 0.2 m, Lr = 0.3 m, S = 0; smooth;	h_u = 0.15 m h_d/h_u = 0, 0.1, 0.2, 0.4	0.2 m	Adana Science and Technology University, Turkey	2019	Water surface profiles at selected times; water depth time series at four locations	Video camera (60 fps)	✗	3D RANS, VOF SPH
Wang et al. [110]	Total; wet bottom	Rectangular channel (rectangular and triangular section) L = 18 m, W = 1 m, Lr = 8.37 m, S = 0; smooth	h_u = 0.4, 0.6 m h_d/h_u = 0.1, 0.2, 0.3, 0.4	1 m	Sichuan University, Chengdu, China	2019	Water surface profiles at selected times; water depth time series at selected locations	Video cameras (48 fps)	✗	
Wu et al. [111]	Total; wet bottom; closed downstream end	Rectangular channel L = 16.38 m, W = 0.4 m, Lr = 5.47 m, S = 0; smooth;	h_u = 0.16, 0.28 m h_d = 0.12 m	0.4 m	Dalian University of Technology, China	2019	Water depth hydrographs at 12 locations; flow velocity time series at four locations	Wave gauges; ADV	✗	2D BOU Hybrid FD–FV (n = 0.01 s m$^{-1/3}$)
Liu et al. [112]	Total; dry and wet bottom	Rectangular channel L = 18 m, W = 1 m, Lr = 8.37 m S = 0, 0.003, 0.02; smooth	h_u = 0.2 m h_d = 0–0.18 m; h_u = 0.4 m h_d = 0–0.36 m	1 m	Sichuan University, Chengdu, China	2020	Video images; water surface and mean velocity profiles; wave front celerity	Video cameras (48 fps)	✗	
Oertel and Süfke [113]	Total; dry bottom	Rectangular channel L = 12.5 m, W = 0.3 m, Lr = 6.5 m S = 0; smooth	h_u = 0.2, 0.3, 0.4 m	0.3 m	Technical University of Applied Sciences, Luebeck, Germany	2020	Water depth at three selected locations; flow velocity vertical profiles	Ultrasonic distance meters; video camera (732 fps); PIV and optical flow methods	✗	
Shugan et al. [114]	Total; dry and wet bottom; first stages	Rectangular channel L = 25 m, W = 0.3 m, Lr = ~11 m, S = 0; smooth	h_u = 0.3, 0.4 m h_d = 0, 0.03, 0.06, 0.09 m	0.3 m	National Cheng Kung University, Taiwan	2020	Water depth time series at 12 locations; water surface profile at selected times; front wave celerity; velocity profiles	Capacitance wave gauges; video camera (30 fps); PIV (video camera, 1000 fps)	✗	

Table 1. *Cont.*

(1) Reference	(2) Dam-Break Type	(3) Setup Characteristics [1]	(4) Initial Conditions [2]	(5) Breach Width	(6) Laboratory	(7) Year	(8) Measured Data	(9) Measuring Technique [3]	(10) Data [4]	(11) Numerical Simulation [5]
Vosoughi et al. [115–117]	Total; silted-up reservoir dry and wet bottom; multiphase flow	Rectangular channel L = 6 m, W = 0.3 m, Lr = 1.52 m S = 0; smooth	h_u = 0.3 m h_d = 0.02, 0.04, 0.05 m (sediment depth: 0–0.24 m)	0.3 m	University of Shiraz, Iran	2020	Video images; water surface profiles; water and sediment depth time series at 16 points	Video cameras (50 fps)	✓	3D NSE, VOF NSE, TFM FV
Wang et al. [118]	Total; dry and wet bottom	Rectangular channel (triangular section) L = 18 m, W = 1 m, Lr = 8.37 m, S = 0; smooth	h_u = 0.2, 0.4, 0.6 m h_d/h_u = 0–0.9	1 m	Sichuan University, Chengdu, China	2020	Water surface profiles at selected times; water depth time series at selected locations; wave front celerity	Video cameras (48 fps)	✗	
Wang et al. [119]	Total; wet bottom	Rectangular channel L = 18 m, W = 1 m, Lr = 8.37 m, S = 0; smooth	h_u = 0.2, 0.4, 0.6 m h_d/h_u = 0.05–0.9	1 m	Sichuan University, Chengdu, China	2020	Water surface profiles at selected times; water level hydrographs at selected locations	Video cameras (48 fps)	✗	2D RANS, VOF FV
Ahmadi and Yamamoto [120]	Partial (trapezoidal and triangular breach); dry bottom	Rectangular channel L = 12 m, W = 0.5 m, Lr = 2.5 m, S = 0; smooth	h_u = 0.25, 0.3 m	0.2,0.3 m	Tokai University, Kanagawa, Japan	2021	Water depth hydrograph at a point located 50 cm upstream of the gate	Video camera	✗	
Ansari et al. [121]	Total; dry and wet bottom	Rectangular channel L = 3.7 m, W = 0.6 m, Lr = 0.6 m, S = 0; smooth	h_u = 0.15 m h_d = 0, 0.015, 0.03, 0.058, 0.07 m	0.6 m	University of Zanjan, Iran	2021	Water surface profiles	Video camera (60fps)	✗	2D (Molecular dynamics software) SPH
Ansari et al. [121]	Total; dry bottom; interaction of two opposite dam-break waves	Rectangular channel L = 3.7 m, W = 0.6 m, Lr = 0.6 m (2 opposite reservoirs at the channel ends), S = 0; smooth	h_{u1} = 0.2 m h_{u2} = 0.2, 0.3 m	0.6 m	University of Zanjan, Iran	2021	Water surface profiles	Video camera (60fps)	✗	2D (Molecular dynamics software) SPH
Birnbaum et al. [122]	Total; dry bottom; three-phase Newtonian suspensions	Rectangular channel L = 1.2 m, W = 0.15 m, Lr = 0.2 m (W = 1m), S = 0; smooth	h_u = 0.04–0.13 m	0.15 m	Columbia University, New York, USA	2021	Wave front position with time	Video cameras (1 fps; 30 fps)	✓	
Espartel and Manica [123]	Total; dry and wet bottom; first stages	Rectangular channel L = 6.71 m, W = 0.24 m, Lr = 0.71 m, S = 0; smooth	h_u = 0.1, 0.2, 0.4 m h_d = 0, 0.02, 0.04, 0.08 m	0.24 m	Universidade Federal do Rio Grande do Sul, Porto Alegre, Brazil	2021	Water surface profiles at selected times	Video cameras (240 fps)	✗	
Kocaman et al. [124]	Partial; dry and wet bottom	Tank L = 1 m, W = 0.5 m, Lr = 0.25 m, S = 0; smooth	h_u = 0.15 m h_d = 0.015, 0.030 m	0.1 m	Iskenderun Technical University, Turkey	2021	Water surface at selected times; water depth time series at five points	Video camera (50 fps); ultrasonic distance meters	✗	3D RANS, VOF FV; 2D SWE FV
Nguyen-Thi et al. [125]	Total; dry and wet bottom; water and three high-viscous Newtonian fluids	Rectangular channel L = 2 m, W = 0.055 m, Lr = 0.28 m, S = 0; smooth	h_u = 0.11 m h_d = 0–0.066 m	0.055 m	Université de Picardie Jules Verne, Amiens, France	2021	Water surface profiles	Video camera (203 fps)	✗	3D RANS, VOF FV
Takagi and Furukawa [126]	Total; dry bottom; different gate opening velocities (0.2–2.5 m/s)	Rectangular channel L = 3 m, W = 0.38 m, Lr = 0.5 m, S = 0; smooth	h_u = 0.5 m	0.38 m	Tokyo Institute of Technology, Japan	2021	Bottom pressure time series at four points along the channel centerline; water surface profiles	Pressure sensors; video camera (2400 fps)	✗	
Wang et al. [127]	Total; dry bottom	Triangular channel L = 18 m, W = 1 m, Lr = 8.37 m, S = 0; smooth	h_u = 0.2, 0.4, 0.6 m h_d/h_u = 0–0.9	1 m	Sichuan University, Chengdu, China	2021	Water surface profiles; water level hydrographs, wave front celerity	Video cameras (48 fps)	✗	
Xu et al. [128]	Total; dry and wet bottom	Rectangular channel L = 13 m, W = 0.25m, Lr = N.A., S = 0.0031; rough	h_u = 0.4 m h_d = 0–0.098 m	0.25 m	University of Queensland, Brisbane, Australia	2021	Shear stress; water depth hydrographs	Shear plate; acoustic distance sensors	✗	(ε = 0.084 m)
Ozmen-Cagatay et al. [129]	Total; dry bottom; closed downstream end; three Newtonian fluids	Rectangular channel L = 1.216 m, W = 0.2 m, Lr = 0.3 m, S = 0; smooth	h_u = 0.15 m	0.2 m	Adana Science and Technology University, Turkey	2022	Water surface profiles, water depth hydrographs	Video camera (60 fps)	✗	2D RANS, VOF FV
Yang et al. [130,131]	Total; dry and wet bottom	Rectangular channel L = 10.72 m, W = 1.485 m, Lr = 4.58 m, S = 0; smooth	h_u = 0.13–0.483 m h_d = 0.02, 0.04, 0.06, 0.08, 0.1, 0.12, 0.14 m	1.485 m	Southwest Jiaotong University, Chengdu, China	2022	Water depth hydrographs; wave front celerity; flow velocity	Wave gauges; ADV	✗	2D RANS, VOF FV

Table 1. *Cont.*

(1) Reference	(2) Dam-Break Type	(3) Setup Characteristics [1]	(4) Initial Conditions [2]	(5) Breach Width	(6) Laboratory	(7) Year	(8) Measured Data	(9) Measuring Technique [3]	(10) Data [4]	(11) Numerical Simulation [5]
Nielsen et al. [132]	Total; dry and wet bottom	Rectangular channel L = 13 m, W = 0.5 m, Lr = 0.625 m, S = 0; smooth and rough (4 different values)	h_u = 0.4 m h_d = 0.018 m	0.5 m	University of Queensland, Brisbane, Australia	2022	Water depth and bottom shear stresses hydrographs; dam-break front celerity	Acoustic transducers; shear plates	✗	
Zhang et al. [133]	Total; dry and wet bottom	Triangular channel (side slope: 45°) L = 18 m, W = 1 m, Lr = 8.37 m, S = 0, 0.003, 0.01, 0.02; smooth	h_u = 0.6 m; 0.4 m h_d/h_u = 0, 0.1, 0.2, 0.4	1 m	Sichuan University, Chengdu, China	2022	Water surface profiles; water depth hydrographs	Video cameras (50 fps)	✗	

Note(s): [1] L = facility length; W = facility width; Lr = reservoir length; Wr = reservoir width (if different from W); S = bottom slope; [2] h_u = upstream water depth; h_d = downstream water depth; [3] ADV = acoustic Doppler velocimeter; PIV = particle image velocimetry; PTV = particle tracking velocimetry; [4] ✗ = not freely available; ✓ = freely available; [5] Approach: 1D = one-dimensional; 2D = two-dimensional; 3D = three-dimensional–Mathematical model: BOU = Boussinesq equations; ETILT = edge-tracked interface locator technique; EUL = Euler equations; NSE = Navier–Stokes equations; RANS = Reynolds-averaged Navier–Stokes equations; SGN = Serre–Green–Naghdi equations; SWE = shallow water equations; VOF = volume of fluid–Numerical method: FD = finite difference; FE = finite element; FV = finite volume; MOC = method of characteristics; MPS = moving particle semi-implicit; SPH = smoothed-particle hydrodynamics; TFM = two-fluid method–n = Manning roughness coefficient; ε = surface roughness; λ = friction factor; γ = Bazin roughness coefficient; N.A. = not available.

Table 2. Experimental investigations of dam-break waves through geometric singularities.

(1) Reference	(2) Dam-Break Type	(3) Setup Characteristics [1]	(4) Initial Conditions [2]	(5) Breach Width	(6) Laboratory	(7) Year	(8) Measured Data	(9) Measuring Technique [3]	(10) Data [4]	(11) Numerical Simulation [5]
Chervet and Dallèves [29]	Total; wet bottom; adverse slope; converging-diverging walls	Rectangular channel L = 23 m, W = 0.3 m Lr = 5, 7.5, 15 m, S = −1, 4, 10% rough channel	h_u = 0.3 m h_d = 0.02 m	0.3 m	Laboratory of Hydraulics, Hydrology and Glaciology, Zurich, Switzerland	1970	Water depth and discharge hydrographs; front position and velocity	Video cameras	✗	1D SWE MOC (n = 0.0077–0.0167 s m$^{-1/3}$)
Matsutomi [134]	Total; dry bottom; adverse slope	Tank with L = 3.9 m, W = 0.3 m, Lr = 1.5 m, S = −0.075, −0.15; rough	h_u = 0.13 m	0.3 m	University of Akita, Japan	1983	Wave front trajectories	N.A.	✗	2D SWE FD (specific resistance law)
Martin [135]	Total; dry and wet bottom	Radial reservoir with variable radius r and diverging walls (θ = 5.71–90°)	h_u = 0.36 m	$r \times \theta$ variable	Dresden Technical University, Germany	1983	Discharge hydrograph at the dam position; water surface profile; water level hydrographs	Photographic film sheeting; oscillograph; photogrammetric plotting	✗	1D SWE MOC
Michouev and Sladkevich [136]	Total; wet bottom; sudden enlargement at the dam	Rectangular channel L = 8.8 m, W = 1.6 m, Lr = 4 m, Wr = 0.4 m, S = 0	h_u = N.A. h_d = 0.1 h_u	0.4 m	State University of Moscow, Russia	1983	Water depth hydrographs at four locations; water depth profiles at three times	N.A.	✗	2D SWE FD
Miller and Chaudhry [137]	Total; dry bottom; 180° curved channel	Rectangular channel L = 11.4 m, W = 0.3 m; S = 0; smooth Reservoir Lr = 1.6 m, Wr = 3.65 m	h_u = 0.1, 0.152, 0.2, 0.254, 0.3 m	0.3 m	State University of Washington, USA	1988	Water depth hydrographs at three points in the channel and five points in the reservoir	Capacitance probes; video camera (60 fps)	✗	1D SWE FD (n = 0.014–0.018 s m$^{-1/3}$)
Townson and Al-Salihi [37]	Total; dry and wet bottom; converging diverging walls (θ = 5°)	Rectangular channel L = 4 m, W = 0.1 m, Lr ~1.9 m, S = 0; smooth	h_u = 0.1 m h_d/h_u = 0.176	0.1 m	University of Strathclyde, Glasgow, UK	1989	Water depth hydrographs; wave front position; water surface profiles	High speed tape recorder; resistance wave probes; pressure transducers	✗	1D SWE (radial) MOC
Bell et al. [138]	Total; dry and wet bottom; 180° curved rectangular channel	Reservoir Lr = 2.29 m, Wr = 3.66 m Rectangular channel W = 0.3 m, S = 0; smooth and rough	h_u = 0.15, 0.2, 0.25, 0.3, 0.35 m h_d = 0, 0.013, 0.025, 0.051, 0.0761 m	0.305 m	State University of Washington, USA	1992	Water depth hydrographs; wave front position	Capacitance probes; video camera (60 fps)	✗	(n = 0.0165, 0.04 s m$^{-1/3}$)
Bellos et al. [139]	Total; dry and wet bottom; gradually variable channel width	Rectangular channel L = 21.2 m, W = 1.4 m, Lr = 8.5 m, S = 0–0.01; smooth	h_u = 0.15–0.3 m h_d = 0, 0.053, 0.101 m	0.6 m	University of Thrace, Xanthi, Greece	1992	Water depth hydrographs; water surface profiles at 10 positions	Wave meters, pressure transducers	✗	2D SWE FD (n = 0.012 s m$^{-1/3}$)

Table 2. *Cont.*

(1) Reference	(2) Dam-Break Type	(3) Setup Characteristics [1]	(4) Initial Conditions [2]	(5) Breach Width	(6) Laboratory	(7) Year	(8) Measured Data	(9) Measuring Technique [3]	(10) Data [4]	(11) Numerical Simulation [5]
Četina and Rajar [140]	Total; dry bottom; sudden enlargement (4 m downstream of the dam)	Rectangular channel L = 20 m, W = 0.4 and 2.8 m, Lr = 8 m, Wr = 1.2 m, S = 0.2%; smooth	h_u = 0.25, 0.35, 0.45 m	0.4 m	University of Skopje, North Macedonia	1994	Water depth time series in 31 points; longitudinal and cross-sectional water surface profiles; flow velocity time series at selected points	Capacitance wave gauges; velocity probes	✗	2D SWE FD (n = 0.0137 s m$^{-1/3}$)
Manciola et al. [44]	Total; wet and dry bottom; adverse slope (−0.084, −0.096, −0.15) (three different gate opening velocities)	Rectangular channel L = 9 m, W = 0.49 m, Lr = 3.366, 5.876 m, S = 0, smooth	h_u =0.2, 0.22, 0.3, 0.35 m h_d = 0, 0.021 m	0.49 m	University of Pavia, Italy	1994	Discharge and water depth hydrographs at the gate section; front celerity hydrographs; wave front propagation	Video cameras (25 fps)	✗	1D SWE FD (n = 0.015 s m$^{-1/3}$)
Aureli et al. [141]	Total; dry and wet bottom; bumps	Rectangular channel L = 7 m, W = 1 m, Lr = 2.25 m, S = 0–0.033; smooth	h_u = 0.292, 0.342, 0.35 m above the bump	1 m	University of Parma, Italy	1999	Water depth and velocity hydrographs	Video camera (25 fps); ADV	✗	1D SWE FD (n = 0.01 s m$^{-1/3}$)
Soares-Frazão and Zech [142]; Soares-Frazão et al. [143]	Total; dry and wet bottom; 90° bend (step at the channel entrance δ = 0.33 m)	Tank L = 2.39 m, W = 2.44 m Channel with 90° bend L = 7.335 m, W = 0.495m S = 0; smooth	h_u = 0.2 m h_d = 0, 0.01 m	0.495 m	Université Catholique de Louvain, Belgium	1999	Water depth time series at six locations; wave front velocity	Water level probes	✓	2D SWE LB (bottom: n = 0.0095 s m$^{-1/3}$; side walls: n = 0.0195 s m$^{-1/3}$)
Soares-Frazão and Zech [142]; Soares-Frazão et al. [143]	Total; dry bottom; 45° bend (step at the channel entrance δ = 0.33 m)	Tank L = 2.39 m, W = 2.44 m Channel with 90° bend L = 8.2 m, W = 0.495m S = 0; smooth	h_u = 0.25 m h_d = 0, 0.01 m	0.495 m	Université Catholique de Louvain, Belgium	1999	Water depth time series at nine locations; wave front velocity	Water level probes	✓	2D SWE LB (bottom: n = 0.0095 s m$^{-1/3}$; side walls: n = 0.0195 s m$^{-1/3}$)
Aureli et al. [144,145]	Total; dry and wet bottom; adverse slope (−8, −9, −10%)	Rectangular channel with adverse slope L = 7 m, W = 1 m, Lr = 2.25 m, S = 0, 1, 2%, smooth and rough	h_u = 0.21, 0.25, 0.292 m h_d = 0, 0.045, 0.05 m	1 m	University of Parma, Italy	2000	Water depth and velocity hydrographs	Video camera (25 fps); ADV	✗	1D SWE FD (n = 0.01, 0.025 s m$^{-1/3}$)
Bento Franco and Betâmio de Almeida [146]; Viseu et al. [147]	Total; wet bottom; sudden enlargement (6.45 m downstream of the dam)	Rectangular channel L = 19.3 m, W = 0.5 m, 2.3 m, Lr = 6.1 m, S = 0; smooth	h_u = 0.504 m h_d = 0.003 m	0.5 m	Istituto Superior Técnico, Lisbon, Portugal	2000	Water depth hydrographs at six points	N.A.	✓	(n = 0.009 s m$^{-1/3}$)
Hiver [148]	Total; dry bottom upstream of the sill, dry and wet bottom downstream; triangular bottom sill	Rectangular channel L = 38 m, W = 1 m, Lr = 15.5 m, S = 0; smooth and rough	h_u = 0.75 m h_d = 0, 0.15 m	1 m	Laboratoire de Recherches Hydrauliques, Châtelet, Belgium	2000	Water depth hydrographs	Gauge measurements	✓	(n = 0.0125 s m$^{-1/3}$)
Soares-Frazão et al. [149]; Soares-Frazão [150]	Total; closed downstream end dry bottom upstream of the sill, wet bottom downstream; triangular bottom sill (±0.14 slopes, 0.065 m high)	Rectangular channel L = 5.6 m, W = 0.5 m, Lr = 2.39 m, S = 0; smooth	h_u = 0.111 m h_d = 0, 0.02, 0.025 m	0.5 m	Université Catholique de Louvain, Belgium	2002	Water surface profiles	Video cameras (25 and 40 fps)	✓	1D SWE FV (n = 0.011 s m$^{-1/3}$)
Soares-Frazão and Zech [151]	Total; dry bottom; 90° bend (step at the channel entrance δ = 0.33 m)	Tank L = 2.39 m, W = 2.44 m Channel with 90° bend L = 7.335 m, W = 0.495m S = 0; smooth	h_u = 0.25 m	0.495 m	Université Catholique de Louvain, Belgium	2002	Water depth profiles; velocity field at the bend	Video camera (200 fps and 40 fps); PIV	✗	Hybrid 1D–2D SWE FV (n = 0.011 s m$^{-1/3}$)
Bukreev [152]	Total; dry and wet bottom; bottom drop (δ = 0.051, 0.072 m)	Channel L = 4.2 m, W = 0.202 Reservoir L = 3.3 m, W = 1 m, S = 0; smooth	h_u = 0.075, 0.102, 0.12, 0.152, 0.154, 0.212 m h_d = N.A.	0.202 m	Russian Academy of Sciences, Novosibirsk	2003	Dimensionless height of water impingement on a vertical wall	Powder coating on end wall	✗	
Bukreev and Gusev [153]	Total; dry and wet bottom; bottom drop (δ = 0.072 m)	Channel L = 4.2 m, W = 0.202 m Reservoir Lr = 3.3 m, W = 1 m, S = 0; smooth	h_u = 0.125 m h_d = 0.022, 0.032, 0.05, 0.056, 0.072, 0.1 m	0.202 m	Russian Academy of Sciences, Novosibirsk	2003	Dimensional and dimensionless hydrographs of water depth for different reservoir and channel depths, water profiles at selected times	Wavemeters; video camera	✗	

Table 2. *Cont.*

(1) Reference	(2) Dam-Break Type	(3) Setup Characteristics [1]	(4) Initial Conditions [2]	(5) Breach Width	(6) Laboratory	(7) Year	(8) Measured Data	(9) Measuring Technique [3]	(10) Data [4]	(11) Numerical Simulation [5]
Soares-Frazão et al. [154]	Total; dry bottom; sudden enlargement	Rectangular channel L = 7.6 m, W = 0.12–0.496 m, Lr = 4 m, S = 0; rough	h_u = 0.2 m	0.12 m	Université Catholique de Louvain, Belgium	2003	Water depth time series at five locations; surface-velocity fields at selected times	Water level gauges; water-level follower; digital imaging	✗	2D SWE FV (n = 0.015 s m$^{-1/3}$)
Bukreev et al. [155]	Total; dry and wet bottom bottom step (δ = 0.06 m)	Channel L = 7.07 m, W = 0.202 m Reservoir Lr = 3.3 m, W = 1–0.202 m, S = 0; smooth	h_u = 0.01–0.22 m h_d = 0, 0.01, 0.09 m	0.202 m	Russian Academy of Sciences, Novosibirsk	2004	Water-level profiles, water depth hydrographs	Wave recorders; video camera	✗	
Bellos [156]	Total; dry and wet bottom; gradually variable channel width	Rectangular channel L = 21.2 m, W = 1.4 m, Lr = 8.5 m, S = −0.005, 0, 0.01; smooth	h_u = 0.1–0.4 m h_d = 0, < 0.02 m; h_d = 0.0635 m for S = −0.005	0.6 m	University of Thrace, Xanthi, Greece	2004	Water depth time series at ten positions	Pressure transducers	✗	2D SWE FD
Natale et al. [157]	Total; dry bottom; sluice gates (gate 1: x = 8.4 m, a = 0.04 m; gate 2: x = 9.3 m, a = 0.02 m)	Rectangular channel L = 9.3 m, W = 0.48 m, Lr = 3.36 m, S = 0; rough	h_u = 0.2 m	0.48 m	University of Pavia, Italy	2004	Water depth profiles	Video camera (25 fps);	✗	1D SWE FV (n = 0.12 s m$^{-1/3}$)
Bukreev [158]	Total; dry and wet bottom; bottom step (δ = 0.038, 0.056 m; l = 0.036, 0.257 m)	Rectangular channel L = 7.2 m, W = 0.2 m, S = 0; smooth	h_u = 0.066, 0.13, 0.15 m h_d = 0.055 m	0.2 m	Russian Academy of Sciences, Novosibirsk	2005	Water-level profiles	Piezometers; wave recorders; video camera	✗	
Bukreev [159]	Partial; dry and wet bottom; bottom step (δ = 0.055 m; l = 0.69 m)	Tank and channel (closed end) L = 7.2 m, W = 0.202 m, Lr = 1.32 m, Wr = 1 m; S = 0; smooth	h_u = 0.145, 0.16 m h_d = N.A.	0.1 m	Russian Academy of Sciences, Novosibirsk	2006	Water-level profiles; depth hydrographs and longitudinal and vertical velocities at three cross sections	Video camera; PIV	✗	
Aureli et al. [14,72]	Partial; dry and wet bottom; bottom sill	Tank L = 2.6 m, W = 1.2 m, Lr = 0.8 m, S = 0; smooth	h_u = 0.15 m h_d = 0.01 m	0.3 m	University of Parma, Italy	2008	Water surface profiles; water depth hydrographs	Video camera (3 fps); ultrasonic distance meters	✓	2D SWE FV (n = 0.007 s m$^{-1/3}$)
Gusev et al. [160]	Total; wet bottom; bottom step (δ = 0.05 m)	Rectangular channel L = 7.06 m, W = 0.202 m Lr = 4.76 m, Wr = 1.0 m, S = 0; smooth	h_u = 0.205 m h_d = 0.01–0.205 m	0.202 m	Russian Academy of Sciences, Novosibirsk	2008	Free-surface hydrographs at two points; velocity of the front behind the step; velocity of the front reflected by the step	Wavemeters	✗	
Bukreev et al. [161]	Partial (vertically); wet bottom; lateral constriction and bottom step (b = 0.06 m, l = 0.38 m, δ = 0.072 m)	Rectangular channel L = 8.3 m, W = 0.20 m, Lr = N.A., S = 0; smooth	$0.08(h_u-\delta) < h_d < 1.1(h_u-\delta)$	0.06 m	Russian Academy of Sciences, Novosibirsk	2008	Dimensionless bore depth and propagation speed	Wavemeters	✗	
Evangelista et al. [162,163]	Total; dry bottom; bottom step (δ = 0.05 m)	Rectangular channel L = 9 m, W = 0.4 m, Lr = N.A., S = 0; smooth	h_u = 0.4 m	0.4 m	University of Cassino and Southern Lazio, Italy	2011	Water surface profiles at two selected times	Video camera (30 fps)	✗	1D SWE FV (n = 0.0125 s m$^{-1/3}$)
Ozmen-Cagatay and Kocaman [164]	Total; dry bottom; trapezoidal bottom sill (δ = 0.075 m, l = 1 m)	Rectangular channel L = 8.9 m, W = 0.3 m, Lr = 4.65 m, S = 0; smooth	h_u = 0.25 m	0.3 m	Cukurova University, Adana, Turkey	2011	Water surface profiles at selected times	Video cameras (50 fps)	✗	2D RANS, VOF FV; 1D SWE FV
Ozmen-Cagatay and Kocaman [165]	Total; dry bottom; trapezoidal contraction (0.95 m long, contraction ratio: 1/3)	Rectangular channel L = 8.9 m, W = 0.3 m, Lr = 4.65 m, S = 0; smooth	h_u = 0.25 m	0.3 m	Cukurova University, Adana, Turkey	2012	Water surface profiles at selected times; water depth hydrographs at seven points	Video cameras (50 fps)	✗	3D RANS, VOF FV

Table 2. *Cont.*

(1) Reference	(2) Dam-Break Type	(3) Setup Characteristics [1]	(4) Initial Conditions [2]	(5) Breach Width	(6) Laboratory	(7) Year	(8) Measured Data	(9) Measuring Technique [3]	(10) Data [4]	(11) Numerical Simulation [5]
Kocaman and Ozmen-Cagatay [166]	Total; dry bottom; triangular obstruction (0.95 m long, contraction ratio: 1/3)	Rectangular channel L = 8.9 m, W = 0.3 m, Lr = 4.65 m, S = 0; smooth	h_u = 0.25 m	0.3 m	Cukurova University, Adana, Turkey	2012	Water surface profiles at selected times; water depth hydrographs at six points	Video cameras (50 fps)	✗	3D RANS, VOF FV
Ozmen-Cagatay et al. [167]	Total; dry bottom; triangular bump (δ = 0.075 m, l = 1 m)	Rectangular channel L = 8.9 m, W = 0.3 m, Lr = 4.65 m, S = 0; smooth	h_u = 0.25 m	0.3 m	Cukurova University, Adana, Turkey	2014	Water surface profiles at selected times; water depth hydrographs at six points	Video cameras (50 fps)	✗	2D RANS, VOF FV; 1D SWE FV
Degtyarev et al. [168]	Total; wet bottom; contraction at the dam location	Rectangular channel L = 10 m, W = 0.254 m Reservoir Lr = 5 m, Wr = 0.38 m, S = 0; smooth	h_u = 0.4 m h_d = 0.04, 0.06, 0.08, 0.1, 0.12, 0.14, 0.16, 0.18, 0.2 m	0.254 m	State University of Novosibirsk, Russia	2014	Water depth hydrographs at three points	Conductive wave meters	✗	1D SWE (n = 0)
Wood and Wang [169]	Total; dry bottom; 90° bend	Rectangular channel with 90° bend L = 6.72m, W = 0.273 m Reservoir Lr = 0.89 m, Wr = 0.89 m, S = 0; smooth	h_u = 0.2794 m	0.29 m	University of Huston, Texas, USA	2015	Water depth hydrographs at four points	Resistance-type water level measurements	✗	2D SWE FD (n = 0.009 s m$^{-1/3}$)
Hooshyaripor and Tahershamsi [90]	Total; dry bottom; reservoir with sloping sides (side angle = 30°, 45°, 60°)	Rectangular channel L = 9.3 m, W = 0.51 m, S = 0; smooth Reservoir Lr = 4.5 m, Wr = 2.25 m	h_u = 0.35 m	0.51 m	Amirkabir University of Technology, Iran	2015	Water depth hydrographs at 11 points; velocity and discharge hydrographs at six locations	Ultrasonic distance meters, ADV	✗	3D RANS, VOF FV (n = 0.011 s m$^{-1/3}$)
Kikkert et al. [170]	Total; dry bottom; sudden contraction at the gate site	Rectangular channel L = 6.6 m, W = 0.3 m, S = 1/20; smooth Reservoir Lr = 7.5 m, Wr = 2 m, S = 0	h_u = 0.35 m	0.3 m	Hong Kong University of Science and Technology	2015	Water depth time series; water depth profiles and wave propagation time	Video cameras (90 fps)	✗	3D RANS, VOF FV (ε = 5 × 10^{-5} m)
Chen et al. [171]	Total; wet bottom; Y-shaped junction	Rectangular channels with junction (Y-shaped; 30°, 45°, 60°, 90°) Side channel (with dam): L = 2.5 m, W = 0.3 m, Lr = 1 m Main channel: L = 5 m, W = 0.3 m S = 0; smooth	h_u = 0.3, 0.4, 0.45 m h_d = N.A.	0.3 m	Sichuan University, Chengdu, China;	2019	Water depth and pressure hydrographs; velocity field	Video cameras; PIV; pressure gauges	✗	3D RANS, VOF FV (n = 0.008 s m$^{-1/3}$)
Kobayashi et al. [172]	Total; wet bottom; meanders	Meandering rectangular channel L = 16.1 m, W = 0.8 m, Lr = 1.5 m, S = 1/600; smooth	h_u = 0.285 m h_d = 0.107, 0.147 m	0.8 m	University of Hiroshima, Japan	2019	Flow depth transversal profiles in eight cross-sections	Wave gauges	✗	1D SWE MOC
Kavand et al. [173]	Total; dry bottom; three 90° bends	Rectangular channel W = 0.2 m, S = 0; smooth and rough	h_u = 0.25, 0.35, 0.45, 0.55 m	0.2 m	University of Ahvaz, Iran	2020	Wave front celerity; wave height al the bend sides	Video camera	✗	(ε = 0, 10, 16, 20 × 10^{-3} m)
Kocaman et al. [174]	Total; dry bottom; triangular and trapezoidal channel contractions	Rectangular channel L = 8.9 m, W = 0.3 m, Lr = 4.65 m, S = 0; smooth	h_u = 0.25 m	0.3 m	Cukurova University, Adana, Turkey	2020	Free surface profiles; flow depth hydrographs	Video cameras (50 fps)	✗	3D RANS, VOF FV; 2D SWE FV
Ansari et al. [121]	Total; dry and wet bottom; triangular bottom sill	Rectangular channel L = 3.7 m, W = 0.6 m, Lr = 0.6 m, S = 0; smooth	h_u = 0.2 m h_d = 0, 0.07 m	0.6 m	University of Zanjan, Iran	2021	Water surface profiles	Video camera (60 fps)	✗	3D RANS SPH
Ismail et al. [175]	Total; wet bottom; Y-shaped junction	Rectangular channels with a Y-shaped junction Side channel (with dam): L = 1.83 m, W = 0.304 m, Lr = 0.91 m, S = 0; smooth Main channel: L = 3.35 m, W = 0.304 m	h_u = 0.25, 0.4, 0.5 m h_d = 0.0425, 0.044, 0.052 m (flow rate and velocity in the main channel: Q = 1.87–2.64 l/s; v = 0.145–0.181 m/s)	0.304 m	University of South Carolina, Columbia, USA	2021	Outflow hydrographs downstream of the junction; water surface elevation at the outlet	Ultrasonic distance meters	✗	

Table 2. *Cont.*

(1) Reference	(2) Dam-Break Type	(3) Setup Characteristics [1]	(4) Initial Conditions [2]	(5) Breach Width	(6) Laboratory	(7) Year	(8) Measured Data	(9) Measuring Technique [3]	(10) Data [4]	(11) Numerical Simulation [5]
Gamero et al. [176]	Total; dry and wet bottom; closed downstream end; Gaussian bottom sill in the reservoir	Rectangular channel L = 15 m, W = 0.405 m, Lr = 9.275 m, S = 0.0015; smooth	h_u = 0.302, 0.3 m h_d = 0, 0.12, 0.18, 0.24 m	0.405 m	University of Córdoba, Spain	2022	Piezometric measures along the centerline of the obstacle; water surface profiles	Piezometers; video cameras (25 fps)	✓	2D VAM Hybrid FV–FD (n = 0.01 s m$^{-1/3}$)
Kobayashi et al. [177]	Total; wet bottom; meanders	Straight rectangular channel L = 16.1 m, W = 0.4 m, Lr = 1.68 m, S = 0; smooth Meandering rectangular channel L = 16.1 m, W = 0.39 m, Lr = 1.66 m, S = 0; smooth	Straight h_u = 0.3 m h_d = 0.02 m Meandering h_u = 0.285 m h_d = 0.107 m	Straight 0.4 m Meand. 0.39 m	University of Hiroshima, Japan	2022	Wave height time series in eight cross-sections; free surface profiles at selected times	Wave gauges	✗	2D SWE; 3D RANS, VOF FV
Vosoughi et al. [178,179]	Total; silted-up reservoir (multiphase flow); dry and wet bottom; semi-circular bottom sill (δ = 0.045 m, l = 0.09 m; δ = 0.075 m, l = 0.15 m)	Rectangular channel L = 6 m, W = 0.3 m, Lr = 1.52 m, S = 0; smooth	h_u = 0.3 m (7 sediment depths: 0.03–0.24 m) h_d = 0, 0.02, 0.04, 0.05 m	0.3 m	University of Shiraz, Iran	2022	Water surface profiles; profile of the saturated sediment layer	Video cameras (50 fps)	✓	3D NSE, VOF FV

Note(s): [1] L = facility length; W = facility width; Lr = reservoir length; Wr = reservoir width (if different from W); S = bottom slope; θ = inclination angle; δ = bottom step/bump height; l = singularity length; b = constriction width; [2] h_u = upstream water depth; h_d = downstream water depth; [3] ADV = acoustic Doppler velocimeter; PIV = particle image velocimetry; [4] ✗ = not freely available; ✓ = freely available; [5] Approach: 1D = one-dimensional; 2D = two-dimensional; 3D = three-dimensional–Mathematical model: NSE = Navier–Stokes equations; RANS = Reynolds-averaged Navier–Stokes equations; SWE = shallow water equations; VOF = volume of fluid–Numerical method: FD = finite difference; FV = finite volume; MOC = method of characteristics; SPH = smoothed particle hydrodynamics–n = Manning roughness coefficient; ε = surface roughness; N.A. = not available.

Table 3. Experimental investigations of the dam-break wave impact against obstacles.

(1) Reference	(2) Dam-Break Type	(3) Setup Characteristics [1]	(4) Initial Conditions [2]	(5) Breach Width	(6) Laboratory	(7) Year	(8) Measured Data	(9) Measuring Technique [3]	(10) Data [4]	(11) Numerical Simulation [5]
Greenspan and Young [180]	Total; dry bottom; impact on containment dykes (θ = 90°, 60°, 30°; variable dyke distance from the gate)	Tank L = 1.22 m, W = 0.23 m, Lr = 0.23 m; S = 0; smooth	$h_u \leq$ 0.2032 m	0.23 m	Massachusetts Institute of Technology, USA	1978	Spillage fraction dependence on dyke inclination	Video recording	✗	1D SWE MOC
Sicard and Nicollet [181]	Total; wet bottom; impact on a vertical wall	Rectangular channel L = 18 m, W = 0.6 m, Lr = 3 m; S = 0; smooth	h_u = N.A. h_d = N.A.	0.6 m	Laboratoire National d'Hydraulique, Chatou, France	1983	Water depth and celerity of the incoming wave; pressure time series on the wall at seven elevations	Piezoresistive pressure transducers	✗	
Ramsden [182]	Total; dry and wet bottom; impact on a vertical wall	Rectangular channel L = 36.6 m, W = 0.396 m, Lr = 8.97 m; S = 0; smooth	h_u = 0.502 m h_d = 0 m; h_u = 0.4801 m h_d = 0.28 m	0.396 m	California Institute of Technology, USA	1996	Impact force; pressure at the wall; position of the wave; 2D profiles near the wall	Force and pressure transducers; contact probes; Argon-ion laser; video camera (300 fps)	✗	
Liu et al. [183]	Total; wet bottom; impact on a vertical porous structure (0.29 m long, 0.37 m high, located 0.02 m downstream of the gate; 2 porous materials)	Tank L = 0.892 m, W = 0.44 m, Lr = 0.28 m; S = 0; smooth	h_u = 0.35, 0.25, 0.15 m h_d = 0.02 m	0.44 m	Cornell University, Ithaca, USA	1999	Water surface profiles at 12 times; water level time series in the center of the porous structure	Camera (10 fps); wave gauge	✗	2D RANS, VOF FD
Gallati and Braschi [55]	Total; dry bottom; impact on obstacle (0.03 × 0.06 m, 0.17 m downstream of the dam)	Tank L = 1.2 m, W = 0.03 m, Lr = 0.3 m, rough	h_u = 0.1 m h_d = 0 m	0.03 m	University of Pavia, Italy	2000	Water surface profiles	Video camera (25 fps)	✗	2D EUL SPH

Table 3. *Cont.*

(1) Reference	(2) Dam-Break Type	(3) Setup Characteristics [1]	(4) Initial Conditions [2]	(5) Breach Width	(6) Laboratory	(7) Year	(8) Measured Data	(9) Measuring Technique [3]	(10) Data [4]	(11) Numerical Simulation [5]
Barakhnin et al. [184]	Total; wet bottom; impact on a reflective vertical wall	Tank $L = l_1 + l_2$, $Lr = l_1$ $50 < l_2/h_d < 90$ l_1 = N.A.	$0.5 \leq (h_u - h_d)/h_d \leq 1.4$ h_d = 0.03, 0.04 m	0.06 m	Russian Academy of Sciences, Novosibirsk	2001	Maximum water level at the wall, splash-up profile, free surface profiles	Video camera (25 fps), resistive wavemeter	✗	1D BOU
Soares-Frazão and Zech [185,186]	Partial; wet bottom; impact on an isolated building (0.4 × 0.8 m)	Rectangular channel L = 36 m; W = 3.6 m, Lr = 6.9 m, S = 0; smooth	h_u = 0.4 m h_d = 0.02 m	1 m	Université Catholique de Louvain, Belgium	2002	Water depth hydrographs at six locations; velocity fields at selected times; flow velocity time series at the gauge points	Resistive level gauges; ADV; video camera (40 fps)	✓	(n = 0.01 s m$^{-1/3}$)
Brufau et al. [187]; Méndez et al. [188]	Partial (asymmetrical); wet bottom; pyramidal obstacle	Tank L = 2.65 m, W = 2.615 m, Lr = 1.3, S = 0; smooth	h_u = 0.5 m h_d = 0.1–0.3 m	0.293 m	University of La Coruña, Spain	2002	Water depth time series at several points	N.A.	✗	2D SWE FV
Ciobataru et al. [189]	Total; dry bottom; impact on pillars (square: 0.12 m × 0.12 m; circular: D = 0.14 m)	Tank L = 16.62 m, W = 0.61 m, Lr = 5.9 m; S = 0; smooth and rough	h_u = 0.1–0.3 m	0.61 m	University of Washington, Seattle, USA	2003	Net force on the structure and velocity hydrographs, free surface profile at mid-channel	Load cell; LDV; PIV	✗	3D NSE ELMMC
Trivellato [190]; Bertolazzi and Trivellato [191]	Total, dry bottom; impact on a vertical wall	Rectangular channel L = 6 m, W = 0.5 m, $0 \leq S \leq 25°$	h_f = 0.04 m u_0 = 2.77 ms^{-1}	0.5 m	University of Trento, Italy	2003	Maximum run-up, pressure at the wall, toe velocity and depth, wall force	Pressure transducers; video camera (25 fps)	✗	2D EUL FV
Campisano et al. [192]	Total; dry bottom; downstream sediment deposit (0.03 m volcanic sand thickness)	Rectangular channel L = 3.9 m, W = 0.15 m, Lr = 1.3 m; S = 0.145%; rough	h_u = 0.10–0.13 m	0.15 m	University of Catania, Italy	2004	Water depth hydrographs, sediment bed profiles	Video camera (25 fps)	✗	1D SWE FD (n = 0.0105 s m$^{-1/3}$)
Gallati and Sturla [63]	Partial; dry bottom; impact on a square obstacle	Tank L = 1.4 m, W = 0.5 m, Lr = 0.4 m, S = 0; smooth	h_u = 0.08 m	0.155 m	University of Pavia, Italy	2004	Images of the flow field in the flood plain at different time steps	Video camera (25 fps)	✗	2D SWE SPH (n = 0.01 s m$^{-1/3}$)
Hu and Kashiwagi [193]	Total; dry bottom; impact on a vertical wall	Tank L = 1.18 m, W = 0.12 m, Lr = 0.68 m; S = 0	h_u = 0.12	0.12 m	Kyushu University, Japan	2004	Pressure hydrograph at the wall	Pressure transducers; video camera	✗	2D NSE CIP, FD
Raad and Bidoe [194]	Total; wet bottom; impact on vertical columns (square: 0.12 m × 0.12 m, 0.75 m high)	Tank L = 1.6 m, W = 0.61 m, Lr = 0.4 m; S = 0; smooth	h_u = 0.3 m h_d = 0.01 m	0.61 m	University of Washington, Seattle, USA	2005	Net force on the structure and velocity hydrographs	Load cell; LDV	✗	3D NSE ELMMC
Arnason [195]	Total; dry bottom; impact on columns (square: 0.12 m × 0.12 m; circular: D = 0.029, 0.0606, 0.14 m)	Tank L = 16.62 m, W = 0.61 m, Lr = 5.9 m; S = 0; smooth and rough	h_u = 0.10–0.40 m	0.61 m	University of Washington, Seattle, USA	2005	Net force on the structure and velocity hydrographs; free surface profiles	Load cell; LDV; video camera; PIV	✗	
Kleefsman et al. [196]; Issa and Violeau [197]; Larese et al. [198]	Total; dry bottom; impact on an obstacle	Tank L = 3.22 m, W = 1.0 m, Lr = 1.228 m; S = 0; smooth	h_u = 0.55 m	1.0 m	MARIN (Maritime Research Institute, The Netherlands)	2005	Water depth, pressure and force hydrographs	Height probes; pressure transducers	✗	3D NSE, VOF FV; 3D NSE SPH, PFEM
Liang et al. [199]	Partial; wet bottom; impact on a column (circular: D = 0.35 m)	Tank L = 25 m, W = 1.6 m, Lr = 2.5 m S = 0; smooth	h_u = 0.235 m h_d = 0.059 m	0.15 m	Delft University of Technology, The Netherlands	2007	Water depth hydrographs; front position and velocity	Video camera (25 fps)	✗	2D SWE FD (n = 0.01 s m$^{-1/3}$)
Aureli et al. [14]	Partial; dry and wet bottom; insubmersible obstacle	Tank L = 2.6 m, W = 1.2 m, Lr = 0.8 m, S = 0; smooth	h_u = 0.15 m h_d = 0.01 m	0.3 m	University of Parma, Italy	2008	Water surface profiles; water depth hydrographs	Video camera (3 fps); ultrasonic distance meters	✓	2D SWE FV (n = 0.007 s m$^{-1/3}$)

Table 3. *Cont.*

(1) Reference	(2) Dam-Break Type	(3) Setup Characteristics [1]	(4) Initial Conditions [2]	(5) Breach Width	(6) Laboratory	(7) Year	(8) Measured Data	(9) Measuring Technique [3]	(10) Data [4]	(11) Numerical Simulation [5]
Nouri [200]; Nistor et al. [201]; Nouri et al. [202]	Total; dry bottom; impact on columns (square: 0.2 m × 0.2 m; circular: D = 0.32 m), constrictions	Rectangular channel L = 10.6 m, W = 2.7 m Lr = 5.58 m, S = 0; rough	h_u = 0.5, 0.75, 0.85, 1.0 m	2.7 m	Canadian Hydraulics Center, Ottawa, Canada	2008	Pressures, water level and impact force hydrographs; point velocities	Capacitance wave gauges; load cells; dynamometer; pressure transducers; ADV	✗	
Bukreev and Zykov [203]	Total; wet bottom; vertical plate	Rectangular channel L = 8.2 m W = 0.2 m Lr > 1.4 m, S = 0; rough	h_u/h_d = 0.186, 0.419, 0.605	0.2 m	Russian Academy of Sciences, Novosibirsk	2008	Water depth and force hydrographs, velocity in the vertical plane	Wavemeters; force transducer; PIV	✗	
Arnason et al. [204]	Total; wet bottom; impact on vertical columns (square: 0.12 m × 0.12 m; circular: D = 0.14 m; 5.2 m downstream of the gate)	Tank L = 16.6 m, W = 0.6 m, Lr = 5.9 m, S = 0; smooth	h_u = 0.10–0.3 m (Δh = 0.025m) h_d = 0.02 m	0.6 m	University of Washington, Seattle, USA	2009	Water depth and velocity hydrographs at different locations; time series of the horizontal force on the columns	Laser induced fluorescence technique; particle image and LDV; load cell	✗	
Cruchaga et al. [205]	Total; dry bottom; obstacles of different shapes	Tank L = 0.456 m, W = 0.228m Lr = 0.114 m, S = 0; smooth	h_u = 0.228 m	0.228 m	University of Santiago, Chile	2009	Water depth profiles at different times	Video camera	✗	2D NSE, ETILT FE
Hu and Sueyoshi [206]	Total; dry bottom; impact on a vertical wall	Tank L = 0.8 m, W = 0.2 m, Lr ~ 0.24 m, S = 0; smooth; closed downstream end	h_u = 0.42 m (estimated)	0.2 m	Kyushu University, Japan	2010	Wave front position; water surface profiles at different times	Video camera	✗	2D NSE CIP, MPS
Yang et al. [75]	Total; dry bottom; impact against a brick (0.22 m × 0.12 m, placed 0.6 m downstream of the gate)	Rectangular channel L = 7 m, W = 0.3 m, Lr = 2 m, S = 0; smooth	$h_u \leq 0.123$ m	0.3 m	Tsinghua University, Beijing, China	2010	Critical reservoir depth h_u causing brick movement	N.A.	✗	3D RANS, VOF FV
Aureli et al. [207]	Partial; dry and wet bottom; insubmersible obstacle	Tank L = 2.6 m, W= 1.2 m, Lr = 0.8 m, S = 0; smooth	h_u = 0.030–0.064 m h_d = 0.0068–0.0157m	0.3 m	University of Parma, Italy	2011	Water depth hydrographs; free surface	Video camera (6.5 fps); ultrasonic distance meters	✗	
Al-Faesly et al. [208]	Total; dry and wet bottom; impact on structural models (square and circular: 0.305 m, placed 4.92 m downstream of the gate); effect of mitigation walls (flat or curved)	Rectangular channel L = 14.56 m, W = 2.7 m, S = 0; smooth	h_u = 0.55, 0.85, 1.15 m h_d = N.A.	2.7 m	University of Ottawa, Canada	2012	Base shear forces and moments on structural models; acceleration and displacement at the top edge; pressures at 10 points; water depth hydrographs on models and channel; wave front velocity	Load cell; accelerometer; displacement transducer; pressure transducers; capacitance wave gauges; free-standing wave gauges; video camera	✗	
Oertel and Bung [87]	Total; dry bottom; submersible obstacle	Rectangular channel L = 22 m, W = 0.3 m, Lr = 13 m, S = 0; smooth	h_u = 0.1, 0.2, 0.3, 0.4 m	0.3 m	Bergische University Wuppertal, Germany	2012	Drag force on the obstacle; water depth profiles and velocity field at selected times	Ultrasonic distance meters; video camera (1000 fps); PIV	✗	2D RANS, VOF FV (ε = 0.0015 × 10^{-3} m)
Lara et al. [209]	Total; wet bottom; impact against a solid square prism (0.12 m × 0.12 m)	Tank L = 1.6 m, W = 0.6 m, Lr = 0.4 m, S = 0; smooth	h_u = 0.3 m h_d = 0.01 m	0.6 m	University of Cantabria, Santander, Spain	2012	Flow velocity time series at a selected point; time history of the net force on the prism	LDV; load cell	✗	3D RANS, VOF FV
Triatmadja and Nurhasanah [210]	Total; wet bottom; impact on a building; effects of a barrier	Rectangular channel L = 24 m, W = 1.45 m, Lr = 8 m, S = 0; smooth	h_u = 0.6, 0.7, 0.8 m h_d = 0.02 m	1.45 m	Gadjah Mada University, Indonesia	2012	Water depth hydrographs; force on the structure	Wave gauges; load cell	✗	
Aguíñiga et al. [211]	Total; wet bottom; impact on a vertical wall placed 2.18 m downstream of the gate	Rectangular channel L = 4.93 m, W = 0.305 m, Lr = 0.305 m, S = 0; smooth	h_u = N.A. h_d = 0.051, 0.076, 0.102 m (bore height: 0.157, 0.203, 0.264 m)	0.305 m	Texas A&M University, Kingsville, USA	2013	Maximum force on the wall	Spring system and video camera	✗	

Table 3. *Cont.*

(1) Reference	(2) Dam-Break Type	(3) Setup Characteristics [1]	(4) Initial Conditions [2]	(5) Breach Width	(6) Laboratory	(7) Year	(8) Measured Data	(9) Measuring Technique [3]	(10) Data [4]	(11) Numerical Simulation [5]
Nakao et al. [212]	Total; wet bottom; model T-girder bridges (placed 7.5 m downstream of the gate)	Rectangular channel L = 30 m, W = 1 m Lr = 12 m, S = 0; smooth	h_u = 0.617 m h_u = 0.1, 0.15, 0.2 m h_d = N.A.	1 m	Public Works Research Institute, Tsukuba, Japan	2013	Tsunami height and reaction force in time; dynamic pressure at the girder	Video cameras; load cells; wave gauges; pressure gauges	✗	
Lobovský et al. [213]	Tank; dry bottom; impact against the downstream end	Tank L = 1.61 m, W = 0.15 m, Lr = 0.6 m, S = 0; smooth	h_u = 0.3, 0.6 m	0.6 m	Technical University of Madrid, Spain	2014	Water surface profiles; wave front propagation; water level hydrographs at four locations; pressure hydrographs at five points	Video camera (300 fps); pressure transducers	✓	
Ratia et al. [214]	Total; wet bottom; closed downstream end; bridge models	Rectangular channel L = 6 m, W = 0.24 m, Lr = 1.56 m, Wr = 0.84 m, S = 0; smooth	h_u = 0.169–0.227 m h_d = 0.009–0.011 m	0.24 m	University of Zaragoza, Spain	2014	Water depth hydrographs in two positions	Water depth gauges	✓	2D SWE FV
Aureli et al. [215]	Partial; dry bottom; impact on a insubmersible obstacle (0.3 m × 0.155 m)	Tank L = 2.6 m, W = 1.2 m, Lr = 0.8 m, S = 0; smooth	h_u = 0.07–0.13 m	0.3 m	University of Parma, Italy	2015	Impact force	Load cell	✓	2D SWE FV (n = 0.007 s m$^{-1/3}$) 3D RANS, VOF FV; 3D NSE SPH
Kocaman and Ozmen-Cagatay [216]	Total; wet bottom; impact on the downstream vertical end	Rectangular channel L = 8.9 m, W = 0.3 m, Lr = 4.65 m, S = 0; smooth	h_u = 0.25 m h_d = 0.025, 0.1 m	0.3 m	Cukurova University, Adana, Turkey	2015	Water surface profiles; water depth hydrographs	Video cameras (50 fps)	✗	2D RANS, VOF FV; 1D SWE FV
Liao et al. [217]	Total; dry bottom; impact on an elastic structure (0.1 m high, 0.4 m downstream of the gate)	Tank L = 0.8 m, W = 0.2 m, Lr = 0.2 m, S = 0; smooth	h_u = 0.2, 0.3, 0.4 m	0.2 m	Kyushu University, Japan	2015	Water surface profiles and deformation of the structure (three markers); longitudinal marker displacement hydrographs	Video camera (1000 fps)	✗	2D NSE, VOF Coupled CIP, FD–FE (interaction fluid–structure)
Liang et al. [218]	Total; wet bottom; bridge	Rectangular channel L = 35.5 m, W = 1 m, Lr = 5.5 m, S = 0; smooth	h_u = 0.4 m h_d = 0.198 m; h_u = 0.204 m h_d = 0.105 m	1 m	Hohai University, Nanjing, China	2016	Water depth and flow velocity time series in seven locations; pressure time series on the bridge piers	Wave gauges; ADV; pressure sensors	✗	2D SWE FV (n = 0.01 s m$^{-1/3}$)
Mohd et al. [219]	Total; dry bottom; impact on a vertical cylinder (square: 0.05 m × 0.05 m; circular D = 0.05 m)	Tank L = 0.8 m, W = 0.2 m, Lr = 0.2 m, S = 0; smooth	h_u = 0.4 m	0.2 m	Kyushu University, Japan	2017	Flow images; wave front celerity; water depth hydrographs	Video cameras	✗	3D LBM
Kamra et al. [220]	Total; dry bottom; impact on the closed downstream end	Tank L = 0.8 m, W = 0.2 m, Lr = 0.2 m; S = 0; smooth	h_u = 0.2 m	0.2 m	Kyushu University, Japan	2018	Water surface profiles; pressure hydrographs; wave front position	Pressure sensors	✗	3D RANS, VOF FV
Liu et al. [221]	Partial; dry bottom; building (0.4 m × 0.2 m × 0.3 m, locked and unlocked door scenarios)	Rectangular channel L = 40 m, W = 2.2 m, Wr = 3.5 m, Lr = 11.5 m; S = 0; smooth	h_u = 0.15, 0.2 m	0.8 m	Tsinghua University, Beijing, China	2018	Water level hydrographs	Pressure gauges; ultrasonic distance meters	✗	
Martínez-Aranda et al. [222]	Partial; dry bottom; obstacles, singularities, and a bridge model	Reservoir and rectangular channel L = 6 m, W = 0.24 m, Lr = 1.57 m; Wr = 0.81 m $S \approx 0$ (in the first 3.26 m downstream of the gate), 0.0404 downstream; smooth	h_u = 0.055, 0.13 m	0.24 m	University of Zaragoza, Spain	2018	Free surface; free surface profiles; flow depth time series	RGB-D sensor	✓	2D SWE FV (n = 0.008–0.012 s m$^{-1/3}$)
Stamataki et al. [223]	Total; dry bottom; building	Rectangular channel L = 20 m, W = 1.2 m, Lr = 2.9 m, S = 1/20; smooth and rough	h_u = 0.1, 0.2 m	1.2 m	University College London, UK	2018	Water depth and hydrodynamic force hydrographs; wave front celerity	Wave gauges; ultrasonic distance meters; load cell; pressure sensors; video camera (250 fps)	✗	2D RANS, VOF FV

Table 3. *Cont.*

(1) Reference	(2) Dam-Break Type	(3) Setup Characteristics [1]	(4) Initial Conditions [2]	(5) Breach Width	(6) Laboratory	(7) Year	(8) Measured Data	(9) Measuring Technique [3]	(10) Data [4]	(11) Numerical Simulation [5]
Tinh et al. [224]	Total; dry and wet bottom; impact on a vertical structure	Rectangular channel $L = 17.6$ m, $W = 0.3$ m, $Lr = 3$ m, $S = 1/20$; smooth	$h_u = 0.15$ m $h_d = 0$; $h_u = 0.2$ m $h_d = 0.05$ m	0.3 m	Tohoku University, Sendai, Japan	2018	Water depth hydrographs; water surface profiles; flow images	Ultrasonic distance meters; video camera	✗	
Demir et al. [225]	Total; dry bottom; impact on the downstream end; interaction with a deformable plate (3 different heights)	Tank $L = 0.6$ m, $W = 0.2$ m, $Lr = 0.15$ m, $S = 0$; smooth	$h_u = 0.3$ m	0.2 m	Technical University of Erzurum, Turkey	2019	Free surface profiles; tip displacement of the plate; pressure in time at the downstream end	Video camera (25 fps); pressure transducers	✗	3D EUL Coupled SPH–FE (interaction fluid–structure)
Ghodoosipour et al. [226,227]	Total; dry and wet bottom; impact on a horizontal transversal pipe ($D = 0.1$ m)	Rectangular channel $L = 30.1$ m, $W = 1.5$ m, $Lr = 21.55$ m, $S = 0$; smooth	$h_u = 0.3, 0.4, 0.5$ m $h_d = 0, 0.03, 0.06, 0.08, 0.12, 0.17$ m	1.5 m	University of Ottawa, Canada	2019	Water depth time series at three locations; wave front celerity; flow velocity at a location; time series of the hydrodynamic force on the pipe	Capacitance wave gauges; ADV; dynamometer; video cameras (70 fps)	✗	
Kamra et al. [228]	Total; dry bottom; impact on a vertical cylinder (square and circular section, square: 0.05 m × 0.05 m, circular: $D = 0.05$ m)	Tank $L = 0.8$ m, $W = 0.2$ m, $Lr = 0.2$ m, $S = 0$; smooth	$h_u = 0.2$ m	0.2 m	Kyushu University, Japan	2019	Flow images; pressure hydrographs	Video camera (1500 fps); piezoresistive pressure sensors	✗	
Mokhtar et al. [229]	Total; wet bottom; impact on a vertical seawall (solid or perforated, located 9 m downstream of the gate)	Rectangular channel $L = 100$ m, $W = 1.5$ m, $Lr = 44$ m, $S = 0$; smooth	$h_u = 0.55, 0.6, 0.65, 0.7, 0.75$ m $h_d = 0.05$ m	1.5 m	National Hydraulic Research Institute, Selangor, Malaysia	2019	Wave depth and pressure hydrographs; flow velocity hydrographs; flow images	Resistance wave gauges; pressure sensors; ADV; video camera (240 fps)	✗	
Dutta et al. [230,231]	Total; dry bottom; impact on a vertical structure	Rectangular channel $L = 6$ m, $W = 0.3$ m, $Lr = 4$ m, $S = 0$; smooth	$h_u = 0.2, 0.25, 0.3, 0.35, 0.4$ m	0.3 m	Indian Institute of Technology, Kharagpur	2020	Flow velocity at two locations; water surface profiles	ADV; video camera	✗	3D RANS, VOF FV
Farahmandpour et al. [232]	Total; dry bottom; impact on a vertical structure	Rectangular channel $L = 10$ m, $W = 2.1$ m $S = 0$; smooth Reservoir (cylindrical, $D = 3$ m)	$h_u = 0.5, 1, 1.25, 1.5, 1.75, 2$ m	3 m	Universiti Teknologi Malaysia	2020	Flow depth time series at two locations; pressure time series on the face of the structure; wave front celerity	Capacitance wave gauges; pressure cells; video cameras	✗	
Kocaman et al. [233]	Partial; dry bottom; insubmersible obstacle (0.15 m × 0.08 m)	Tank $L = 1$ m, $W = 0.5$ m, $Lr = 0.25$ m, $S = 0$; smooth	$h_u = 0.15$ m	0.1 m	Iskenderun Technical University, Turkey	2020	Wave front; water depth time series at five gauge points	Video camera (300 fps); ultrasonic distance meters	✗	3D RANS, VOF FV
Pratiwi et al. [234]	Partial; dry bottom; insubmersible oblique obstacle	Rectangular channel $L = 10$ m, $W = 1$ m $S = 0$; smooth Reservoir $Lr = 2$ m, $Wr = 5.2$ m	$h_u = 0.4$ m	1 m	Institut Teknologi Bandung, Indonesia	2020	Water depth and flow velocity at five locations	Ultrasonic distance meters; current meters	✗	
Shen et al. [235]	Total; dry bottom; impact on a vertical wall	Rectangular channel $L = 4$ m, $W = 0.4$ m, $Lr = 1$ m $S = 0$; smooth	$h_u = 0.3$ m	0.4 m	Zhejiang University, Hangzhou, China	2020	Pressure time series at five elevations on the vertical wall; water depth at the wall; flow images	Pressure transducers; capacitance wave gauge; video cameras (100 and 200 fps)	✗	
Ansari et al. [121]	Total; dry bottom; circular cylinder, square cylinder, and cubic obstacle	Rectangular channel $L = 3.7$ m, $W = 0.6$ m, $Lr = 0.6$ m, $S = 0$; smooth	$h_u = 0.2$ m	0.6 m	University of Zanjan, Iran	2021	Water surface profiles	Video camera (60fps)	✗	3D (Molecular dynamics software) SPH
Memarzadeh et al. [236]	Total; dry and wet bottom; impact against an overtoppable vertical wall (0.33 m from the gate)	Rectangular channel $L = 1$ m, $W = 0.5$ m, $Lr = 0.32$ m, $S = 0$; smooth	$h_u = 0.25$ m	0.5 m	Shahid Bahonar University, Kerman, Iran	2021	Water surface profiles at selected times	Video camera	✗	3D NSE SPH; 3D RANS, VOF FV ($\varepsilon = 0.3 \times 10^{-5}$ m)

Table 3. *Cont.*

(1) Reference	(2) Dam-Break Type	(3) Setup Characteristics [1]	(4) Initial Conditions [2]	(5) Breach Width	(6) Laboratory	(7) Year	(8) Measured Data	(9) Measuring Technique [3]	(10) Data [4]	(11) Numerical Simulation [5]
Del Gaudio et al. [237]	Total; dry bottom; impact on the end vertical wall	Rectangular channel $L = 3$ m, $W = 0.4$ m, $Lr = 1.5$ m, $S = 0$; smooth	$h_u = 0.2$ m	0.4 m	University of Naples Federico II, Italy	2022	Water surface profiles at selected times; pressure time series at six locations on the end wall	Video cameras (164 fps); pressure transducers	✗	1D SWE FV ($C/g^{1/2} = 22$)
Fang et al. [238]	Total; dry and wet bottom; effect of front buildings on the wave impact on buildings	Rectangular channel $L = 17.3$ m, $W = 0.8$ m, $Lr = 0.625$ m, $S = 0$; smooth	$h_u = 0.35, 0.5, 0.65$ m	0.8 m	Tongji University, Shanghai, China	2022	Water depth time series at fourlocations; flow velocity at a gauge point; impact force on the building; pressure distribution on the impact front	Ultrasonic distance meters; ADV; multiaxial dynamometer; uniaxial force transducers	✗	
Garoosi et al. [239]	Total; dry and wet bottom; closed downstream end; impact on a vertical wall	Rectangular channel $L = 0.7$ m, $W = 0.4$ m, $Lr = 0.25$ m, $S = 0$; smooth (dry bottom case); $L = 1$ m, $W = 0.4$ m, $Lr = 0.25$ m, $S = 0$; smooth (wet bottom case)	$h_u = 0.15$ m (dry bottom case); $h_u = 0.20$ m (wet bottom case) $h_d = 0.02$ m	0.4 m	École Polytechnique de Montréal, Canada	2022	Water surface profiles; impact pressures on the downstream wall	Video camera (480 fps); pressure sensors	✓	2D NSE, VOF FV; 2D NSE MPS
Lin et al. [240]	Total; wet bottom; movable boulder (placed 1.87 m from the gate)	Rectangular channel $L = 25$ m, $W = 0.3$ m $Lr = 0.25$ m, $S = 0$; smooth	$h_u = 0.23$–0.35 m $h_d = 0.03$–0.06 m	0.3 m	Tainan Hydraulics Laboratory, Taiwan	2022	Images of the bore impact on the boulder; boulder transportation process and boulder final posture	Video camera (1000 fps); inertial measurement unit	✗	
Liu et al. [241]	Total; dry bottom; impact on a vertical wall (placed 0.85 m from the gate)	Tank $L = 1.2$ m, $W = 0.44$ m, $Lr = 0.25$ m, $S = 0$; smooth	$h_u = 0.2, 0.25, 0.3$ m	0.44 m	University of Ottawa, Canada	2022	Images of the wave propagation; water depth time series at the vertical wall; dynamic pressure time series at ten points on the wall	Video camera (60 fps); ultrasonic distance meters; pressure transducers	✓	
Wang et al. [242]	Total; dry bottom; impact on flood barriers (kinetic umbrellas, placed 1.11 m from the gate)	Tank $L = 3$ m, $W = 0.56$ m, $Lr = 0.616$ m, $S = 0$; smooth	$h_u = 0.1, 0.15, 0.2$ m	0.616 m	Princeton University, USA	2022	Hydrodynamic force time history; flow images	Resistive load cell; video cameras	✗	3D NSE Coupled SPH–FE (interaction fluid–structure)
Xie and Shimozono [243]	Total; dry bottom; closed downstream end; impact on a vertical wall	Rectangular channel $L = 1.52$ m, $W = 0.42$ m, $Lr = 0.51$ m, $S = 0$; smooth	$h_u = 0.08$–0.14 m	0.42 m	University of Tokyo, Japan	2022	Dam-break wave front celerity; dam-break wave front slope; impact pressure on a vertical wall	Video camera (500 fps); pressure sensors	✓	
Yang et al. [131]	Total; dry and wet bottom; impact on a circular pier ($D = 0.08$ m) located 4 m downstream of the gate	Rectangular channel $L = 10.72$ m, $W = 1.485$ m, $Lr = 4.58$ m, $S = 0$; smooth	$h_u = 0.13$–0.483 m (dry bottom cases); $h_u = 0.13$–0.487 m (wet bottom cases); $h_d = 0.02, 0.04, 0.06, 0.08, 0.1, 0.12, 0.14$ m	1.485 m	Southwest Jiaotong University, Chengdu, China	2022	Water depth hydrographs at five locations; forces and moments on the pier; pressure time series on 16 points on the front, back, and lateral sides of the pier	Wave gauges; load cell; pressure sensors	✗	

Note(s): [1] L = facility length; W = facility width; Lr = reservoir length; Wr = reservoir width (if different from W); S = bottom slope; θ = inclination angle; D = diameter; [2] h_u = upstream water depth; h_d = downstream water depth; [3] ADV = acoustic Doppler velocimeter; LDV = laser Doppler velocimeter; PIV = particle image velocimetry; [4] ✗ = not freely available; ✓ = freely available; [5] Approach: 1D = one-dimensional; 2D = two-dimensional; 3D = three-dimensional–Mathematical model: BOU = Boussinesq equations; ETILT = edge-tracked interface locator technique; EUL = Euler equations; LBM = lattice Boltzmann method; NSE = Navier–Stokes equations; RANS = Reynolds-averaged Navier–Stokes equations; SWE = shallow water equations; VOF = volume of fluid–Numerical method: CIP = constrained interpolation profile; ELMMC = Eulerian–Lagrangian marker and micro cell method; FD = finite difference; FE = finite element; FV = finite volume; MOC = method of characteristics; MPS = moving particle semi-implicit; PFEM = particle finite element method; SPH = smoothed particle hydrodynamics–n = Manning roughness coefficient; ε = surface roughness; C = Chézy's resistance factor; g = gravity acceleration; N.A. = not available.

Table 4. Experimental investigations of the dam-break wave propagation in idealized urban areas.

(1) Reference	(2) Dam-Break Type	(3) Setup Characteristics [1]	(4) Initial Conditions [2]	(5) Breach Width	(6) Laboratory	(7) Year	(8) Measured Data	(9) Measuring Technique [3]	(10) Data [4]	(11) Numerical Simulation [5]
Shige-eda and Akiyama [59]	Partial (asymmetric); dry bottom; impact on square pillars (0.06 m × 0.06 m)	Tank L = 4.8 m, Wr = 2.98 m Lr = 1.93 m, S = 0; smooth	h_u = 0.2 m	0.5 m	Kyushu Institute of Technology, Kitakyushu, Japan	2003	Wave front position, flow depths and surface velocity hydrographs at four positions, forces on selected pillars	Digital video tape recorder; particle tracking velocimetry; load cells	✗	2D SWE FV (n < 0.07 s m$^{-1/3}$)
Soares-Frazão et al. [244]; Soares-Frazão and Zech [245]	Partial; wet bottom; three urban district layouts (blocks: 0.3 m × 0.3 m; streets: 0.1 wide)	Trapezoidal channel L = 35.8 m, W = 3.6 m, Lr = 6.75 m, S = 0; smooth	h_u = 0.40 m h_d = 0.011 m	1 m	Université Catholique de Louvain, Belgium	2006	Water levels time series at 64 points; water surface profiles; surface velocity measurements	Resistive water level gauges; digital imaging technique; Voronoï PTV technique	✗	2D SWE FV (n = 0.01 s m$^{-1/3}$)
Szydlowski and Twarog [246]	Partial; dry bottom; urban district layout with aligned buildings (0.1 m sides)	Tank L = 6.75 m, W = 3 m, Lr = 3.0 m, Wr = 3.5 m, S = 0; smooth	h_u = 0.21 m	0.5 m	Gdansk University of Technology, Poland	2006	Water depth time series at 11 locations	Pressure transducers; depth-control gauge	✗	2D SWE FV (n = 0.018 s m$^{-1/3}$)
Yoon [247] Kim et al. [248]	Partial; dry bottom; 0.2 m × 0.2 m block arranged as two 3 × 3 groups	Plane L =30 m, W = 30 m, Lr = 5 m, S = 0; smooth	h_u = 0.3, 0.45 m	1 m	Urban Flood Disaster Management Research Center, Seoul, South Korea	2007	Water depth time series at 17 points	Capacitance wave gauges	✗	2D SWE (with porosity) FV (ε = 0.3–3 × 10^{-3} m)
Albano et al. [249]	Total; dry bottom; two fixed buildings (0.3 m × 0.15 m × 0.3 m); three floating bodies (0.118 m × 0.045 m × 0.043 m, mass: 0.025 kg)	Rectangular channel L = 2.5 m, W = 0.5 m, Lr = 0.5 m, S = 0; smooth	h_u = 0.1 m	0.5 m	Basilicata University, Italy	2016	Water depth time series at two locations (in front of the fixed obstacles); displacement of movable bodies	Resistive water depth gauges; cameras	✗	3D EUL (Euler-Newton equations for the rigid body dynamics) SPH
Norin et al. [250]	Total; dry bottom; staggered 0.1 m × 0.1 m parallelepipeds	Rectangular channel L = 7 m, W = 1.39 m, Lr = N.A., S = 0; smooth	h_u = 0.225 m	1.39 m	Scientific Research Institute of Power Structures, Russia	2017	Water level time series at two points; flow velocity profiles	Water level gauges; flow meters	✗	2D SWE FV
Guinot et al. [251,252]	Total; dry bottom; blocks (0.5 m × 0.75 m); two configurations	Rectangular channel L = 20 m, W = 1 m, Lr = N.A., S = 0; smooth	h_u = 0.35 m	1 m	Université Catholique de Louvain, Belgium	2018	Water depth time series at selected locations	Ultrasonic distance meters	✗	1D SWE (with porosity) FV
Kusuma et al. [253]	Partial; dry bottom; blocks (0.1 m × 0.1 m); four configurations (1, 3, 5, 8 blocks)	Rectangular channel L = 10 m, W = 1 m, S = 0; smooth Reservoir Lr = 2 m, Wr = 4 m	h_u = 0.2, 0.3, 0.4 m	1 m	Institut Teknologi Bandung, Indonesia	2019	Water depth profiles at selected times; water depth and flow velocity hydrographs at selected locations	Wave probe and piezometers; current meter	✗	–
Chumchan and Rattanadecho [254]	Partial; dry bottom; blocks (0.085 m sides); two configurations	Tank L = 0.984 m, W = 0.484 m, Lr = 0.24 m, S = 0; smooth	h_u = 0.15 m	0.1 m	Thammasat University, Pathumthani, Thailand	2020	Flow images; wave front	Video camera (240 fps)	✗	3D RANS, VOF FV, LB
Dong et al. [255]	Partial; dry bottom; idealized urban street; six configurations (with buildings, greenbelt sections, sidewalks, and an underground sewer system)	Rectangular channel L = 20.5 m, W = 3 m, Lr = 4.5 m, S = 0; smooth	h_u = 0.09, 0.19, 0.29 m	1 m	North China University of Water Resources and Electric Power, China	2021	Water hydrographs at seven points; flow velocity time series at three points; drainage discharge time series at inlets	Ultrasonic distance meters; electromagnetic velocity meter; electromagnetic flowmeters	✗	2D SWE FV (n = 0.009–0.011 s m$^{-1/3}$)

Note(s): [1] L = facility length; W = facility width; Lr = reservoir length; Wr = reservoir width (if different from W); S = bottom slope; [2] h_u = upstream water depth; h_d = downstream water depth; [3] PTV = particle tracking velocimetry; [4] ✗ = not freely available; [5] Approach: 1D = one-dimensional; 2D = two-dimensional; 3D = three-dimensional–Mathematical model: EUL = Euler equations; RANS = Reynolds-averaged Navier–Stokes equations; SWE = shallow water equations; VOF = volume of fluid–Numerical method: FV = finite volume; LB = lattice Boltzmann; SPH = smoothed particle hydrodynamics–n = Manning roughness coefficient; ε = surface roughness; N.A. = not available.

Table 5. Experimental investigations of the propagation of tsunami bores (generated by the removal of a gate) in the swash zone.

(1) Reference	(2) Dam-Break Type	(3) Setup Characteristics [1]	(4) Initial Conditions [2]	(5) Breach Width	(6) Laboratory	(7) Year	(8) Measured Data	(9) Measuring Technique [3]	(10) Data [4]	(11) Numerical Simulation [5]
Yeh and Ghazali [256], Yeh et al. [257]	Total; wet bottom; sloping beach starting 0.4 m downstream of the gate	Tank $L = 9$ m, $W = 1.2$ m, $Lr = 2.97$ m, $S_b = 7.5°$; smooth	$h_u/h_d = 2.31$ $h_d = 0.0975$ m (fully developed bore); $h_u/h_d = 1.72$ $h_d = 0.0975$ m (undular bore)	1.2 m	University of Washington, Seattle, USA	1988	Longitudinal profile of the bore; maximum run-up height; bore celerity	Video camcorder and photo camera (laser-induced fluorescence); water sensors	✗	
Petroff et al. [258]	Total; wet bottom; sloping beach; prismatic movable obstacles of different sizes and orientations	Rectangular channel $L = 20$ m, $W = 0.6$ m, $Lr = 7$ m; $S_b = 0.1$; smooth and rough	$h_u = 0.3$ m $h_d = 0.02$ m	0.61 m	University of Washington, Seattle, USA	2001	Advection distance of obstacles	Video camera (18 fps)	✗	(beach roughened with sand: $d_{50} = 0.84 \times 10^{-3}$ m)
Anh [259]	Total; dry bottom; adverse slope; Vetiver hedge 0.5 m thick (160–530 stems/m^2)	Tank $L > 12.5$ m, $W = 0.4$ m, $Lr = 6$ m, $S_b = 1/30$; smooth	$h_u = 0.35$–0.5 m	0.4 m	Delft University of Technology, The Netherlands	2007	Water depth hydrographs; overtopping discharge	Pressure transducers, water depth gauges	✗	
Barnes et al. [260]	Total; wet bottom; sloping beach starting 4 m downstream of the gate	Rectangular channel $L = 20$ m, $W = 0.45$ m, $Lr = 1$ m, $S_b = 0.1$; smooth and rough	$h_u = 0.65$ m $h_d = 0.065$ m	0.45 m	University of Aberdeen, UK	2009	Flow depth, bottom shear stress, and flow velocity time series	Acoustic displacement sensors; shear plate; PIV	✗	
De Leffe et al. [261]	Total; dry bottom; sloping beach starting 1.15 m downstream of the gate	Rectangular channel $L = 8$ m, $W = 1$ m, $Lr = 2.25$ m, $S_b = 0.1$; smooth	$h_u = 0.25$ m	1 m	École Centrale Nantes, France	2010	Flow depth time series at 2 gauge points	N.A.	✗	1D, 2D SWE SPH ($n = 0.001$ s m$^{-1/3}$)
O'Donoghue et al. [262]	Total; wet bottom; sloping beach starting 3.8 m downstream of the gate	Rectangular channel $L = 20$ m, $W = 0.45$ m, $Lr = 1$ m, $S_b = 0.1$; smooth and rough	$h_u = 0.65$ m $h_d = 0.06$ m	0.45 m	University of Aberdeen, UK	2010	Water depth time series at 25 locations; runup; flow velocity profiles at five cross-sections	Capacitance depth gauges; PIV	✗	1D SWE FV ($\lambda = 0.064$, $\lambda = 0.16$)
Kikkert et al. [263]	Total; wet bottom; sloping beach starting 4.82 m downstream of the gate	Rectangular channel $L = 20$ m, $W = 0.45$ m; $Lr = 1$ m, $S_b = 1/10$; rough	$h_u = 0.6$ m $h_d = 0.062$ m	0.45 m	University of Aberdeen, UK	2012	Flow depth time series and velocity profiles at six cross-sections	Laser induced fluorescence and video camera; PIV	✗	
Adegoke et al. [264]	Total; dry and wet bottom; sloping beach starting 2.7 m downstream of the gate	Rectangular channel $L = 4.7$ m, $W = 0.4$ m, $Lr = 1$ m, S_b = N.A.; smooth	$h_u = 0.15$–0.55 m $h_d = 0.05, 0.10, 0.15$ m	0.4 m	Liverpool John Moores University, UK	2014	Wave front velocity	Video Camera (40 fps); wave probes; pressure transducers	✗	
Rahman et al. [265]	Total; dry bottom; building model (cubic, $l = 0.08$ m) placed 4 m from the gate; effect of solid and perforated sea walls (at various distances from the building model)	Rectangular channel $L = 17.5$ m, $W = 0.6$ m $Lr = 5$ m, $S = 0$; smooth	$h_u = 0.15, 0.2, 0.25, 0.3$ m	0.6 m	University of Malaya, Kuala Lumpur, Malaysia	2014	Wave height time series at four positions; force time series on the building model	Wave probes; load cell	✗	
Hartana and Murakami [266]	Total; wet bottom; adverse slope starting 5.5 m from the gate building models (0.2 × 0.2 × 0.26 m): solid and with 40% opening ratio	Rectangular channel $L = 12$ m, $W = 0.4$ m $Lr = 5$ m, $S = 0$, $S_b = 1/40$; smooth	$h_u = 0.15, 0.2, 0.25, 0.3$ m $h_d = 0.05$ m	0.4 m	University of Mataram, Indonesia	2015	Water depth hydrographs at three locations; flow velocity hydrographs at two locations; pressure time series at 15 points on the building faces	Video cameras; wave gauges; propeller current meters; pressure gauges	✗	3D NSE, VOF FV, FE

Table 5. *Cont.*

(1) Reference	(2) Dam-Break Type	(3) Setup Characteristics [1]	(4) Initial Conditions [2]	(5) Breach Width	(6) Laboratory	(7) Year	(8) Measured Data	(9) Measuring Technique [3]	(10) Data [4]	(11) Numerical Simulation [5]
Chen et al. [267]	Total; two-dam-break systems (1 m apart) wet bottom; adverse slope of starting 3.006 m downstream of the first gate; swash-swash interaction	Rectangular channel L = 12.5 m, W = 0.3 m, Lr = 2.443 m, S = 0, S_b =1/10; smooth, rough adverse slope	h_{u1} = 0.35 m, h_{u2} = 0.5 m, h_d =0.035 m; h_{u1} = 0.4 m, h_{u2} = 0.4 m, h_d =0.04 m (time delay between the opening of the two gates: 1.5–6.5 s)	0.3 m	Hong Kong University of the Science and Technology	2016	Water depth hydrographs at five locations; velocity profiles and water surface elevation	Acoustic distance sensors; PIV	✗	
Chen et al. [268]	Total; dry bottom (wet bottom in the foreshore area); impact on a wharf model (three deck heights and eight wharf slopes)	Reservoir area 77 m^2, capacity 50 m^3 Rectangular channel L = 14 m, W = 1.2 m, S = 0, S_b = 30°; smooth	h_u = 0.3, 0.4, 0.5, 0.6 m (h_d = 0.05 m); (different gate openings)	1.2 m	University of Auckland, New Zealand	2016	Water level hydrographs at two locations; bore velocities; time series of uplift pressures at eight points on the wharf	Wave gauges; video camera (210 fps); pressure sensors	✗	
Chen et al. [269]	Total; dry bottom (wet bottom in the foreshore area); impact on a wharf model and a protective vertical wall (four positions and three wall heights)	Reservoir Lr = 11 m, Wr = 7.3 m Rectangular channel L = 14 m, W = 1.2 m, S = 0, S_b = 30°; smooth	h_u = 0.3, 0.4, 0.6 m (h_d = 0.05 m); (different gate openings)	1.2 m	University of Auckland, New Zealand	2017	Water level hydrographs at three locations; bore velocities; pressure time series on the wharf and the wall	Wave gauges; pressure sensors	✗	
Esteban et al. [270]	Total; dry bottom (wet bottom in the foreshore area); sloping beach; impact on different overtoppable structures (high vertical wall, low block, dyke)	Rectangular channel L = 14 m, W = 0.41 m, Lr = 4.5 m; S = 0, S_b = 1/10; smooth	h_u = 0.3, 0.4, 0.6 m (h_d = 0, 0.1, 0.2 m)	0.41 m	Waseda University, Tokyo, Japan	2017	Wave depth hydrographs at six locations; overtopping flow velocity; bore impact images	Wave gauges; electromagnetic current meters; video camera	✗	
Dai et al. [271]	Total; wet bottom; sloping beach starting 3.006 m downstream of the first gate	Rectangular channel L = 12.5 m, W = 0.3 m, Lr = 1.006 m, Wr = 0.279 m; S = 0, S_b = 1/10; smooth, rough adverse slope	h_u = 0.5 m h_d =0.05 m	0.3 m	Hong Kong University of the Science and Technology	2017	Flow depth and velocity hydrographs at five locations; entrained air	Combined laser-induced fluorescence and PIV; phase detection optical probe system; bubble image velocimetry	✗	
Tar et al. [272]	Total; wet bottom; sloping beach; impact on a oil storage tank model and protective multiple flexible pipes	Rectangular channel L = 44 m, W = 0.7 m, Lr = 7.9 m; S = 0, S_b = 1/40 and 1/100; smooth	h_u = 0.65 m h_d = 0.4 m	0.7 m	University of Osaka, Japan	2017	Flow velocity upstream and downstream of the flexible pipes; hydrodynamic force on the tank model; flow images	Electromagnetic velocity meters; load cell; video camera	✗	3D RANS, VOF FV
Chen et al. [273]	Total; dry bottom (wet bottom in the foreshore area); impact on the piles of a wharf model; protective effect of a vertical wall (four positions and three wall heights)	Reservoir Lr = 11 m, Wr = 7.3 m Rectangular channel L = 14 m, W = 1.2 m, S = 0, S_b = 30°; smooth	h_u = 0.3, 0.4, 0.6 m (h_d = 0.05 m); (different gate openings)	1.2 m	University of Auckland, New Zealand	2018	Water level time series at three locations; bore velocities; pressure time series on the piles and deck	Wave gauges; pressure sensors	✗	
Chen et al. [274]	Total; dry bottom; impact on a bridge model (four different contraction ratios)	Reservoir: 50 m^3 Rectangular channel L = 14 m, W = 1.2 m, S = 0, S_b = 30°; smooth	h_u = 0.3, 0.4, 0.6 m (different gate openings)	1.2 m	University of Auckland, New Zealand	2018	Force and momentum acting on the bridge; pressure time series on the bridge deck; wave height time series	Load cell; pressure transducers; capacitance wave gauges; video camera (30 fps)	✗	

Table 5. *Cont.*

(1) Reference	(2) Dam-Break Type	(3) Setup Characteristics [1]	(4) Initial Conditions [2]	(5) Breach Width	(6) Laboratory	(7) Year	(8) Measured Data	(9) Measuring Technique [3]	(10) Data [4]	(11) Numerical Simulation [5]
Ishii et al. [275]	Total; dry bottom (wet bottom in the foreshore area); sloping beach starting 4.45 m downstream of the upstream end; impact on a vertical structure	Tank $L = 9$ m, $W = 4$ m, Lr = N.A., $S = 0$, $S_b \approx 8.5^{\circ}$; smooth	h_u = N.A. ($h_d = 0.2$ m)	4 m	Waseda University, Tokyo, Japan	2018	Flow vortices behind the structure	Load cell; wave gauges, PIV	✗	3D RANS, VOF FV
Lu et al. [276]	Total; dry and wet (in the foreshore area) bottom; sloping beach starting 1.8 m downstream of the gate	Rectangular channel $L = 6.5$ m, $W = 0.4$ m, $Lr = 1.5$ m, $S = 0$, $S_b = 1/7.5$; smooth	$h_u = 0.08$–0.24 m ($h_d = 0, 0.02, 0.04, 0.06, 0.08$ m)	0.4 m	Zhejiang University, Hangzhou, China	2018	Wave front position; maximum run-up; flow images	Video camera (150 fps)	✗	
Chen et al. [277]	Total; dry bottom; sloping beach starting 0.76 m downstream of the dam; run-up height of balls with different diameters and densities	Rectangular channel $L = 4.4$ m, $W = 0.3$ m, $Lr = 1.29$ m, $S = 0$; $S_b = 15$–90°; smooth	$h_u = 0.06, 0.1, 0.14, 0.18, 0.22$ m	0.3 m	University of Fuzhou, China	2020	Water surface profiles; ball climbing height	Video camera	✗	
Chen et al. [278]	Total; dry bottom; impact on a container model	Rectangular channel $L = 4.4$ m, $W = 0.3$ m, $Lr = 1.27$ m, $S = 0$, -1°; smooth	$h_u = 0.13, 0.14, 0.15, 0.16, 0.17$ m	0.3 m	University of Fuzhou, China	2020	Tsunami wave height; shift of the container model; flow images	Water level gauge; video camera	✗	
Elsheikh et al. [279,280]	Total; dry bottom; interaction with a transverse canal located 3 m downstream of the gate (three different depths and widths)	Rectangular channel $L = 15.56$ m, $W = 0.38$ m, $Lr = 7.76$ m, $S = 0$; smooth	$h_u = 0.2, 0.3, 0.4$ m	0.38 m	University of Ottawa, Canada	2020	Wave front motion and wave height over the canal; wave profiles; water level hydrographs at four locations; flow velocity time series at three points	Video cameras; capacitance wave gauge and ultrasonic distance meters; ADV	✗	3D RANS, VOF FV
Barranco and Liu [281]	Total; wet bottom; sloping beach starting 11.1 m downstream of the gate	Rectangular channel $L = 36$ m, $W = 0.9$ m, $Lr = 2, 4, 8, 17.6$ m; $S = 0$; $S_b = 1/10$; smooth	$h_u = 0.128, 0.157, 0.188, 0.221, 0.256, 0.292, 0.329, 0.368, 0.408$ m $h_d = 0.1$ m	0.9 m	National University of Singapore	2021	Water depth time series at seven locations; run-up on the adverse slope	Capacitance gauges; ultrasonic distance meters; video camera (100fps)	✓	2D SWE (non-hydrostatic) FD
Chen and Wang [282]	Total; dry bottom; sloping beach starting 0.76 m downstream of the gate energy dissipation effect of grasses; run-up height of steel balls	Rectangular channel $L = 4.4$ m, $W = 0.3$ m, $Lr = 1.29$ m, $S = 0$, $S_b = 30^{\circ}$; smooth, rough reach (artificial grasses)	$h_u = 0.06, 0.1, 0.14, 0.18, 0.22$ m	0.3 m	University of Fuzhou, China	2022	Wave maximum height at a location; wave celerity; ball climbing height	Water level gauges	✗	
Liu et al. [283]	Total; dry bottom; sloping channel starting 0.45 m downstream of the gate; impact on a vertical wall placed 0.85 m from the gate	Tank $L = 1.2$ m, $W = 0.44$ m, $Lr = 0.25$ m, $S = 0$; $S_b = 5^{\circ}, 10^{\circ}, 15^{\circ}$; smooth	$h_u = 0.25$ m	0.44 m	University of Ottawa, Canada	2022	Wave runup on the vertical wall; images of the wave propagation; free surface profiles at selected times; time history of the wave front	Ultrasonic distance meters; video camera (60 fps)	✗	
Liu et al. [284]	Total; dry bottom; sloping channel starting 0.45 m downstream of the gate; impact on a vertical wall placed 0.85 m from the gate	Tank $L = 1.2$ m, $W = 0.44$ m, $Lr = 0.25$ m, $S = 0$, $S_b = 5^{\circ}, 10^{\circ}, 15^{\circ}$; smooth	$h_u = 0.3$ m	0.44 m	University of Ottawa, Canada	2022	Dynamic pressure time series at five points on the wall	Pressure transducers	✗	3D RANS, VOF FV; 3D NSE SPH

Table 5. *Cont.*

(1) Reference	(2) Dam-Break Type	(3) Setup Characteristics [1]	(4) Initial Conditions [2]	(5) Breach Width	(6) Laboratory	(7) Year	(8) Measured Data	(9) Measuring Technique [3]	(10) Data [4]	(11) Numerical Simulation [5]
Rajaie et al. [285]	Total; wet bottom; sloping channel starting 4.3 m downstream of the gate insub-mersible structure	Rectangular channel L = 30 m, W = 1.5 m, Lr = 21.55 m, S = 0, S_b = 5%; smooth, rough reach (sand bed)	h_u = 0.25, 0.3, 0.35, 0.4 m h_d = 0.03, 0.1 m	1.5 m	University of Ottawa, Canada	2022	Water depth time series at two locations and in front of the structure; flow velocity time series at a gauge point	Capacitance wave gauges and ultrasonic distance meters; ADV	✗	
von Häfen et al. [286]	Total; dry bottom; sloping beach starting 10 m downstream of the (swing) gate composite bathymetry (horizontal inland)	Rectangular channel L = 100 m, W = 2 m, Lr = 80 m, S = 0, S_b =5%, followed by a horizontal bottom; smooth	h_u = 0.4, 0.5, 0.6 m	2 m	Technische Universität Braunschweig, Germany	2022	Water depth time series at four locations	Capacitance wave gauges	✗	3D RANS, LSM FD; 2D SWE (non-hydrostatic) FD (ε = 0.001 m)

Note(s): [1] L = facility length; W = facility width; Lr = reservoir length; Wr = reservoir width (if different from W); S = bottom slope; S_b = beach (adverse) slope; l = obstacle characteristic length; [2] h_u = upstream water depth; h_d = downstream water depth; [3] ADV = acoustic Doppler velocimeter; PIV = particle image velocimetry; [4] ✗ = not freely available; ✓ = freely available; [5] Approach: 1D = one-dimensional; 2D = two-dimensional; 3D = three-dimensional–Mathematical model: LSM = level set method; NSE = Navier–Stokes equations; RANS = Reynolds-averaged Navier–Stokes equations; SWE = shallow water equations; VOF = volume of fluid–Numerical method: FD = finite difference; FE = finite element; FV = finite volume; SPH = smoothed-particle hydrodynamics–n = Manning roughness coefficient; ε = surface roughness; λ = friction factor; N.A. = not available.

Table 6. Experimental investigations on green water events using dam-break waves.

(1) Reference	(2) Dam-Break Type	(3) Setup Characteristics [1]	(4) Initial Conditions [2]	(5) Breach Width	(6) Laboratory	(7) Year	(8) Measured Data	(9) Measuring Technique [3]	(10) Data [4]	(11) Numerical Simulation [5]
Buchner [287]	Total; dry bottom; impact on a rigid panel	Tank L = 3.22 m, W = 1 m, Lr = 1.2 m, S = 0; smooth	h_u = 0.6 m	1 m	Delft University of Technology, The Netherlands	2002	Water depth hydrographs at four locations; time series of impact loads on the panel in different areas	Force and pressure transducers	✗	
Hernández-Fontes et al. [288,289]	Total; wet bottom; vessel structure located 0.505 m downstream of the gate	Tank L = 1 m, W = 0.355 m, Lr = 0.3 m, S = 0; smooth; f = 0.006, 0.024, 0.042 m	h_u = 0.18, 0.2, 0.21, 0.22, 0.24 m h_d/h_u = 0.6	0.355 m	Federal University of Rio de Janeiro, Brazil	2017	Water elevation hydrographs at two locations; video-images of green water flow	Conductive wave probes; video cameras (500 fps)	✗	
Hernández-Fontes et al. [290]	Total; wet bottom; vessel structure located 1.258 m downstream of the gate	Tank L = 1.95 m, W = 0.5 m, Lr = 0.3 m, S = 0; smooth; f = 0.006–0.042 m	h_u = 0.18, 0.21, 0.24 m h_d/h_u = 0.6	0.5 m	Federal University of Rio de Janeiro, Brazil	2019	Freeboard exceedance time series; vertical load on the structure deck	Load cells; video cameras (500 fps);	✓	1D SWE
Hernández-Fontes et al. [291]	Total; wet bottom; vessel structure located 0.505 m downstream of the gate	Tank L = 1 m, W = 0.355 m, Lr = 0.3 m, S = 0; smooth; f = 0.03–0.042 m	h_d = 0.108, 0.12 m h_d/h_u = 0.8, 0.7, 0.6, 0.5, 0.4	0.355 m	Federal University of Rio de Janeiro, Brazil	2020	Water elevation hydrographs at four locations; freeboard exceedance time series; vertical load on the structure deck; video-images of green water flow	Conductive wave probes); load cells; video cameras (500 fps)	✗	
Hernández-Fontes et al. [292]	Total; wet bottom; vessel structure located 1.455 m downstream of the gate	Tank L = 1.95 m, W = 0.5 m, Lr = 0.3 m, S = 0; smooth; f = 0.006–0.042 m	h_u = 0.18, 0.2, 0.21, 0.22, 0.24, 0.27, 0.3 m h_d/h_u = 0.6, 0.5, 0.4	0.5 m	Federal University of Rio de Janeiro, Brazil	2020	Water elevation hydrographs at five locations; freeboard exceedance time series; vertical load on the structure deck; video-images of green water flow	Conductive wave probes); load cells; video cameras (500 fps)	✗	
Hernández-Fontes et al. [293]	Total; wet bottom; vessel structure located 0.505 m downstream of the gate	Tank L = 1 m, W = 0.355 m, Lr = 0.3 m, S = 0; smooth; f = 0.03 m	h_u = 0.3 m h_d = 0.12 m	0.355 m	Federal University of Rio de Janeiro, Brazil	2020	Water elevation hydrographs at five locations; water surface profiles; video-images of green water flow	Conductive wave probes; video cameras (250 fps)	✗	

Table 6. *Cont.*

(1) Reference	(2) Dam-Break Type	(3) Setup Characteristics [1]	(4) Initial Conditions [2]	(5) Breach Width	(6) Laboratory	(7) Year	(8) Measured Data	(9) Measuring Technique [3]	(10) Data [4]	(11) Numerical Simulation [5]
Wang and Dong [294]	Total; wet bottom; interaction with a floating box (0.3 m × 0.595 m × 0.1 m, placed 0.75 m or 1.2 m from the gate)	Tank $L = 2$ m, $W = 0.6$ m, $Lr = 0.5$ m, $S = 0$; smooth; $f = 0.07$ m	$h_u = 0.25, 0.3, 0.35$ m $h_d = 0.15$ m	0.6 m	Ocean University, Qingdao, China	2022	Pressure hydrographs at two points on the box upstream face; water surface hydrographs at two locations; motion of the floating structure	Pressure probes; wave gauges; motion capture system	✗	

Note(s): [1] L = facility length; W = facility width; Lr = reservoir length; Wr = reservoir width (if different from W); S = bottom slope; f = freeboard; [2] h_u = upstream water depth; h_d = downstream water depth; [3] –; [4] ✗ = not freely available; ✓ = freely available; [5] Approach: 1D = one-dimensional–Mathematical model: SWE = shallow water equations; N.A. = not available.

Table 7. Experimental investigations of dam-break waves of non-Newtonian fluids.

(1) Reference	(2) Dam-Break Type	(3) Setup Characteristics [1]	(4) Initial Conditions [2]	(5) Breach Width	(6) Laboratory	(7) Year	(8) Measured Data	(9) Measuring Technique [3]	(10) Data [4]	(11) Numerical Simulation [5]
Chanson et al. [295]; Chanson et al. [296]	Total; dry bottom; thixotropic fluid (bentonite suspension)	Rectangular channel $L = 2$ m, $W = 0.34$ m, $Lr = h_u/\sin(15^\circ)$, $S = 15^\circ$; rough	$h_u = 0.0472$–0.0784 m	0.34 m	Laboratory of Materials and Structures in Civil Engineering, Champs sur Marne, France	2004	Free surface; wave front propagation; wave front profiles	Video cameras (25 fps)	✗	
Jánosi et al. [64]	Total; dry and wet bottom; polyethylene-oxide; different concentrations	Tank $L = 9.93$ m, $W = 0.15$ m, $Lr = 0.38$ m, $S = 0$; smooth	$h_u = 0.11$–0.25 m $h_d = 0$–0.005 m	0.15 m	Eötvös University, Budapest, Hungary	2004	Water profiles; front position and velocity	Video cameras	✗	
Komatina and Đorđević [297]	Total; dry bottom; mixture of water and copper tailings; different volumetric concentrations of the solid phase	Rectangular channel $L = 4.5$ m, $W = 0.15$ m, $Lr = 2$ m, $Wr = 0.155$ m, $S = 0$–0.01; smooth	$h_u = 0.1$–0.3 m	0.155 m	University of Belgrade, Serbia & Montenegro	2004	Flow depth profiles at different times	Video camera (5 fps)	✗	1D SWE FD
Cochard and Ancey [298]; Cochard [299]; Cochard and Ancey [300]	Total; dry bottom; viscoplastic fluid (Carbopol Ultrez 10)	Plane $L = 6$ m, $W = 1.8$ m, $S = 0$–18°; smooth Reservoir $Wr = 1.8$ m, Mass = 120 kg	N.A.	1.6 m	EPFL, Lausanne, Switzerland	2006	Free surface and flow depth profiles at different times	Video camera	✗	
Balmforth et al. [301]	Total; dry bottom; Newtonian and non-Newtonian fluids (corn syrup and aqueous suspensions of xanthan gum, kaolin, Carbopol, and cornstarch)	Rectangular channel $L > 1$ m, $W = 0.1$ m, $Lr = 0.4$ m, $S = 0$; smooth	$h_u = 0.02$–0.0435 m	0.1 m	N.A.	2007	Wave front position	Video camera	✗	
Ancey and Cochard [302]	Total; dry bottom; viscoplastic (Herschel–Bulkley) fluid (Carbopol Ultrez 10)	Rectangular channel $L = 4$ m, $W = 0.3$ m, $Lr = 0.51$ m, $S = 6, 12, 18, 24^\circ$; smooth	Mass in the reservoir: 23–43 kg	0.3 m	EPFL, Lausanne, Switzerland	2009	Free surface and flow depth profiles at selected times; front position with time	Video camera	✗	
Cochard and Ancey [303]	Partial; dry bottom; viscoplastic (Herschel–Bulkley) fluid (Carbopol Ultrez 10)	Plane $L = 5.5$ m, $W = 1.8$ m, $S = 0$–18°; smooth Reservoir $Lr = 0.51$ m, $Wr = 0.3$ m	$h_u = 0.3$–0.36 m	0.3 m	EPFL, Lausanne, Switzerland	2009	Free surface at selected times	Video camera (45 fps)	✗	

Table 7. *Cont.*

(1) Reference	(2) Dam-Break Type	(3) Setup Characteristics [1]	(4) Initial Conditions [2]	(5) Breach Width	(6) Laboratory	(7) Year	(8) Measured Data	(9) Measuring Technique [3]	(10) Data [4]	(11) Numerical Simulation [5]
Brondani Minussi and de Freitas Maciel [304]	Total; dry bottom; viscoplastic (Herschel–Bulkley) fluid (Carbopol 940, different concentrations)	Rectangular channel L = 1.91 m, W = 0.32 m, Lr = 0.5 m, S = 0; smooth	h_u = 0.07, 0.1, 0.13 m	0.32 m	Paulista State University, Ilha Solteira, Brazil	2012	Free surface at selected times; wave front position	Video camera	✗	2D NSE, VOF FV
Bates and Ancey [305]	Total; dry bottom; viscoplastic (Herschel–Bulkley) fluid; contact with a stationary layer of the same fluid	Rectangular channel L = 3.5 m, W = 0.1 m, Lr = 0.3 m; S = 12°, 16°, 20°, 24°; smooth	Fluid mass: 3 kg	0.1 m	EPFL, Lausanne, Switzerland	2017	Wave front position; water surface profiles; velocity field	PIV; video cameras	✓	1D (lubrification theory) GM
Jing et al. [306]	Total; dry bottom; mudflow (three different grain sizes)	Rectangular channel L = 6 m, W = 0.3 m; S = 0.02; smooth Reservoir Lr = 2 m, Wr = 0.6 m	h_u = 0.30 m	0.3 m	University of Mining and Technology, Beijing, China	2019	Flow depth, velocity and pressure hydrographs at four locations	Video cameras (300 fps); pressure sensors	✗	
Modolo et al. [307]	Total; dry bottom; Bingham fluid (different solutions)	Tank L = 1.52 m, W = 0.05 m, Lr = 0.4 m, S = 0; smooth and rough	h_u = 0.24 m	0.05 m	Federal University of Rio de Janeiro, Brazil	2019	Flow images; flow depth profiles at selected times	Video camera; PIV	✗	
Tang et al. [308]	Total; dry bottom; mud flow (Herschel–Bulkley fluid)	Rectangular channel L = 3 m, W = 0.23 m, Lr = 0.48 m, S = 0, 5°, 10°; smooth	Mud volume: 38.6, 36.3, 34 l	0.23 m	Sichuan University, Chengdu, China	2022	Flow depth and bottom pressure hydrographs at two locations	Pressure sensors; laser sensors	✗	2D NSE, VOF FD

Note(s): [1] L = facility length; W = facility width; Lr = reservoir length; Wr = reservoir width (if different from W); S = bottom slope; [2] h_u = upstream water depth; h_d = downstream water depth; [3] PIV = particle image velocimetry; [4] ✗ = not freely available; ✓ = freely available; [5] Approach: 1D = one-dimensional; 2D = two-dimensional–Mathematical model: NSE = Navier–Stokes equations; SWE = shallow water equations; VOF = volume of fluid–Numerical method: FD = finite difference; FV = finite volume; GM = Galerkin method; N.A. = not available.

Table 8. Experimental investigations of dam-breaks in cascade reservoirs.

(1) Reference	(2) Dam-Break Type	(3) Setup Characteristics [1]	(4) Initial Conditions [2]	(5) Breach Width	(6) Laboratory	(7) Year	(8) Measured Data	(9) Measuring Technique [3]	(10) Data [4]	(11) Numerical Simulation [5]
Yang et al. [309]	Two total dam-breaks; three different distances between the two dams (7.8, 9.8, 11.8 m); dry bottom	Rectangular channel L = 20 m, W = 0.5 m, S = 12°; smooth	h_u = 0.184–0.531 m	0.5 m	Sichuan University, Chengdu, China	2011	Water depth hydrographs in 10 positions	Water probes; high resolution camera	✗	
Chen et al. [310]	Total dam-break; pressure load on a downstream dam; dry bottom	Upstream reservoir Lr = 2 m, Wr = 0.4 m, S = 0 Rectangular channel L = 10 m, W = 0.4 m S = 4, 8, 12°; smooth	h_u = 0.1–0.3 m (upstream reservoir); h_d = 0–0.3 m (downstream reservoir)	0.4 m	Sichuan University, Chengdu, China	2014	Pressure hydrographs 20 positions at five different elevations on the downstream dam	Pressure sensors	✗	
Liu et al. [311]	Two total dam-breaks; dry bottom (dam height: 0.4 m)	Reservoirs Lr = 2 m, W = 0.8 m Rectangular channel L = 12 m, W = 0.4 m, S = 1/12.5, smooth;	$h_{u1} = h_{u2}$ = 0.3 m (downstream dam breaks 0, 2, 4 s after the upstream one); $h_{u1} = h_{u2}$ = 0.2 m (downstream dam breaks due to overtopping)	0.4 m	Changjiang River Scientific Research Institute, Wuhan, China	2017	Water depth hydrographs at six locations	Ultrasonic distance meters	✗	

Table 8. *Cont.*

(1) Reference	(2) Dam-Break Type	(3) Setup Characteristics [1]	(4) Initial Conditions [2]	(5) Breach Width	(6) Laboratory	(7) Year	(8) Measured Data	(9) Measuring Technique [3]	(10) Data [4]	(11) Numerical Simulation [5]
Zhang and Xu [312]	Three dams; total break of the upstream dam; dry bottom; retarding effects of an intermediate intact dam (dam height: 0–0.6 m)	Upstream reservoir Lr = 2.97 m, Wr = 1.93 m Rectangular channel L = 20 m, W = 0.5 m, S = 12°, smooth;	h_u = 0.1–0.3 m (upstream dam); h_u = 0.1–0.5 m (downstream dam); h_u = 0–0.6 m (intermediate dam)	0.5 m	Sichuan University, Chengdu, China	2017	Pressure time series on the face of the intermediate dam; flow images	Pressure sensors; video cameras	✓	
Luo et al. [313]	Two dams; total break of the upstream dam; dry bottom; flow in the downstream reservoir	Rectangular channel L = 10 m, W = 0.4 m, S = 4°; smooth	h_u = 0.2 m (upstream dam); h_u = 0.15, 0.3 m (downstream dam)	0.4 m	Sichuan University, Chengdu, China	2019	Flow images; water depth and pressure hydrographs at three points	N.A	✗	3D NSE SPH
Luo et al. [313]	Three dam-breaks; dry bottom	Rectangular channel L = 15.6 m, W = 0.5 m, S = 4°; smooth	h_u = 0.5 m (upstream dam); h_u = 0.5 m (downstream dams)	0.5 m	Sichuan University, Chengdu, China	2019	Flow images; water depth hydrograph at six points	N.A	✗	3D NSE SPH
Kocaman and Dal [314]	Two dams; total break of the upstream dam on the reservoir of the downstream one; overtopping of the downstream dam	Rectangular channel L = 2.5 m, W = 0.25 m, Lr = 0.75 m (both dams) S = 1/5, smooth	h_u = 0.15 m (both dams)	0.25	Iskenderun Technical University, Turkey	2020	Water depth hydrographs; images of the free surface profiles	Video cameras (120 and 50 fps)	✗	3D NSE SPH

Note(s): [1] L = facility length; W = facility width; Lr = reservoir length; Wr = reservoir width (if different from W); S = bottom slope; [2] h_u = upstream water depth; h_d = downstream water depth; [3] –; [4] ✗ = not freely available; ✓ = freely available; [5] Approach: 3D = three-dimensional–Mathematical model: NSE = Navier–Stokes equations–Numerical method: SPH = smoothed particle hydrodynamics; N.A. = not available.

Table 9. Experimental investigations of dike-break induced flows on a lateral floodplain.

(1) Reference	(2) Dike-Break type	(3) Setup Characteristics [1]	(4) Initial Conditions [2]	(5) Breach Width	(6) Laboratory	(7) Year	(8) Measured Data	(9) Measuring Technique [3]	(10) Data [4]	(11) Numerical Simulation [5]
Bechteler et al. [315]	Sudden trapezoidal opening (1V:1.11H slope)	Rectangular channel L = 30 m, W = 2 m, S = 0 Floodplain L = 5 m, W = 10 m, S = 0; smooth	h_u = 0.2 m	0.5 m	University of German Federal Armed Forces, Munich Germany	1992	Pressure hydrographs at 29 locations; flooded area perimeter	Pressure transducers; video camera	✗	2D SWE FV (n = 0.001 s m$^{-1/3}$)
Liem and Köngeter [316]	N.A.	Rectangular channel L = N.A., W = N.A., S = N.A. Floodplain L = 8.5 m, W = 3.5 m, S = 0.05; smooth	N.A.	0.6 m	Aachen University of Technology, Germany	1999	Water levels hydrographs in 72 points; front wave propagation	Electrode system; capacity sensors	✗	(n = 0.01 s m$^{-1/3}$)
Aureli and Mignosa [317,318]	Sudden opening	Rectangular channel L = 10 m, W = 0.3 m, S = 0.001 Floodplain L = 1.5 m, W = 2.6 m; smooth	Steady flow 5–15 l/s	0.28 m	University of Parma, Italy	2002	Water depth hydrographs at nine locations; transverse velocity profiles; discharge flowing through the breach	Ultrasonic distance meters; ADV; triangular weir	✗	2D SWE FD (n = 0.01 s m$^{-1/3}$)
Sarma and Das [319]	Sudden opening	Compound channel L = 9.2 m, W = N.A., S = N.A. Floodplain L = 2 m, W = 2.5 m, S = 0; smooth	N.A.	N.A.	Indian Institute of Technology, Guwahati, India	2003	Wave front in the flooding plane at three times	N.A.	✗	(n = 0.013 s m$^{-1/3}$)
Briechle et al. [320]; Briechle [321]; Harms et al. [322]	Sudden opening	Rectangular channel L = N.A., W = 1 m, S = N.A. Floodplain L = 3.5 m, W = 4 m, S = N.A.; smooth	Steady flow: 300 l/s h_u = 0.3–0.5 m	0.5 m	Aachen University of Technology, Germany	2004	Water depth hydrographs; wave front position and velocity	Ultrasonic distance meters; Video cameras (>50 fps)	✗	2D SWE DG (n = 0.0083 s m$^{-1/3}$)

Table 9. *Cont.*

(1) Reference	(2) Dike-Break type	(3) Setup Characteristics [1]	(4) Initial Conditions [2]	(5) Breach Width	(6) Laboratory	(7) Year	(8) Measured Data	(9) Measuring Technique [3]	(10) Data [4]	(11) Numerical Simulation [5]
Oertel and Schlenkhoff [323]; Oertel [324]	N.A.	Rectangular channel $W \approx 0.6$ m Floodplain $L = 4$ m, $W = 5.6$ m, $S = 0$; smooth	N.A.	0.5 m	Bergische University Wuppertal, Germany	2008	Water depth contour maps	Ultrasonic distance meters	✗	
Roger et al. [325]	Sudden opening	Rectangular channel L = N.A., $W = 1$ m, S = N.A. Floodplain $L = 4$ m, $W = 3.5$ m, $S = 0$; smooth	Steady flow: 100–300 l/s $h_u = 0.3$–0.5 m	0.3–0.7 m	Aachen University of Technology, Germany	2009	Surface profiles at different times; breach discharge	Ultrasonic distance meters; LDA	✗	2D SWE DG, FV ($n = 0.005$–0.02 s m$^{-1/3}$)
Sun et al. [326]	Sudden breaching	Rectangular channel $L = 40$ m, $W = 1$ m, $Lr = 15.5$ m Floodplain $L = 25$ m, $W = 2.5$ m, $S = 0$; smooth	Steady flow: 80 l/s	1 m	University of Tsinghua, Beijing, China	2017	Water depth and flow velocity hydrographs at several locations;	Pressure gauges; ADV	✗	2D SWE FD ($n = 0.012$ s m$^{-1/3}$)
Al-Hafidh et al. [327]	Sudden breaching	Rectangular channel $L = 11$ m, $W = 0.4$ m, $S = 0$ Floodplain $L = 1.83$ m, $W = 4.87$ m; smooth	Different inflow hydrographs	0.2, 0.4, 0.8 m	University of South Carolina, USA	2022	Water depth hydrographs at eight locations	Ultrasonic distance meters	✗	
Yoon et al. [328]	Gradual trapezoidal breaching (sliding opening) 1V:0.3H	Rectangular channel $L = 30$ m, $W = 5$ m, $S = 0$; Floodplain $L = 25$ m, $W = 30$ m; smooth	$h_u = 0.3, 0.35, 0.4, 0.45, 0.5, 0.55$ m	0.5, 1, 1.5, 2, 2.5, 3 m	Institute of Civil Engineering and Building Technology, Korea	2022	Water depth hydrographs; propagation of the wave front	Wave height meters	✗	

Note(s): [1] L = facility length; W = facility width; Lr = reservoir length; Wr = reservoir width (if different from W); S = bottom slope; [2] h_u = water depth in the channel; [3] ADV = acoustic Doppler velocimeter; LDA = laser Doppler anemometer; [4] ✗ = not freely available; [5] Approach: 2D = two-dimensional–Mathematical model: SWE = shallow water equations–Numerical method: DG = discontinuous Galerkin; FD = finite difference; FV = finite volume–n = Manning roughness coefficient; N.A. = not available.

Table 10. Experimental investigations of collapses of storage tanks and bunds or dike overtopping.

(1) Reference	(2) Dam-Break Type	(3) Setup Characteristics [1]	(4) Initial Conditions [2]	(5) Bund Characteristics	(6) Laboratory	(7) Year	(8) Measured Data	(9) Measuring Technique [3]	(10) Data [4]	(11) Numerical Simulation [5]
Greenspan and Johansson [329]	Total and partial (orifice over a 30° arc: 0.0254 m wide, $h = 0.076$ m high); dry bottom	Cylindrical tank $D = 0.19$ m, $S = 0$; smooth	0.05 m $< h_u < 0.22$ m	Circular: bund radius = 0.127, 0.178, 0.229, 0.279 m; bund inclination = 30°, 60°, 90°; bund height = 0.033, 0.038, 0.051, 0.064 m	Massachusetts Institute of Technology, USA	1981	Overtopping fraction (as a function of the dike characteristics)	Needle depth gauge; video camera	✗	
Sharifi [330]	Total; dry bottom; three configurations: unconfined flow, barrier flow, and confined flow (wall height = $0.25h_u$)	Cylindrical tank $D = 0.087$ m, $S = 0$; smooth	$h_u = 0.5D$, $0.75D$, D	Circular: bund radius = 0.175, 0.24, 0.258, 0.3, 0.34 m; bund inclination = 40°, 90° bund height = 0.022, 0.032, 0.044 0.065 m	Imperial College of Science and Technology, London, UK	1987	Water depth hydrographs at eight positions; wave front propagation	Light-sensitive photodiodes; video camera (128 fps)	✗	
Maschek et al. [331]	Total; dry and wet bottom; symmetric and asymmetric water column (off-centeredness = 0.055, 0.0825, 0.11 m); effect of obstacles in the flow: rings, rods, and particles	Cylindrical tanks Inner $D = 0.11, 0.19$ m; Outer $D = 0.44$ m; $S = 0$; smooth	$h_u = 0.05, 0.1, 0.2, 0.22, 0.23$ m $h_d = 0, 0.01, 0.03, 0.05, 0.1$ m	Circular: bund height = 0.02, 0.03 m	Karlsruhe Nuclear Research Centre, Germany	1992	Arrival time at the wall; time of maximum height; maximum height at the container wall; time of maximum height; maximum height at pool center	Video camera	✗	

Table 10. *Cont.*

(1) Reference	(2) Dam-Break Type	(3) Setup Characteristics [1]	(4) Initial Conditions [2]	(5) Bund Characteristics	(6) Laboratory	(7) Year	(8) Measured Data	(9) Measuring Technique [3]	(10) Data [4]	(11) Numerical Simulation [5]
Cleaver et al. [332]; Cronin and Evans [333]	Total; dry bottom; different bunding arrangements; impact on an additional cylindrical tank (D = 3.5 m)	Quarter of cylinder tank D = 3.5 m; S = 0; smooth	h_u = 1.45, 1.6, 1.75 m	Circular: bund radius = 5, 7.1, 10 m; bund inclination = 30°, 45°, 90°; bund height = 0.05, 0.1, 0.2 m; Square: bund distance = 6.27, 4.43, 8.89 m; bund inclination = 90°; bund height = 0.05, 0.2 m	Advantica Technologies Ltd. (for Health and Safety Executive), Loughborough, UK	2001	Time of water arrival at 60 positions; water head in the tank; overtopping volume	Video camera (125 fps); pressure transducer; depth resistance probes; calibrated container	✗	
Atherton [334]	Total; dry bottom; different bunding arrangements	Quarter of cylinder tank, D = 0.6 m; S = 0; smooth	h_u = 0.12, 0.3, 0.6 m	Circular: bund radius = 0.315–1.9 m; bund inclination = 90°; bund height = 0.006–0.72 m; Triangular and Rectangular: bund distance = 0.441, 1.247 m; bund inclination = 90°; bund height = 0.012, 0.12 m	Liverpool John Moores University, UK	2005	Dynamic pressure vertical profiles on the bund; wave heights; fluid mass overtopping the bund	Piezotronic pressure transducers; resistive wave gauges; water balance; video camera	✗	
Atherton [335]	Partial; (orifice: 0.019–0.084 m diameter; slot: 0.157 m wide, 0.007–0.18 m high)	Quarter of cylinder tank, D = 0.6 m; S = 0; smooth	h_u = 0.12, 0.3, 0.6 m	Circular: bund radius = 0.497–1.407 m; bund inclination = 90°; bund height = 0.006–0.24 m	Liverpool John Moores University, UK	2008	Dynamic pressure vertical profiles on the bund; wave heights; fluid mass overtopping the bund	Piezotronic pressure transducers; resistive wave gauges; water balance; video camera	✗	
Zhang et al. [336]	Total; dry bottom; straight and curved dikes	Cylindrical tank D = 0.1, 0.2 m; S = 0; smooth	h_u > 0.3 m (for D = 0.1 m); h_u > 0.2 m (for D = 0.2 m)	Circular: bund radius = 0.1–0.15 m; bund inclination = 90°; bund height = 0.022–0.051 m; Square: bund distance (equivalent radius) = 0.11–0.23 m; bund inclination = 90°; bund height = 0.022–0.045 m	Mary Kay O'Connor Process Safety Center, Texas A&M University, USA	2017	Fluid mass overtopping the bund	Balance; video camera	✗	
Zhang et al. [336]	Total; dry bottom; straight and curved dikes	Cylindrical tank D = 0.229 m; S = 0; smooth	h_u = 0.5 m	Square: bund distance (equivalent radius) = 0.516 m; bund inclination = 90°; bund height = 0.08–0.098 m	Mary Kay O'Connor Process Safety Center, Texas A&M University, USA	2017	Fluid mass overtopping the bund	Balance; video camera	✗	
Megdiche [337]	Total and partial (slot: 0.157 m wide, 0.018–0.09 m high); dry bottom; different bunding arrangements; viscous fluid (olive oil)	Quarter of cylinder tank, D = 0.6 m; S = 0; smooth	h_u = 0.12, 0.3, 0.6 m	Circular: bund radius = 0.497–1.9 m; bund inclination = 90°; bund height = 0.03–0.72 m; Triangular, square, and rectangular: bund distance = 0.324–1.095 m; bund inclination = 90°; bund height = 0.012, 0.12 m	Liverpool John Moores University, UK	2018	Dynamic pressure vertical profiles on the bunds; wave heights; fluid mass overtopping the bund	Piezotronic pressure transducers; resistive wave gauges; water balance, video camera	✗	3D RANS, VOF FV
Zhao et al. [338]	Total; dry bottom; bunds with different shapes, inclinations and breakwaters	Cylindrical tank D = 0.27 m; S = 0; smooth	Various tank filling ratios	Circular: bund radius = 0.272 m; bund inclination = 45°, 60°, 90°, 120°; bund height = 0.1 m; Square: (0.483 m × 0.483 m) and rectangular (0.341 m × 0.683 m): bund inclination = 45°, 60°, 90°, 120°; bund height = 0.1 m	Nanjing Tech University, China	2022	Dynamic pressure time series at selected points on the bunds overtopping fraction	Pressure sensors; balance; video camera	✗	3D RANS, VOF FV

Note(s): [1] D = diameter of the cylindrical tank; S = bottom slope; [2] h_u = upstream water depth; h_d = downstream water depth; [3] –; [4] ✗ = not freely available; [5] Approach: 3D = three-dimensional–Mathematical model: RANS = Reynolds-averaged Navier–Stokes equations; VOF = volume of fluid–Numerical method: FV = finite volume; N.A. = not available.

3. Discussion and Advances

Tables 1–10 show the considerable amount of experimental investigations performed, covering a broad spectrum of dam-break flow conditions. The first laboratory tests dated back even more than 100 years ago [19,20], but more than 70% of the dam-break experiments reviewed here were carried out in the last 20 years, suggesting an increasing interest worldwide in experimental research on dam-break flows.

Basic features of dam-break flows (wave profile, wavefront motion, etc.) have been the most investigated, especially in the past, as indicated by Table 1. To this end, capacitive or resistive probes and pressure gauges were used in most investigations before 2000. The former probes are very easy to implement in laboratory facilities, but they are intrusive devices locally disturbing the flow [150]. Accordingly, non-intrusive pressure gauges or ultrasonic distance meters have sometimes been preferred (e.g., [46,95,207]), even if they may show spurious dynamic oscillations in fast transient flows, especially when the slope of the free surface is high [207]. The limitation of such gauges is that they provide a local flow depth measure. Therefore, since earlier times, there has been an interest in measurements over extended areas of the flow. Martin and Moyce [23] and Dressler [13] were pioneers in using video cameras to record images of a dam-break flow from which quantitative information about the wave motion, especially wave profiles at fixed times, can be extracted. In particular, Dressler [13] used five electrically synchronized cameras with an impressive acquisition speed for the time (1800 frames per second). Experimental wave profiles are of utmost relevance to understanding the characteristics of the flow and verifying the capability of the classic analytical solutions of the dam-break problem (i.e., Ritter's [339] and Stoker's [340] solutions) or numerical solutions of dam-break models to predict the dam-break wave profile (e.g., [52,76]). Recent optical and image-processing techniques overcome the limitations of the punctual gauges and the cumbersome post-treatment of analogic video records, allowing for the accurate non-intrusive measurement of the free surface on an area of selected extension with a suitable time rate [4].

In addition to the wave profile, the velocity field is a flow characteristic of interest in dam-break experiments. Earlier investigations focused only on the wavefront velocity, tracking its position on flow images (e.g., [26,29]). Local flow velocity has often been measured using acoustic Doppler velocimetry (ADV) [186], despite the disturbances induced by the measuring device on the flow. Moreover, the fast variation in time of the free-surface elevation implies that the probe's position below the free surface does not remain constant, making it difficult to interpret the measures [186]. Velocity fields are efficiently measured on selected regions of the flow using non-intrusive imaging techniques (e.g., particle imaging velocimetry-PIV or particle tracking velocimetry-PTV) based on the tracking of particles floating on the flow surface [245] or buoyant within the flow [82]. The measurement of the surface velocity field provides insight into the flow features because it allows the reconstruction of flow trajectories on the flow surface. Moreover, surface velocities are a good approximation of depth-averaged flow velocities in fast transient shallow flows, where the turbulent velocity profile is not yet established. The measurement of the vertical velocity field (e.g., [82]) further enhances the understanding of the flow dynamics allowing the validity of the basic assumptions of the numerical simulation tools to be checked.

Experimental investigations on the effect of geometrical singularities and the impact of dam-break flows against obstacles have gradually taken hold alongside research on basic features of the dam-break flow. Experiments listed in Table 2 concern flows in more complex geometries than a simple straight channel due to the presence of contractions, bottom sills, bends, etc. The variety of cases reported in Table 2 demonstrates the need for an in-depth understanding of complex flow features generated by geometric singularities, each isolated in well-defined experimental situations. Accordingly, such test cases should not be considered scale physical models of real situations but prototype cases highlighting specific flow features. The availability of experimental data allows for modelers to check the treatment of each type of singularity in numerical models. The key challenge for new numerical approaches is indeed to reproduce the effects of the geometric singularities in the best possible way, especially in the context of shallow water models, because most of the considered singularities induce local deviations from the hydrostatic pressure assumption.

Table 3 reports test cases and laboratory investigations of the effects of isolated obstacles or structures on a dam-break flow. The experimental analysis of such situations is of practical interest. Indeed, natural and artificial obstacles are commonly present in real-field applications, and flooding propagation in flood-prone areas can be strongly influenced by such singularities, which may act as barriers to the flow. Obstacles of different sizes, shapes, and orientations were considered in the dam-break experiments. Prismatic blocks, vertical columns (of a square, rectangular, circular, but also pyramidal shape [187]), and solid walls (simulating protective barriers) were mainly used as obstacles, but occasionally also bridge models [214,218,222]. Furthermore, the obstacle's position and distance from the gate are crucial to defining the test conditions. A few tests involved obstacles overtopped by the flow [87,236] or deformable structures [217,225], which induce complex flow features and wave-structure interactions, respectively. Moreover, the impact of the dam-break wave against a structure was investigated in detail by some studies (e.g., [215]). Other applications concern the effect of mitigation walls placed in front of model structures for protection purposes [208,210] or the performance of new flood protection structures [242]; others concern permeable structures (i.e., buildings with openings [221,266] or perforated walls [229], and even the presence of movable obstacles carried away by the flow [240,258]. Despite this variety of cases investigated in the literature, the experimental analysis of flood scenarios in which a structure is destroyed by the flow is lacking. Flow depth and velocity were typically measured at selected locations to describe the features of the flow, especially near the obstacles (e.g., [186]). In recent years, the use of imaging techniques to capture wave propagation and measure the free surface over an extended area (e.g., [63,207]) has become widespread. The hydrodynamic load acting on the whole structure (e.g., [215,223]) or impact pressures at selected gauge points on the structure faces (e.g., [218,228]) were also measured. These data are valuable for the validation of numerical models used for evaluating hydrodynamic forces and other hydraulic variables useful for the structural design and verification of structural reliability.

Flood inundation of urban areas (possibly induced by a dam-break or a tsunami invading a city) is a research topic that arouses considerable interest nowadays due to the high exposure of residential or industrial settlements close to waterways, dams, or coastal areas [341,342]. Table 4 lists the studies on urban flooding conducted through experimental modelling. In these models, dam-break experiments were performed using idealized urban districts constituted by arrays of solid blocks with different configurations and orientations, which simulate the idealized layouts of buildings and cannot be considered scale models of existing urban areas [245]. Complex flow processes occur in these experiments, with multiple flow paths (dictated by the arrangement of buildings and streets) and high flow velocities. Hydraulic variables describing flow dynamics and directly involved in flood impact assessment were typically measured. Accordingly, flow depth and velocity time series were usually provided at selected locations both inside and around the city layout (e.g., [245,252,253]). Moreover, the measurement of hydrodynamic loads on buildings has received less attention [59]. More insight into urban flooding could come from considering quasi-realistic urban district models [255], taking into account additional events associated with urban floods [342], such as the penetration of water into buildings through openings [221], the flow exchange between the streets and the sewer system, the transport of cars or urban debris [249], and the diffusion of pollutants. Experimental data from such experiments would better support the validation of urban flood simulation models, which have become increasingly sophisticated in recent years [341]. Among these numerical models, the coarse-grid ones (for example based on the porosity approach [248,252]) can provide accurate results preserving computational efficiency. Models of that type require such experiments in an idealized urban environment for their validation.

Wave runup prediction on sloping beaches is one of the main concerns in the swash zone studies. Wave runup and overtopping on coastal structures have historically been investigated through physical models, since storm waves occur infrequently, and field measurements are expensive and difficult during storms. Laboratory investigations of waves normally incident on structures and beaches were usually conducted in wave flumes. More in-depth investigations of the wave dynamics require large basins equipped with more complex and expensive facilities due to the nearshore non-uniformity. In laboratory investigations, a single bore was often generated by lifting a gate separating the initially quiescent water on the beach from the deeper water behind the gate, exploiting the strict similarity between tsunami and dam-break waves. Table 5 includes only experiments in which single bores were generated through a dam-break. In addition to basic investigations of the characteristics of waves propagating on simple sloping beaches, advanced ones considered the presence of vegetation, moving objects, wharves, overtoppable or insubmersible structures, floating tanks, bridges, crossing canals, vertical walls, or composite bathymetries. Flow depth hydrographs and velocity profiles at selected positions were often measured, as well as the wavefront position in time and the maximum run-up height, thanks to the analysis of images acquired through video cameras. Wave profiles at selected times were measured less frequently (e.g., [283]). Time series of pressure and force against structures hit by the bore were measured in several studies (e.g., [269,274]), whereas data related to overtopping phenomena are seldom recorded [270]. Only a few works focused on bottom shear stress data, flow vortices behind structures, and phenomena related to moving objects transported by the flow (e.g., [277,278]).

Vessels and offshore structures can be affected by extreme waves causing green water run-up and wave impingement, with consequent extensive damage and failure to superstructures, deck plating, hatches, and topside equipment. Moreover, green water represents a serious concern for the safety of personnel. In past years, a significant resemblance was recognized between the green water event and dam-break flow [343]. Therefore, as shown by Table 6, many studies applied the dam-break theory to green water predictions [344], and the use of dam-break solutions has become the standard design analysis approach to estimate the front velocity in green water phenomena. For at least two decades, researchers have tried to obtain experimental data useful for the validation of theories and numerical codes through laboratory investigations of green water phenomena caused by dam-breaks (e.g., [287,343,345]), taking advantage of the relative simplicity of dam-break setups. Conditions characterized by different freeboards were usually considered. The interaction between a dam-break flow and a floating box was also investigated [294]. Loads and pressures exerted on structures are of primary interest in such applications and were acquired through force and pressure transducers, respectively. Moreover, measures of free surface elevation at selected positions were often performed using conventional wave probes. In recent years, the availability of high-speed video cameras has allowed a more in-depth investigation of the initial phases of the phenomenon (e.g., [291,292]).

The dam-break flow of non-Newtonian fluids has recently received considerable attention due to environmental and industrial applications, such as the flood hazard assessments associated with tailings dam failures. Experimental data are even more valuable given the complex rheological behavior of such fluids and the scarcity of analytical solutions available. As shown in Table 7, experiments were conducted in simple laboratory facilities (planforms or rectangular channels), sometimes with steep bottom slopes [302]. Typically, the liquids used were aqueous suspensions or mudflows with viscoplastic behavior, but also Bingham fluids are used [307]. Moreover, the test conditions usually considered are characterized by a total dam-break and dry downstream bottom, with few exceptions [64,303]. Non-intrusive imaging techniques were preferably used to record data.

Table 8 lists experimental investigations of dam-break flows in cascade reservoirs. This line of research has recently been developed, motivated by the significant number of cascade dams built in recent years along several rivers. In this case, the channel bottom downstream of the dam is always assumed to be dry in the experiments. The influence of the initial reservoir levels and the distances between the dams are analyzed to highlight the attenuation effect on the dam-break flood in case the downstream dam does not fail. Flow depths time series were typically measured at selected positions. Sometimes pressure time series were recorded on the upstream face of the dam hit by the flood wave [310]. Numerical simulations accompanying the experimental investigations were usually performed through 3D models.

In experimental investigations of a dike-break induced flow presented in Table 9 the laboratory facilities consisted of an initially dry, smooth lateral floodplain linked to a straight main channel. A gate on the side wall of the channel was lifted to simulate the dike failure and induce the flooding of the lateral floodplain. In most cases, the gate opening was sudden, and the breach was rectangular or trapezoidal. Seldom was the gate removed gradually to represent the typical progressive failure of earth-fill embankments [328]. The flow in the main channel was typically assumed to be steady, even if a river embankment realistically fails during a flood event [327]. The wavefront propagation and flood depth time series at different positions in the floodplain were usually measured with non-intrusive devices. The experimental data acquired are particularly useful for validating 2D depth-averaged numerical models.

Dangerous liquids for industrial applications are often stored in large tanks built above ground. The failure of such storage tanks can lead to disastrous consequences for people, assets, and the surrounding environment, due to the sudden and uncontrolled release of large volumes of impounded materials, sometimes potentially flammable. Many major incidents occurred in recent years due to natural disasters, atmospheric phenomena, maintenance or operational errors, equipment failures, or corrosion [337]. A containment system formed by dikes or bunds is crucial (and required by technical regulations) to mitigate the risk associated with these catastrophic events. The bund system must be designed to prevent massive liquid overflow and withstand the dynamic pressures generated by the wave impact. Table 10 shows that in the past 40 years, several experimental studies have dealt with the problem of predicting the bund overtopping fraction as a function of the level of the impounded liquid, as well as the bund shape and characteristic parameters (distance from the tank, height, inclination, presence of breakwaters, etc.). To this end, small-scale experimental investigations have typically been conducted in dam-break setups by suddenly releasing fixed volumes of water or oil stored in cylindrical or quarter-cylinder-shaped tanks. Few medium-scale experimental investigations have so far been conducted [332,333]. Sometimes, the liquid release resulting from the opening of fractures or holes in the tank walls is considered [335]. In recent years, computational fluid dynamics (CFD) has increasingly been used in this context, since it gives the possibility to study in detail the phenomenon. The accuracy of the CFD models is assessed through validation against experimental data [337,338].

In over half of the total entries of Tables 1–10, a numerical analysis was coupled with the laboratory investigation, and the experimental data were immediately used to validate the numerical models. The information about those numerical simulations provided in Column 11 of the tables indicates that the most adopted solution approach is the two-dimensional one (approximately 50% of cases); the one- and three-dimensional approaches are equally adopted in about 25% of cases. One- and two-dimensional numerical models are usually based on the depth-averaged shallow water equations (SWE), while three-dimensional models on the Reynolds averaged Navier–Stokes equations (RANS), coupled with the volume of fluid (VOF) technique for the tracking of the free surface. In the examined studies, the most used numerical method for the solution of the governing equations is the finite volume method (over 50% of cases), followed by the finite difference method (over 20% of cases, especially before 2000). The mesh-free particle methods, such as the smoothed-particle hydrodynamics (SPH) and the moving particle semi-implicit (MPS) ones, have spread rapidly more recently and were used in about 15% of the cases considered.

The impact of scale effects in open channel flow physical models deserves special attention [346,347]. Some analyses in the literature have confirmed that the Froude number is dominant in dam-break flows over a fixed bed (e.g., [150]). However, the validation of numerical simulation tools designed for real applications against small-scale laboratory tests must take into account the distortions introduced by the scale effects.

In all experiments listed in the previous tables, the waves were generated by the sudden removal of a gate. In most cases, a lift gate moving upward is used, but there are also examples of downward-moving lift gates [82] or flap gates (e.g., [104]). Experimental and numerical studies have compared gate-opening modalities, investigating the gate motion effect on the dam-break flow [105,348]. However, none of these systems mimics exactly the instantaneous disappearance of the gate assumed in the classic theoretical approach to the dam-break problem, nor a real dam collapse. Technical regulations on dam-break flood risk assessment adopted worldwide prescribe that the structural failure of concrete gravity and arch dams is assumed to occur practically instantaneously (e.g., [349,350]). If this assumption is made, the gate opening time in dam-break experiments should be short enough to represent a 'nearly-instantaneous' dam collapse. To this end, suitable criteria for the gate opening timing have been presented in the literature [50,351] and are usually checked at the beginning of the experimental investigations. However, in the hydraulics laboratory, the question of the 'instantaneous' removal of the gate (or of the actual non-presence of the gate itself) remains a subject of debate, and often suggestive and imaginative hypotheses are formulated in the breaks between the experimental tests to remove the gate as quickly as possible.

The large number of articles reviewed here demonstrates that an impressive amount of experimental work has been carried out on dam-break flows, considering a variety of test conditions covering a wide range of flow situations. Many aspects of the physical process having practical implications have been investigated, including the effects of obstacles and structures that interfere with the flow. Nevertheless, the dam-break flow remains a topic of current research that continues to attract considerable interest, also from an experimental point of view [352]. Non-intrusive techniques appear preferable in dam-break flow measurements as they do not disturb the flow. In particular, digital imagery enables the acquisition of flow data (such as free surface profiles or flow depth and velocity fields at selected times) over an extended area, but requires optical access and often free surface seeding or the use of a coloring agent, as well as laborious calibration procedures [4].

The literature review shows that no systematic experimental investigations have been conducted on floods caused by a partial dam collapse in the vertical direction, producing a breach in the upper portion of the dam. To the authors' knowledge, this dam-break scenario was only hypothesized in a historical study based on a physical model [5,353]. Therefore, it could be considered for future research to collect experimental data to support the development of hybrid 3D-2D numerical models simulating the breach outflow with a 3D model and the downstream flooding with a depth-averaged 2D model (e.g., [354]). Furthermore, the movement of the pieces of a breached dam within the flow has never been experimentally studied since the release of the impounded water is usually simulated by removing (and not breaking) a retaining plate. This aspect related to the collapse of a concrete or masonry dam could also be the subject of future experimental research; after all, the modeling capabilities of current CFD software include the possibility of handling moving objects which dynamically interact with the flow and rigid body interactions (e.g., [355]).

4. Conclusions

This paper provides a comprehensive review of the state-of-the-art experimental investigations on unsteady, rapidly varying flows generated by the sudden removal of a retaining structure. Only experiments performed in schematic laboratory setups with a fixed, non-erodible bottom were considered. This review, based on journal papers, reports, theses, and documents published until the end of 2022, was carried out with passion and dedication by four researchers who share the experience of conducting physical experimentation of dam-break phenomena for over twenty years. Although the authors tried to conduct an extensive and meticulous review, it may not be exhaustive. Some studies on the subject, especially the older ones, may be missing since they were published in journals of local diffusion or not written in a vehicular language, or those in which the experimental data are marginal and do not represent the focus of the research.

A large number of references was reviewed and divided into tables according to the investigation's purposes. These tables report extensive information on test conditions, datasets, measuring techniques, relevant bibliographic references, and data availability.

This review may guide researchers to compare existing datasets and identify remaining knowledge gaps deserving additional experimental investigation. Moreover, it may help modelers select suitable test cases for validating their numerical models and testing new numerical approaches. Indeed, most experiments aimed at collecting benchmark data were expressly designed to highlight specific computational difficulties for numerical schemes. This review may also support practitioners looking for new technical solutions for mitigating the destructive effects of dam-break flood waves.

Unfortunately, most datasets are not directly accessible in digital format as supplemental material linked to the original works. Therefore, we hope a public repository will soon be made available, where experimental data can be freely uploaded to form a comprehensive open-access database for all researchers interested in dam-break flows.

An impressive amount of laboratory investigations was carried out on dam-break flows, and a variety of test conditions were considered in the literature. However, experimental studies on flows caused by dam breaches with height and width lower than those of the dam and on the movement of the blocks resulting from the dam collapse are still lacking and may be the subject of future research.

Author Contributions: F.A., A.M., G.P. and S.S.-F. contributed equally to the conceptualization and implementation of the research, to the revision of the existing literature, and to the writing of the manuscript. All authors have read and agreed to the published version of the manuscript.

Funding: This work was supported by the Italian Ministry of University and Research through the PRIN 2017 Project RELAID (REnaissance of LArge Italian Dams), project number 2017T4JC5K. The support from Italian Ministry of University and Research and University of Pavia (Italy) within the program "Dipartimenti di Eccellenza 2023–2027" is acknowledged.

Data Availability Statement: No new data were created or analyzed in this study. Data sharing is not applicable to this article.

Conflicts of Interest: The authors declare no conflict of interest.

References

1. Friedrich, H.; Ravazzolo, D.; Ruiz-Villanueva, V.; Schalko, I.; Spreitzer, G.; Tunnicliffe, J.; Weitbrecht, V. Physical Modelling of Large Wood (LW) Processes Relevant for River Management: Perspectives from New Zealand and Switzerland. *Earth Surf. Process. Landforms* **2022**, *47*, 32–57. [CrossRef]
2. Nezu, I.; Sanjou, M. PIV and PTV Measurements in Hydro-Sciences with Focus on Turbulent Open-Channel Flows. *J. Hydro-environ. Res.* **2011**, *5*, 215–230. [CrossRef]
3. Soares-Frazão, S. Review of Imaging-Based Measurement Techniques for Free Surface Flows Involving Sediment Transport and Morphological Changes. *J. Hydroinform.* **2020**, *22*, 958–971. [CrossRef]
4. Gomit, G.; Chatellier, L.; David, L. Free-Surface Flow Measurements by Non-Intrusive Methods: A Survey. *Exp. Fluids* **2022**, *63*, 94. [CrossRef]
5. Aureli, F.; Maranzoni, A.; Petaccia, G. Review of Historical Dam-Break Events and Laboratory Tests on Real Topography for the Validation of Numerical Models. *Water* **2021**, *13*, 1968. [CrossRef]
6. Testa, G.; Zuccalà, D.; Alcrudo, F.; Mulet, J.; Soares-Frazão, S. Flash Flood Flow Experiment in a Simplified Urban District. *J. Hydraul. Res.* **2007**, *45* (Suppl. 1), 37–44. [CrossRef]
7. Levin, L. Mouvement Non Permanent sur les Cours d'Eau à la Suite de Rupture de Barrage [Unsteady River Flow Caused by a Dam-Break]. *Rev. Gén. Hydr.* **1952**, *18*, 297–315. (In French)
8. Escande, L.; Nougaro, J.; Castex, L.; Barthet, H. Influence de Quelques Paramètres sur une Onde de Crue Subite à l'Aval d'un Barrage [The Influence of Certain parameters on a Sudden Flood Wave Downstream from a Dam]. *Houille Blanche* **1961**, *5*, 565–575. Available online: https://www.shf-lhb.org/articles/lhb/pdf/1961/07/lhb1961043.pdf (accessed on 25 January 2023). (In French) [CrossRef]

9. Özgenç Aksoy, A.; Doğan, M.; Oğuzhan Güven, S.; Tanır, G.; Güney, M.Ş. Experimental and Numerical Investigation of the Flood Waves due to Partial Dam Break. *Iran. J. Sci. Technol. Trans. Civ. Eng.* **2022**, *46*, 4689–4704. [CrossRef]
10. Chanson, H. Application of the Method of Characteristics to the Dam Break Wave Problem. *J. Hydraul. Res.* **2009**, *47*, 41–49. [CrossRef]
11. Hager, W.H.; Lauber, G. Hydraulische Experimente zum Talsperrenbruchproblem [Hydraulic Experiments on the Dam-Break Problem]. *Schweiz. Ing. Archit.* **1996**, *114*, 515–524. (In German) [CrossRef]
12. Hager, W.H.; Chervet, A. Geschichte der Dammbruchwelle [Hystory of dam-break wave]. *Wasser Energ. Luft* **1996**, *88*, 49–54. Available online: https://www.e-periodica.ch/digbib/view?pid=wel-004%3A1996%3A88%3A%3A61#67 (accessed on 25 January 2023). (In German)
13. Dressler, R. Comparison of Theories and Experiments for the Hydraulic Dam-Break Wave. In Proceedings of the International Association of Scientific Hydrology, Assemblée Générale, Rome, Italy, 14–25 September 1954; Volume 3, pp. 319–328.
14. Aureli, F.; Maranzoni, A.; Mignosa, P.; Ziveri, C. Dam-Break Flows: Acquisition of Experimental Data through an Imaging Technique and 2D Numerical Modeling. *J. Hydraul. Eng.* **2008**, *134*, 1089–1101. [CrossRef]
15. Toro, E.F.; Garcia-Navarro, P. Godunov-Type Methods for Free-Surface Shallow Flows: A Review. *J. Hydraul. Res.* **2007**, *45*, 736–751. [CrossRef]
16. Hui Pu, J.; Shao, S.; Huang, Y.; Hussain, K. Evaluations of SWEs and SPH Numerical Modelling Techniques for Dam Break Flows. *Eng. Appl. Comput. Fluid Mech.* **2013**, *7*, 544–563. [CrossRef]
17. Teng, J.; Jakeman, A.J.; Vaze, J.; Croke, B.F.W.; Dutta, D.; Kim, S. Flood Inundation Modelling: A Review of Methods, Recent Advances and Uncertainty Analysis. *Environ. Model. Softw.* **2017**, *90*, 201–216. [CrossRef]
18. Bates, P.D. Flood Inundation Prediction. *Annu. Rev. Fluid Mech.* **2022**, *54*, 287–315. [CrossRef]
19. Schoklitsch, A. Über Dambruchwellen [On Waves Produced by Broken Dams]. In *Sitzungsberichte, Mathematisch-Naturwissenschaftliche Klasse*; Akademie der Wissenschaften in Wien: Vienna, Austria, 1917; Volume 126 (IIa), pp. 1489–1514. (In German)
20. Trifonov, E.K. Experimental Investigation of Positive Wave's Propagation along Dry Bottom. *Sci. Research Inst. Hydrotechnics Trans.* **1933**, *10*, 169–188. (In Russian)
21. Trifonov, E.K. Étude Expérimental de la Propagation d'une Onde Positive le Long d'un Fond Sec [Experimental Investigation of the Propagation of a Positive Wave over a Dry Bottom]. *Bull. Congr. Navig.* **1935**, *10*, 66–77. (In French)
22. Eguiazaroff, J.B. Réglage du Niveau de l'Eau dans les Biefs des Rivières Canalisées et Réglage du Débit en Aval du Dernier Barrage Selon que la Puissance Hydraulique Est, ou Non, Utilisée [Regulation of the Water Level in the Reaches of Canalized Rivers and Regulation of the Flow Below the Last Lock Dam According to Whether the Water Power Is or Is Not Used]. In Proceedings of the 16th International Congress of Navigation, Brussels, Belgium, 3–10 September 1935. (In French)
23. Martin, J.C.; Moyce, W.J. An Experimental Study of the Collapse of Liquid Columns on a Rigid Horizontal Plane. *Philos. Trans. R. Soc. Lond. Ser. A* **1952**, *244*, 312–324. [CrossRef]
24. U.S. Army Engineer Waterways Experiment Station (WES). *Floods Resulting from Suddenly Breached Dams. Conditions of Minimum Resistance. Hydraulic Model Investigation*; Miscellaneous Paper No. 2-374, Report I; U.S. Corps of Engineers: Vicksburg, MS, USA, 1960.
25. U.S. Army Engineer Waterways Experiment Station (WES). *Floods Resulting from Suddenly Breached Dams. Conditions of High Resistance. Hydraulic Model Investigation*; Miscellaneous Paper No. 2-374, Report II; U.S. Corps of Engineers: Vicksburg, MS, USA, 1961.
26. Faure, J.; Nahas, N. Étude Numérique et Expérimentale d'Intumescences à Forte Courbure du Front [A Numerical and Experimental Study of Steep-Fronted Solitary Waves]. *La Houille Blanche* **1961**, *47*, 576–587. (In French) [CrossRef]
27. Estrade, J. Contribution à l'Étude de la Suppression d'un Barrage—Phase Initial de l'Écoulement (Contribution to the Study of Dam-Break—Initial Flow Phase). *Dir. Etudes Rech. Bullet* **1967**, *1*, 5–90. (In French)
28. Nakagawa, H.; Nakamura, S.; Ichihashi, K. Generation and Development of a Hydraulic Bore Due to the Breaking of a Dam (1). *Bull. Disas. Prev. Res. Inst. Kyoto Univ.* **1969**, *19*, 1–17. Available online: https://core.ac.uk/download/pdf/39254633.pdf (accessed on 25 January 2023).
29. Chervet, A.; Dallèves, P. Calcul de l'Onde de Submersion Consécutive à la Rupture d'un Barrage. Deuxième Partie: Comparaisons Entre Observations et Calculs [Calculation of a Dam-Break Wave. Second Part: Comparison Between Observations and Numerical Results]. *Schweiz. Bauztg.* **1970**, *88*, 425–432. (In French) [CrossRef]
30. Cunge, J.-A. Calcul de Propagation des Ondes de Rupture de Barrage [Dam-Break Wave Propagation Modelling]. *Houille Blanche* **1970**, *56*, 25–33. [CrossRef]
31. Cavaillé, Y. Contribution à l'Étude de l'Écoulement Variable Accompagnant la Vidange Brusque d'une Retenue [Contribution to the Analysis of Dam-Break Flow]. In *Aeronautical Documentation and Technical Information Service (No. 410)*; Service de Documentation et D'information Technique de L'aéronautique: Paris, France, 1965. (In French)
32. Maxworthy, T. Experiments on Collisions Between Solitary Waves. *J. Fluid Mech.* **1976**, *76*, 177–186. [CrossRef]
33. Xanthopoulos, T.; Koutitas, C. Numerical simulation of a two dimensional flood wave propagation due to dam failure. *J. Hydraul. Res.* **1976**, *14*, 321–331. [CrossRef]
34. Barr, D.I.H.; Das, M.M. Numerical Simulation of Dam-Burst and Reflections, with Verification against Laboratory Data. *Proc. Inst. Civ. Eng.* **1980**, *69*, 359–373. [CrossRef]

35. Barr, D.I.H.; Das, M.M. Simulation of Surges After Removal of a Separating Barrier Between Shallower and Deeper Bodies of Water. *Proc. Inst. Civ. Eng.* **1981**, *71*, 911–919. [CrossRef]
36. Memos, C.D.; Georgakakos, A.; Vomvoris, S. Some Experimental Results of the Two-Dimensional Dam-Break Problem. In Proceedings of the 20th Congress of International Association for Hydraulic Research, Moscow, Russia, 5–9 September 1983; Volume 2, pp. 555–563.
37. Townson, J.M.; Al-Salihi, A.H. Models of Dam-Break Flow in R-T Space. *J. Hydraul. Eng.* **1989**, *115*, 561–575. [CrossRef]
38. Menendez, A.N.; Navarro, F. An Experimental Study on the Continuous Breaking of a Dam. *J. Hydraul. Res.* **1990**, *28*, 753–772. [CrossRef]
39. Iverson, R.M.; Costa, J.E.; LaHusen, R.G. *Debris-Flow Flume at H.J. Andrews Experimental Forest, Oregon;* Open-File Report 92-483; U.S. Geological Survey, Department of the Interior: Reston, VA, USA, 1992. [CrossRef]
40. Logan, M.; Iverson, R.M.; Obryk, M.K. *Video Documentation of Experiments at the USGS Debris-Flow Flume 1992–2017 (Ver. 1.4, January 2018)*; Open-File Report 2007-1315; U.S. Geological Survey, Department of the Interior: Reston, VA, USA, 2018. [CrossRef]
41. Antunes Do Carmo, J.S.; Seabra Santos, F.J.; Almeida, A.B. Numerical Solution of the Generalized Serre Equations with the MacCormack Finite-Difference Scheme. *Int. J. Numer. Methods Fluids* **1993**, *16*, 725–738. [CrossRef]
42. Tingsanchali, T.; Rattanapitikon, W. 2-D Mathematical Modelling for Dam Break Wave Propagation in Supercritical and Subcritical Flows. In Proceedings of the 25th IAHR World Congress, Tokyo, Japan, 30 August–9 September 1993; pp. 25–32.
43. Braschi, G.; Dadone, F.; Gallati, M. Plain Flooding: Near Field and Far Fields Simulations. In *Modelling of Flood Propagation over Initially Dry Areas, Proceedings of the Specialty Conference, Milan, Italy, 29 June–1 July 1994*; Molinaro, P., Natale, L., Eds.; American Society of Civil Engineers: New York, NY, USA, 1994; pp. 45–59.
44. Manciola, P.; Mazzoni, A.; Savi, F. Formation and Propagation of Steep Wave: An Investigative Experimental Interpretation. In *Modelling of Flood Propagation over Initially Dry Areas, Proceedings of the specialty Conference, Milan, Italy, 29 June–1 July 1994*; Molinaro, P., Natale, L., Eds.; American Society of Civil Engineers: New York, NY, USA, 1994; pp. 283–297.
45. Aguirre-Pe, J.; Plachco, F.P.; Quisca, S. Tests and Numerical One-Dimensional Modelling of a High-Viscosity Fluid Dam-Break Wave. *J. Hydraul. Res.* **1995**, *33*, 17–26. [CrossRef]
46. Fraccarollo, L.; Toro, E.F. Experimental and Numerical Assessment of the Shallow Water Model for Two-Dimensional Dam-Break Type Problems. *J. Hydraul. Res.* **1995**, *33*, 843–864. [CrossRef]
47. Jovanović, M.; Djordjević, D. Experimental Verification of the MacCormack Numerical Scheme. *Adv. Eng. Softw.* **1995**, *23*, 61–67. [CrossRef]
48. Koshizuka, S.; Oka, Y. Moving-Particle Semi-Implicit Method for Fragmentation of Incompressible Fluid. *Nucl. Sci. Eng.* **1996**, *123*, 421–434. [CrossRef]
49. Koshizuka, S.; Tamako, H.; Oka, Y. A Particle Method for Incompressible Viscous Flow with Fluid Fragmentation. *Comput. Fluid Dyn. J.* **1995**, *4*, 29–46.
50. Lauber, G.; Hager, W.H. Experiments to Dambreak Wave: Horizontal Channel. *J. Hydraul. Res.* **1998**, *36*, 291–307. [CrossRef]
51. Lauber, G.; Hager, W.H. Experiments to Dambreak Wave: Sloping Channel. *J. Hydraul. Res.* **1998**, *36*, 761–773. [CrossRef]
52. Stansby, P.K.; Chegini, A.; Barnes, T.C.D. The initial stages of dam-break flow. *J. Fluid Mech.* **1998**, *374*, 407–424. [CrossRef]
53. Blaser, F.; Hager, W.H. Positive Front of Dambreak Wave on Rough Bottom. In Proceedings of the 28th IAHR World Congress, Graz, Austria, 22–27 August 1999; Technical University Graz: Graz, Austria, 1999. Available online: https://www.iahr.org/library/infor?pid=13612 (accessed on 14 July 2022).
54. Nsom, B.; Debiane, K.; Piau, J.-M. Bed Slope Effect on the Dam Break Problem. *J. Hydraul. Res.* **2000**, *38*, 459–464. [CrossRef]
55. Gallati, M.; Braschi, G. Simulazione Lagrangiana di Flussi con Superficie Libera in Problemi di Idraulica [Lagrangian Simulation of Free Surface Flows in Hydraulic Problems]. *L'Acqua* **2000**, *5*, 7–18. Available online: https://www.idrotecnicaitaliana.it/wp-content/uploads/2020/04/Gallati-et-al-LAcqua-n.-5-2000.pdf (accessed on 19 October 2022). (In Italian)
56. Liem, R.; Schramm, J.; Köngeter, J. Evaluating the Implementation of Shallow Water Equations within Numerical Models Focusing the Propagation of Dambreak Waves. *WIT Trans. Modelling Simul.* **2001**, *30*, 231–240. [CrossRef]
57. Briechle, S.; Köngeter, J. Experimental Data for Dike-Break Waves. In *River Flow 2002, Proceedings of the International Conference on Fluvial Hydraulics, Louvain-la-Neuve, Belgium, 4–6 September 2002*; Bousmar, D., Zech, Y., Eds.; Balkema: Lisse, The Netherlands, 2002; Volume 1, pp. 467–473.
58. Soares-Frazão, S.; Zech, Y. Undular Bores and Secondary Waves–Experiments and Hybrid Finite-Volume Modelling. *J. Hydraul. Res.* **2002**, *40*, 33–43. [CrossRef]
59. Shige-eda, M.; Akiyama, J. Numerical and Experimental Study on Two-Dimensional Flood Flows with and without Structures. *J. Hydraul. Eng.* **2003**, *129*, 817–821. [CrossRef]
60. Stelling, G.S.; Duinmeijer, S.P.A. A Staggered Conservative Scheme for Every Froude Number in Rapidly Varied Shallow Water Flows. *Int. J. Numer. Methods Fluids* **2003**, *43*, 1329–1354. [CrossRef]
61. Duinmeijer, A. *Verification of Delft FLS*; Delft University of Technology: Delft, The Netherlands, 2002.
62. Chegini, A.H.N.; Pender, G.; Slaouti, A.; Tait, S.J. Velocity Measurements in Dam-Break Flow Using Imaging System. In *River Flow 2004, Proceedings of the International Conference on Fluvial Hydraulics, Naples, Italy, 23–25 June 2004*; Greco, M., Carravetta, A., Della Morte, R., Eds.; Balkema: Lisse, The Netherlands, 2004; Volume 2, pp. 859–866.

63. Gallati, M.; Sturla, D. SPH Simulation of Dam-Break Flow in Shallow Water Approximation. In *River Flow 2004, Proceedings of the International Conference on Fluvial Hydraulics, Naples, Italy, 23–25 June 2004*; Greco, M., Carravetta, A., Della Morte, R., Eds.; Balkema: Lisse, The Netherlands, 2004; Volume 2, pp. 919–928.
64. Jánosi, M.I.; Jan, D.; Szabó, K.G.; Tél, T. Turbulent Drag Reduction in Dam-Break Flows. *Exp. Fluids* **2004**, *37*, 219–229. [CrossRef]
65. Bukreev, V.I.; Gusev, A.V. Initial Stage of the Generation of Dam-Break Waves. *Dokl. Phys.* **2005**, *50*, 200–203. [CrossRef]
66. Eaket, J.; Hicks, F.E.; Peterson, A.E. Use of Stereoscopy for Dam Break Flow Measurement. *J. Hydraul. Eng.* **2005**, *131*, 24–29. [CrossRef]
67. Piau, J.-M.; Debiane, K. Consistometers Rheometry of Power-Law Viscous Fluids. *J. Non-Newton. Fluid Mech.* **2005**, *127*, 213–224. [CrossRef]
68. Barnes, M.P.; Baldock, T.E. Bed Shear Stress Measurements in Dam Break and Swash Flows. In Proceedings of the International Conference on Civil and Environmental Engineering, Hiroshima, Japan, 28–29 September 2006. Available online: https://www.researchgate.net/publication/37620092_Bed_shear_stress_measurements_in_dam_break_and_swash_flows (accessed on 14 June 2022).
69. Bateman, A.; Granados, A.; Medina, V.; Velasco, D.; Nalesso, M. Experimental Procedure to Obtain 2D Time-Space High-Speed Water Surfaces. In *River Flow 2006, International Conference on Fluvial Hydraulics, Lisbon, Portugal, 6–8 September 2006*; Ferreira, R., Alves, E., Leal, J., Cardoso, A., Eds.; Taylor & Francis: London, UK, 2006; Volume 2, pp. 1879–1888.
70. Cruchaga, M.A.; Celentano, D.J.; Tezduyar, T.E. Collapse of a Liquid Column: Numerical Simulation and Experimental Validation. *Comput. Mech.* **2007**, *39*, 453–476. [CrossRef]
71. Maranzoni, A.; Pilotti, M.; Tomirotti, M.; Valerio, G. Modellazione Numerica e Sperimentale dei Primi Istanti di Moto Conseguenti alla Rapida Rimozione di uno Sbarramento [Numerical and Experimental Modelling of the Early Stages of a Dam-Break Flow]. In *AIMETA 2007, Proceedings of 18th Congress of Italian Association of Theoretical and Applied Mechanics, Brescia, 11–14 September 2007*; Carini, A., Mimmi, G., Piva, R., Eds.; Starrylink: Brescia, Italy, 2007.
72. Aureli, F.; Maranzoni, A.; Mignosa, P. Experimental Modeling of Rapidly Varying Flows on Wet Bed and in Presence of Submersible Obstacles. In *River Flow 2004, Proceedings of the International Conference on Fluvial Hydraulics, Naples, Italy, 23–25 June 2004*; Greco, M., Carravetta, A., Della Morte, R., Eds.; Balkema: Lisse, The Netherlands, 2004; Volume 2, pp. 849–858.
73. Mohamed, A. Characterization of Tsunami-Like Bores in Support of Loading Structures. Master's Thesis, University of Hawaii, Manoa, HI, USA, 2008. Available online: https://scholarspace.manoa.hawaii.edu/items/a7b0cee7-82fe-47ad-abbd-03a5a07705ee (accessed on 19 October 2022).
74. Ancey, C.; Cochard, S.; Andreini, N. The Dam-Break Problem for Viscous Fluids in the High-Capillary-Number Limit. *J. Fluid Mech.* **2009**, *624*, 1–22. [CrossRef]
75. Yang, C.; Lin, B.; Jiang, C.; Liu, Y. Predicting Near-Field Dam-Break Flow and Impact Force using a 3D Model. *J. Hydraul. Res.* **2010**, *48*, 784–792. [CrossRef]
76. Ozmen-Cagatay, H.; Kocaman, S. Dam-Break Flows During Initial Stage Using SWE and RANS Approaches. *J. Hydraul. Res.* **2010**, *48*, 603–611. [CrossRef]
77. Çağatay, H.; Kocaman, S. Experimental Study of Tailwater Level Effects on Dam-Break Flood Wave Propagation. In *River Flow 2008, Proceedings of the International Conference on Fluvial Hydraulics, Çeşme, Izmir, Turkey, 3–5 September 2008*; Altinakar, M.S., Kokpinar, M.A., Aydin, I., Cokgor, S., Kirkgoz, S., Eds.; Kubaba Congress Department and Travel Services: Ankara, Turkey, 2008; Volume 1, pp. 635–644.
78. Duarte, R.; Ribeiro, J.; Boillat, J.-L.; Schleiss, A. Experimental Study on Dam-Break Waves for Silted-Up Reservoirs. *J. Hydraul. Eng.* **2011**, *137*, 1385–1393. [CrossRef]
79. Boillat, J.-L.; Ribeiro, J.; Duarte, R.; Darbre, G. Dam Break in Case of Silted-Up Reservoirs. In *River Flow 2008, Proceedings of the International Conference on Fluvial Hydraulics, Çeşme, Izmir, Turkey, 3–5 September 2008*; Altinakar, M.S., Kokpinar, M.A., Aydin, I., Cokgor, S., Kirkgoz, S., Eds.; Kubaba Congress Department and Travel Services: Ankara, Turkey, 2008; Volume 1, pp. 689–695.
80. Ribeiro, J.; Boillat, J.-L.; Duarte, R.; Darbre, G. Front Wave Propagation in Case of Dam Break. Comparison of Water Filled and Silted-Up Reservoirs. In Proceedings of the 33th IAHR World Congress, Vancouver, BC, Canada, 9–14 August 2009; pp. 6010–6017.
81. Marra, D.; Earl, T.; Ancey, C. Experimental Investigations of Dam Break Flows down an Inclined Channel. In Proceedings of the 34th IAHR World Congress, Brisbane, Australia, 26 June–1 July 2011; pp. 623–630.
82. Aleixo, R.; Soares-Frazão, S.; Zech, Y. Velocity-Field Measurements in a Dam-Break Flow Using a PTV Voronoï Imaging Technique. *Exp. Fluids* **2011**, *50*, 1633–1649. [CrossRef]
83. Aleixo, R.; Zech, Y.; Soares-Frazão, S. Turbulence Measurements in Dam-Break Flows. In *River Flow 2012, Proceedings of the International Conference on Fluvial Hydraulics, San José, Costa Rica, 5–7 September 2012*; Murillo Muñoz, R.E., Ed.; CRC Press: London, UK, 2012; pp. 311–318.
84. Aleixo, R.; Soares-Frazão, S.; Zech, Y. Before the Dam Breaks: Analysis of the Flow Behind a Downward Moving Gate. In *River Flow 2016, Proceedings of the International Conference on Fluvial Hydraulics, Iowa City, IA, USA, 11–14 July 2016*; Constantinescu, G., Garcia, M., Hanes, D., Eds.; CRC Press: London, UK, 2016; pp. 436–442.
85. Aleixo, R.; Soares-Frazão, S.; Zech, Y. Statistical Analysis Methods for Transient Flows–The Dam-Break Case. *J. Hydraul. Res.* **2018**, *57*, 688–701. [CrossRef]
86. Feizi Khankandi, A.; Tahershamsi, A.; Soares-Frazão, S. Experimental Investigation of Reservoir Geometry Effect on Dam-Break Flow. *J. Hydraul. Res.* **2012**, *50*, 376–387. [CrossRef]

87. Oertel, M.; Bung, D.B. Initial Stage of Two-Dimensional Dam-Break Waves: Laboratory versus VOF. *J. Hydraul. Res.* **2012**, *50*, 89–97. [CrossRef]
88. LaRocque, L.A.; Imran, J.; Chaudhry, M.H. Experimental and Numerical Investigations of Two-Dimensional Dam-Break Flows. *J. Hydraul. Eng.* **2013**, *139*, 569–579. [CrossRef]
89. Miani, M.; Consoli, S.; Rossi, E.; Galliano, D.; Annunziato, A. *Physical and Numerical Simulation of a Scaled Tsunami Wave*; Technical Report EUR 26079 EN; Joint Research Centre–Institute for the Protection and Security of the Citizen; Publications Office of the European Union: Luxembourg, 2013. [CrossRef]
90. Hooshyaripor, F.; Tahershamsi, A. Effect of Reservoir Side Slopes on Dam-Break Flood Waves. *Eng. Appl. Comput. Fluid Mech.* **2015**, *9*, 458–468. [CrossRef]
91. Jiang, Z.; Baldock, T.E. Direct Bed Shear Measurements under Loose Bed Swash Flows. *Coast. Eng.* **2015**, *100*, 67–76. [CrossRef]
92. McMullin, N. Numerical and Experimental Modelling of Dam Break Interaction with a Sediment Bed. Ph.D. Thesis, University of Nottingham, Nottingham, UK, 2015. Available online: https://core.ac.uk/download/pdf/33573992.pdf (accessed on 24 January 2023).
93. Mrokowska, M.M.; Rowiński, P.M.; Kalinowska, M.B. Evaluation of Friction Velocity in Unsteady Flow Experiments. *J. Hydraul. Res.* **2015**, *53*, 659–669. [CrossRef]
94. Aleixo, R.; Ozeren, Y.; Altinakar, M. PIV-PTV Measurements of a Tailings Dam-Break Flow. In *Hydrodynamic and Mass Transport at Freshwater Aquatic Interfaces*; Rowiński, P.M., Marion, A., Eds.; Springer: Cham, Switzerland, 2016; pp. 255–268. [CrossRef]
95. Elkholy, M.; LaRocque, L.A.; Chaudhry, M.H.; Imran, J. Experimental Investigations of Partial-Breach Dam-Break Flows. *J. Hydraul. Eng.* **2016**, *142*, 04016042. [CrossRef]
96. Javadian, M.; Kaveh, R.; Mahmoodinasab, F. A Study on Experimental Model of Dam Break Problem and Comparison Experimental Results with Analytical Solution of Saint-Venant Equations. *Int. J. Adv. Biotechnol. Res.* **2016**, *7*, 1239–1245.
97. Hooshyaripor, F.; Tahershamsi, A.; Razi, S. Dam Break Flood Wave under Different Reservoir's Capacities and Lengths. *Sādhanā* **2017**, *42*, 1557–1569. [CrossRef]
98. Liu, H.; Liu, H.; Guo, L.; Lu, S. Experimental Study on Dam-Break Hydrodynamic Characteristics under Different Conditions. *J. Disaster Res.* **2017**, *12*, 198–207. [CrossRef]
99. Liu, H.; Liu, H. Experimental Study on the Dam-Break Hydrographs at the Gate Location. *J. Ocean Univ. China* **2017**, *16*, 697–702. [CrossRef]
100. Cordero, S.; Cagninei, A.; Poggi, D. Dam-Break on an Idealised Hill Side: Preliminary Results of a Physical Model. *E3S Web Conf.* **2018**, *40*, 05002. [CrossRef]
101. Liu, W.; Wang, B.; Chen, Y.; Wu, C.; Liu, X. Assessing the Analytical Solution of One-Dimensional Gravity Wave Model Equations Using Dam-Break Experimental Measurements. *Water* **2018**, *10*, 1261. [CrossRef]
102. Hamid, H.; Khan, M.; Mehmood, M.; Ahmed, F. Experimental Modeling of Dam Break Flood Wave Depth Propagation on a Fixed Bed. *Tech. J. UET Taxila* **2018**, *23*, 1–4.
103. Hamid, H.; Khan, F.A.; Khan, M.; Ajmal, M.; Mahmood, M.; Aslam, M.S.; Tufail, M. Dam Break Wave Propagation on a Non-Erodible Bed–Comparison of Experimental and Numerical Results. *Int. J. Emerg. Trends Eng. Res.* **2021**, *9*, 733–740. [CrossRef]
104. Stolle, J.; Ghodoosipour, B.; Derschum, C.; Nistor, I.; Petriu, E.; Goseberg, N. Swing Gate Generated Dam-Break Waves. *J. Hydraul. Res.* **2018**, *57*, 675–687. [CrossRef]
105. von Häfen, H.; Goseberg, N.; Stolle, J.; Nistor, I. Gate-Opening Criteria for Generating Dam-Break Waves. *J. Hydraul. Eng.* **2019**, *145*, 04019002. [CrossRef]
106. Liu, W.; Wang, B.; Wang, H.; Zhang, J.; Chen, Y.; Peng, Y.; Liu, X.; Yang, S. Experimental and Numerical Modeling of Dam-Break Flows in Wet Downstream Conditions. In Proceedings of the 38th IAHR World Congress, Panama City, Panama, 1–6 September 2019. [CrossRef]
107. Melis, M.; Poggi, D.; Giovanni; Fasanella, O.D.; Cordero, S.; Katul, G.G. Resistance to Flow on a Sloping Channel Covered by Dense Vegetation Following a Dam Break. *Water Resour. Res.* **2019**, *55*, 1040–1058. [CrossRef]
108. Turhan, E.; Ozmen-Cagatay, H.; Tantekin, A. Modeling Flood Shock Wave Propagation with the Smoothed Particle Hydrodynamics (SPH) Method: An Experimental Comparison Study. *Appl. Ecol. Environ. Res.* **2019**, *17*, 3033–3047. [CrossRef]
109. Turhan, E.; Ozmen-Cagatay, H.; Kocaman, S. Experimental and Numerical Investigation of Shock Wave Propagation Due to Dam-Break Over a Wet Channel. *Pol. J. Environ. Stud.* **2019**, *28*, 2877–2898. [CrossRef]
110. Wang, B.; Zhang, J.; Chen, Y.; Peng, Y.; Liu, X.; Liu, W. Comparison of Measured Dam-Break Flood Waves in Triangular and Rectangular Channels. *J. Hydrol.* **2019**, *575*, 690–703. [CrossRef]
111. Wu, K.F.J.; Liu, Z.; Sun, J. Numerical Simulation of Undular Bore Using a Shock-Capturing Boussinesq Model. In Proceedings of the 29th International Ocean and Polar Engineering Conference, Honolulu, HI, USA, 16–21 June 2019; pp. 3568–3573. Available online: https://onepetro.org/ISOPEIOPEC/proceedings-abstract/ISOPE19/All-ISOPE19/ISOPE-I-19-003/20847?redirectedFrom=PDF (accessed on 27 June 2022).
112. Liu, W.; Wang, B.; Guo, Y.; Zhang, J.; Chen, Y. Experimental Investigation on the Effects of Bed Slope and Tailwater on Dam-Break Flows. *J. Hydrol.* **2020**, *590*, 125256. [CrossRef]
113. Oertel, M.; Süfke, F. Two-Dimensional Dam-Break Wave Analysis: Particle Image Velocimetry Versus Optical Flow. *J. Hydraul. Res.* **2020**, *58*, 326–334. [CrossRef]

114. Shugan, I.V.; Chen, Y.-Y.; Hsu, C.-J. Experimental and Theoretical Study on Flood Bore Propagation and Forerunner Generation in Dam-Break Flow. *Phys. Wave Phenom.* **2020**, *28*, 274–284. [CrossRef]
115. Vosoughi, F.; Rakhshandehroo, G.; Nikoo, M.R.; Sadegh, M. Experimental Study and Numerical Verification of Silted-Up Dam Break. *J. Hydrol.* **2020**, *590*, 125267. [CrossRef]
116. Vosoughi, F.; Nikoo, M.R.; Rakhshandehroo, G.; Gandomi, A.H. Experimental Videos in Studying the Influences of Dry- and Wet-Bed Downstream Conditions on Dam Break Multiphase Flood Waves in a Reservoir with 7.5 cm Sediment Depth (25% silted-up). Mendeley Data, V3. Available online: https://doi.org/10.17632/p6bsdz7xch.3 (accessed on 21 June 2022).
117. Vosoughi, F.; Nikoo, M.R.; Rakhshandehroo, G.; Gandomi, A.H. Experimental Dataset on Water Levels, Sediment Depths and Wave Front Celerity Values in the Study of Multiphase Shock Wave for Different Initial Up- and Down-Stream Conditions. *Data Brief* **2021**, *36*, 107082. [CrossRef]
118. Wang, B.; Liu, X.; Zhang, J.; Guo, Y.; Chen, Y.; Peng, Y.; Liu, W.; Yang, S.; Zhang, F. Analytical and Experimental Investigations of Dam-Break Flows in Triangular Channels with Wet-Bed Conditions. *J. Hydraul. Eng.* **2020**, *146*, 04020070. [CrossRef]
119. Wang, B.; Liu, W.; Wang, W.; Zhang, J.; Chen, Y.; Peng, Y.; Liu, X.; Yang, S. Experimental and numerical investigations of similarity for dam-break flows on wet bed. *J. Hydrol.* **2020**, *583*, 124598. [CrossRef]
120. Ahmadi, S.M.; Yamamoto, Y. A New Dam-Break Outflow-Rate Concept and Its Installation to a Hydro-Morphodynamics Simulation Model Based on FDM (An Example on Amagase Dam of Japan). *Water* **2021**, *13*, 1759. [CrossRef]
121. Ansari, A.; Khavasi, E.; Ghazanfarian, J. Experimental and SPH Studies of Reciprocal Wet-Bed Dam-Break Flow over Obstacles. *Int. J. Mod. Phys. C* **2021**, *32*, 2150098. [CrossRef]
122. Birnbaum, J.; Lev, E.; Llewellin, E.W. Rheology of Three-Phase Suspensions Determined via Dam-Break Experiments. *Proc. R. Soc. A* **2021**, *477*, 20210394. [CrossRef]
123. Espartel, L.; Manica, R. Experiments on Initial Stages of Development of Dam-Break Waves. *Rev. Bras. Recur. Hidr.* **2021**, *26*, e6. [CrossRef]
124. Kocaman, S.; Evangelista, S.; Guzel, H.; Dal, K.; Yilmaz, A.; Viccione, G. Experimental and Numerical Investigation of 3D Dam-Break Wave Propagation in an Enclosed Domain with Dry and Wet Bottom. *Appl. Sci.* **2021**, *11*, 5638. [CrossRef]
125. Nguyen-Thi, L.-Q.; Nguyen, V.-D.; Pierens, X.; Coorevits, P. An Experimental and Numerical Study of the Influence of Viscosity on the Behavior of Dam-Break Flow. *Theor. Comput. Fluid Dyn.* **2021**, *35*, 345–362. [CrossRef]
126. Takagi, H.; Furukawa, F. Stochastic Uncertainty in a Dam-Break Experiment with Varying Gate Speeds. *J. Mar. Sci. Eng.* **2021**, *9*, 67. [CrossRef]
127. Wang, B.; Zhang, F.; Liu, X.; Guo, Y.; Zhang, J.; Peng, Y. Approximate Analytical Solution and Laboratory Experiments for Dam-Break Wave Tip Region in Triangular Channels. *J. Hydraul. Eng.* **2021**, *147*, 06021015. [CrossRef]
128. Xu, B.; Zhang, S.; Nielsen, P.; Wüthrich, D. Measurements of Bed Shear Stresses Near the Tip of Dam-Break Waves on a Rough Bed. *Exp. Fluids* **2021**, *62*, 49. [CrossRef]
129. Ozmen-Cagatay, H.; Turhan, E.; Kocaman, S. An Experimental Investigation of Dam-Break Induced Flood Waves for Different Density Fluids. *Ocean Eng.* **2022**, *243*, 110227. [CrossRef]
130. Yang, S.; Yang, W.; Zhang, C.; Qin, S.; Wei, K.; Zhang, J. Experimental and Numerical Study on the Evolution of Wave Front Profile of Dam-Break Waves. *Ocean Eng.* **2022**, *247*, 110681. [CrossRef]
131. Yang, S.; Tan, Z.; Yang, W.; Imani, H.; Song, D.; Luo, J.; Zhang, J. Experimental Study on Hydrodynamic Interaction between Dam-Break Waves and Circular Pier. *Ocean Eng.* **2022**, *266*, 113093. [CrossRef]
132. Nielsen, P.; Xu, B.; Wüthrich, D.; Zhang, S. Friction Effects on Quasi-Steady Dam-Break Wave Propagation on Horizontal Beds. *J. Fluid Mech.* **2022**, *939*, A21. [CrossRef]
133. Zhang, F.; Wang, B.; Guo, Y. Experimental Study of the Dam-Break Waves in Triangular Channels with a Sloped Wet Bed. *Ocean Eng.* **2022**, *255*, 111399. [CrossRef]
134. Matsutomi, H. Numerical Computations of Two-Dimensional Inundation of Rapidly Varied Flows due to Breaking of Dams. In Proceedings of the 20th IAHR Congress, Moscow, Russia, 5–9 September 1983; Volume 2, pp. 479–493.
135. Martin, H. Dam-Break Wave in Horizontal Channels with Parallel and Divergent Side Walls. In Proceedings of the 20th IAHR Congress, Moscow, Russia, 5–9 September 1983; Volume 2, pp. 494–505.
136. Michouev, A.V.; Sladkevich, M. Écoulement Dans le Bief Aval Dans le Cas d'une Rupture Partielle d'un Barrage [Flow Downstream in the Event of a Partial Dam-Break]. In Proceedings of the 20th IAHR Congress, Moscow, Russia, 5–9 September 1983; Volume 2, pp. 512–519. (In French)
137. Miller, S.; Chaudhry, M.H. Dam-Break Flows in Curved Channel. *J. Hydraul. Eng.* **1989**, *115*, 1465–1478. [CrossRef]
138. Bell, S.W.; Elliot, R.C.; Chaudhry, M.H. Experimental Results of Two-Dimensional Dam-Break Flows. *J. Hydraul. Res.* **1992**, *30*, 225–252. [CrossRef]
139. Bellos, C.V.; Soulis, V.; Sakkas, J.G. Experimental Investigation of Two-Dimensional Dam-Break Induced Flows. *J. Hydraul. Res.* **1992**, *30*, 47–63. [CrossRef]
140. Četina, M.; Rajar, R. Two-Dimensional Dam-Break Flow Simulation in a Sudden Enlargement. In *Modelling of Flood Propagation Over Initially Dry Areas, Proceedings of the Specialty Conference Co-Sponsored by ASC-CNR/GNDCI-ENEL Spa, Milan, Italy, 29 June–1 July 1994*; Molinaro, P., Natale, L., Eds.; American Society of Civil Engineers: New York, NY, USA, 1994; pp. 268–282.

141. Aureli, F.; Mignosa, P.; Tomirotti, M. Dam-break flows in presence of abrupt bottom variations. In Proceedings of the 28th IAHR World Congress, Graz, Austria, 22–27 August 1999; Technical University Graz: Graz, Austria, 1999. Available online: https://www.iahr.org/library/infor?pid=13679 (accessed on 14 July 2022).
142. Soares-Frazão, S.; Zech, Y. Effects of a Sharp Bend on Dam-Break Flow. In Proceedings of the 28th IAHR World Congress, Graz, Austria, 22–27 August 1999; Technical University Graz: Graz, Austria, 1999. Available online: https://www.iahr.org/library/infor?pid=13631 (accessed on 17 February 2023).
143. Soares-Frazão, S.; Sillen, X.; Zech, Y. Dam-Break Flow through Sharp Bends Physical Model and 2D Boltzmann Model Validation. In Proceedings of the 1st CADAM Workshop, Wallingford, UK, 2–3 March 1998.
144. Aureli, F.; Mignosa, P.; Tomirotti, M. Numerical Simulation and Experimental Verification of Dam-Break Flows with Shocks. *J. Hydraul. Res.* **2000**, *38*, 197–206. [CrossRef]
145. Aureli, F.; Belicchi, M.; Maione, U.; Mignosa, P.; Tomirotti, M. Fenomeni di Moto Vario Conseguenti al Crollo di Opere di Ritenuta. Parte II: Indagini Sperimentali e Modellazione Numerica in Presenza di Onde di Shock [Dam-Break Flows. Part II: Experimental and Numerical Modelling in the Presence of Shocks]. *L'Acqua* **1998**, *5*, 27–36. (In Italian)
146. Bento Franco, A.; Betâmio de Almeida, A. Dam-break in a channel. In *Concerted Action on Dam Break Modelling: Objectives Project, Report, Test Cases, Meeting Proceedings*; Soares Frazão, S., Morris, M., Zech, Y., Eds.; Université Catholique de Louvain Louvain-la-Neuve, Belgium, 2000.
147. Viseu, T.; Bento Franco, A.; Betâmio de Almeida, A. Numerical and Computational Results of the 2D BIPLAN Model. In Proceedings of the 1st CADAM Workshop, Wallingford, UK, 2–3 March 1998.
148. Hiver, J. Adverse-Slope and Slope (Bump). In *Concerted Action on Dam Break Modelling: Objectives, Project, Report, Test Cases, Meeting Proceedings*; Soares Frazão, S., Morris, M., Zech, Y., Eds.; Université Catholique de Louvain: Louvain-la-Neuve, Belgium, 2000.
149. Soares-Frazão, S.; de Bueger, C.; Dourson, V.; Zech, Y. Dam-Break Wave Over a Triangular Bottom Sill. In *River Flow 2002, Proceedings of the International Conference on Fluvial Hydraulics, Louvain-la-Neuve, Belgium, 4–6 September 2002*; Bousmar, D., Zech, Y., Eds.; Balkema: Lisse, The Netherlands, 2002; Volume 1, pp. 437–442.
150. Soares-Frazão, S. Experiments of Dam-Break Wave Over a Triangular Bottom Sill. *J. Hydraul. Res.* **2007**, *45* (Suppl. 1), 19–26. [CrossRef]
151. Soares-Frazão, S.; Zech, Y. Dam Break in Channels with 90° Bend. *J. Hydraul. Eng.* **2002**, *128*, 956–968. [CrossRef]
152. Bukreev, V.I. Water Impingement on a Vertical Wall Due to Discontinuity Decay Above a Drop. *J. Appl. Mech. Tech. Phys.* **2003**, *44*, 59–63. [CrossRef]
153. Bukreev, V.I.; Gusev, A.V. Gravity Waves Due to Discontinuity Decay Over an Open-Channel Bottom Drop. *J. Appl. Mech. Tech. Phys.* **2003**, *44*, 506–515. [CrossRef]
154. Soares-Frazão, S.; Lories, D.; Taminiau, D.; Zech, Y. Dam-Break Flow in a Channel with a Sudden Enlargement. In Proceedings of the 30th IAHR Congress, Thessaloniki, Greece, 25–27 August 2003.
155. Bukreev, V.I.; Gusev, A.V.; Malysheva, A.A.; Malysheva, I.A. Experimental Verification of the Gas-Hydraulic Analogy with Reference to the Dam-Break Problem. *Fluid Dyn.* **2004**, *39*, 801–809. [CrossRef]
156. Bellos, C. Experimental Measurements of Flood Wave Created by a Dam Break. *Eur. Water* **2004**, *7*, 3–15.
157. Natale, L.; Petaccia, G.; Savi, F. Mathematical Simulation of the Effects of Bridges and Structures on Flood Waves Propagation. In *River Flow 2004, Proceedings of the International Conference on Fluvial Hydraulics, Naples, Italy, 23–25 June 2004*; Greco, M., Carravetta, A., Della Morte, R., Eds.; Balkema: Lisse, The Netherlands, 2004; Volume 2, pp. 895–901.
158. Bukreev, V.I. On the Water Depth in the Breach during a Partial Dam Break. *Fluid Dyn.* **2005**, *40*, 769–776. [CrossRef]
159. Bukreev, V.I. On the discharge characteristic at the dam site after dam break. *J. Appl. Mech. Tech. Phys.* **2006**, *47*, 679–687. [CrossRef]
160. Gusev, A.V.; Ostapenko, V.V.; Malysheva, A.A.; Malysheva, I.A. Open-Channel Waves Generated by Propagation of a Discontinuous Wave over a Bottom Step. *J. Appl. Mech. Tech. Phys.* **2008**, *49*, 23–33. [CrossRef]
161. Bukreev, V.I.; Degtyarev, V.V.; Chebotnikov, A.V. Experimental Verification of Methods for Calculating Partial Dam-Break Waves. *J. Appl. Mech. Tech. Phys.* **2008**, *49*, 754–761. [CrossRef]
162. Evangelista, S.; Di Cristo, C.; De Marinis, G. Experiments and Simulations of Dam Break over Fixed and Granular Beds. In Proceedings of the 34th IAHR World Congress, Brisbane, Australia, 26 June–1 July 2011; pp. 3450–3457. Available online: https://www.researchgate.net/publication/273060021_Experiments_and_simulations_of_dam_break_over_fixed_and_granular_beds#fullTextFileContent (accessed on 23 December 2022).
163. Evangelista, S. Esperimenti e Simulazioni Numeriche di Dam Break su Letto Fisso e Mobile [Experiments and Numerical Simulations of Dam Break on Fixed and Movable Beds]. *L'Acqua* **2014**, *3*, 9–21. (In Italian)
164. Ozmen-Cagatay, H.; Kocaman, S. Dam-Break Flow in the Presence of Obstacle: Experiment and CFD Simulation. *Eng. Appl. Comput. Fluid Mech.* **2011**, *5*, 541–552. [CrossRef]
165. Ozmen-Cagatay, H.; Kocaman, S. Investigation of Dam-Break Flow Over Abruptly Contracting Channel with Trapezoidal-Shaped Lateral Obstacles. *J. Fluids Eng.* **2012**, *134*, 081204. [CrossRef]
166. Kocaman, S.; Ozmen-Cagatay, H. The Effect of Lateral Channel Contraction on Dam Break Flows: Laboratory Experiment. *J. Hydrol.* **2012**, *432–433*, 145–153. [CrossRef]
167. Ozmen-Cagatay, H.; Kocaman, S.; Guzel, H. Investigation of Dam-Break Flood Waves in a Dry Channel with a Hump. *J. Hydro-Environ. Res.* **2014**, *8*, 304–315. [CrossRef]

168. Degtyarev, V.V.; Ostapenko, V.V.; Kovyrkina, O.A.; Zolotykh, A.V. Comparison of Theory and Experiment in Simulation of Dam Break in a Rectangular Channel with a Sudden Change in Cross-Sectional Area. *J. Appl. Mech. Tech. Phys.* **2014**, *55*, 999–1004. [CrossRef]
169. Wood, A.; Wang, K.H. Modeling Dam-Break Flows in Channels with 90 Degree Bend Using an Alternating-Direction Implicit Based Curvilinear Hydrodynamic Solver. *Comput. Fluids* **2015**, *114*, 254–264. [CrossRef]
170. Kikkert, G.A.; Liyanage, T.; Shang, C. Dam-Break Generated Flow from an Infinite Reservoir into a Positively Inclined Channel of Limited Width. *J. Hydro-Environ. Res.* **2015**, *9*, 519–531. [CrossRef]
171. Chen, S.; Li, Y.; Tian, Z.; Fan, Q. On Dam-Break Flow Routing in Confluent Channels. *Int. J. Environ. Res. Public Health* **2019**, *16*, 4384. [CrossRef] [PubMed]
172. Kobayashi, D.; Uchida, T.; Kawahara, Y. The Characteristics of Dam Break Flows in a Meandering Channel. In Proceedings of the 38th IAHR World Congress, Panama City, Panama, 1–6 September 2019. [CrossRef]
173. Kavand, R.; Ghomeshi, M.; Daryaee, M. Experimental Study of the Effect of Bed Roughness on Wave Characteristics Resulting from Dam Break in Curve Channel. *J. Hydraul.* **2020**, *14*, 111–122. (In Arabic) [CrossRef]
174. Kocaman, S.; Güzel, H.; Evangelista, S.; Ozmen-Cagatay, H.; Viccione, G. Experimental and Numerical Analysis of a Dam-Break Flow through Different Contraction Geometries of the Channel. *Water* **2020**, *12*, 1124. [CrossRef]
175. Ismail, H.; Larocque, L.A.; Bastianon, E.; Chaudhry, M.H.; Imran, J. Propagation of Tributary Dam-Break Flows through a Channel Junction. *J. Hydraul. Res.* **2021**, *59*, 214–223. [CrossRef]
176. Gamero, P.; Cantero-Chinchilla, F.N.; Bergillos, R.J.; Castro-Orgaz, O.; Dey, S. Shallow-Water Lee-Side Waves at Obstacles: Experimental Characterization and Turbulent Non-Hydrostatic Modeling Using Weighted-Averaged Residual Equations. *Environ. Model. Softw.* **2022**, *155*, 105422. [CrossRef]
177. Kobayashi, D.; Uchida, T. Experimental and Numerical Investigation of Breaking Bores in Straight and Meandering Channels with Different Froude Numbers. *Coast. Eng. J.* **2022**, *64*, 442–457. [CrossRef]
178. Vosoughi, F.; Nikoo, M.R.; Rakhshandehroo, G.; Adamowski, J.F.; Gandomi, A.H. Downstream Semi-Circular Obstacles' Influence on Floods Arising from the Failure of Dams with Different Levels of Reservoir Silting. *Phys. Fluids* **2022**, *34*, 013312. [CrossRef]
179. Vosoughi, F.; Nikoo, M.R.; Rakhshandehroo, G.; Alamdari, N.; Gandomi, A.H.; Al-Wardy, M. The Application of Bayesian Model Averaging Based on Artificial Intelligent Models in Estimating Multiphase Shock Flood Waves. *Neural Comput. Appl.* **2022**, *34*, 20411–20429. [CrossRef]
180. Greenspan, H.P.; Young, R.E. Flow Over a Containment Dyke. *J. Fluid Mech.* **1978**, *87*, 179–192. [CrossRef]
181. Sicard, J.; Nicollet, G. Effets d'une onde de rupture sur un barrage aval [Effects of a dam-break wave on a downstream dam]. In Proceedings of the 20th IAHR Congress, Moscow, Russia, 5–9 September 1983; Volume 2, pp. 564–570.
182. Ramsden, J.D. Forces on a Vertical Wall Due to Long Waves, Bores, and Dry-Bed Surges. *J. Waterw. Port Coast. Ocean. Eng.* **1996**, *122*, 134–141. [CrossRef]
183. Liu, P.L.-F.; Lin, P.; Chang, K.-A.; Sakakiyama, T. Numerical Modeling of Wave Interaction with Porous Structures. *J. Waterw. Port Coast. Ocean Eng.* **1999**, *125*, 322–330. [CrossRef]
184. Barakhnin, V.B.; Krasnoshchekova, T.V.; Potapov, I.N. Reflection of a Dam–Break Wave at a Vertical Wall. Numerical Modeling and Experiment. *J. Appl. Mech. Tech. Phys.* **2001**, *42*, 269–275. [CrossRef]
185. Soares-Frazão, S.; Zech, Y. Dam Break Flow Experiment: The Isolated Building Test Case. In Proceedings of the 2nd IMPACT Workshop, Mo i Rana, Norway, 12–13 September 2002.
186. Soares-Frazão, S.; Zech, Y. Experimental Study of Dam-Break Flow against an Isolated Obstacle. *J. Hydraul. Res.* **2007**, *45* (Suppl. 1), 27–36. [CrossRef]
187. Brufau, P.; Garcia-Navarro, P.; Vásquez-Cendón, M.E.; Méndez, A.; Puertas, J. Numerical Model Validation with Experimental Data on Dam Break Problems Involving Wetting/Drying Fronts over Initially Dry Bed Adverse Slopes. In *River Flow 2002, Proceedings of the International Conference on Fluvial Hydraulics, Louvain-la-Neuve, Belgium, 4–6 September 2002*; Bousmar, D., Zech, Y., Eds.; Balkema: Lisse, The Netherlands, 2002; Volume 1, pp. 487–494.
188. Méndez, A.; Pena, L.; Brufau, P.; Puertas, J. An Experimental Approach to the Dam Break Problem. Real Data to Check Numerical Models. In Proceedings of the 29th Congress of International Association for Hydraulic Research, Beijing, China, 16–21 September 2001; Li, G., Ed.; Tsinghua University Press: Beijing, China, 2001; Theme C; pp. 184–189.
189. Ciobotaru, R.M.; Bidoae, R.; Raad, P.E. Interaction of a Single Large Wave with a Tall Fixed Structure: A Numerical Study. In Proceedings of the FEDSM'03, 4th ASME-JSME Joint Fluids Engineering Conference, Honolulu, HI, USA, 6–10 July 2003; pp. 437–442. [CrossRef]
190. Trivellato, F. Experimental and Numerical Investigation of Bore Impact on a Wall. *WIT Trans. Built Environ.* **2003**, *71*, 3–12. [CrossRef]
191. Bertolazzi, E.; Trivellato, F. *Numerical Modeling of Clear Water Bore Impact*; University of Trento: Trento, Italy, 2003. Available online: https://e.bertolazzi.dii.unitn.it/files/2002-DIMS-18.pdf (accessed on 19 October 2022).
192. Campisano, A.; Creaco, E.; Modica, C. Experimental and Numerical Analysis of the Scouring Effects of Flushing Waves on Sediment Deposits. *J. Hydrol.* **2004**, *299*, 324–334. [CrossRef]
193. Hu, C.; Kashiwagi, M. A CIP-based method for numerical simulations of violent free-surface flows. *J. Mar. Sci. Technol.* **2004**, *9*, 143–157. [CrossRef]

194. Raad, P.E.; Bidoae, R. The Three-Dimensional Eulerian–Lagrangian Marker and Micro Cell Method for the Simulation of Free Surface Flows. *J. Comput. Phys.* **2005**, *203*, 668–699. [CrossRef]
195. Arnason, H. Interactions between an Incident Bore and a Free-Standing Coastal Structure. Ph.D. Thesis, University of Washington, Seattle, WA, USA, 2005.
196. Kleefsman, K.M.T.; Fekken, G.; Veldman, A.E.P.; Iwanowski, B.; Buchner, B. A Volume-of-Fluid Based Simulation Method for Wave Impact Problems. *J. Comput. Phys.* **2005**, *206*, 363–393. [CrossRef]
197. Issa, R.; Violeau, D. 3D Dambreaking-Test-Case 2. EDF-Ercoftac-SPH European Research Interest Community SIG. 2006. Available online: https://docplayer.net/63169198-Ercoftac-test-case-2-3d-dambreaking-release-1-1-march-reza-issa-and-damien-violeau-electricite-de-france.html (accessed on 31 January 2023).
198. Larese, A.; Rossi, R.; Oñate, E.; Idelsohn, S.R. Validation of the Particle Finite Element Method (PFEM) for Simulation of Free Surface Flows. *Eng. Comput.* **2008**, *25*, 385–425. [CrossRef]
199. Liang, D.; Falconer, R.A.; Jiang, C.; Wang, X. Numerical Simulation of Flood Flows Due to Levee Breaches. In Proceedings of the 32nd IAHR Congress, Venice, Italy, 1–6 July 2007; CORILA: Venice, Italy, 2007. Available online: https://www.iahr.org/library/infor?pid=15794 (accessed on 31 January 2023).
200. Nouri, Y. The Impact of Hydraulic Bores and Debris on Free Standing Structures. Master's Thesis, University of Ottawa, Ottawa, ON, Canada, 2008.
201. Nistor, I.; Nouri, Y.; Palermo, D.; Cornett, A. Experimental Investigation of the Impact of a Tsunami-Induced Bore on Structures. In *Coastal Engineering 2008, Proceedings of the 31st International Conference, Hamburg, Germany, 31 August–5 September 2008*; McKee Smith, J., Ed.; World Scientific Publishing: Singapore, 2008; pp. 3324–3336. [CrossRef]
202. Nouri, Y.; Nistor, I.; Palermo, D.; Cornett, A. Experimental Investigation of Tsunami Impact on Free Standing Structures. *Coast. Eng. J.* **2010**, *52*, 43–70. [CrossRef]
203. Bukreev, V.I.; Zykov, V.V. Bore Impact on a Vertical Plate. *J. Appl. Mech. Tech. Phys.* **2008**, *49*, 926–933. [CrossRef]
204. Arnason, H.; Petroff, C.; Yeh, H. Tsunami bore impingement onto a vertical column. *J. Disaster Res.* **2009**, *4*, 391–403. [CrossRef]
205. Cruchaga, M.A.; Celentano, D.J.; Tezduyar, T.E. Computational modeling of the collapse of a liquid column over an obstacle and experimental validation. *J. Appl. Mech.* **2009**, *76*, 021202. [CrossRef]
206. Hu, C.; Sueyoshi, M. Numerical Simulation and Experiment on Dam Break Problem. *J. Mar. Sci. Appl.* **2010**, *9*, 109–114. [CrossRef]
207. Aureli, F.; Maranzoni, A.; Mignosa, P.; Ziveri, C. An Image Processing Technique for Measuring Free Surface of Dam-Break Flows. *Exp. Fluids* **2011**, *50*, 665–675. [CrossRef]
208. Al-Faesly, T.; Palermo, D.; Nistor, I.; Cornett, A. Experimental Modeling of Extreme Hydrodynamic Forces on Structural Models. *Int. J. Prot. Struct.* **2012**, *3*, 477–505. [CrossRef]
209. Lara, J.L.; del Jesus, M.; Losada, I.J. Three-Dimensional Interaction of Waves and Porous Coastal Structures: Part II: Experimental Validation. *Coast. Eng.* **2012**, *64*, 26–46. [CrossRef]
210. Triatmadja, R.; Nurhasanah, A. Tsunami Force on Buildings with Openings and Protection. *J. Earthq. Tsunami* **2012**, *6*, 1250024. [CrossRef]
211. Aguíñiga, F.; Jaiswal, M.; Sai, J.O.; Cox, D.T.; Gupta, R.; van de Lindt, J.W. Experimental Study of Tsunami Forces on Structures. In *Coastal Hazards*; Huang, W., Wang, K., Cehng, Q.J., Eds.; American Society of Civil Engineers: Resto, VA, USA, 2013; pp. 111–118. [CrossRef]
212. Nakao, H.; Zhang, G.; Sumimura, T.; Hoshikuma, J. Numerical Assessment of Tsunami-Induced Effect on Bridge Behavior. In Proceedings of the 29th US–Japan Bridge Engineering Workshop, Tsukuba, Japan, 11–13 November 2013. Available online: https://www.pwri.go.jp/eng/ujnr/tc/g/pdf/29/29-1-3_Nakao.pdf (accessed on 1 February 2023).
213. Lobovský, L.; Botia-Vera, E.; Castellana, F.; Mas-Soler, J.; Souto-Iglesias, A. Experimental Investigation of Dynamic Pressure Loads During Dam Break. *J. Fluids Struct.* **2014**, *48*, 407–434. [CrossRef]
214. Ratia, H.; Murillo, J.; García-Navarro, P. Numerical Modelling of Bridges in 2D Shallow Water Flow Simulations. *Int. J. Numer. Methods Fluids* **2014**, *75*, 250–272. [CrossRef]
215. Aureli, F.; Dazzi, S.; Maranzoni, A.; Mignosa, P.; Vacondio, R. Experimental and Numerical Evaluation of the Force Due to the Impact of a Dam-Break Wave on a Structure. *Adv. Water Resour.* **2015**, *76*, 29–42. [CrossRef]
216. Kocaman, S.; Ozmen-Cagatay, H. Investigation of Dam-Break Induced Shock Waves Impact on a Vertical Wall. *J. Hydrol.* **2015**, *525*, 1–12. [CrossRef]
217. Liao, K.; Hu, C.; Sueyoshi, M. Free Surface Flow Impacting on an Elastic Structure: Experiment versus Numerical Simulation. *Appl. Ocean Res.* **2015**, *50*, 192–208. [CrossRef]
218. Liang, Q.; Chen, K.-C.; Hou, J.; Xiong, Y.; Wang, G.; Qiang, J. Hydrodynamic Modelling of Flow Impact on Structures under Extreme Flow Conditions. *J. Hydrodynam. B* **2016**, *28*, 267–274. [CrossRef]
219. Mohd, N.; Kamra, M.M.; Sueyoshi, M.; Hu, C. Lattice Boltzmann Method for Free Surface Impacting on Vertical Cylinder: A Comparison with Experimental Data. *Evergreen* **2017**, *4*, 28–37. [CrossRef]
220. Kamra, M.M.; Mohd, N.; Liu, C.; Sueyoshi, M.; Hu, C. Numerical and Experimental Investigation of Three-Dimensionality in the Dam-Break Flow Against a Vertical Wall. *J. Hydrodyn.* **2018**, *30*, 682–693. [CrossRef]
221. Liu, L.; Sun, J.; Lin, B.; Lu, L. Building Performance in Dam-Break Flow–An Experimental Study. *Urban Water J.* **2018**, *15*, 251–258. [CrossRef]

222. Martínez-Aranda, S.; Fernández-Pato, J.; Caviedes-Voullième, D.; García-Palacín, I.; García-Navarro, P. Towards Transient Experimental Water Surfaces: A New Benchmark Dataset for 2D Shallow Water Solvers. *Adv. Water Resour.* **2018**, *121*, 130–149. [CrossRef]
223. Stamataki, I.; Zang, J.; Buldakov, E.; Kjeldsen, T.; Stagonas, D. Study of Dam Break Flow Interaction with Urban Settlements over a Sloping Channel. *E3S Web Conf.* **2018**, *40*, 06006. [CrossRef]
224. Tinh, N.X.; Mitobe, Y.; Tanaka, H. Laboratory Experiment Study on the Building Relative Angle against Tsunami Waves. *Tohoku J. Nat. Disaster Sci.* **2018**, *54*, 7–12.
225. Demir, A.; Dincer, A.E.; Bozkus, Z.; Tijsseling, A.S. Numerical and Experimental Investigation of Damping in a Dam-Break Problem with Fluid-Structure Interaction. *J. Zhejiang Univ. Sci. A* **2019**, *20*, 258–271. [CrossRef]
226. Ghodoosipour, B.; Stolle, J.; Nistor, I.; Mohammadian, A.; Goseberg, N. Experimental Study on Extreme Hydrodynamic Loading on Pipelines. Part 1: Flow Hydrodynamics. *J. Mar. Sci. Eng.* **2019**, *7*, 251. [CrossRef]
227. Ghodoosipour, B.; Stolle, J.; Nistor, I.; Mohammadian, A.; Goseberg, N. Experimental Study on Extreme Hydrodynamic Loading on Pipelines Part 2: Induced Force Analysis. *J. Mar. Sci. Eng.* **2019**, *7*, 262. [CrossRef]
228. Kamra, M.M.; Al Salami, J.; Sueyoshi, M.; Hu, C. Experimental Study of the Interaction of Dambreak with a Vertical Cylinder. *J. Fluids Struct.* **2019**, *86*, 185–199. [CrossRef]
229. Mokhtar, Z.A.; Mohammed, T.A.; Yusuf, B.; Lau, T.L. Experimental Investigation of Tsunami Bore Impact Pressure on a Perforated Seawall. *Appl. Ocean Res.* **2019**, *84*, 291–301. [CrossRef]
230. Dutta, D.; Kumar, L.; Saud Afzal, M.; Rathore, P. Hydrodynamic Study of the Flows Caused by Dam Break Around a Rectangular Obstacle. In Proceeding of ICRACEM 2020, 1st Online International Conference on Recent Advances in Computational and Experimental Mechanics, Kharagpur, India, 4–6 September 2020; SM-20-046.
231. Dutta, D.; Kumar, L.; Saud Afzal, M.; Rathore, P. Hydrodynamic Study of the Flows Caused by Dam Break Around a Rectangular Obstacle. In *Recent Advances in Computational and Experimental Mechanics, Vol II. Lecture Notes in Mechanical Engineering*; Springer: Singapore, 2022; pp. 159–169. [CrossRef]
232. Farahmandpour, O.; Marsono, A.K.; Forouzani, P.; Tap, M.M.; Abu Bakar, S. Experimental Simulation of Tsunami Surge and Its Interaction with Coastal Structure. *Int. J. Prot. Struct.* **2020**, *11*, 258–280. [CrossRef]
233. Kocaman, S.; Evangelista, S.; Viccione, G.; Güzel, H. Experimental and Numerical Analysis of 3D Dam-Break Waves in an Enclosed Domain with a Single Oriented Obstacle. *Environ. Sci. Proc.* **2020**, *2*, 35. [CrossRef]
234. Pratiwi, V.; Kusuma, M.S.B.; Kardhana, H.; Farid, M. Experimental Study of Dam Break Flow Generated by the Flap Gate in Horizontal Channel. *Int. J. GEOMATE* **2020**, *19*, 19–26. [CrossRef]
235. Shen, J.; Wei, L.; Wu, D.; Liu, H.; Huangfu, J. Spatiotemporal Characteristics of the Dam-Break Induced Surge Pressure on a Vertical Wall. *Coast. Eng. J.* **2020**, *62*, 566–581. [CrossRef]
236. Memarzadeh, R.; Sheybanifard, H.; Zounemat-Kermani, M. Numerical and Experimental Study of Abrupt Wave Interaction with Vertical and Inclined Rectangular Obstacles. *J. Appl. Fluid Mech.* **2021**, *14*, 921–933. [CrossRef]
237. Del Gaudio, A.; De Paola, F.; Di Cristo, C.; La Forgia, G.; Leopardi, A.; Vacca, A. Experimental Investigation and Numerical Evaluation of the Free Surface of a Dam Break Wave in the Presence of an Obstacle. *Environ. Sci. Proc.* **2022**, *21*, 23. [CrossRef]
238. Fang, Q.; Liu, S.; Zhong, G.; Liang, J.; Zhen, Y. Experimental Investigation of Extreme Flood Loading on Buildings Considering the Shadowing Effect of the Front Building. *J. Hydraul. Eng.* **2022**, *148*, 04022007. [CrossRef]
239. Garoosi, F.; Mellado-Cusicahua, A.N.; Shademani, M.; Shakibaeinia, A. Experimental and Numerical Investigations of Dam Break Flow over Dry and Wet Beds. *Int. J. Mech. Sci.* **2022**, *215*, 106946. [CrossRef]
240. Lin, J.-H.; Chang, Y.-W.; Chen, G.-Y. Boulder Transportation on the Flat Bed by Dam Break. *J. Earthq. Tsunami* **2022**, *16*, 2241002. [CrossRef]
241. Liu, S.; Nistor, I.; Mohammadian, A.; Azimi, A.H. Experimental Investigation on the Impact of Dam-Break Induced Surges on a Vertical Wall. *Fluids* **2022**, *7*, 258. [CrossRef]
242. Wang, S.; Garlock, M.; Deike, L.; Glisic, B. Feasibility of Kinetic Umbrellas as Deployable Flood Barriers During Landfalling Hurricanes. *J. Struct. Eng.* **2022**, *148*, 04022047. [CrossRef]
243. Xie, W.; Shimozono, T. Water Surge Impingement onto a Vertical Wall: Laboratory Experiments and Stochastic Analysis on Impact Pressure. *Ocean Eng.* **2022**, *248*, 110422. [CrossRef]
244. Soares-Frazão, S.; Spinewine, B.; Duthoit, A.; Deswijsen, J.F.; Zech, Y. Dam-Break Flow Experiments in Simplified City Layouts. In *River Flow 2006, Proceedings of the International Conference on Fluvial Hydraulics, Lisbon, Portugal, 6–8 September 2006*; Ferreira, R., Alves, E., Leal, J., Cardoso, A., Eds.; Taylor & Francis: London, UK, 2006; Volume 1, pp. 513–521.
245. Soares-Frazão, S.; Zech, Y. Dam-Break Flow Through an Idealised City. *J. Hydraul. Res.* **2008**, *46*, 648–658. [CrossRef]
246. Szydłowski, M.; Twaróg, B. Numerical Investigation of Flooding of Real-Topography Developed Areas Following River Embankment Failure. *Task Q.* **2006**, *10*, 321–338.
247. Yoon, K. *Experimental Study on Flood Inundation Considering Urban Characteristics (FFC06-05)*; Urban Flood Disaster Management Research Center: Seoul, Korea, 2007.
248. Kim, B.; Sanders, B.F.; Famiglietti, J.S.; Guinot, V. Urban Flood Modeling with Porous Shallow-Water Equations: A Case Study of Model Errors in the Presence of Anisotropic Porosity. *J. Hydrol.* **2015**, *523*, 680–692. [CrossRef]
249. Albano, R.; Sole, A.; Mirauda, D.; Adamowski, J. Modelling Large Floating Bodies in Urban Area Flash-Floods via a Smoothed Particle Hydrodynamics Model. *J. Hydrol.* **2016**, *541*, 344–358. [CrossRef]

250. Norin, S.V.; Belikov, V.V.; Aleksyuk, A.I. Simulating Flood Waves in Residential Areas. *Power Technol. Eng.* **2017**, *51*, 52–57. [CrossRef]
251. Guinot, V.; Soares-Frazão, S.; Delenne, C. Experimental Validation of Transient Source Term in Porosity-Based Shallow Water Models. *E3S Web Conf.* **2018**, *40*, 06033. [CrossRef]
252. Guinot, V.; Delenne, C.; Soares-Frazão, S. Self-Similar Solutions of Shallow Water Equations with Porosity. *J. Hydraul. Res.* **2022**, *61*, 109–119. [CrossRef]
253. Kusuma, M.S.B.; Setiawati, T.; Farid, M. Experimental Model of Dam Break Flow Around Several Blockages Configurations. *Int. J. GEOMATE* **2019**, *16*, 26–32. [CrossRef]
254. Chumchan, C.; Rattanadecho, P. Experimental and Numerical Investigation of Dam Break Flow Propagation Passed through Complex Obstacles Using LES Model Based on FVM and LBM. *Songklanakarin J. Sci. Technol.* **2020**, *42*, 564–572. [CrossRef]
255. Dong, B.; Xia, J.; Zhou, M.; Deng, S.; Ahmadian, R.; Falconer, R.A. Experimental and Numerical Model Studies on Flash Flood Inundation Processes Over a Typical Urban Street. *Adv. Water Resour.* **2021**, *147*, 103824. [CrossRef]
256. Yeh, H.H.; Ghazali, A. On Bore Collapse. *J. Geophys. Res. Atmos.* **1988**, *93*, 6930–6936. [CrossRef]
257. Yeh, H.H.; Ghazali, A.; Marton, I. Experimental Study of Bore Run-Up. *J. Fluid Mech.* **1989**, *206*, 563–578. [CrossRef]
258. Petroff, C.M.; Moore, A.L.; Arnason, H. Particle Advection by Turbulent Bores–Orientation Effects. In Proceedings of the International Tsunami Symposium, Seattle, WA, 7–10 August 2001; NOAA/PMEL: Seattle, WA, USA, 2001; pp. 897–904.
259. Anh, V.M. Reduction of Wave Overtopping by Vetiver Grass. Master's Thesis, Delft University, Delft, The Netherlands, 2007.
260. Barnes, M.P.; O'Donoghue, T.; Alsina, J.M.; Baldock, T.E. Direct Bed Shear Stress Measurements in Bore-Driven Swash. *Coast. Eng.* **2009**, *56*, 853–867. [CrossRef]
261. De Leffe, M.; Le Touzé, D.; Alessandrini, B. SPH Modeling of Shallow-Water Coastal Flows. *J. Hydraul. Res.* **2010**, *48* (Suppl. S1), 118–125. [CrossRef]
262. O'Donoghue, T.; Pokrajac, D.; Hondebrink, L.J. Laboratory and Numerical Study of Dambreak-Generated Swash on Impermeable Slopes. *Coast. Eng.* **2010**, *57*, 513–530. [CrossRef]
263. Kikkert, G.A.; O'Donoghue, T.; Pokrajac, D.; Dodd, N. Experimental Study of Bore-Driven Swash Hydrodynamics on Impermeable Rough Slopes. *Coast. Eng.* **2012**, *60*, 149–166. [CrossRef]
264. Adegoke, P.B.; Atherton, W.; Al Khaddar, R.M. A Novel Simple Method for Measuring the Velocity of Dam-Break Flow. *WIT Trans. Ecol. Environ.* **2014**, *184*, 23–34. [CrossRef]
265. Rahman, S.; Akib, S.; Khan, M.T.R.; Shirazi, S.M. Experimental Study on Tsunami Risk Reduction on Coastal Building Fronted by Sea Wall. *Sci. World J.* **2014**, *2014*, 729357. [CrossRef]
266. Hartana; Murakami, K. Numerical and Experimental Simulation of Two-Phase Tsunami Flow Through Buildings with Openings. *J. Earthq. Tsunami* **2015**, *9*, 1550007. [CrossRef]
267. Chen, B.-T.; Kikkert, G.A.; Pokrajac, D.; Dai, H.-J. Experimental Study of Bore-Driven Swash–Swash Interactions on an Impermeable Rough Slope. *Coast. Eng.* **2016**, *108*, 10–24. [CrossRef]
268. Chen, C.; Melville, B.W.; Nandasena, N.A.K.; Shamseldin, A.Y.; Wotherspoon, L. Experimental Study of Uplift Loads Due to Tsunami Bore Impact on a Wharf Model. *Coast. Eng.* **2016**, *117*, 126–137. [CrossRef]
269. Chen, C.; Melville, B.W.; Nandasena, N.A.K.; Shamseldin, A.Y.; Wotherspoon, L. Mitigation Effect of Vertical Walls on a Wharf Model Subjected to Tsunami Bores. *J. Earthq. Tsunami* **2017**, *11*, 1750004. [CrossRef]
270. Esteban, M.; Glasbergen, T.; Takabatake, T.; Hofland, B.; Nishizaki, S.; Nishida, Y.; Stolle, J.; Nistor, I.; Bricker, J.; Takagi, H.; et al. Overtopping of Coastal Structures by Tsunami Waves. *Geosciences* **2017**, *7*, 121. [CrossRef]
271. Dai, H.-J.; Kikkert, G.A.; Chen, B.-T.; Pokrajac, D. Entrained Air in Bore-Driven Swash on an Impermeable Rough Slope. *Coast. Eng.* **2017**, *121*, 26–43. [CrossRef]
272. Tar, T.; Kato, N.; Suzuki, H.; Nagai, Y.; Ohnishi, K.; Okubayashi, T. Experimental and Numerical Study on the Reduction of Tsunami Flow Using Multiple Flexible Pipes. *J. Loss Prev. Process Ind.* **2017**, *50*, 364–385. [CrossRef]
273. Chen, C.; Melville, B.W.; Nandasena, N.A.K. Investigations of reduction Effect of Vertical Wall on Dam-Break-Simulated Tsunami Surge Exerted on Wharf Piles. *J. Earthq. Tsunami* **2018**, *12*, 1840006. [CrossRef]
274. Chen, C.; Melville, B.W.; Nandasena, N.A.K.; Farvizi, F. An Experimental Investigation of Tsunami Bore Impacts on a Coastal Bridge Model with Different Contraction Ratios. *J. Coast. Res.* **2018**, *34*, 460–469. [CrossRef]
275. Ishii, H.; Shibayama, T.; Stolles, J. Physical and Numerical Modelling of the Flow Structure behind Structure in Tsunami-like Flow. *Coast. Eng. Proc.* **2018**, *1*, currents17. [CrossRef]
276. Lu, S.; Liu, H.; Deng, X. An Experimental Study of the Run-Up Process of Breaking Bores Generated by Dam-Break under Dry- and Wet-Bed Conditions. *J. Earthq. Tsunami* **2018**, *12*, 1840005. [CrossRef]
277. Chen, C.; Wang, F.; Lin, H. Experimental Study on Dam-Break-Like Tsunami Surge Impacts on Small Balls Climbing on Different Slopes. *J. Earthq. Tsunami* **2020**, *14*, 2050022. [CrossRef]
278. Chen, C.; Chen, J.; Lin, P.; Chen, C.; Chen, H. Experimental Study of Dam-Break-Like Tsunami Bore Impact Mechanism on a Container Model. *Pol. Marit. Res.* **2020**, *27*, 53–59. [CrossRef]
279. Elsheikh, N.; Azimi, A.H.; Nistor, I.; Mohammadian, A. Experimental Investigations of Hydraulic Surges Passing over a Rectangular Canal. *J. Earthq. Tsunami* **2020**, *14*, 2040004. [CrossRef]
280. Elsheikh, N.; Nistor, I.; Azimi, A.H.; Mohammadian, A. Tsunami-Induced Bore Propagating over a Canal—Part 1: Laboratory Experiments and Numerical Validation. *Fluids* **2022**, *7*, 213. [CrossRef]

281. Barranco, I.; Liu, P.L.-F. Run-Up and Inundation Generated by Non-Decaying Dam-Break Bores on a Planar Beach. *J. Fluid Mech.* **2021**, *915*, A81. [CrossRef]
282. Chen, C.; Wang, F. An Experimental Study of Energy Dissipation of Sea Grass in a Dam Break Flume. *Glob. NEST J.* **2022**, *24*, 129–134. [CrossRef]
283. Liu, S.; Nistor, I.; Mohammadian, A.; Azimi, A. Experimental Investigation of Beach Slope Effects on the Kinematic Behaviors of Dam Break Flow. In Proceedings of the 39th IAHR World Congress, Granada, Spain, 19–24 June 2022; Ortega-Sánchez, M., Ed.; IAHR: Madrid, Spain, 2022; pp. 6194–6202. [CrossRef]
284. Liu, S.; Nistor, I.; Mohammadian, A.; Azimi, A.H. Experimental and Numerical Investigation of Beach Slope Effects on the Hydrodynamic Loading of Tsunami-like Surges on a Vertical Wall. *J. Mar. Sci. Eng.* **2022**, *10*, 1580. [CrossRef]
285. Rajaie, M.; Azimi, A.H.; Nistor, I.; Rennie, C.D. Experimental Investigations on Hydrodynamic Characteristics of Tsunami-Like Hydraulic Bores Impacting a Square Structure. *J. Hydraul. Eng.* **2022**, *148*, 04021061. [CrossRef]
286. von Häfen, H.; Krautwald, C.; Bihs, H.; Goseberg, N. Dam-Break Waves' Hydrodynamics on Composite Bathymetry. *Front. Built Environ.* **2022**, *8*, 877378. [CrossRef]
287. Buchner, B. Green Water on Ship-Type Offshore Structures. Ph.D. Thesis, Delft University of Technology, Delft, The Netherlands, 2002. Available online: https://repository.tudelft.nl/islandora/object/uuid:e383a7ef-43b5-4137-93d1-d5ab8fc75568/datastream/OBJ/download (accessed on 8 February 2023).
288. Hernández-Fontes, J.V.; Vitola, M.A.; Silva, M.C.; Esperança, P.D.T.T.; Sphaier, S.H. Use of Wet Dam-Break to Study Green Water Problem. In Proceedings of the 36th International Conference on Offshore Mechanics and Arctic Engineering, Trondheim, Norway, 25–30 June 2017; American Society of Mechanical Engineers: New York, NY, USA, 2017; OMAE2017-62113, V07AT06A065. [CrossRef]
289. Hernández-Fontes, J.V.; Vitola, M.A.; Silva, M.C.; Esperança, P.D.T.T.; Sphaier, S.H. On the Generation of Isolated Green Water Events Using Wet Dam-Break. *J. Offshore Mech. Arct. Eng.* **2018**, *140*, 051101. [CrossRef]
290. Hernández-Fontes, J.V.; Vitola, M.A.; Esperança, P.D.T.T.; Sphaier, S.H. Assessing Shipping Water Vertical Loads on a Fixed Structure by Convolution Model and Wet Dam-Break Tests. *Appl. Ocean Res.* **2019**, *82*, 63–73. [CrossRef]
291. Hernández-Fontes, J.V.; Vitola, M.A.; Esperança, P.D.T.T.; Sphaier, S.H.; Silva, R. Patterns and Vertical Loads in Water Shipping in Systematic Wet Dam-Break Experiments. *Ocean Eng.* **2020**, *197*, 106891. [CrossRef]
292. Hernández-Fontes, J.V.; Esperança, P.D.T.T.; Silva, R.; Mendoza, E.; Sphaier, S.H. Violent Water-Structure Interaction: Overtopping Features and Vertical Loads on a Fixed Structure Due to Broken Incident Flows. *Mar. Struct.* **2020**, *74*, 102816. [CrossRef]
293. Hernández-Fontes, J.V.; Hernández, I.D.; Silva, R.; Mendoza, E.; Esperança, P.D.T.T. A Simplified and Open-Source Approach for Multiple-Valued Water Surface Measurements in 2D Hydrodynamic Experiments. *J. Braz. Soc. Mech. Sci. Eng.* **2020**, *42*, 623. [CrossRef]
294. Wang, D.; Dong, S. Experimental Investigation on the Interactions Between Dam-Break Flow and a Floating Box. *J. Ocean Univ. China* **2022**, *21*, 633–646. [CrossRef]
295. Chanson, H.; Coussot, P.; Jarny, S.; Tocquer, L. *A Study of Dam Break Wave of Thixotropic Fluid: Bentonite Surges Down an Inclined Plane*; Rep. No. CH54/04; Department of Civil Engineering, The University of Queensland: Brisbane, Australia, 2004. Available online: https://espace.library.uq.edu.au/view/UQ:9437 (accessed on 12 December 2022).
296. Komatina, D.; Đorđević, D. Numerical Simulation of Hyper-Concentrated Flows. In *River Flow 2004, Proceedings of the International Conference on Fluvial Hydraulics, Naples, Italy, 23–25 June 2004*; Greco, M., Carravetta, A., Della Morte, R., Eds.; Balkema: Lisse, The Netherlands, 2004; Volume 2, pp. 1111–1120.
297. Cochard, S.; Ancey, C. Accurate Measurements of Free-Surface in the Dam-Break Problem. In *River Flow 2006, Proceedings of the International Conference on Fluvial Hydraulics, Lisbon, Portugal, 6–8 September 2006*; Ferreira, R., Alves, E., Leal, J., Cardoso, A., Eds.; Taylor & Francis: London, UK, 2006; Volume 2, pp. 1863–1872.
298. Cochard, S. Measurements of Time-Dependent, Free-Surface Viscoplastic Flows Down Steep Slopes. Ph.D. Thesis, École Polytechnique Fédéral de Lausanne, Lausanne, Switzerland, 2007.
299. Cochard, S.; Ancey, C. Tracking The Free Surface of Time-Dependent Flows: Image Processing for the Dam-Break Problem. *Exp. Fluids* **2008**, *44*, 59–71. [CrossRef]
300. Chanson, H.; Jarny, S.; Coussot, P. Dam Break Wave of Thixotropic Fluid. *J. Hydraul. Eng.* **2006**, *132*, 280–293. [CrossRef]
301. Balmforth, N.J.; Craster, R.V.; Perona, P.; Rust, A.C.; Sassi, R. Viscoplastic Dam Breaks and the Bostwick Consistometer. *J. Non-Newton. Fluid Mech.* **2007**, *142*, 63–78. [CrossRef]
302. Ancey, C.; Cochard, S. The Dam-Break Problem for Herschel–Bulkley Viscoplastic Fluids Down Steep Flumes. *J. Non-Newton. Fluid Mech.* **2009**, *158*, 18–35. [CrossRef]
303. Cochard, S.; Ancey, C. Experimental Investigation of the Spreading of Viscoplastic Fluids on Inclined Planes. *J. Non-Newton. Fluid Mech.* **2009**, *158*, 73–84. [CrossRef]
304. Brondani Minussi, R.; de Freitas Maciel, G. Numerical Experimental Comparison of Dam Break Flows with Non-Newtonian Fluids. *J. Braz. Soc. Mech. Sci. Eng.* **2012**, *34*, 167–178. [CrossRef]
305. Bates, B.M.; Ancey, C. The Dam-Break Problem for Eroding Viscoplastic Fluids. *J. Non-Newton. Fluid Mech.* **2017**, *243*, 64–78. [CrossRef]
306. Jing, X.; Chen, Y.; Xie, D.; Williams, D.J.; Wu, S.; Wang, W.; Yin, T. The Effect of Grain Size on the Hydrodynamics of Mudflow Surge from a Tailings Dam-Break. *Appl. Sci.* **2019**, *9*, 2474. [CrossRef]

307. Modolo, A.V.F.; Loureiro, B.V.; Soares, E.J.; Thompson, R.L. Influence of the Plastic Number on the Evolution of a Yield Stress Material Subjected to a Dam Break. *J. Appl. Fluid Mech.* **2019**, *12*, 1967–1978. [CrossRef]
308. Tang, J.B.; Lin, P.Z.; Cui, P. Depth-Resolved Numerical Model of Dam Break Mud Flows with Herschel-Bulkley Rheology. *J. Mt. Sci.* **2022**, *19*, 1001–1017. [CrossRef]
309. Xue, Y.; Xu, W.-L.; Luo, S.-J.; Chen, H.-Y.; Li, N.-W.; Xu, L.-J. Experimental Study of Dam-Break Flow in Cascade Reservoirs with Steep Bottom Slope. *J. Hydrodyn.* **2011**, *23*, 491–497. [CrossRef]
310. Chen, H.Y.; Xu, W.L.; Deng, J.; Xue, Y.; Li, J. Experimental Investigation of Pressure Load Exerted on a Downstream Dam by Dam-Break Flow. *J. Hydraul. Eng.* **2014**, *140*, 199–207. [CrossRef]
311. Liu, H.J.; Li, Y.; Wang, X.G.; Xuan, G.X.; Zhu, L. Experimental Study of the Flood Superposition Effect Due to Cascade Dam Break. In Proceedings of the 37th IAHR World Congress, Kuala Lumpur, Malaysia, 13–18 August 2017; Ghani, A.A., Chan, N.W., Ariffin, J., Wahab, A.K.A., Harun, S., Kassim, A.H.M., Karim, O.A., Eds.; IAHR: Madrid, Spain, 2017; pp. 2299–2306.
312. Zhang, Y.; Xu, W. Retarding Effects of an Intermediate Intact Dam on the Dam-Break Flow in Cascade Reservoirs. *J. Hydraul. Res.* **2017**, *55*, 438–444. [CrossRef]
313. Luo, J.; Xu, W.; Tian, Z.; Chen, H. Numerical Simulation of Cascaded Dam-Break Flow in Downstream Reservoir. *Proc. Inst. Civ. Eng.–Water Manag.* **2019**, *172*, 55–67. [CrossRef]
314. Kocaman, S.; Dal, K. A New Experimental Study and SPH Comparison for the Sequential Dam-Break Problem. *J. Mar. Sci. Eng.* **2020**, *8*, 905. [CrossRef]
315. Bechteler, W.; Kulisch, H.; Nujic, M. 2-D Dam-Break Flooding Waves – Comparison Between Experimental and Calculated Results. In *Floods and Flood Management*; Saul, A.J., Ed.; Springer: Dordrecht, The Netherlands, 1992; Volume 15, pp. 247–260. [CrossRef]
316. Liem, R.; Köngeter, J. The Influence of Initial Flow Conditions on the Propagation of Dam Break Wave. In Proceedings of the 28th IAHR World Congress, Graz, Austria, 22–27 August 1999. Available online: https://www.iahr.org/library/infor?pid=13676 (accessed on 3 October 2022).
317. Aureli, F.; Mignosa, P. Rapidly Varying Flows Due to Levee-Breaking. In *River Flow 2002, Proceedings of the International Conference on Fluvial Hydraulics, Louvain-la-Neuve, Belgium, 4–6 September 2002*; Bousmar, D., Zech, Y., Eds.; Balkema: Lisse, The Netherlands, 2002; Volume 1, pp. 459–466.
318. Aureli, F.; Mignosa, P. Comparison Between Experimental and Numerical Results of 2D Flows Due to Levee Breaking. In Proceedings of the 29th IAHR World Congress, Beijing, China, 16–21 September 2001; Li, G., Ed.; Tsinghua University Press: Beijing, China, 2001; Theme C, pp. 252–258.
319. Sarma, A.K.; Das, M.M. Analytical Solution of a Flood Wave Resulting from Dike Failure. *Proc. Inst. Civ. Eng.—Water Marit. Eng.* **2003**, *156*, 41–45. [CrossRef]
320. Briechle, S.; Joeppen, A.; Köngeter, J. Physical Model Tests for Dike-Break Induced, Two-Dimensional Flood Wave Propagation. In *River Flow 2004, Proceedings of the International Conference on Fluvial Hydraulics, Naples, Italy, 23–25 June 2004*; Greco, M., Carravetta, A., Della Morte, R., Eds.; Balkema: Lisse, The Netherlands, 2004; Volume 2, pp. 959–966.
321. Briechle, S.R. Die flachenhafte Ausbreitung der Flutwellenach Versagen von Hochwasserschutzeinrichtungen an Fliesgewassern [Flood Wave Generated by the Collapse of a Flood Protection]. Ph.D. Thesis, University of Aachen, Aachen, Germany, 2006. Available online: https://d-nb.info/98060897x/34 (accessed on 19 October 2022). (In German)
322. Harms, M.; Briechle, S.; Köngeter, J.; Schwanenberg, D. Dike-Break Induced Flow: Validation of Numerical Simulations and Case Study. In *River Flow 2004, Proceedings of the International Conference on Fluvial Hydraulics, Naples, Italy, 23–25 June 2004*; Greco, M., Carravetta, A., Della Morte, R., Eds.; Balkema: Lisse, The Netherlands, 2004; Volume 2, pp. 937–944.
323. Oertel, M.; Schlenkhoff, A. Flood Wave Propagation and Flooding of Underground Facilities. In *River Flow 2008, Proceedings of the International Conference on Fluvial Hydraulics, Çeşme, Izmir, Turkey, 3–5 September 2008*; Altinakar, M.S., Kokpinar, M.A., Aydin, I., Cokgor, S., Kirkgoz, S., Eds.; Kubaba Congress Department and Travel Services: Ankara, Turkey, 2008; Volume 1, pp. 595–600.
324. Oertel, M. Analyse der Flutung Unterirdischer Bauwerke in Flussnahen Urbanen Regionen Nach Versagen von Hochwasserschutzeinrichtungen [Analyzing Flooding Processes of Underground Facilities in Urban River Areas after Malfunction of Flood Protection Measures]. Ph.D. Thesis, University of Wuppertal, Wuppertal, Germany, 2008. Available online: https://www.hydro.uni-wuppertal.de/fileadmin/bauing/hydro/berichte/Nr.15_Oertel.pdf (accessed on 19 October 2022). (In German)
325. Roger, S.; Dewals, B.J.; Erpicum, S.; Schwanenberg, D.; Schüttrumpf, H.; Köngeter, J.; Pirotton, M. Experimental and Numerical Investigations of Dike-Break Induced Flows. *J. Hydraul. Res.* **2009**, *47*, 349–359. [CrossRef]
326. Sun, J.; Lu, L.; Lin, B.; Liu, L. Processes of Dike-Break Induced Flows: A Combined Experimental and Numerical Model Study. *Int. J. Sediment Res.* **2017**, *32*, 465–471. [CrossRef]
327. Al-Hafidh, I.A.I.; Calamak, M.; LaRocque, L.A.; Chaudhry, M.H.; Imran, J. Experimental Investigation of Flood Management by an Instantaneous Levee Breach. *J. Hydraul. Eng.* **2022**, *148*, 04021056. [CrossRef]
328. Yoon, K.S.; Rehman, K.; Yoo, H.J.; Lee, S.O.; Hong, S.H. Large Scale Laboratory Experiment: The Impact of the Hydraulic Characteristics of Flood Waves Caused by Gradual Levee Failure on Inundation Areas. *Water* **2022**, *14*, 1446. [CrossRef]
329. Greenspan, H.P.; Johansson, A.V. An Experimental Study of Flow over an Impounding Dike. *Stud. Appl. Math.* **1981**, *64*, 211–223. [CrossRef]
330. Sharifi, T. An Experimental Study of Catastrophic Failure of Liquid Storage Tanks. Ph.D. Thesis, Imperial College of Science and Technology, University of London, London, UK, 1987. Available online: https://spiral.imperial.ac.uk/bitstream/10044/1/46527/2/Sharifi-T-1987-PhD-Thesis.pdf (accessed on 7 February 2023).

331. Maschek, W.; Roth, A.; Kirstahler, M.; Meyer, L. *Simulation Experiments for Centralized Liquid Sloshing Motions*; Report KfK-5090; Kernforschungszentrum Karlsruhe: Karlsruhe, Germany, 1992. Available online: https://core.ac.uk/download/pdf/197569152.pdf (accessed on 7 February 2023).
332. Cleaver, R.P.; Cronin, P.S.; Evans, J.A.; Hirst, I.L. An Experimental Study of Spreading Liquid Pools. In *Institution of Chemical Engineers Symposium Series No. 148*; Institution of Chemical Engineers: Yarraville, Australia, 2001; pp. 167–179. Available online: https://www.icheme.org/media/10158/xvi-paper-13.pdf (accessed on 7 February 2023).
333. Cronin, P.S.; Evans, J.A. *A Series of Experiments to Study the Spreading of Liquid Pools with Different Bund Arrangements*; Contract Research Report 405/2002; Advantica Technologies Limited for the Health and Safety Executive: Leicestershire, UK, 2002. Available online: https://www.hse.gov.uk/research/crr_pdf/2002/crr02405.pdf (accessed on 7 February 2023).
334. Atherton, W. *An Experimental Investigation of Bund Wall Overtopping and Dynamic Pressures on the Bund Wall Following Catastrophic Failure of a Storage Vessel*; Research Report 333; Liverpool John Moores University for the Health and Safety Executive: Liverpool, UK, 2005.
335. Atherton, W. An Empirical Investigation of Catastrophic and Partial Failures of Bulk Storage Vessels and Subsequent Bund Wall Overtopping and Dynamic Pressures. Ph.D. Thesis, Liverpool John Moores University, Liverpool, UK, 2008. Available online: https://researchonline.ljmu.ac.uk/id/eprint/5866/1/446356.pdf (accessed on 7 February 2023).
336. Zhang, B.; Liu, Y.; Zhu, W.; Gopalaswami, N.; Mannan, M.S. Experimental Study of Bund Overtopping Caused by a Catastrophic Failure of Tanks. *Ind. Eng. Chem. Res.* **2017**, *56*, 12227–12235. [CrossRef]
337. Megdiche, I. Evaluation of the Mechanical Strength of Bund Walls under the Catastrophic Failure of Storage Tanks via Fluid Structure Interaction. Ph.D. Thesis, Liverpool John Moores University, Liverpool, UK, 2018. Available online: https://researchonline.ljmu.ac.uk/id/eprint/10217/7/2019MegdichePhD.pdf (accessed on 7 February 2023).
338. Zhao, S.; Huo, J.; Xu, R.; Liu, Y.; Jing, M.; Zhang, B. Prevention of Bund Overtopping after a Catastrophic Tank Failure Accident: Effects of Bund Design, Liquids and Scale-Up. *Process Saf. Environ. Prot.* **2022**, *166*, 41–56. [CrossRef]
339. Ritter, A. Die Fortpflanzung der Wasserwellen [Propagation of Water Waves]. *Z. Ver. Dtsch. Ing.* **1892**, *36*, 947–954. (In German)
340. Stoker, J.J. *Water Waves: The Mathematical Theory with Applications*; Wiley: New York, NY, USA, 1957.
341. Mignot, E.; Dewals, B. Hydraulic Modelling of Inland Urban Flooding: Recent Advances. *J. Hydrol.* **2022**, *609*, 127763. [CrossRef]
342. Mignot, E.; Li, X.; Dewals, B. Experimental Modelling of Urban Flooding: A Review. *J. Hydrol.* **2019**, *568*, 334–342. [CrossRef]
343. Buchner, B. The Impact of Green Water on FPSO Design. In *OTC 95, Proceedings of the 27th Offshore Technology Conference, Houston, Texas, 1–4 May 1995*; OTC 7698; OnePetro: Richardson, TX, USA, 1995; Volume 2, pp. 45–57. [CrossRef]
344. Schønberg, T.; Rainey, R.T.C. A Hydrodynamic Model of Green Water Incidents. *Appl. Ocean Res.* **2002**, *24*, 299–307. [CrossRef]
345. Chuang, W.-L.; Chang, K.-A.; Mercier, R. Review of Experimental Modeling of Green Water in Laboratories. In *ISOPE 2019, Proceedings of the 29th International Ocean and Polar Engineering Conference, Honolulu, HI, USA, 16–21 June 2019*; Chung, J.S., Akselsen, O.M., Jin, H.W., Kawai, H., Lee, Y., Matskevitch, D., Van, S.H., Wan, D., Wang, A.M., Yamaguchi, S., Eds.; International Society of Ocean and Polar Engineers: Cupertino, CA, USA, 2019; Volume 3, pp. 2466–2472.
346. Henderson, F.M. *Open Channel Flow*; Macmillan: New York, NY, USA, 1966; pp. 491–493.
347. Heller, V. Scale Effects in Physical Hydraulic Engineering Models. *J. Hydraul. Res.* **2011**, *49*, 293–306. [CrossRef]
348. Ye, Z.; Zhao, X.; Deng, Z. Numerical Investigation of the Gate Motion Effect on a Dam Break Flow. *J. Mar. Sci. Technol.* **2016**, *21*, 579–591. [CrossRef]
349. FEMA. *Federal Guidelines for Inundation Mapping of Flood Risks Associated with Dam Incidents and Failures*; FEMA P-946; U.S. Department of Homeland Security: Washington, DC, USA, 2013. Available online: https://www.fema.gov/sites/default/files/2020-08/fema_dam-safety_inundation-mapping-flood-risks.pdf (accessed on 23 January 2023).
350. NZSOLD. *New Zealand Dam Safety Guidelines*; New Zealand Society on Large Dams (NZSOLD): Wellington, New Zealand, 2015. Available online: https://nzsold.org.nz/wp-content/uploads/2019/10/nzsold_dam_safety_guidelines-may-2015-1.pdf (accessed on 23 January 2023).
351. Vischer, D.L.; Hager, W.H. *Dam Hydraulics*; Wiley: Chichester, UK, 1998; pp. 287–288.
352. Tan, T.; Ma, Y.; Zhang, J.; Niu, X.; Chang, K.-A. Experimental Study on Flow Kinematics of Dam-Break Induced Surge Impacting onto a Vertical Wall. *Phys. Fluids* **2023**, *35*, 025127. [CrossRef]
353. De Marchi, G. Sull'Onda di Piena che Seguirebbe al Crollo della Diga di Cancano. Prove su Modello per il Tronco Fluviale dalla Diga a Ponte Cepina–Calcolo per il Tronco da Ponte Cepina a Tirano [On the Dam-Break Wave Resulting from the Collapse of the Cancano Dam. Physical Model for the River Reach from the Dam to Ponte Cepina–Numerical Model for the River Reach from Ponte Cepina to Tirano]. *L'Energia Elettr.* **1945**, *22*, 319–340. (In Italian)
354. FLOW-3D Modeling Capabilities. Hybrid Shallow Water/3D Flow. Available online: https://www.flow3d.com/modeling-capabilities/hybrid-shallow-water-3d-flow/ (accessed on 10 March 2023).
355. FLOW-3D Modeling Capabilities. Moving Objects. Available online: https://www.flow3d.com/modeling-capabilities/moving-objects/ (accessed on 10 March 2023).

water

MDPI

Review

Three-Dimensional Numerical Modelling of Real-Field Dam-Break Flows: Review and Recent Advances

Andrea Maranzoni [1,*] and Massimo Tomirotti [2]

[1] Department of Engineering and Architecture, University of Parma, 43124 Parma, Italy
[2] Department of Civil, Environmental, Architectural Engineering and Mathematics, University of Brescia, 25121 Brescia, Italy; massimo.tomirotti@unibs.it
* Correspondence: andrea.maranzoni@unipr.it

Abstract: Numerical modelling is a valuable and effective tool for predicting the dynamics of the inundation caused by the failure of a dam or dyke, thereby assisting in mapping the areas potentially subject to flooding and evaluating the associated flood hazard. This paper systematically reviews literature studies adopting three-dimensional hydrodynamic models for the simulation of large-scale dam-break flooding on irregular real-world topography. Governing equations and numerical methods are analysed, as well as recent advances in numerical techniques, modelling accuracy, and computational efficiency. The dam-break case studies used for model validation are highlighted. The advantages and limitations of the three-dimensional dam-break models are compared with those of the commonly used two-dimensional depth-averaged ones. This review mainly aims at informing researchers and modellers interested in numerical modelling of dam-break flow over real-world topography on recent advances and developments in three-dimensional hydrodynamic models so that they can better direct their future research. Practitioners can find in this review an overview of available three-dimensional codes (research, commercial, freeware, and open-source) and indications for choosing the most suitable numerical method for the application of interest.

Keywords: dam break; flooding; numerical modelling; real-world topography; review; three-dimensional models

check for updates

Citation: Maranzoni, A.; Tomirotti, M. Three-Dimensional Numerical Modelling of Real-Field Dam-Break Flows: Review and Recent Advances. *Water* **2023**, *15*, 3130. https://doi.org/10.3390/w15173130

Academic Editor: Anargiros I. Delis

Received: 22 June 2023
Revised: 25 August 2023
Accepted: 27 August 2023
Published: 31 August 2023

1. Introduction

Dam-break floods are caused by the uncontrolled release of water stored in a reservoir due to a total or partial collapse of a constructed or natural dam. Such phenomena can result in potentially catastrophic consequences and many more casualties than other kinds of floods and disasters (e.g., [1,2]). Zhang et al. [3] documented 1443 failures of constructed dams and 1044 failures of landslide dams worldwide over the past two centuries and provided a list of the 20 most significant dam failures (each causing more than 500 casualties), resulting in more than 44,000 fatalities. Among these 20 selected dam disasters, the most catastrophic one is the 1975 Banqiao dam failure (China), which led to the inundation of an area of approximately 12,000 km^2 and the loss of more than 26,000 lives, followed by the disasters of the Vajont dam (Italy) in 1963, the South Fork dam (US) in 1889, and the Machhu-II dam (India) in 1979, each with more than 2000 casualties [1,3,4]. In addition, dam-break flooding can cause huge economic losses and extensive environmental damage [5]. For example, the 1976 Teton dam disaster (US) resulted in only 11 fatalities, but the flood covered an area of 77 km^2 and reached up to 250 km downstream of the dam, causing more than USD 400 million of damage [6].

Accordingly, the assessment of the flooding hazard associated with hypothetical dam-break scenarios has gained great importance and attracted considerable attention in the last decades, both in engineering practice (e.g., [7,8]) and research (e.g., [9–12]), for dam-break flood risk management, emergency response, and flood hazard mitigation planning.

In principle, a dam-break problem can be studied using either an experimental [13] or a mathematical modelling approach [14,15], or both together (e.g., [16]). However, numerical modelling has become attractive and increasingly popular due to its flexibility and cost-effectiveness and the continuous growth of computing capacities, which allows the processing of a large amount of data and solving mathematical models of great complexity and significant predictive capability [17]. Moreover, the current availability of several commercial and freeware Computational Fluid Dynamics (CFD) software codes, equipped with friendly interfaces and including effective pre- and post-processing tools, facilitates the diffusion of numerical models in flood hazard analyses [18]. The high-cost and time-consuming complex operations required to carry out accurate laboratory experimentation have determined that the physical modelling approach is currently mainly adopted to obtain experimental data useful for the validation of numerical models [13,14].

Usually, dam-break flood hazard assessment and mapping are performed using depth-averaged two-dimensional (2D) models, which solve the 2D shallow water equations (SWEs) through finite difference, finite volume, or finite element methods (e.g., [15,19–21]) to predict relevant hydraulic quantities associated with flood hazard (namely, maximum flood depth and velocity). Despite the simplifying assumptions underlying the 2D SWEs, there are many computational challenges in numerically solving these equations, such as shock-capturing capability, treatment of wet and dry fronts, treatment of bottom and friction source terms, reproduction of flow regime transitions, and preservation or achievement of stationary or steady-flow conditions. In any case, the SWEs strictly apply to flows with no significant curvature of the free surface and negligible vertical acceleration (and hence with nearly hydrostatic pressure distribution) over small bottom slopes (e.g., [21,22]). However, in gravity-driven geophysical flows, the terrain may be very steep, at least locally, especially in mountain regions or near topographic singularities [23–25]. Moreover, the hydrostatic pressure assumption is violated in curvilinear flows, which can also occur as a result of dam failures, especially during the first stages of the motion (e.g., [26,27]) or in case the flooding wave propagates in the presence of bends (e.g., [28]), contractions (e.g., [29,30]), bottom singularities (e.g., [31,32]), or obstacles and structures (e.g., [33–35]). For example, Figure 1 shows two pictures of the impact of a dam-break wave against a prismatic block [34]. Such a physical process is characterised by marked three-dimensional (3D) features.

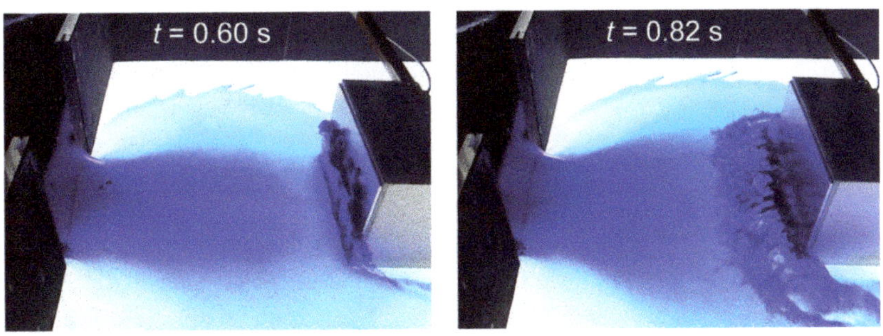

Figure 1. Three-dimensional features of a dam-break wave impacting a structure. The pictures were taken during the laboratory investigation performed by Aureli et al. [34]. Time t starts from the sudden gate removal.

Local 3D effects cannot be reproduced by the 2D SWEs [36], with consequent possible limitations in the predictive capability of 2D shallow-water models and inaccuracies in the prediction of the relevant hydraulic variables, such as flood inundation extent, maximum flood depths, and impact loads on structures [37]. More general formulations of the 2D depth-averaged SWEs have been proposed in the literature to overcome this drawback without resorting to more computationally expensive 3D models. In these enhanced formulations, some of the restrictive assumptions of the classic shallow-water model are relaxed while retaining its robustness and simplicity. For example, steep-slope SWEs (SSSWEs) were introduced to simulate shallow flows over steep terrain (e.g., [24,25,38]). Boussinesq-type models (e.g., [23,39,40]), as well as vertically averaged and moment (VAM) equations (e.g., [41]) or depth-averaged equations incorporating an "enhanced" gravity (e.g., [42,43]), can be used to simulate non-hydrostatic flows, preserving the vertical momentum balance and including the effect of the vertical flow acceleration.

Recent advances in computing performance have fostered the application of 3D numerical models [44], which offer an improved predictive capability [45] and a more accurate description of the flow features, especially where vertical flow acceleration cannot be neglected and the pressure distribution is far from hydrostatic [19]. Three-dimensional models with different degrees of complexity (and, consequently, different computational costs) can be used, namely [40] direct numerical simulation (DNS) models, which numerically solve the Navier–Stokes equations (NSEs), resolving all turbulence spatial and temporal scales; large-eddy simulation (LES) models, which solve the filtered NSEs, thus ignoring the smallest length scales; and models that solve the Reynolds-averaged Navier–Stokes equations (RANS), coupled with a closure turbulence model [46]. DNS and LES models are significantly more computationally expensive than the RANS ones, especially at the high Reynolds numbers encountered in environmental free-surface flows. Therefore, the application of DNS and LES models in this field is feasible only in limited domains [19]. Consequently, RANS solvers are the most common choice for large-scale real-world dam-break flood simulations. However, since the flow depth does not explicitly appear in the basic equations of 3D models, great care has to be devoted to spatial discretization in the vertical direction, and additional computational effort is required for free surface tracking [37]. Several free-surface-tracking techniques have been proposed in the literature. The best known are the Volume of Fluid (VOF; [47]) and the Level Set [48] methods (or a combination of the two in coupled Level Set–VOF methods; e.g., [49]). The VOF model is the most popular in dam-break flow simulations [50]. An alternative strategy to solve the 3D governing equations, avoiding the construction of a computational mesh and the adoption of complex free-surface-tracking algorithms, is based on meshless particle-based methods, such as Smoothed Particle Hydrodynamics (SPH; [51]), which has also been applied in the modelling of dam-break flow (e.g., [52,53]), even on real-world topography, and other environmental applications [54,55].

However, the high computational cost of 3D models is still a significant limitation, especially in large-scale field studies or when high spatial accuracy is required. This limitation has hindered the diffusion of 3D models in the past for real-field applications, favouring the 2D depth-averaged ones. Hence, high-performance computing, such as parallel or Graphics Processing Unit (GPU) computing, is today a valuable support to enhance the computational efficiency of 3D models and reduce computational time.

Dam-break models (both 2D and 3D) are usually validated against experimental data of laboratory test cases [13], which, however, are typically schematic (e.g., [44,56]) or include isolated singularities (e.g., [50,57–59]). Instead, real dam-break events involve irregular topography and constitute far more challenging benchmarks for numerical models. However, well-documented historical dam-break events are scarce and often characterised by uncertain available information [14], making their use arduous for complete validation of the numerical models (especially for 3D ones, in which several parameters are involved). Nonetheless, they provide valuable information about the different degrees of reliability of the models in reproducing dam-break flow features over real-world topography.

Table 1 provides a comparative summary of the advantages and shortcomings of 3D and depth-averaged 2D models. Given the increasing diffusion of 3D CFD models and their improved predictive capabilities compared with the 2D ones, this review focuses on 3D modelling of large-scale real-field dam-break floods.

Table 1. Summary and comparison of the advantages and shortcomings of 3D and 2D models.

Model	Advantages	Shortcomings
2D depth-averaged	Easy to build and implement Computationally cheap Few parameters to calibrate (roughness) Robust and stable	Limitations due to the shallow-water assumptions (hydrostatic distribution of pressure and small bottom slopes)
3D	High accuracy (mild restrictive assumptions) Reproduction of non-hydrostatic effects	Laborious to build and implement Complex calculations Computationally expensive Several parameters involved

Figure 2 shows a sketch of the methodological framework of this study. The systematic review is restricted to publications focusing on 3D CFD modelling of dam-break flow and providing application examples of numerical simulation of dam-break flooding over irregular real-world topography. A comparative analysis is performed on the documents collected, and the information relevant to 3D CFD modelling is reported in tabular form (Section 2). The results of the analysis allow statistical information on key items of the reviewed documents (such as year of publication, software status, model type, and numerical scheme) to be obtained. The discussion of the review findings is centred on the following aspects: improvements in simulation accuracy, model validation and calibration, improvements in computational efficiency, and improvements in result visualization (Section 3). Finally, conclusions are drawn, outlining implications that can facilitate practical applications and future research on 3D CFD models for the simulation of large-scale dam-break flooding on irregular real-world topography (Section 4).

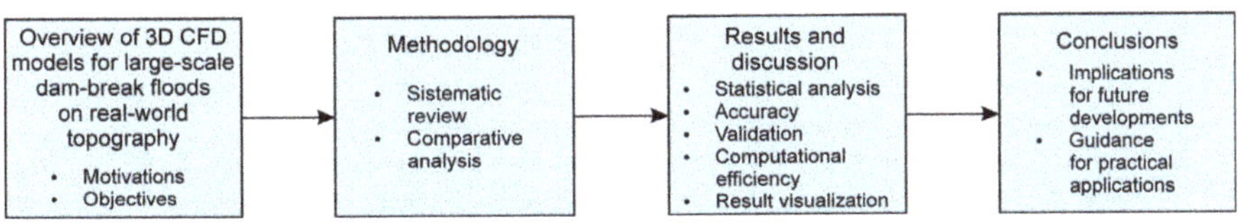

Figure 2. Schematic of the methodological framework of this review study.

1.1. Motivations of the Present Review

There is a vast literature overviewing, describing, comparing, and evaluating flood inundation numerical models for flood risk assessment and water resources management (e.g., [15,18,19]). For instance, Teng et al. [15] conducted a comprehensive review of different modelling approaches (empirical, hydrodynamic, and simplified conceptual models), highlighting their advantages and drawbacks and discussing the sources of uncertainty in flood inundation modelling. Mudashiru et al. [18] provided a similar overview, focusing on flood hazard mapping. Bates [19] surveyed recent advances in floodplain inundation modelling, discussing the ability of the numerical models to reproduce the physical aspects of the flooding process. Luo et al. [60] and Mignot and Dewals [61] limited the area of focus, reviewing specifically urban flood simulation models and analysing their hydrologic and hydrodynamic component modules. More recently, Kumar et al. [62] and Avila-Aceves et al. [63] offered a broad overview of different modelling approaches and numerical techniques for simulating large-scale inundations, providing insight into their strengths and weaknesses.

Only a few of the previously mentioned reviews touch on the use of 3D hydrodynamic models for flood inundation simulations [15,19]. However, to our knowledge, no literature reviews exist on the specific topic of 3D hydrodynamic modelling of large-scale dam-break flows on real-world topography. Recent advances in 3D dam-break modelling, the growing attention given to 3D CFD models in recent times due to enhanced computational efficiency and the prospect of using 3D models extensively in engineering practice in the near future for dam-break flood risk assessment are the main motivations for this review.

1.2. Objectives of the Present Review

This review aims to inform researchers new to the field and more experienced researchers of the recent advances in 3D CFD dam-break modelling on real-world topography, thereby enabling them to have a general, updated overview of this topic. The comparative analysis of the relevant references selected can help modellers, practitioners, emergency response agencies, and dam owners remain up to date on the latest developments in dam-break simulation tools and choose the most suitable numerical model for the specific application of interest. Finally, this review study provides state-of-the-art information to direct future research on 3D dam-break numerical modelling for real field applications.

2. Review and Comparative Analysis

A systematic and careful search was performed in the most well-known scientific databases (based on the keywords: "dam-break", "3D numerical modelling", "real topography", "complex topography", and similar) to ensure wide coverage of the existing literature on 3D numerical modelling of large-scale dam-break flows over real-world topography. Mainly the academic literature (i.e., journal or conference articles) was considered; the grey literature was not included.

The documents retrieved in the literature survey are arranged in Table 2, including studies on dam spillway flows or overtopping flows reporting the numerical simulation of the subsequent flood propagation downstream. Table entries are organised in chronological order. References that contain multiple real-world case studies are repeated for each of these and appear in different rows of the table (i.e., one row for each case study).

Table 2 provides the following, most significant information.

(1) *Relevant references retrieved.* Multiple references are reported in the same table row when details providing a complete description of the case study analysed can be obtained from various articles. In the case of duplicate studies, the most complete one was considered, and subordinately, the earliest one.

(2) *Code/software name,* if available. Besides well-known commercial (e.g., TELEMAC-3D [64]; FLOW-3D [65]) or open-source CFD software codes (e.g., OpenFOAM [66]; DualSPHysics [67]) widely used in many hydrodynamic applications, numerical codes (sometimes with no name) developed by universities or research centres for research purposes (e.g., [68]) appear in the table.

(3) *Basic equations of the simulation models*: in order of increasing complexity, the Euler equations, the Navier–Stokes equations, and the RANS equations for incompressible flow. In the last case, a closure turbulence model is coupled with the governing equations. Lattice Boltzmann methods (based on the Boltzmann equation and simulating the flow through collision models of fictitious particles moving on a discrete lattice grid) are used more rarely in dam-break flooding simulations. The table also reports the rheological model coupled with the hydrodynamic equations in mudflow modelling of tailings dam breaks and the erosional model used in the analysis of geomorphic dam-break flows over an erodible bottom.

(4) *Numerical methods* used for spatial and temporal discretization of the model equations. The indication of the numerical scheme used for the simulations is accompanied by the specification of the discretization technique (finite volume, finite element, SPH, etc.).

(5) *Dam-break case studies* considered to demonstrate the applicability of the numerical models. Some case studies concern historical dam-break events, while others concern hypothetical dam failures. The former case studies, if well documented, can be used for model validation and to evaluate the model's ability to reproduce real-field dam-break flooding. Case studies of the latter type concern model applications to real-world situations, mainly aimed at assessing dam-break flood hazards in potentially floodable areas.

(6) *Details about the computational domain and the spatial resolution* adopted in the numerical simulation of the case studies considered (i.e., the number of grid cells in mesh-based models or fluid particles in meshless models). The mesh type (structured or unstructured) and the shape of the grid elements are specified for mesh-based models.

(7) *Outcomes of the numerical modelling*. Typically, model results include flooded areas, flood depth contour maps and flow velocity fields at selected times, contour maps of the maximum values of flood depth and velocity magnitude, time series of flood depth and flow velocity at given locations, and time series of flow discharge at selected cross-sections.

(8) *Focus of the studies*. The focus may be on the model validation, the prediction or reconstruction of the inundation dynamics, or the 3D effects due to flow curvature. Different aspects are sometimes examined contextually.

(9) *Computational efficiency*. The simulation (physical) time and the corresponding computational run-time are reported for the case studies (when these data are available), along with strategies implemented to improve computational efficiency.

(10) *Publication year*. This bibliographic information is useful to place each contribution in time, outlining the evolution of the numerical models over time and research trends. If multiple references are associated with a single table row, the publication year of the oldest one is reported.

(11) *Status of the CFD model*. This can be commercial, freeware, open-source, or research.

Table 2. Overview of studies on 3D hydrodynamic modelling of large-scale dam-break flooding over real-world topography.

(1) Reference	(2) Model Name	(3) Model Type [1]	(4) Numerical Method [2]	(5) Case Study	(6) Computational Domain and Elements	(7) Output Data	(8) Focus of the Study	(9) Computational Efficiency [3]	(10) Year	(11) Status
Roubtsova and Kahawita [69] *	N/A	Navier–Stokes and continuity equations	Weakly compressible SPH	Historical 1963 overtopping of the Vajont dam (Italy) Volume of the rockslide: 270 million m^3 Stored water volume: 115 million m^3 (reservoir water level provided) Overtopping water volume: 30 million m^3	Modelled area extent: N/A (the reservoir and the Vajont River downstream) Number of particles: N/A Particle spacing: N/A	Water surface at selected times; transverse water surface profiles at a selected cross-section in the reservoir	Performance of the numerical technique Reconstruction of the event Comparison with field observations	Simulation time: 220 s Run time/simulation time: ~74	2006	Research
Cleary et al. [70]; Prakash et al. [71]	N/A	Navier–Stokes and continuity equations	Weakly compressible SPH	Historical 1928 St. Francis dam break (California) Water volume: 47 million m^3	Modelled area extent: the reservoir and a valley stretch downstream of the dam Number of particles: 1.4×10^6 Particle spacing: 4 m	Flow fields (velocity magnitude) and flooded areas at selected times; motion of wall fragments; flow discharge hydrograph at the dam site; flood arrival times and maximum flood depths at selected locations	Flooding dynamics for different collapse scenarios Comparison with field data Modelling of the motion of dam wall blocks 3D effects Sensitivity on particle resolution (4 m; 6 m; 8 m)	Simulation time: 25 min Run time/simulation time: N/A	2010	Research
Cleary et al. [70]; Ye et al. [72]; Cleary et al. [73]	N/A	Navier–Stokes and continuity equations	Weakly compressible SPH	Hypothetical Geheyan dam break (China) Water volume: 3.12 billion m^3 Different dam failure scenarios	Modelled area extent: the reservoir and a valley stretch downstream of the dam Number of particles: 1.3×10^6 (fluid), 1.9×10^6 (boundaries) Particle spacing: 15 m (fluid), 30 m (boundaries)	Flow fields (velocity magnitude) and flooded areas at selected times (different views); discharge hydrograph at the dam site; flow discharge hydrographs at selected sections; flood depth hydrographs at selected locations	Flooding dynamics 3D effects Effect of different dam failure scenarios Modelling of the motion of dam wall blocks	Simulation time: 60 min Run time/simulation time: N/A	2010	Research
Lee et al. [74] *	N/A	Navier–Stokes and continuity equations	Weakly compressible and truly incompressible SPH	Ski-jump spillway of the Goulours dam (France)	Modelled area extent: The reservoir (assumed to be of prismatic shape) and ~250 m-long valley reach downstream of the dam (according to a 1:20 scale physical model) Number of particles: 9.366×10^5 (wall particles: 2.169×10^5; fictitious particles: 2.196×10^5) Particle spacing (initial): 0.2 m	Spillway flow dynamics; flooded areas at selected times	Qualitative reconstruction of the spillway process Spillway flow features	Simulation time: 16 s Run time/simulation time: ~2.7×10^4	2010	Research

Table 2. *Cont.*

(1) Reference	(2) Model Name	(3) Model Type [1]	(4) Numerical Method [2]	(5) Case Study	(6) Computational Domain and Elements	(7) Output Data	(8) Focus of the Study	(9) Computational Efficiency [3]	(10) Year	(11) Status
Caboussat et al. [75]	N/A	Incompressible Navier–Stokes equations coupled with VOF	Finite element; implicit time splitting scheme (advection and diffusion steps)	Historical 1959 Malpasset dam break (France) Water volume: 50 million m^3	Modelled area extent: 17.5 km × 9 km Unstructured grid of tetrahedral cells (diffusion step) Number of cells: 1.716×10^6 Spatial resolution: 5 m Structured grid of cubic cells (advection step) Number of cells: N/A Spatial resolution: 2 m	Flooded areas and flow velocity fields at selected times; maximum flood depths and arrival times at selected points	Comparison with physical model data	Simulation time: >8 min Run time/ simulation time: 600	2011	Research
Caboussat et al. [75]	N/A	Incompressible Navier–Stokes equations coupled with VOF	Finite element; implicit time splitting scheme (advection and diffusion steps)	Hypothetical Grande-Dixence dam break (Switzerland) Water volume: 400 million m^3	Modelled area extent: 28.9 km × 5.75 km Unstructured grid of tetrahedral cells (diffusion step) Number of cells: 13.876×10^6 Spatial resolution: 50 m Structured grid of cubic cells (advection step) Number of cells: N/A Spatial resolution: 10 m	Flooded areas and flow velocity fields at selected times; flood depth contour maps at selected times	Inundation dynamics	Simulation time: 10 min Run time/ simulation time: N/A	2011	Research
Vassilevski et al. [76]	N/A	Incompressible Navier–Stokes equations coupled with grid level set function (for free surface tracking); Herschel–Bulkley rheological relation for viscoplastic fluids	Finite difference/ finite volume; Chorin–Temam-Yanenko time splitting scheme	Hypothetical Sayano–Shushenskaya partial dam-break (Russia) Water volume: N/A	Modelled area extent: the reservoir and a valley stretch downstream of the dam Structured octree staggered grid	Flood depth hydrographs at selected points; time series of the bottom pressure at the base of the spillway	Dam-break flow	Simulation time: 100 s Run time/ simulation time: N/A	2012	Research
Vacondio et al. [77] *	DualSPHysics	Navier–Stokes and continuity equations	Weakly compressible SPH	Historical 1963 overtopping of the Vajont dam (Italy) Volume of the rockslide: 310 million m^3 Stored water volume: N/A (reservoir water level provided)	Modelled area extent: the reservoir and a valley stretch downstream of the dam Number of particles: 3.954×10^6 (bottom particles: 2.144×10^6; rockslide particles: 1.274×10^5; fluid particles: 1.683×10^6) Particle size: 5 m	Water surface elevation at selected times; maximum run-up on the reservoir side; water surface elevation in the residual lake; flow velocity field; overflow hydrograph	Reconstruction of the wave generated by the Vajont rockslide and of the dam-overtopping phenomenon Comparison with field observations	Simulation time: 21 min Run time/ simulation time: 177 Parallelization on GPU	2013	Open-source

Table 2. *Cont.*

(1) Reference	(2) Model Name	(3) Model Type [1]	(4) Numerical Method [2]	(5) Case Study	(6) Computational Domain and Elements	(7) Output Data	(8) Focus of the Study	(9) Computational Efficiency [3]	(10) Year	(11) Status
Zhainakov and Kurbanaliev [78]; Jainakov et al. [79]	OpenFOAM	RANS coupled with VOF; standard k-ε turbulence model	Finite volume; PIMPLE algorithm; explicit Euler first-order time discretization method	Hypothetical Andijan dam break (Uzbekistan) Water volume: N/A	Modelled area extent: 6 km × 4 km Structured mesh of hexahedral cells Number of cells: 120 × 120 × 80	Contour maps of the water volume fraction at selected times	Flood wave propagation	Simulation time: 240 s Run time/simulation time: 135	2013	Open-source
Zhainakov and Kurbanaliev [78]; Jainakov et al. [79]	OpenFOAM	RANS coupled with VOF; standard k-ε turbulence model	Finite volume; PIMPLE algorithm; explicit Euler first-order time discretization method	Hypothetical Papan dam break (Kyrgyzstan) Water volume: N/A	Modelled area extent: 5 km × 5 km Structured mesh of hexahedral cells Number of cells: 50 × 60 × 30	Contour maps of the water volume fraction at selected times	Flood wave propagation	Simulation time: 260 s Run time/simulation time: 69	2013	Open-source
Džebo et al. [80]	Tis Isat	Navier–Stokes and continuity equations	Weakly compressible SPH	Hypothetical break of the embankment of the reservoir of the Kolarjev vrh pumped-storage hydropower plant (Slovenia) Water volume: 3.1 million m^3	Modelled area extent: the reservoir and a 4.5 km long valley stretch downstream Number of particles: (a) 21.890 × 10^3; (b) 174.884 × 10^3 Particle size: (a) 5 m; (b) 2.5 m	Water surface elevation at selected times; transverse water surface profiles at given cross-sections; flow depth hydrographs at selected gauge points	Flooding dynamics Comparison with 2D depth-averaged model predictions and physical model experimental data Effects of different bottom roughness values and spatial resolutions	Simulation time: 200 s Run time/simulation time: (a) 30; (b) 981	2014	Research
Marsooli and Wu [50]	N/A	RANS coupled with VOF; Smagorinsky eddy viscosity turbulence model	Finite volume; PISO algorithm; CICSAM scheme	Flash flood in the Toce River 1:100 physical model (Italy) Controlled impulsive inflow	Modelled area extent: 5 km long reach of the Toce River (5.5 km × 1.2 km) Unstructured mesh of hexahedral cells Number of cells: 3.1 × 10^6	Flow-depth hydrographs at selected points	Comparison with experimental physical model data Comparison with 2D depth-averaged model predictions	Simulation time: 3 min Run time/simulation time: 4100	2014	Research
Zhou et al. [81]	N/A	RANS coupled with VOF; k-ε turbulence model	Finite volume; PISO algorithm	Flash flood in the Toce River 1:100 physical model (Italy) Controlled impulsive inflow	Modelled area extent: 5 km long reach of the Toce River (50 m × 11 m in the physical model; presence of an idealised urban district) Structured mesh of prismatic cells Number of cells: 8.904 × 10^3 Spatial resolution: 1 m (horizontal) 10^{-2} m (vertical)	Water depth hydrographs at selected gauge points	Validation (comparison with physical model experimental data)	Simulation time: 60 s Run time/simulation time: N/A	2014	Research

Table 2. *Cont.*

(1) Reference	(2) Model Name	(3) Model Type [1]	(4) Numerical Method [2]	(5) Case Study	(6) Computational Domain and Elements	(7) Output Data	(8) Focus of the Study	(9) Computational Efficiency [3]	(10) Year	(11) Status
Zhou et al. [81]	N/A	RANS coupled with VOF; k-ε turbulence model	Finite volume; PISO algorithm	Hypothetical Dongwushi dam break (China) Water volume: 161.5 million m^3 (Hypothetical dam break of four other dams in the same Haihe River basin, China)	Modelled area extent: upper reach of the valley of the Fuyang River Unstructured mesh of hexahedral cells Number of cells: 79.513×10^3	VOF spatial distribution at selected times; flow velocity field at selected times; flow discharge hydrographs at selected cross-sections (including the dam site); flood depth spatial distribution at selected times	Flood wave propagation Dam-break risk analysis	Simulation time: ~47 h Run time/ simulation time: N/A	2014	Research
Biscarini et al. [82]	OpenFOAM	RANS coupled with VOF; k-ε turbulence model	Finite volume; PISO algorithm; MULES scheme	Historical 1959 Malpasset dam break (France) Water volume: 50 million m^3	Modelled area extent: 17.5 km × 10 km Unstructured mesh Number of cells: 2.203×10^6 Spatial resolution: N/A	Arrival time at selected points; flood hydrographs at selected points; flooded area at selected times; transverse free surface profiles at a river bend cross-sections; velocity fields in selected areas	Comparison with experimental (field and physical model) data 3D effects at sharply curved river bends	Simulation time: 40 min Run time/ simulation time: N/A	2016	Open-source
TELEMAC Modelling System [83]	TELEMAC-3D	Navier–Stokes and continuity equations (Boussinesq approximation)	Finite element; three-fractional-step algorithm	Historical 1959 Malpasset dam break (France) Water volume: 50 million m^3	Modelled area extent: 17 km × 9 km Unstructured horizontal mesh of triangular elements Number of cells: (a) 26×10^3; (b) 104×10^3 Vertical mesh: 2 or 6 layers regularly spaced in the vertical direction	Flood depth contour maps at selected times; flood depth hydrographs at selected locations	Flood wave propagation	Simulation time: 4000 s Run time/ simulation time: N/A	2016	Freeware
Amicarelli et al. [84]	SPHERA	Continuity equations of the fluid and solid incompressible phases + volume balance equation; momentum equations of the fluid and solid phases; momentum equation for the mixture	Weakly-compressible SPH; Leapfrog time integration scheme	Erosional dam-break demonstrative ICOLD benchmark Water volume: N/A (reservoir water level provided)	Modelled area extent: 24.627 km × 9.855 km Mobile bottom downstream of the dam (granular material of fixed characteristics) Number of particles: 6.8×10^5 Particle spacing: 4 m	3D distribution of the particles and velocity fields at selected times; maps of maximum values of mixture depth and specific flow rate; water and bed-load flow rate and mixture depth hydrographs at selected cross-sections	Dynamics of the phenomenon	Simulation time: 25 min Run time/ simulation time: N/A	2017	Open-source

Table 2. *Cont.*

(1) Reference	(2) Model Name	(3) Model Type [1]	(4) Numerical Method [2]	(5) Case Study	(6) Computational Domain and Elements	(7) Output Data	(8) Focus of the Study	(9) Computational Efficiency [3]	(10) Year	(11) Status
Wang et al. [85]	N/A	RANS coupled with VOF; k-ε turbulence model	Finite volume; PISO algorithm	Flash flood in the Toce River 1:100 physical model (Italy) Controlled impulsive inflow	Modelled area extent: 5 km long reach of the Toce River (50 m × 11 m in the physical model; two idealised urban district configurations) Unstructured mesh of polyhedral cells Number of cells: ~2×10^5 Spatial resolution: 0.1 m	Computational time; computational error; water depth hydrographs at selected gauge points	Validation (comparison with physical model experimental data) Dam-break flooding of an urban area Comparison of computational performance of different mesh types (polyhedral, tetrahedral, hexahedral)	Simulation time: 60 s Run time/simulation time: ~20	2017	Research
Wang et al. [85]	N/A	RANS coupled with VOF; k-ε turbulence model	Finite volume; PISO algorithm	Hypothetical dam break of an urban reservoir (SZ City, China) Water volume: 94 million m^3	Modelled area extent: 40.12 km^2 area Unstructured mesh of polyhedral cells Number of cells: 4.229×10^6	VOF spatial distribution at selected times; flood depth and flow velocity hydrographs at selected sites in the urban area; velocity and vorticity fields; maximum flood depth and flow velocity contour maps	Dam-break flooding of an urban area	Simulation time: 5 h Run time/simulation time: N/A	2017	Research
Wang et al. [86]	DualSPHysics	Navier–Stokes and continuity equations	Weakly compressible SPH	Historical 2015 Fundão tailings dam break (Brazil) Released tailings volume: 32 million m^3	Modelled area extent: N/A (the pond and the area around) Number of particles: 2.988×10^6 (fluid) 18.132×10^6 (boundaries) Particle spacing: 3 m	Flow fields (velocity magnitude) and flooded areas at selected times; flow depth, velocity, and impact pressure time series at a selected location	Tailings flow dynamics Comparison with field data	Simulation time: 30 min Run time/simulation time: N/A Parallelization on GPU	2018	Open-source
Wang et al. [86]	DualSPHysics	Navier–Stokes and continuity equations	Weakly compressible SPH	Hypothetical dam break of an operating overhead tailings pond (China) Pond capacity: 33 million m^3	Modelled area extent: N/A (the pond and the area around) Number of particles: 4.463×10^6 (fluid) 3.9×10^6 (boundaries) Particle spacing: 3 m	Flow fields (velocity magnitude) and flooded areas at selected times; flow depth, velocity, and impact pressure time series at a selected location	Tailings flow dynamics	Simulation time: 10 min Run time/simulation time: N/A Parallelization on GPU	2018	Open-source

Table 2. *Cont.*

(1) Reference	(2) Model Name	(3) Model Type [1]	(4) Numerical Method [2]	(5) Case Study	(6) Computational Domain and Elements	(7) Output Data	(8) Focus of the Study	(9) Computational Efficiency [3]	(10) Year	(11) Status
Zhang et al. [87]	N/A	Navier–Stokes and continuity equations (Boussinesq approximation)	Finite element; θ time-stepping method	Hypothetical dike-break flooding on a realistic topography (fixed inflow velocity)	Modelled area extent: 100 m × 100 m Unstructured horizontal mesh of triangular elements; spatial resolution: 5 m Vertical mesh: 1 layer Unstructured mesh of tetrahedral cells Number of cells: 3.114×10^3	Velocity fields and flooded areas at selected times	Flood wave propagation 3D effects	Simulation time: 7 min Run time/simulation time: N/A	2018	Research
Zhang et al. [87]	N/A	Navier–Stokes and continuity equations (Boussinesq approximation)	Finite element; θ time-stepping method	Hypothetical dike-break flooding in a realistic urban area (flow velocity through the breach: 0.1 m/s)	Modelled area extent: 5.495 km × 2.5 km Unstructured horizontal mesh of triangular elements; spatial resolutions: (a) 30–50 m; (b) 60–100 m; (c) 120–200 m Vertical mesh: 1 layer Unstructured mesh of tetrahedral cells Number of cells: (a) 9376; (b) 3024; (c) 816	Flooded areas, flood depth contour maps, and velocity fields at selected times; flood depth hydrographs at selected points	3D effects Sensitivity to the mesh resolution	Simulation time: 6 h Run time/simulation time: (a) 17.5; (b) 3.5; (c) 0.3	2018	Research
Chen et al. [88]	LS-DYNA	Navier–Stokes and continuity equations; material with fluid-elastoplastic properties	SPH	Historical 1985 Stava tailings dam break (Italy) Released tailings volume: 185×10^3 m^3	Modelled area extent: the pond and a 4.2 km long stretch of the valley Number of particles: 11.119×10^3 (fluid) Particle spacing: 2.5 m	Flow fields (velocity magnitude) and flooded areas at selected times; average velocity profile of the debris flow front; velocity field near the check dam; arrival time at an observation point; final deposition zones; impact force hydrographs (considering a single or multiple check dams)	Debris flow dynamics Fluid–structure interactions Comparison with other numerical results Effect of the presence of hypothetical check dams (rigid indestructible dams or concrete destructible dams) placed at different positions	Simulation time: N/A Run time/simulation time: N/A	2019	Commercial
Kurbanaliev et al. [89]	OpenFOAM	RANS coupled with VOF; standard k-ε turbulence model	Finite volume; PISO algorithm; explicit Euler first-order time discretization method	Hypothetical dam-break flow in the Willow Creek Mountain area (California) Water volume: N/A	Modelled area extent: ~8 km × 3 km Mesh of hexahedral cells Number of cells: 0.45×10^6	Maps of the water volume fraction at selected times; flood depth hydrographs at selected points	Flood wave propagation	Simulation time: 400 s Run time/simulation time: ~45	2019	Open-source

Table 2. *Cont.*

(1) Reference	(2) Model Name	(3) Model Type [1]	(4) Numerical Method [2]	(5) Case Study	(6) Computational Domain and Elements	(7) Output Data	(8) Focus of the Study	(9) Computational Efficiency [3]	(10) Year	(11) Status
Issakhov and Zhandaulet [90]	N/A	RANS coupled with VOF; three incompressible phases for the simulation of mixed water–mud flow: two Newtonian fluids (air and water) and a non-Newtonian liquid; realizable k-ω turbulence model	Finite volume; PISO algorithm	Hypothetical Mynzhylky erosional dam break (Kazakhstan) Water volume: 50×10^3 m^3	Modelled area extent: 17×10^3 m^2, 1.317 km long river reach downstream of the dam Homogeneous mud layer of fixed thickness downstream of the dam Structured mesh of tetrahedral cells Number of cells: 2.433×10^6 Spatial resolution: 0.5 m	Flood depth hydrographs at selected points; water surfaces and inundated areas at selected times (for different mud layer thicknesses)	Flood wave (with mud) propagation Effect of the initial mud layer thickness	Simulation time: 60 s Run time/simulation time: N/A	2020	N/A
Munoz and Constantinescu [37]	STAR-CCM+	RANS coupled with VOF; realizable k-ε turbulence model	Finite volume; SIMPLE algorithm	Hypothetical Coralville dam-break (USA) Water volume: N/A (reservoir water level provided)	Modelled area extent: 18 km long river reach downstream of the dam and floodplains Lake: unstructured grid with polyhedral cells; spatial resolution: 100 m River and floodplains: unstructured grid with prismatic cells; multi-resolution Number of cells: 18×10^6	Flooded areas at different times; free surface profile along the river at peak flood extent; discharge hydrographs at selected river sections; unit discharge transverse profiles in selected cross-sections at peak flood extent; details of the velocity field	Flood wave propagation 3D effects Comparison with 2D depth-averaged model predictions Recalibration of the 2D model parameter to improve the agreement between 2D and 3D model results	Simulation time: 5 h Run time/simulation time: 144 Parallelization using MPI	2020	Commercial
Munoz and Constantinescu [37]	STAR-CCM+	RANS coupled with VOF; realizable k-ε turbulence model	Finite volume; SIMPLE algorithm	Hypothetical Saylorville dam break (USA) Water volume: N/A (reservoir water level provided)	Modelled area extent: 18 km long river reach downstream of the dam and floodplains Lake: unstructured grid with polyhedral cells; spatial resolution: N/A River and floodplains: unstructured grid with prismatic cells; multi-resolution Number of cells: 40×10^6	Flooded areas at different times; free surface profile along the river at the end of the simulation; discharge hydrographs at selected river sections	Flood wave propagation 3D effects Comparison with 2D depth-averaged model predictions	Simulation time: 3.75 h Run time/simulation time: 230 Parallelization using MPI	2020	Commercial
Wang et al. [91]	DualSPHysics	Navier–Stokes and continuity equations; generalised Herschel–Bulkley–Papanastasiou rheological model	Weakly compressible SPH	Hypothetical Yujiaquan tailings dam break (China) Pond capacity: 52.55 million m^3	Modelled area extent: the pond and the area around (~2 km × 2 km) Number of particles: 3.495×10^6 (fluid) 0.936×10^6 (boundaries) Particle spacing: 2 m	Flow fields (velocity magnitude) and flooded areas at selected times	Tailings flow dynamics	Simulation time: 10 min Run time/simulation time: N/A Parallelization on GPU	2020	Open-source

Table 2. *Cont.*

(1) Reference	(2) Model Name	(3) Model Type [1]	(4) Numerical Method [2]	(5) Case Study	(6) Computational Domain and Elements	(7) Output Data	(8) Focus of the Study	(9) Computational Efficiency [3]	(10) Year	(11) Status
Yu et al. [92]	OpenFOAM	RANS coupled with VOF; standard k-ε turbulence model; Bingham–Papanastasiou rheological model	Finite volume; PISO algorithm	Historical 2019 Feijão (Brumadinho) tailings dam break (Brazil) Pond capacity: 12.7 million m^3 Released tailings volume: 11.7 million m^3	Modelled area extent: N/A (suitable area around the reservoir) Unstructured mesh of hexahedral cells Number of cells: 3.242×10^6 Spatial resolution: 10 m (horizontal) 3 m (vertical)	Flow velocity magnitude contour maps at selected times; wave front motion; free surface average velocity hydrograph; flooded area	Tailings flow dynamics Comparison with field data	Simulation time: 2500 s Run time/simulation time: N/A Parallelization using MPI (analysis of the speed-up of different numbers of processors)	2020	Open-source
Yu et al. [92]	OpenFOAM	RANS coupled with VOF; standard k-ε turbulence model; Bingham–Papanastasiou rheological model	Finite volume; PISO algorithm	Hypothetical A'xi gold tailings dam break (China) Pond capacity: 3.6 million m^3	Modelled area extent: N/A (suitable area around the reservoir) Unstructured mesh of hexahedral cells; number of cells: 6.657×10^6 Spatial resolution: 3 m	Flow velocity magnitude contour maps at selected times; wave front motion; free surface average velocity hydrograph; flooded area	Tailings flow dynamics	Simulation time: 800 s Run time/simulation time: N/A Parallelization using MPI (analysis of the speed-up of different numbers of processors)	2020	Open-source
Zhuang et al. [93]	FLOW-3D	RANS coupled with VOF; RNG k-ε turbulence model	Finite volume	Historical landslide dam break consequent to the 2000 Yigong landslide (China) Water volume: N/A (water depth of 60 m at the barrier lake)	Modelled area extent: ~33 km long stretch of the Yigong River valley Mesh details: N/A	Flood depth contour maps at selected times; flow depth and velocity hydrographs at selected points; flow discharge at selected sections	Landslide and following landslide dam-break coupled 3D simulations Comparison with field data Flood wave propagation	Simulation time: 3 h 20 min Run time/simulation time: N/A	2020	Commercial
Amicarelli et al. [94]	SPHERA	Euler and continuity equations	Weakly-compressible SPH	Hypothetical Alpe Gera dam break (Italy) Water volume: 68.1 million m^3	Modelled area extent: 7.9 km × 9.9 km Number of particles: N/A Particle spacing: N/A	Flooded areas; velocity fields at selected times; maximum flood depth contour map; discharge and flood depth hydrographs at selected sections	Urban flood features Comparison with experimental laboratory data Adoption of a flooding damage model	Simulation time: 50 min Run time/simulation time: N/A	2021	Free and open-source
Karam et al. [95]	FLOW-3D	RANS coupled with VOF; RNG k-ε turbulence model	Finite volume	Hypothetical Attabad Lake landslide dam break (Pakistan) Water volume: 305 million m^3	Modelled area extent: N/A (stretch of the downstream valley) Multiple mesh blocks of hexahedral cells Number of cells: N/A	Flow depth hydrographs at selected sites; flow discharge hydrographs at selected cross-sections; flood inundation maps and velocity fields at selected times; flood arrival times at selected locations	Flood wave propagation	Simulation time: ~1 h 19 min Run time/simulation time: N/A	2021	Commercial

Table 2. *Cont.*

(1) Reference	(2) Model Name	(3) Model Type [1]	(4) Numerical Method [2]	(5) Case Study	(6) Computational Domain and Elements	(7) Output Data	(8) Focus of the Study	(9) Computational Efficiency [3]	(10) Year	(11) Status
Miliani et al. [96]	N/A	Lattice Boltzmann equation with the Bhatnagar–Gross–Krook (BGK) collisional operator; interface tracking method	Lattice Boltzmann algorithm	Flash flood in the Toce River 1:100 physical model (Italy) Controlled impulsive inflow Two case studies considered: the presence of actual buildings and an idealised array of buildings	Modelled area extent: 5 km long reach of the Toce River (5.5 km × 1.2 km)	Video animations of the numerical results; flood depth hydrographs at selected gauge points	Flood wave propagation	Simulation time: N/A Run time/ simulation time: N/A	2021	Research
Ai et al. [97]	N/A	RANS coupled with a free-surface equation; non-hydrostatic and hydrostatic versions; standard k-ε turbulence model	Coupled finite volume–finite difference; explicit projection method	Flash flood in the Toce River 1:100 physical model (Italy) Controlled impulsive inflow (high-inflow and low-inflow hydrographs) Array of aligned buildings simulating a simplified urban district	Modelled area extent: 5 km long reach of the Toce River (5.5 km × 1.2 km) Unstructured mesh of prismatic cells with triangular basis in a vertical boundary-fitted coordinate system Number of cells: 1.041×10^5 (2.082×10^4 triangular elements on the bottom and 5 layers along the vertical)	Flow depth hydrographs at selected points	Non-hydrostatic effects Comparison with experimental physical model data Comparison with hydrostatic 3D model predictions	Simulation time: 1 min Run time/ simulation time: 29.4 (high-inflow hydrograph) Run time/ simulation time: 24 (low-inflow hydrograph)	2022	Research
Issakhov et al. [98]	ANSYS Fluent	RANS coupled with VOF; three incompressible phases for the simulation of mixed water–mud flow: two Newtonian fluids (air and water) and a non-Newtonian liquid; realizable k-ω turbulence model	Finite volume; PISO algorithm	Hypothetical erosional dam-break flow along the Kargalinka River (Kazakhstan) Water volume: 333.5×10^3 m^3	Modelled area extent: N/A (a stretch of the river) Homogeneous mud layer of fixed thickness downstream of the dam Structured mesh of uniform cells Number of cells: 0.985×10^6	Water surfaces and inundated areas at different times; flood depth hydrographs at selected points (for different mud layer thicknesses)	Flood wave (with mud) propagation Effect of the initial mud layer thickness	Simulation time: 34.5 s Run time/ simulation time: N/A	2022	Commercial
Yang et al. [99]	ANSYS CFX	RANS coupled with VOF; standard k-ε turbulence model; Bingham rheological model	Finite volume; PISO algorithm	Hypothetical Dagangding tailings dam break (China) Pond capacity: 3.59 million m^3 (Theoretical inflow discharge at the dam site)	Modelled area extent: N/A (selected area downstream of the dam) Unstructured mesh with tetrahedral and pentahedral cells Number of cells: 0.543×10^6	Flow fields (velocity magnitude) and flooded areas at selected times; wave front advancement and celerity in time; final deposition area and depth distribution; longitudinal and transverse profiles of the final deposit	Tailings flow dynamics	Simulation time: 2000 s Run time/ simulation time: N/A	2022	Commercial

Table 2. *Cont.*

(1) Reference	(2) Model Name	(3) Model Type [1]	(4) Numerical Method [2]	(5) Case Study	(6) Computational Domain and Elements	(7) Output Data	(8) Focus of the Study	(9) Computational Efficiency [3]	(10) Year	(11) Status
Zhuang et al. [100]	DAN3D	Hydrodynamic equations; rheological models (Bingham model)	SPH	Historical 2017 Tonglüshan tailings dam break (China) Pond capacity: 15.78 million m^3; moved slurry volume: 0.5 million m^3	Modelled area extent: ~2 km × 1.5 km area around the tailings pond Number of particles: 4×10^3 Particle spacing: N/A	Flow depth maps at different times; final deposition area and slurry depth distribution; maximum velocity magnitude map	Propagation of the tailings slurry Comparison with field data concerning the final deposition distribution of the tailings slurry Sensitivity analysis on the solid concentration of the tailings slurry	Simulation time: 300 s Run time/ simulation time: N/A	2022	Freeware

Notes: [1] RANS = Reynolds-Averaged Navier–Stokes equations; RNG = Re-normalization Group; VOF = Volume of Fluid. [2] CICSAM = Compressive Interface Capturing Scheme for Arbitrary Meshes; MULES = Multidimensional Universal Limiter with Explicit Solution; PIMPLE = combination of PISO and SIMPLE; PISO = Pressure Implicit with Split Operators; SIMPLE = Semi-Implicit Method for Pressure-Linked Equations; SPH = Smoothed-Particle Hydrodynamics. [3] GPU = Graphics Processing Unit; MPI = Message Passing Interface. N/A = not available; * dam spillway or overtopping flow.

3. Results and Discussion

A total of 34 documents relating to the period from 2006 to the present were retrieved from the literature review. However, 39 entries appear in Table 2, as nine of the references examined present two different case studies, hence being repeated in two different table rows, and four references are mentioned together with other works that contain identical case studies.

Figure 3 summarises the main statistical information derived from the analysis of the papers reviewed. About a third of these (12) were published in the last three years (2020–2022), confirming a recent growing interest in 3D modelling of real-field dam-break flows (Figure 3a). In most case studies examined (43.6%), in-house research codes were adopted for the numerical analyses; commercial CFD software packages and open-source or freeware codes were used in 17.9% and 35.9% of cases, respectively (Figure 3b). The type of code was unspecified in only one case. Mesh-based methods appear in Table 2 more times (25) than particle-based ones (14) (Figure 3c). Among the former methods, the VOF-based finite volume ones are the most commonly used, whereas, among the latter, the SPH methods are prevalent. This state of the art was also observed by other researchers (e.g., [101]) who, in addition, showed how numerical results obtained from these two methods are in good agreement [102]. The Lattice Boltzmann method appears in Table 2 in only one case. This modern numerical technique still has limited application to dam-break problems [103,104]. Finite volume techniques are the most used in the reviewed papers for discretizing the governing equations in mesh-based models (Figure 3d). Indeed, the finite volume method is particularly suitable for solving conservation law equations because it exploits the integral formulation of the equations to capture their weak solutions [105]. Finite element methods appear fewer times in Table 2, since this numerical technique still constitutes a developing research area in CFD [106] and free surface flow modelling [107], despite its ability to treat complex geometries and reach high-order accuracy.

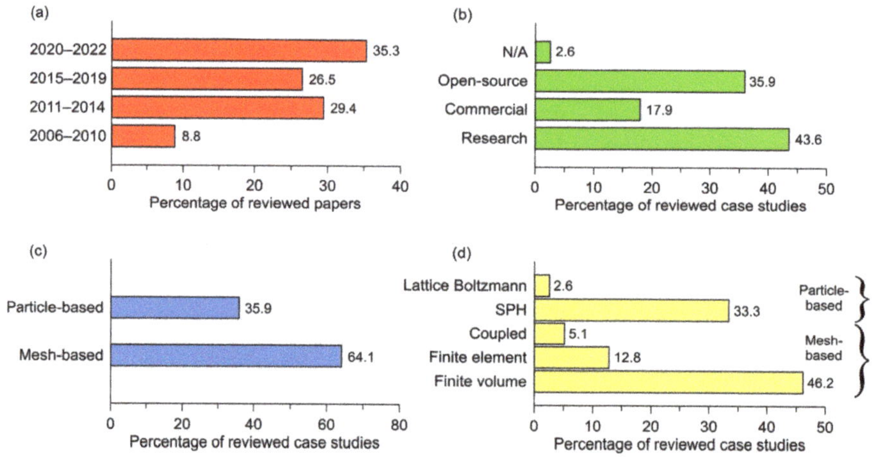

Figure 3. Statistical information from the database of the reviewed papers on 3D modelling of large-scale real-world dam-break floods: (**a**) year of publication; (**b**) status of the software; (**c**) model type; (**d**) numerical scheme.

The catastrophic scenario of the sudden and total dam collapse is mostly considered in the case studies reviewed, thus neglecting the breach development dynamics.

Three-dimensional models employed in dam-break modelling are also commonly used to simulate floods in rivers (e.g., [45,108]) or inundations in floodplains [19] or in urban areas (e.g., [36,109]).

3.1. Improvements in Simulation Accuracy

Three-dimensional hydrodynamic modelling improves the mathematical description of dam-break flows compared to routinely adopted 2D depth-averaged modelling, because 3D models overcome the intrinsic limitations of 2D ones. Therefore, 3D modelling is of wide applicability and general interest in dam-break problems. Indeed, 3D models calculate the pressure field and include the vertical fluid acceleration, thereby inherently taking into account the effects of flow curvature. Moreover, 3D models involve the vertical velocity component and describe the vertical variation of flow velocity. Conversely, 2D depth-averaged models introduce hypotheses on the vertical pressure distribution (assumed hydrostatic in SWE models) or the shape of the vertical velocity profile in non-hydrostatic flow models [23]. Finally, 3D models can include vertical turbulence and spiral flows [60] and can correctly simulate the impact of flows against obstacles or structures [109,110]. These enhanced modelling capabilities of 3D models can ensure more accurate flow predictions, especially around structures, at topographic singularities where sudden changes in the bottom surface occur, and in urban flooding simulations. Reproducing small-scale eddies, vertical turbulence, and residual circulation can be crucial for accurately predicting both near-field and far-field features of the flooding dynamics and the flow variables involved in flood hazard assessments.

The capabilities of 3D models were verified and validated in most of the articles reviewed, but the analysis of 3D flow effects is the focus of only a few of them (e.g., [70,71,82,87]). For example, Biscarini et al. [82] discussed the 3D effects induced on a dam-break flow by a sharp river bend. Even fewer are the articles that, based on a real-field case study, compare the results obtained through a 3D model with those obtained through a standard 2D depth-averaged shallow-water model, analysing the differences [37,50,80]. For example, Munoz and Constantinescu [37], studying the flood inundations induced by the hypothetical failure of two flood-protection dams in the United States, found that their 2D depth-averaged model underpredicted the wave propagation speed and the inundation extent compared with a 3D model. Most frequently, in the literature, the comparison of 3D and 2D model performance in predicting dam-break flow was made on the basis of schematic test cases characterised by a simple geometry (e.g., [27,111]).

In 3D models based on the RANS equations, the effect of turbulence is introduced through a closure turbulence model. The classic k-ε model is the most adopted in the references reviewed (entries [37,82,85,95] in Table 2). However, no systematic sensitivity analyses on the turbulence model type were performed for real-world dam-break case studies. The prediction capabilities of different turbulence models were typically compared again on the basis of dam-break test cases characterised by a simple and schematic geometry (e.g., [110,112–116]), exploiting laboratory experimental data.

Some references analysed in this review consider mudflows resulting from a tailings dam failure. In these studies, the tailings slurry is assumed as a homogeneous non-Newtonian viscoplastic fluid, and a suitable constitutive equation is added to the set of governing equations to characterise its rheological properties [88,91,92,99,100]. In some other studies among those reviewed, non-Newtonian rheological models are used to describe the behaviour of moving sediment layers in the path of the dam-break wave [90,98], thereby simulating the geomorphic effects produced by a dam-break flood on an erodible bed. An erosional dam break (with bed-load transport) on real-world topography was modelled by Amicarelli et al. [84] through an SPH model applied to a mixture of water and a non-cohesive granular material.

Recently, topographic data accuracy and spatial resolution have been significantly improved thanks to highly accurate terrain surveying techniques, such as scanning airborne laser altimetry (LIDAR), which can reach a horizontal resolution even below 1 m and a vertical accuracy of 10^{-1} m [19], and unmanned aerial vehicle (UAV) photogrammetry, which is very effective in terms of timeliness, repeatability, and high resolution [91]. Hence, highly accurate and high-resolution topographic digital models are widespread nowadays, even in urban areas [85], and allow for very accurate dam-break flow modelling.

On the other hand, despite the implicit greater descriptive capacity (and complexity) of 3D models, accuracy improvement in 3D numerical predictions compared with simpler, low-dimensional models may be illusory if topographic and input data are limited and inaccurate and reliable real-field validation data are scarce [19,117].

3.2. Model Validation and Calibration

Dam-break models are usually validated by comparing numerical results with experimental data [13]. To this end, real-field data of historical dam-break events or experimental data obtained from small-scale physical models with an irregular bottom can be very useful [14], especially if the numerical model will be used in practical engineering applications for dam-break inundation mapping and flood hazard assessment. Validation against real-field data can also help discriminate among potential competing models able to return plausible and realistic results.

Based on the articles reviewed, the most used real-field test case for validating 3D dam-break numerical models is the historical event of the 1959 Malpasset dam break (entries [75,82,83] in Table 2). An extensive database is available for this test case, including field data collected immediately after the event and data measured during a laboratory investigation in a reduced-scale physical model [118,119]. However, other historical dam-break events are well documented [14]; for some of these, the test case and the experimental data (from field surveys or consultation of historical documents) are available in digital format [120,121].

Another test case frequently used for validating 3D dam-break numerical models is the Toce River test case (entries [50,81,85,96] in Table 2), which concerns a dam-break-like flash flood induced by imposing an inflow discharge hydrograph in a 1:100 physical model of a 5 km long stretch of the Toce River valley in northern Italy [122]. An idealised urban district was inserted in this physical model to simulate the dam-break flooding of an urban area. Other experimental data from reduced-scale physical models are available in the literature [123,124] and could be considered for validating 3D dam-break models.

In complex 3D models, a large number of model parameters requiring calibration appear, mainly concerning numerical schemes and turbulence models. Accordingly, an optimization process involving many parameters should be performed, which might be challenging and computationally expensive due to the long runtime of each model execution. Therefore, calibration parameters are usually set based on expert judgement, according to the suggestions of software users' guides or following the choices made in similar numerical studies.

Uncertainty analysis is widely recognised as desirable, if not indispensable, in environmental system modelling to associate uncertainty estimates with model predictions [15]. In real-world applications of dam-break inundation modelling, the main uncertainty sources are topographic input data, model parameters (including the mesh resolution), and initial conditions defining the dam-break scenario (i.e., breach parameters and reservoir filling conditions). Modern remote sensing acquisition technologies (such as LIDAR) have drastically reduced the uncertainty in terrain description, and the availability of high-resolution topographic data makes it possible to include in the computational domain obstacles, structures, and topographic details that may significantly influence the inundation process [60]. The typical way to individually quantify the effects of model or scenario parameters on numerical predictions is to perform a sensitivity analysis based on a sampling approach (e.g., [125,126]). Monte Carlo methods are usually adopted to this end, considering a sample of the appropriate size to ensure convergence. However, exhaustive sensitivity analyses based on large sets of parameter values (and, consequently, many model runs) may be prohibitive using 3D hydrodynamic models. To overcome this limitation, Rizzo et al. [10] (who, however, used a 2D depth-averaged shallow-water model) proposed a probabilistic method based on a limited set of dam-break scenarios, each of which had a (conditional) probability associated with it. As regards the spatial resolution, Zhang et al. [87] performed a sensitivity analysis on the mesh size for a realistic dike-break flooding case, concluding that the spatial resolution significantly impacts the accuracy of model results when the terrain is irregular, as in real-field applications.

3.3. Improvements in Computational Efficiency

The main reason for the limited use of 3D models in large-scale dam-break flood simulation over real-world topography is that they are time-consuming due to their high computational cost, which depends on various factors, such as the complexity of the numerical method adopted to solve the governing equations, the extent of the computational domain, the simulation (physical) time (usually set long enough to reconstruct the salient features of the flooding process), and the mesh type and resolution (in mesh-based models) or the particle size (in particle-based models). This problem—which is exacerbated in real-world applications—does not concern, as a rule, dam-break tests characterised by a schematic geometry (i.e., verification against analytical solutions or validation against laboratory data), because such simple tests do not require, in general, significant computational resources.

A viable option to overcome the limitation related to computational efficiency without resorting to approximate models [127] is to accelerate 3D model calculations via high-performance computing and GPU technology to reduce model running times [19]. In particular, parallel computing is a valid and widely used method to improve computational efficiency, possibly exploiting GPU computing power and processing capabilities.

Reducing the computational time of model executions also allows larger domains to be considered with high spatial resolution, extending the simulation of the flooding dynamics for a longer simulation time. Moreover, a large set of dam-break scenarios can be used in sensitivity analyses for assessing model uncertainty.

Table 2 shows that parallelization techniques based on Message Passing Interface (MPI; [39,72]) and the GPU implementation of the simulation model [77,86,91] were used in 3D modelling of real-field dam-break flows. The ratio of model run time to simulation time is highly variable, ranging from 10^2 to 10^4, depending on the number of processors and the total number of computational cells or particles. Yu et al. [92] analysed the gain in parallel speed-up as a function of the number of processors.

3.4. Improvements in Result Visualization

Geographic information systems (GISs; [72,73]) and 3D virtual geographic environment (VGE) systems have recently become attractive tools for the 3D dynamic visualization of model results, improving the communication of the dam-break flooding dynamics and the potential consequences (e.g., [128]). The availability of 3D numerical results facilitates the development of 3D virtual reality environments and visualizations, which, currently, are usually based on 2D hydraulic model data and results extrapolated into 3D (e.g., [129,130]).

4. Conclusions

Three-dimensional numerical modelling of dam-break flow has developed considerably in the last decade thanks to the significant increase in the available computing resources, which has made 3D modelling a viable alternative to routinely used 2D depth-averaged modelling in large-scale real-field applications, despite its higher computational cost. Even if 3D numerical models have only recently become a real and feasible option in large-scale dam-break modelling on real-world topography, we found in the literature a noticeable number of contributions concerning this fluid dynamic application, including standard dam-break water flows as well as geomorphic and tailings dam-break flows. Mesh-based models are mostly used for solving the governing hydrodynamic equations, which are the Navier–Stokes equations or the RANS equations coupled with a free-surface tracking technique (e.g., the VOF method) and a closure turbulence model. However, in recent years, particle-based models based on the SPH technique have become widespread in computational fluid dynamics and in the simulation of dam-break flows. Indeed, this method benefits from both being mesh-free and not requiring computationally expensive free-surface tracking techniques. The application of the Lattice Boltzmann method to large-scale 3D dam-break flood modelling is a relatively new research field that deserves further exploration [96].

This paper systematically reviews the state-of-the-art 3D numerical modelling of large-scale dam-break floods on irregular real-world topography. The literature survey is mainly based on journal and conference papers published until July 2023, excluding the grey literature. We aimed to conduct an exhaustive and meticulous review based on comprehensive search parameters and inclusive keywords. Nevertheless, we may have missed studies published in journals of local diffusion or not specifically focused on 3D numerical modelling.

The references reviewed are organised into a table (Table 2) reporting extensive information on numerical models, case studies analysed, research focus, numerical results, and computational efficiency. Recent developments and key improvements in modelling and computational aspects, such as model accuracy and efficiency, are discussed. Regarding computational efficiency, code running on Graphics Processing Units and massive parallelization ensure a significant computational time reduction. This general improvement in computational efficiency allows for high spatial resolutions, even in large-scale applications. However, model calibration remains challenging due to the large number of parameters involved in 3D models, especially when coupled with a turbulence model. The 3D visualization of the numerical results can improve the quality of the communication of the dam-break flood hazard to managers and stakeholders, thus contributing to the mitigation of dam-break flood consequences.

Compared to the less computationally demanding 2D depth-averaged models, the 3D ones allow for a more detailed and accurate prediction of the flooding dynamics, inherently including the fluid vertical acceleration and 3D effects due to the flow curvature induced by the irregular topography. Freeware and commercial CFD software (relatively user-friendly) are available nowadays for dam-break flow analysis, even concerning large-scale problems on real-world topography.

Future research may concern coupled 2D–3D models as a valid compromise between simulation accuracy and computational efficiency. This modelling option is currently already introduced in some CFD software (e.g., [131]) in order to simulate large-scale flows in which the shallow water assumptions are approximately valid with a 2D depth-averaged model, and near-field flows and localised flow features (near structures, obstacles, or significant topographic irregularities) with a 3D model. For instance, hybrid 2D–3D models could be used to simulate dam-break flooding generated by a partial dam failure, with the breach involving only the upper part of the dam. In this case, the weir-type outflow could be modelled by the 3D model and the wave propagation in the downstream area by the 2D depth-averaged model. The portions of the computational domain in which to apply the two different models can be previously identified and efficiently linked.

This review can guide researchers, modellers, and practitioners to compare existing 3D dam-break numerical models, choose the most suitable model for the application of interest, and select state-of-the-art numerical approaches. This review can also support modellers and researchers, providing a basis for future research in 3D simulation models and computational techniques for dam-break flow modelling, with special attention to large-scale real-field applications.

Author Contributions: A.M. and M.T. contributed equally to the conceptualization and implementation of the research, the revision of the existing literature, the analysis of results, and the writing of the manuscript. All authors have read and agreed to the published version of the manuscript.

Funding: A.M. was supported by the Italian Ministry of University and Research through the PRIN 2017 Project RELAID (REnaissance of LArge Italian Dams), project number 2017T4JC5K.

Data Availability Statement: No new data were created or analysed in this study. Data sharing is not applicable to this article.

Conflicts of Interest: The authors declare no conflict of interest.

References

1. Costa, J.E. *Floods from Dam Failures*; Open-File Report 85-560; US Geological Survey: Denver, CO, USA, 1985. Available online: https://pubs.usgs.gov/of/1985/0560/report.pdf (accessed on 21 July 2023).
2. Charles, J.A.; Tedd, P.; Warren, A. *Lessons from Historical Dam Incidents*; Project SC080046/R1; Environment Agency: Bristol, UK, 2011. Available online: https://assets.publishing.service.gov.uk/media/603369e7e90e07660cc43890/_Lessons_from_Historical_Dam_Incidents_Technical_Report.pdf (accessed on 21 July 2023).
3. Zhang, L.; Peng, M.; Chang, D.; Xu, Y. *Dam Failure Mechanisms and Risk Assessment*; Wiley: Singapore, 2016.
4. Graham, W.J. *A Procedure for Estimating Loss of Life Caused by Dam Failure*; DSO-99-06; US Department of Interior, Bureau of Reclamation, Dam Safety Office: Denver, CO, USA, 1999. Available online: https://www.usbr.gov/ssle/damsafety/TechDev/DSOTechDev/DSO-99-06.pdf (accessed on 21 July 2023).
5. ASDSO (Association of State Dam Safety Officials). Lesson Learned from Dam Incidents and Failures—Case Studies. Available online: https://damfailures.org/case-study/ (accessed on 21 July 2023).
6. International Water Power and Dam Construction. Learning Historical Dam Safety Lessons. Available online: https://www.waterpowermagazine.com/features/featurelearning-historical-dam-safety-lessons-4958949/ (accessed on 21 July 2023).
7. FEMA. *Federal Guidelines for Inundation Mapping of Flood Risks Associated with Dam Incidents and Failures*; FEMA P-946; Federal Emergency Management Agency, US Department of Homeland Security: Washington, DC, USA, 2013. Available online: https://www.fema.gov/sites/default/files/2020-08/fema_dam-safety_inundation-mapping-flood-risks.pdf (accessed on 25 May 2023).
8. CDSO. *Guidelines for Mapping Flood Risk Associated with Dams*; CDSO_GUD_DS_05_v1.0; Central Dam Safety Organization, Central Water Commission, Dam Safety Rehabilitation Directorate: New Delhi, India, 2018. Available online: https://damsafety.cwc.gov.in/ecm-includes/PDFs/Guidelines_for_Mapping_Flood_Risks_Associated_with_Dams.pdf (accessed on 25 May 2023).
9. Morris, M.W. *CADAM Concerted Action on Dam Break Modeling*; Final Report SR 571; HR Wallingford: Oxford, UK, 2000; Available online: https://eprints.hrwallingford.com/447/1/CADAM.pdf (accessed on 25 May 2023).
10. Rizzo, C.; Maranzoni, A.; D'Oria, M. Probabilistic Mapping and Sensitivity Assessment of Dam-Break Flood Hazard. *Hydrol. Sci. J.* **2023**, *68*, 700–718. [CrossRef]
11. Maranzoni, A.; D'Oria, M.; Rizzo, C. Quantitative Flood Hazard Assessment Methods: A Review. *J. Flood Risk Manag.* **2023**, *16*, e12855. [CrossRef]
12. Ferrari, A.; Vacondio, R.; Mignosa, P. High-Resolution 2D Shallow Water Modelling of Dam Failure Floods for Emergency Action Plans. *J. Hydrol.* **2023**, *618*, 129192. [CrossRef]
13. Aureli, F.; Maranzoni, A.; Petaccia, G.; Soares-Frazão, S. Review of Experimental Investigations of Dam-Break Flows over Fixed Bottom. *Water* **2023**, *15*, 1229. [CrossRef]
14. Aureli, F.; Maranzoni, A.; Petaccia, G. Review of Historical Dam-Break Events and Laboratory Tests on Real Topography for the Validation of Numerical Models. *Water* **2021**, *13*, 1968. [CrossRef]
15. Teng, J.; Jakeman, A.J.; Vaze, J.; Croke, B.F.W.; Dutta, D.; Kim, S. Flood Inundation Modelling: A Review of Methods, Recent Advances and Uncertainty Analysis. *Environ. Model. Softw.* **2017**, *90*, 201–216. [CrossRef]
16. De Marchi, G. Sull'Onda di Piena che Seguirebbe al Crollo della Diga di Cancano [On the Dam-Break Wave Resulting from the Collapse of the Cancano Dam]. *L'Energia Elettr.* **1945**, *22*, 157–169. (In Italian)
17. Antunes do Carmo, J.S. Physical Modelling vs. Numerical Modelling: Complementarity and Learning. *Preprints* **2020**, 2020070753. Available online: https://www.preprints.org/manuscript/202007.0753/v2/download (accessed on 21 July 2023).
18. Mudashiru, R.B.; Sabtu, N.; Abustan, I.; Balogun, W. Flood Hazard Mapping Methods: A Review. *J. Hydrol.* **2021**, *603*, 126846. [CrossRef]
19. Bates, P.D. Flood Inundation Prediction. *Annu. Rev. Fluid Mech.* **2022**, *54*, 287–315. [CrossRef]
20. Toro, E.F.; Garcia-Navarro, P. Godunov-Type Methods for Free-Surface Shallow Flows: A Review. *J. Hydraul. Res.* **2007**, *45*, 736–751. [CrossRef]
21. Castro-Orgaz, O.; Hager, W.H. *Shallow Water Hydraulics*; Springer: Cham, Switzerland, 2019.
22. Toro, E.F. *Shock-Capturing Methods for Free-Surface Shallow Flows*; John Wiley & Sons: Chichester, UK, 2001.
23. Castro-Orgaz, O.; Hager, W.H. *Non-Hydrostatic Free Surface Flows*; Springer: Cham, Switzerland, 2017.
24. Maranzoni, A.; Tomirotti, M. New Formulation of the Two-Dimensional Steep-Slope Shallow Water Equations. Part I: Theory and Analysis. *Adv. Water Resour.* **2022**, *166*, 104255. [CrossRef]
25. Maranzoni, A.; Tomirotti, M. New Formulation of the Two-Dimensional Steep-Slope Shallow Water Equations. Part II: Numerical Modeling, Validation, and Application. *Adv. Water Resour.* **2023**, *177*, 104403. [CrossRef]
26. Stansby, P.K.; Chegini, A.; Barnes, T.C.D. The Initial Stages of Dam-Break Flow. *J. Fluid Mech.* **1998**, *374*, 407–424. [CrossRef]
27. Ozmen-Cagatay, H.; Kocaman, S. Dam-Break Flows During Initial Stage Using SWE and RANS Approaches. *J. Hydraul. Res.* **2010**, *48*, 603–611. [CrossRef]
28. Soares-Frazão, S.; Zech, Y. Dam Break in Channels with 90° Bend. *J. Hydraul. Eng.* **2002**, *128*, 956–968. [CrossRef]
29. Kocaman, S.; Ozmen-Cagatay, H. The Effect of Lateral Channel Contraction on Dam Break Flows: Laboratory Experiment. *J. Hydrol.* **2012**, *432–433*, 145–153. [CrossRef]
30. Khoshkonesh, A.; Nsom, B.; Bahmanpouri, F.; Dehrashid, F.A.; Adeli, A. Numerical Study of the Dynamics and Structure of a Partial Dam-Break Flow Using the VOF Method. *Water Resour. Manage.* **2021**, *35*, 1513–1528. [CrossRef]

31. Soares-Frazão, S. Experiments of Dam-Break Wave Over a Triangular Bottom Sill. *J. Hydraul. Res.* **2007**, *45* (Suppl. S1), 19–26. [CrossRef]
32. Ozmen-Cagatay, H.; Kocaman, S.; Guzel, H. Investigation of Dam-Break Flood Waves in a Dry Channel with a Hump. *J. Hydro-Environ. Res.* **2014**, *8*, 304–315. [CrossRef]
33. Soares-Frazão, S.; Zech, Y. Experimental Study of Dam-Break Flow against an Isolated Obstacle. *J. Hydraul. Res.* **2007**, *45* (Suppl. S1), 27–36. [CrossRef]
34. Aureli, F.; Dazzi, S.; Maranzoni, A.; Mignosa, P.; Vacondio, R. Experimental and Numerical Evaluation of the Force Due to the Impact of a Dam-Break Wave on a Structure. *Adv. Water Resour.* **2015**, *76*, 29–42. [CrossRef]
35. Khoshkonesh, A.; Daliri, M.; Riaz, K.; Dehrashid, F.A.; Bahmanpouri, F.; Di Francesco, S. Dam-Break Flow Dynamics over a Stepped Channel with Vegetation. *J. Hydrol.* **2022**, *613*, 128395. [CrossRef]
36. Zhang, T.; Feng, P.; Maksimović, Č.; Bates, P.D. Application of a Three-Dimensional Unstructured-Mesh Finite-Element Flooding Model and Comparison with Two-Dimensional Approaches. *Water Resour. Manage.* **2016**, *30*, 823–841. [CrossRef]
37. Munoz, D.H.; Constantinescu, G. 3-D Dam Break Flow Simulations in Simplified and Complex Domains. *Adv. Water Resour.* **2020**, *137*, 103510. [CrossRef]
38. Fernandez-Feria, R. Dam-Break Flow for Arbitrary Slopes of the Bottom. *J. Eng. Math.* **2006**, *54*, 319–331. [CrossRef]
39. Castro-Orgaz, O.; Cantero-Chinchilla, F.N. Non-Linear Shallow Water Flow Modelling over Topography with Depth-Averaged Potential Equations. *Environ. Fluid Mech.* **2020**, *20*, 261–291. [CrossRef]
40. Lu, X.; Dong, B.; Zhang, X. A Two-Dimensional Depth-Integrated Non-Hydrostatic Numerical Model for Nearshore Wave Propagation. *Ocean Model.* **2015**, *96*, 187–202. [CrossRef]
41. Cantero-Chinchilla, F.N.; Bergillos, R.J.; Gamero, P.; Castro-Orgaz, O.; Cea, L.; Hager, W.H. Vertically Averaged and Moment Equations for Dam-Break Wave Modeling: Shallow Water Hypotheses. *Water* **2020**, *12*, 3232. [CrossRef]
42. Denlinger, R.P.; Iverson, R.M. Granular Avalanches Across Irregular Three-Dimensional Terrain: 1. Theory and Computation. *J. Geophys. Res. Earth Surf.* **2004**, *109*, F01014. [CrossRef]
43. Denlinger, R.P.; O'Connell, D.R. Computing Nonhydrostatic Shallow-Water Flow over Steep Terrain. *J. Hydraul. Eng.* **2008**, *134*, 1590–1602. [CrossRef]
44. Biscarini, C.; Di Francesco, S.; Manciola, P. CFD Modelling Approach for Dam Break Flow Studies. *Hydrol. Earth Syst. Sci.* **2010**, *14*, 705–718. [CrossRef]
45. Lane, S.N.; Bradbrook, K.F.; Richards, K.S.; Biron, P.A.; Roy, A.G. The Application of Computational Fluid Dynamics to Natural River Channels: Three-Dimensional Versus Two-Dimensional Approaches. *Geomorphology* **1999**, *29*, 1–20. [CrossRef]
46. Versteeg, H.K.; Malalasekera, W. *An Introduction to Computational Fluid Dynamics. The Finite Volume Method*; Longman Scientific & Technical: Harlow, UK, 1995.
47. Hirt, C.W.; Nichols, B.D. Volume of Fluid (VOF) Methods for the Dynamics of Free Boundaries. *J. Comput. Phys.* **1981**, *39*, 201–225. [CrossRef]
48. Osher, S.; Fedkiw, R.P. Level Set Methods: An Overview and Some Recent Results. *J. Comput. Phys.* **2001**, *169*, 463–502. [CrossRef]
49. Sussman, M.; Puckett, E.G. A Coupled Level Set and Volume-of-Fluid Method for Computing 3D and Axisymmetric Incompressible Two-Phase Flows. *J. Comput. Phys.* **2000**, *162*, 301–337. [CrossRef]
50. Marsooli, R.; Wu, W. 3-D Finite-Volume Model of Dam-Break Flow over Uneven Beds Based on VOF Method. *Adv. Water. Resour.* **2014**, *70*, 104–117. [CrossRef]
51. Monaghan, J.J. Simulating Free Surface Flows with SPH. *J. Comput. Phys.* **1994**, *110*, 399–406. [CrossRef]
52. Xu, X. An Improved SPH Approach for Simulating 3D Dam-Break Flows with Breaking Waves. *Comput. Methods Appl. Mech. Eng.* **2016**, *311*, 723–742. [CrossRef]
53. Xu, X.; Jiang, Y.-L.; Yu, P. SPH Simulations of 3D Dam-Break Flow Against Various Forms of the Obstacle: Toward an Optimal Design. *Ocean Eng.* **2021**, *229*, 108978. [CrossRef]
54. Cleary, P.W.; Prakash, M. Discrete-Element Modelling: Methods and Applications in the Environmental Sciences. *Phil. Trans. R. Soc. Lond. A* **2004**, *362*, 2003–2030. [CrossRef] [PubMed]
55. Wu, J.; Bao, K.; Zhang, H. Research Progress on Dam-Break Floods. In Proceedings of the 2nd IEEE International Conference on Emergency Management and Management Sciences, Beijing, China, 8–10 August 2011; pp. 334–338. [CrossRef]
56. Ferrari, A.; Fraccarollo, L.; Dumbser, M.; Toro, E.F.; Armanini, A. Three-Dimensional Flow Evolution after a Dam Break. *J. Fluid Mech.* **2010**, *663*, 456–477. [CrossRef]
57. Xie, Z.; Stoesser, T.; Xia, J. Simulation of Three-Dimensional Free-Surface Dam-Break Flows over a Cuboid, Cylinder, and Sphere. *J. Hydraul. Eng.* **2021**, *147*, 6021009. [CrossRef]
58. Issakhov, A.; Imanberdiyeva, M. Numerical Simulation of the Movement of Water Surface of Dam Break Flow by VOF Methods for Various Obstacles. *Int. J. Heat Mass Transf.* **2019**, *136*, 1030–1051. [CrossRef]
59. Issakhov, A.; Borsikbayeva, A. The Impact of a Multilevel Protection Column on the Propagation of a Water Wave and Pressure Distribution During a Dam Break: Numerical Simulation. *J. Hydrol.* **2021**, *598*, 126212. [CrossRef]
60. Luo, P.; Luo, M.; Li, F.; Qi, X.; Huo, A.; Wang, Z.; He, B.; Takara, K.; Nover, D.; Wang, Y. Urban Flood Numerical Simulation: Research, Methods and Future Perspectives. *Environ. Model. Softw.* **2022**, *156*, 105478. [CrossRef]
61. Mignot, E.; Dewals, B. Hydraulic Modelling of Inland Urban Flooding: Recent Advances. *J. Hydrol.* **2022**, *609*, 127763. [CrossRef]

62. Kumar, V.; Sharma, K.V.; Caloiero, T.; Mehta, D.J.; Singh, K. Comprehensive Overview of Flood Modeling Approaches: A Review of Recent Advances. *Hydrology* **2023**, *10*, 141. [CrossRef]
63. Avila-Aceves, E.; Plata-Rocha, W.; Monjandin-Armenta, S.A.; Rangel-Peraza, J.G. Geospatial Modelling of Floods: A Literature Review. *Stoch. Environ. Res. Risk Assess.* **2023**. [CrossRef]
64. TELEMAC-3D—3D Hydrodynamics. Available online: http://www.opentelemac.org/index.php/presentation?id=18 (accessed on 27 March 2023).
65. FLOW-3D. Available online: https://www.flow3d.com/products/flow-3d/ (accessed on 27 March 2023).
66. About OpenFOAM. Available online: https://www.openfoam.com/ (accessed on 27 March 2023).
67. DualSPHysics: From Fluid Dynamics to Multiphysics Problems. Available online: https://dual.sphysics.org/ (accessed on 5 April 2023).
68. Amicarelli, A.; Manenti, S.; Albano, R.; Agate, G.; Paggi, M.; Longoni, L.; Mirauda, D.; Ziane, L.; Viccione, G.; Todeschini, S.; et al. SPHERA v. 9.0.0: A Computational Fluid Dynamics Research Code, Based on the Smoothed Particle Hydrodynamics Mesh-Less Method. *Comput. Phys. Commun.* **2020**, *250*, 107157. [CrossRef]
69. Roubtsova, V.; Kahawita, R. The SPH Technique Applied to Free Surface Flows. *Comput. Fluids* **2006**, *35*, 1359–1371. [CrossRef]
70. Cleary, P.W.; Prakash, M.; Rothauge, K. Combining Digital Terrain and Surface Textures with Large-Scale Particle-Based Computational Models to Predict Dam Collapse and Landslide Events. *Int. J. Image Data Fusion* **2010**, *1*, 337–357. [CrossRef]
71. Prakash, M.; Rothauge, K.; Cleary, P.W. Modelling the Impact of Dam Failure Scenarios on Flood Inundation Using SPH. *Appl. Math. Model.* **2014**, *38*, 5515–5534. [CrossRef]
72. Ye, F.; Wang, H.; Ouyang, S.; Tang, X.; Li, Z.; Prakash, M. Spatio-Temporal Analysis and Visualization Using SPH for Dam-Break and Flood Disasters in a GIS Environment. In Proceedings of the 2012 International Symposium on Geomatics for Integrated Water Resource Management, Lanzhou, China, 19–21 October 2012. [CrossRef]
73. Cleary, P.W.; Prakash, M.; Mead, S.; Tang, X.; Wang, H.; Ouyang, S. Dynamic Simulation of Dam-Break Scenarios for Risk Analysis and Disaster Management. *Int. J. Image Data Fusion* **2012**, *3*, 333–363. [CrossRef]
74. Lee, E.-S.; Violeau, D.; Issa, R.; Ploix, S. Application of Weakly Compressible and Truly Incompressible SPH to 3-D Water Collapse in Waterworks. *J. Hydraul. Res.* **2010**, *48* (Suppl. S1), 50–60. [CrossRef]
75. Caboussat, A.; Boyaval, S.; Masserey, A. On the Modeling and Simulation of Non-Hydrostatic Dam Break Flows. *Comput. Visual. Sci.* **2011**, *14*, 401–417. [CrossRef]
76. Vassilevski, Y.V.; Nikitin, K.D.; Olshanskii, M.A.; Terekhov, K.M. CFD Technology for 3D Simulation of Large-Scale Hydrodynamic Events and Disasters. *Russ. J. Numer. Anal. Math. Model.* **2012**, *27*, 399–412. [CrossRef]
77. Vacondio, R.; Mignosa, P.; Pagani, S. 3D SPH Numerical Simulation of the Wave Generated by the Vajont Rockslide. *Adv. Water Res.* **2013**, *59*, 146–156. [CrossRef]
78. Zhainakov, A.Z.; Kurbanaliev, A.Y. Verification of the Open Package OpenFOAM on Dam Break Problems. *Thermophys. Aeromech.* **2013**, *20*, 451–461. [CrossRef]
79. Jainakov, A.; Kurbanaliev, A.; Oskonbaev, M. Large-Scale Modeling of Dam Break Induced Flows. In *Dam Engineering*; Tosun, H., Ed.; IntechOpen: London, UK, 2019; pp. 59–72. [CrossRef]
80. Džebo, E.; Žagar, D.; Krzyk, M.; Četina, M.; Petkovšek, G. Different Ways of Defining Wall Shear in Smoothed Particle Hydrodynamics Simulations of a Dam-Break Wave. *J. Hydraul. Res.* **2014**, *52*, 453–464. [CrossRef]
81. Zhou, Z.; Wang, X.; Sun, R.; Ao, X.; Sun, X.; Song, M. Study of the Comprehensive Risk Analysis of Dam-Break Flooding Based on the Numerical Simulation of Flood Routing. Part II: Model Application and Results. *Nat. Hazards* **2014**, *72*, 675–700. [CrossRef]
82. Biscarini, C.; Di Francesco, S.; Ridolfi, E.; Manciola, P. On the Simulation of Floods in a Narrow Bending Valley: The Malpasset Dam Break Case Study. *Water* **2016**, *8*, 545. [CrossRef]
83. TELEMAC Modelling System. *3D Hydrodynamics, TELEMAC-3D Software*; Version 7.0, Validation Document; EDF R&D: Paris, France, 2016; Available online: http://www.opentelemac.org/index.php/component/jdownloads/summary/44-v7p0/1302-telemac3d-validation-v7p0?Itemid=54 (accessed on 30 March 2023).
84. Amicarelli, A.; Kocak, B.; Sibilla, S.; Grabe, J. A 3D Smoothed Particle Hydrodynamics Model for Erosional Dam-Break Floods. *Int. J. Comput. Fluid Dyn.* **2017**, *31*, 413–434. [CrossRef]
85. Wang, X.; Chen, W.; Zhou, Z.; Zhu, Y.; Wang, C.; Liu, Z. Three-Dimensional Flood Routing of a Dam Break Based on a High-Precision Digital Model of a Dense Urban Area. *Nat. Hazards* **2017**, *86*, 1147–1174. [CrossRef]
86. Wang, K.; Yang, P.; Hudson-Edwards, K.A.; Lyu, W.; Yang, C.; Jing, X. Integration of DSM and SPH to Model Tailings Dam Failure Run-Out Slurry Routing Across 3D Real Terrain. *Water* **2018**, *10*, 1087. [CrossRef]
87. Zhang, T.; Peng, L.; Feng, P. Evaluation of a 3D Unstructured-Mesh Finite Element Model for Dam-Break Floods. *Comput. Fluids* **2018**, *160*, 64–77. [CrossRef]
88. Chen, H.-X.; Li, J.; Feng, S.-J.; Gao, H.-Y.; Zhang, D.-M. Simulation of Interactions Between Debris Flow and Check Dams on Three-Dimensional Terrain. *Eng. Geol.* **2019**, *251*, 48–62. [CrossRef]
89. Kurbanaliev, A.I.; Maksutov, A.R.; Obodoeva, G.S.; Oichueva, B.R. Using OpenFOAM Multiphase Solver InterFoam for Large Scale Modeling. In *Proceedings of the 27th World Congress on Engineering and Computer Science, San Francisco, CA, USA, 22–24 October 2019*; International Association of Engineers: Hong Kong, China; pp. 366–370. Available online: https://www.iaeng.org/publication/WCECS2019/WCECS2019_pp366-370.pdf (accessed on 25 August 2023).

90. Issakhov, A.; Zhandaulet, Y. Numerical Study of Dam Break Waves on Movable Beds for Complex Terrain by Volume of Fluid Method. *Water Resour. Manage.* **2020**, *34*, 463–480. [CrossRef]
91. Wang, K.; Yang, P.; Yu, G.; Yang, C.; Zhu, L. 3D Numerical Modelling of Tailings Dam Breach Run Out Flow over Complex Terrain: A Multidisciplinary Procedure. *Water* **2020**, *12*, 2538. [CrossRef]
92. Yu, D.; Tang, L.; Chen, C. Three-Dimensional Numerical Simulation of Mud Flow from a Tailing Dam Failure Across Complex Terrain. *Nat. Hazards Earth Syst. Sci.* **2020**, *20*, 727–741. [CrossRef]
93. Zhuang, Y.; Yin, Y.; Xing, A.; Jin, K. Combined Numerical Investigation of the Yigong Rock Slide-Debris Avalanche and Subsequent Dam-Break Flood Propagation in Tibet, China. *Landslides* **2020**, *17*, 2217–2229. [CrossRef]
94. Amicarelli, A.; Manenti, S.; Paggi, M. SPH Modelling of Dam-break Floods, with Damage Assessment to Electrical Substations. *Int. J. Comput. Fluid Dyn.* **2021**, *35*, 3–21. [CrossRef]
95. Karam, W.; Khan, F.A.; Alam, M.; Ali, S. Simulation of Dam-Break Flood Wave and Inundation Mapping: A Case Study of Attabad Lake. *Int. J.* **2021**, *9*, 703–714. [CrossRef]
96. Miliani, S.; Montessori, A.; La Rocca, M.; Prestininzi, P. Dam-Break Modeling: LBM as the Way Towards Fully 3D, Large-Scale Applications. *J. Hydraul. Eng.* **2021**, *147*, 4021017. [CrossRef]
97. Ai, C.; Ma, Y.; Ding, W.; Xie, Z.; Dong, G. Three-Dimensional Non-Hydrostatic Model for Dam-Break Flows. *Phys. Fluids* **2022**, *34*, 22105. [CrossRef]
98. Issakhov, A.; Borsikbayeva, A.; Abylkassymova, A.; Issakhov, A.; Khikmetov, A. Numerical Modeling of the Dam-Break Flood over Natural Rivers on Movable Beds. *Int. J. Nonlinear Sci. Numer. Simul.* **2022**. [CrossRef]
99. Yang, Y.; Zhou, X.; Chen, X.; Xie, C. Numerical Simulation of Tailings Flow from Dam Failure over Complex Terrain. *Materials* **2022**, *15*, 2288. [CrossRef] [PubMed]
100. Zhuang, Y.; Jin, K.; Cheng, Q.; Xing, A.; Luo, H. Experimental and Numerical Investigations of a Catastrophic Tailings Dam Break in Daye, Hubei, China. *Bull. Eng. Geol. Environ.* **2022**, *81*, 9. [CrossRef]
101. Jiang, H.; Zhao, B.; Dapeng, Z.; Zhu, K. Numerical Simulation of Two-Dimensional Dam Failure and Free-Side Deformation Flow Studies. *Water* **2023**, *15*, 1515. [CrossRef]
102. Oertel, M.; Bung, D.B. Comparison of 2D Dam-Break Waves with VOF and SPH Method. In *Proceedings of the 35th IAHR World Congress, Chengdu, China, 8–13 September 2013*; Tsinghua University Press: Beijing, China. Available online: https://www.iahr.org/library/infor?pid=14676 (accessed on 24 August 2023).
103. Purbasari, R.J.; Suryanto, A.; Anam, S. Numerical Simulations of Dam-Break Flows by Lattice Boltzmann Method. *AIP Conf. Proc.* **2021**, *2021*, 60027. [CrossRef]
104. Maquignon, N.; Smaoui, H.; Sergent, P.; Bader, B. A Simplified and Stable Lattice Boltzmann Shallow Water Model. *J. Phys. Conf. Ser.* **2022**, *2202*, 12055. [CrossRef]
105. LeVeque, R.J. *Finite Volume Methods for Hyperbolic Problems*; Cambridge University Press: Cambridge, UK, 2002.
106. Dick, E. Introduction to Finite Element Methods in Computational Fluid Dynamics. In *Computational Fluid Dynamics*, 3rd ed.; Wendt, J.F., Ed.; Springer: Berlin, Germany, 2009; pp. 235–274.
107. Hervouet, J.-M. *Hydrodynamics of Free Surface Flows: Modelling with the Finite Element Method*; Wiley: Chichester, UK, 2007.
108. Munoz, D.H.; Constantinescu, G. A Fully 3-D Numerical Model to Predict Flood Wave Propagation and Assess Efficiency of Flood Protection Measures. *Adv. Water Resour.* **2018**, *122*, 148–165. [CrossRef]
109. Rong, Y.; Zhang, T.; Zheng, Y.; Hu, C.; Peng, L.; Feng, P. Three-Dimensional Urban Flood Inundation Simulation Based on Digital Aerial Photogrammetry. *J. Hydrol.* **2020**, *584*, 124308. [CrossRef]
110. Peng, L.; Zhang, T.; Li, J.; Feng, P. Three-Dimensional Numerical Study of Dam-Break Flood Impacting Problem with VOF Method and Different Turbulence Closures. *Water Resour. Manage.* **2023**, *37*, 3875–3895. [CrossRef]
111. Pu, J.H.; Shao, S.; Huang, Y.; Hussain, K. Evaluations of SWEs and SPH Numerical Modelling Techniques for Dam Break Flows. *Eng. Appl. Comput. Fluid Mech.* **2013**, *7*, 544–563. [CrossRef]
112. Issakhov, A.; Zhandaulet, Y.; Nogaeva, A. Numerical Simulation of Dam Break Flow for Various Forms of the Obstacle by VOF Method. *Int. J. Multiph. Flow* **2018**, *109*, 191–206. [CrossRef]
113. Park, I.-R.; Kim, K.-S.; Kim, J.; Van, S.-H. Numerical Investigation of the Effects of Turbulence Intensity on Dam-Break Flows. *Ocean Eng.* **2012**, *42*, 176–187. [CrossRef]
114. Larocque, L.A.; Imran, J.; Chaudhry, M.H. 3D Numerical Simulation of Partial Breach Dam-Break Flow Using the LES and k–ϵ Turbulence Models. *J. Hydraul. Res.* **2013**, *51*, 145–157. [CrossRef]
115. Yang, S.; Yang, W.; Qin, S.; Li, Q. Comparative Study on Calculation Methods of Dam-Break Wave. *J. Hydraul. Res.* **2019**, *57*, 702–714. [CrossRef]
116. Simsek, O.; Islek, H. 2D and 3D Numerical Simulations of Dam-Break Flow Problem with RANS, DES, and LES. *Ocean Eng.* **2023**, *276*, 114298. [CrossRef]
117. Neal, J.; Villanueva, I.; Wright, N.; Willis, T.; Fewtrell, T.; Bates, P. How Much Physical Complexity is Needed to Model Flood Inundation? *Hydrol. Process.* **2012**, *26*, 2264–2282. [CrossRef]
118. Goutal, N. The Malpasset Dam Failure. An Overview and Test Case Definition. In Proceedings of the 4th CADAM Meeting, Zaragoza, Spain, 18–19 November 1999.
119. Hervouet, J.-M.; Petitjean, A. Malpasset Dam-Break Revisited with Two-Dimensional Computations. *J. Hydraul. Res.* **1999**, *37*, 777–788. [CrossRef]

120. Alcrudo, F.; Mulet, J. Description of the Tous Dam Break Case Study (Spain). *J. Hydraul. Res.* **2007**, *45* (Suppl. S1), 45–57. [CrossRef]
121. Pilotti, M.; Maranzoni, A.; Tomirotti, M.; Valerio, G. 1923 Gleno Dam Break: Case Study and Numerical Modeling. *J. Hydraul. Eng.* **2011**, *137*, 480–492. [CrossRef]
122. Testa, G.; Zuccalà, D.; Alcrudo, F.; Mulet, J.; Soares-Frazão, S. Flash Flood Flow Experiment in a Simplified Urban District. *J. Hydraul. Res.* **2007**, *45* (Suppl. S1), 37–44. [CrossRef]
123. Pilotti, M.; Milanesi, L.; Bacchi, V.; Tomirotti, M.; Maranzoni, A. Dam-Break Wave Propagation in Alpine Valley with HEC-RAS 2D: Experimental Cancano Test Case. *J. Hydraul. Eng.* **2020**, *146*, 5020003. [CrossRef]
124. Güney, M.S.; Tayfur, G.; Bombar, G.; Elci, S. Distorted Physical Model to Study Sudden Partial Dam Break Flows in an Urban Area. *J. Hydraul. Eng.* **2014**, *140*, 5014006. [CrossRef]
125. Norton, J. An Introduction to Sensitivity Assessment of Simulation Models. *Environ. Model. Softw.* **2015**, *69*, 166–174. [CrossRef]
126. Pianosi, F.; Beven, K.; Freer, J.; Hall, J.W.; Rougier, J.; Stephenson, D.B.; Wagener, T. Sensitivity Analysis of Environmental Models: A Systematic Review with Practical Workflow. *Environ. Model. Softw.* **2016**, *79*, 214–232. [CrossRef]
127. Castro-Orgaz, O.; Hager, W.H.; Katopodes, N.D. Variational Models for Nonhydrostatic Free-Surface Flow: A Unified Outlook to Maritime and Open-Channel Hydraulics Developments. *J. Hydraul. Eng.* **2023**, *149*, 4023014. [CrossRef]
128. Yu, D.; Tang, L.; Ye, F.; Chen, C. A Virtual Geographic Environment for Dynamic Simulation and Analysis of Tailings Dam Failure. *Int. J. Digit. Earth* **2021**, *14*, 1194–1212. [CrossRef]
129. Macchione, F.; Costabile, P.; Costanzo, C.; De Santis, R. Moving to 3-D Flood Hazard Maps for Enhancing Risk Communication. *Environ. Model. Softw.* **2019**, *111*, 510–522. [CrossRef]
130. Spero, H.R.; Vazquez-Lopez, I.; Miller, K.; Joshaghani, R.; Cutchin, S.; Enterkine, J. Drones, Virtual Reality, and Modeling: Communicating Catastrophic Dam Failure. *Int. J. Digit. Earth* **2022**, *15*, 585–605. [CrossRef]
131. FLOW-3D Modeling Capabilities. Hybrid Shallow Water/3D Flow. Available online: https://www.flow3d.com/modeling-capabilities/hybrid-shallow-water-3d-flow/ (accessed on 27 April 2023).

water MDPI

Article

Dam Break Modeling in a Cascade of Small Earthen Dams: Case Study of the Čižina River in the Czech Republic

Jaromír Říha, Stanislav Kotaška * and Lubomír Petrula *

Faculty of Civil Engineering, Institute of Water Structures, Brno University of Technology, Veveří 331/95, 602 00 Brno, Czech Republic; riha.j@fce.vutbr.cz

* Correspondence: kotaska.s@fce.vutbr.cz (S.K.); petrula.l@fce.vutbr.cz (L.P.); Tel.: +420-541-147-761 (S.K.); +420-541-147-756 (L.P.)

Received: 26 June 2020; Accepted: 14 August 2020; Published: 17 August 2020

Abstract: Failures of small dams can pose a serious threat to people and property even if the size of the schemes is relatively low. In many cases, small dams are situated in a cascade along streams, meaning that the failure of the uppermost dam may cause the dams downstream to fail. In this paper, a cascade of three small reservoirs, Lichnov II (14.6 m high), Lichnov III (10 m high), and Pocheň (8.5 m high), is the subject of the dam break analyses carried out via various methods such as empirical formulae, analogy, and hydraulic modeling. The dam-break flood routing was simulated using a shallow water flow hydraulic model. The simulations confirm that the attenuation effect of the peak discharge is governed by the flood volume, slope, and morphology of the floodplain and increases with the distance from the breached dam following an approximately exponential trend. When estimating peak discharge, empirical formulae derived for a single dam break should be applied carefully as they may underestimate the peak outflow by up to 10% in the case of a dam cascade. The attenuation volume of small reservoirs is small when compared to the flood volume, meaning that the attenuation of the peak discharge usually varies between 5–10%.

Keywords: dam break; dam overtopping; internal erosion; cascade of dams; erodibility

1. Introduction

Every dam represents a certain threat due to the possibility of its failure, which may lead to a dam-break flood in the area downstream. Statistics show that the most frequent causes of earth-fill dam failures are overtopping and internal erosion [1–3]. The safety of small dams, ponds, and dry reservoirs is still topical. In the Czech Republic, there are about 20,000 small dams. During extreme floods in 1997, 2002, and 2010, dozens of collapsed small dams were reported [4]. Therefore, dam break studies, including dam-break flood propagation modeling are also carried out for systems of small dams. An example of the application of dam break modeling to a cascade of small earthen dams is presented in this case study of the Čižina River in the Czech Republic.

For Emergency Action Plans, it is necessary to specify the characteristics of the dam-break flood and the extent of the endangered area. The knowledge concerning such flood-prone areas aids in the development of warning systems and evacuation plans. Various hydraulic and sediment transport models can be used for this purpose [5–8].

Numerous authors have dealt with the issue of dam breach modeling. In the 1970s, Cristofano [1] collected data from historical failures of earth-fill dams occurring up to 1965, based on which he performed a simulation of the onset of erosion. In the early eighties, the BRDAM model [9] was used to simulate the erosion of an earth dam in the case of overtopping or internal erosion.

Subsequent authors [10] improved the model by specifying the place of origin, simulating the erosion at the lowest point of the dam crest. The surface erosion proceeding on the downstream slope was modeled. The model used continuity equations, sediment transport equations, and equations for the formation of the dam breach opening via surface erosion. Nogueria [11] subsequently presented a 1D model using Saint–Venant's equations to describe unsteady flow and Exner's, Meyer–Peter, and Muller's formula for sediment transport. Almost at the same time, Fread [12] published the numerical models BREACH and Dam Break Forecasting model (DAMBRK), which allowed the simulation of a dam breach via both overtopping and internal erosion, including the effect of lateral slope due to a landslide. That same year, Fread [13,14] introduced the National Weather Service dam-break flood forecasting model (NWS DAMBRK) numerical model. It was based on complex hydrodynamics and erosion processes, and marked the first inclusion of soil strength characteristics within such a model. An improved model, BEED [15], also simulated erosion processes during a dam break. The existing empirical and numerical procedures were summarized by Wahl [3], who proposed an inventory for the development and further improvement of numerical models. These works were followed by Říha and Daněček [16], Tingsanchali, and Chinnarasri [17], and others using different approaches to derive a breach outflow hydrograph using different erosion formulae. Later, a group of authors at HR Wallingford [18–20] developed the numerical model named HR BREACH now Embankment Breach (EMBREA), which allows the input of parameter intervals characterizing the material properties of soils, and then performs Monte–Carlo simulations to find the most unfavorable result of the simulation. In 2005, Hanson developed the Simplified Breach Analysis (SIMBA) now Windows Dam Analysis Modules (WINDAM) model for the Agricultural Research Service and Natural Resources Conservation Service in the US [21]. Currently, the newest version of the program (WINDAM C) is able to model the overtopping and internal erosion of embankment dams, as well as head-cut erosion and the failure of protective layers. In 2012, the AREBA software was developed by [22]. The software is capable of simulating a dam break via overtopping or internal erosion in homogeneous and composite embankments. The numerical module Breach_Macchione [23] was adopted to the code Basement in 2012 by Voltz et al. [24]. The model is based on the application of two-dimensional shallow-water equations to breaching processes occurring in non-cohesive embankments due to overtopping. Wu [25] developed A Simplified Physically-Based Dam/Levee Breach (DL Breach), which simulates dam and levee breaching via overtopping and internal erosion in homogeneous and composite materials. It can simulate the failure of cover surfaces and can perform Monte–Carlo simulations. Recently, researchers have incorporated simplified modules into River Analysis System (HEC-RAS), A modelling system for Rivers and Channels (MIKE11 DB), WOLF, Rubar 20TS, and other software for modeling water flow in inundation areas [5,26–28]. Recent work by King and Simonovic [29] presents a deterministic Monte–Carlo simulation concept used for the assessment of dam safety management control.

In the case of the breaching of dams in a cascade, the resulting dam-break flood depends on numerous factors like the rate of breaching of the upper dam, the ratio of the volumes of the reservoirs in the cascade, and the ability of the lower reservoirs to attenuate the flood wave. The distance between individual reservoirs and the characteristics of the floodplain are also important factors. Therefore, the dam-break flood routing along a cascade of reservoirs may vary according to the failure mode of the uppermost dam and to the current water level and free volume of the reservoirs in the cascade. Usually, the most unfavorable scenario is used for Emergency Action Plans. Only a few studies deal with dam breaks in a cascade of reservoirs [30–33]. Liu et al. [31] describe a procedure for the numerical simulation of a dam breach due to overtopping in a cascade together with risk analysis. This includes the 1D flow open channel modeling of flood propagation. Stoyanova and Coombs [32] use the methodology applied in the AREBA software. Both simple two-reservoir and more complex four-reservoir cascade scenarios are dealt with. Cai et al. [30] applied Bayesian networks for the quantification of the flood risk due to dam breaks in a cascade. Studies [30–33] focused on large dams with the possibility of outflow discharge regulation.

The lessons learned during past extreme floods [2,4,34] show that on average, about ten small dams fail in the Czech Republic during extreme flood events and that such dams are more vulnerable to breaching than large dams, which is primarily due to their insufficient spillway capacity. As a rule, such schemes have a flood attenuation volume, which is relatively small when compared to the volume of the arriving extreme flood. Their fixed spillways usually have a limited capacity due to the low importance of the dams.

This study aims to fill a gap in the research concerning the breaching of small embankment dams in a cascade. It deals with a cascade of three relatively new small embankment dams in the Čižina River catchment in the Moravian–Silesian Region of the Czech Republic. One of the aims was to demonstrate the effect of the distance between dams on dam-break flood attenuation. In our case study, the two upper reservoirs are quite close to one another, while the lower one is located more remotely downstream along the Čižina River. A complete set of technical data about the dams was available. The lowest dam, Pocheň, has already experienced overtopping twice but it resisted in both cases and did not fail.

Based on the results of the analysis, typical dam-break flood routing characteristics are summarized for a cascade of small dams. The method proposed in the study may be used for the development of Emergency Action Plans containing dam break and flood routing analyses for the affected area below a dam. The results of such analyses contain flood arrival times and maps with flood zones where water velocities and depths are included.

2. Case Study

2.1. Study Area

The case study covers an area of the Čižina River catchment in the Moravian–Silesian Region of the Czech Republic (Figure 1) that extends for about 103 km^2.

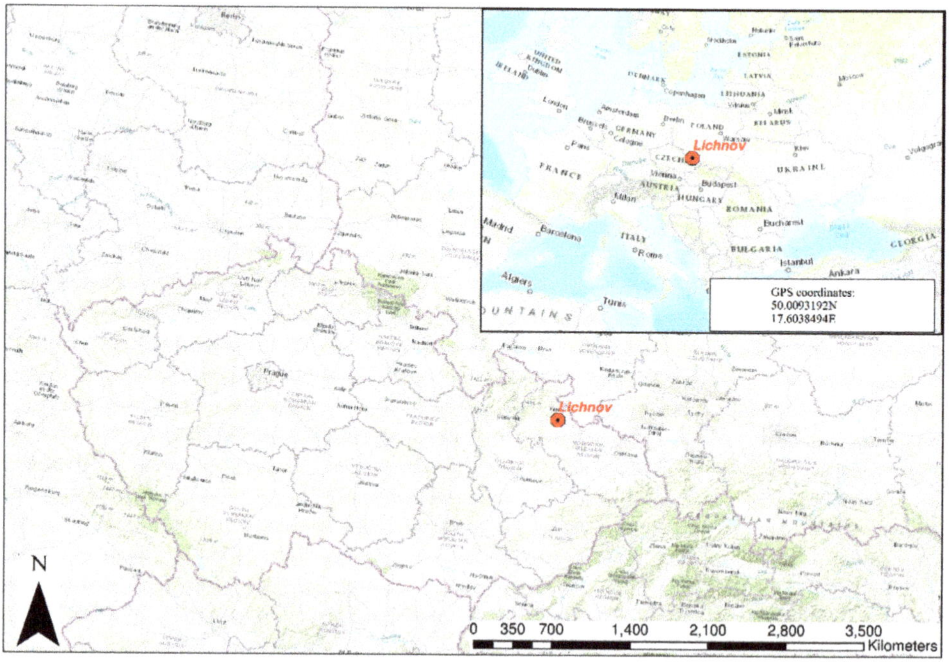

Figure 1. Location of the area.

In the catchment area, three small reservoirs named Lichnov II, Lichnov III, and Pocheň are arranged in a cascade (Figure 2). The cascade was selected because the distance between the dams differs significantly, allowing the demonstration of the impact of distance on dam-break flood attenuation. The two upper reservoirs are quite close to one another, while the lower one (Pocheň) is about 15 km downstream from the Lichnov III dam. A complete set of technical data about the dams was available. The lowest dam, Pocheň, has already experienced overtopping twice, but it resisted in both cases and did not fail.

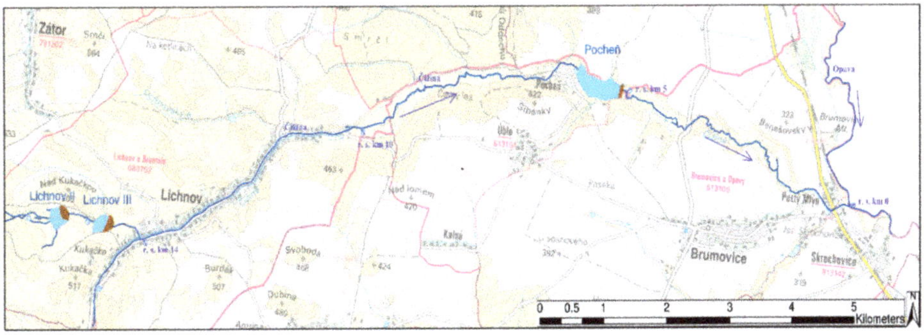

Figure 2. Map of the area.

2.2. *Description of the Small Dams*

The main characteristics of the studied dams are described in Table 1.

Table 1. Description of the small dams.

Basic Characteristic of Dam	Lichnov II	Lichnov III	Pocheň	Units
Dam height	14.6	10	8.5	(m)
Crest width	5	4	4	(m)
Upstream slope	2	3	3	(1/X)
Downstream slope	2.8	2.6	2.3	(1/X)
Material classification	Gravel clay	Loam gravel	Gravel clay	(-)

The Lichnov II dam is located about 1 km above the village of Lichnov on Tetřeví Stream, which is a tributary of the Čižina River. The dam was built in 2015. The main purpose of the reservoir is to provide flood protection to the villages downstream. The dam is homogenous and is made of sandy to gravelly clay. The maximum height of the dam above the local terrain is 16 m, and the dam crest is 5 m wide. A longitudinal section along the dam crest is shown in Figure 3, while a typical cross-section is displayed in Figure 4. The dam crest elevation at the left abutment is 0.85 m lower than the dam crest due to the presence of an emergency spillway. This place where overtopping may start, and water flows rapidly along the downstream slope was considered to be the potential point of origin of a breach due to overtopping. At the toe of the right abutment, extensive leakage through disintegrated colluvial soil below the base of the dam was identified via technical surveillance during the first trial filling of the reservoir. It is considered that there is a threat of backward erosion piping and a possible breach at this location.

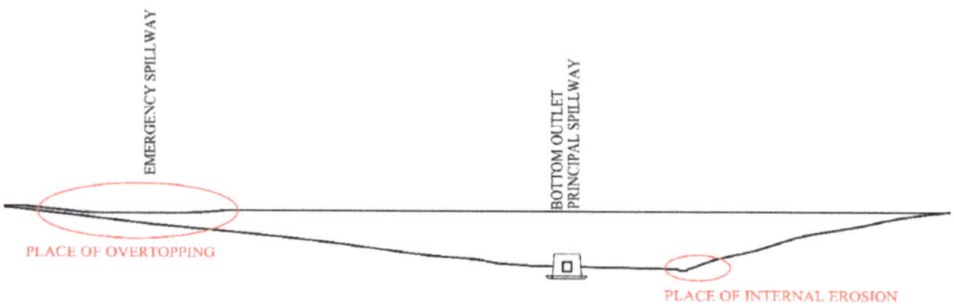

Figure 3. Longitudinal section of the Lichnov II dam with indicated breach locations.

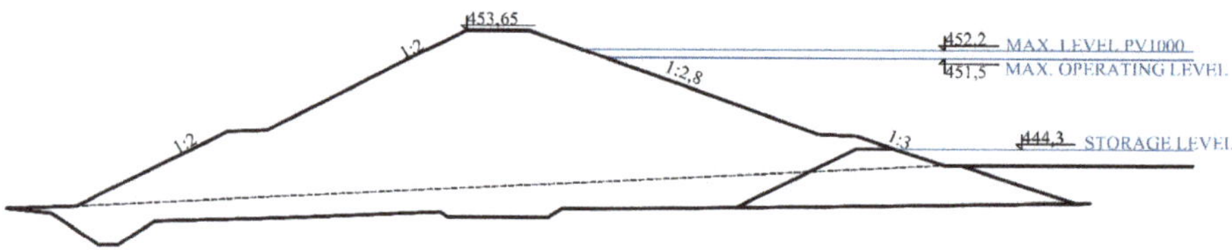

Figure 4. Typical cross-section of the Lichnov II dam.

The length of Tetřeví Stream between the Lichnov II and Lichnov III reservoirs is relatively small, about 500 m. The valley is relatively steep and narrow and is covered by forest and meadows.

The small detention dam Lichnov III is located approximately 400 m upstream of the village of Lichnov. It was built in 2018. Its purpose is to protect the village and to transform incoming flood waves from the Lichnov II dam and from the area adjacent to the reservoir. The dam body is homogeneous. It is made of sandy to gravelly clay and is very similar to that of Lichnov II. The dam height is about 12 m; see Table 1 for its main characteristics and parameters. As a dam-break flood arriving from the Lichnov II dam would be sudden, it is not probable that a gradually evolving piping process would develop, and so only overtopping was considered as a potential failure mode. The place of overtopping was similar to that of Lichnov II, i.e., at the auxiliary crest spillway at the left abutment (Figure 5).

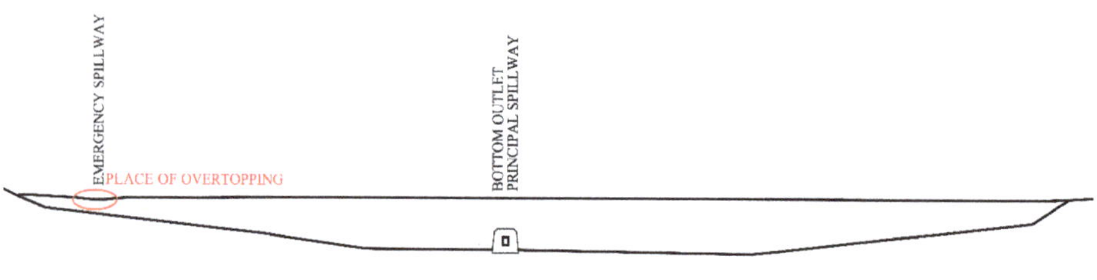

Figure 5. Longitudinal section of the Lichnov III dam with indicated possible failure locations.

About 600 m downstream of the Lichnov III dam, Tetřeví Stream flows into the Čižina River, which passes through the village of Lichnov. Numerous civil structures such as bridges and culverts are located there, and the area along the 3.5 km long river reach is urbanized with numerous buildings. Further, on the downstream of the village, the Čižina River continues into a region of fields and meadows covered with trees. This section of the Čižina River stretches for about 6 km, after which the river enters the Pocheň reservoir.

The small dam named Pocheň is located on the Čižina River about 5 km upstream of its junction with the larger Opava River. The Pocheň dam was built between 1973 and 1975 as a source of water for agricultural activities in nearby areas. During the extreme flood event in the year 1996, the dam was seriously damaged by overtopping but not breached. Due to this damage, reconstruction of the dam body took place, and the spillway capacity was improved over the following years. The dam height is about 9.5 m, and the dam is constructed from local gravelly clay. The geodetic survey indicated the lowest crest elevation is approximately in the middle of the crest length (see Figure 6). Due to the short duration of the extreme load, a progressive piping failure would not develop, and so only the overtopping failure mode was taken into account.

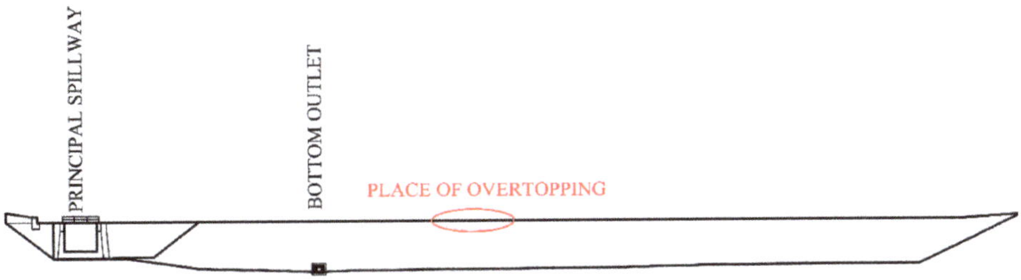

Figure 6. Longitudinal section of the Pocheň dam with an indicated place of overtopping.

Downstream of the Pocheň reservoir, the Čižina River passes through an area of forest, flows below several road and railway bridges, and then enters the Opava River close to the village of Skrochovice.

3. Methods

Both empirical formulae and comparison with real incidents were applied for the first preliminary estimates of the dam break peak discharges. The obtained values provide an idea about the range of the peak outflows and provide a general framework for numerical modeling. Due to the lack of relevant incidents and reliable data from real breached dams, these equations suffer from a great deal of inaccuracy and are mainly useful for the preliminary estimation of parameters, which can then be determined more precisely by numerical models.

The dam break simulation was carried out using the numerical model of breaching due to overtopping and internal erosion (the latter phenomenon at the uppermost dam only), which is normally applied to individual dams. The peak discharges were compared with the results of empirical formulae and with real cases involving the failure of small dams.

Two-dimensional shallow-water flow hydraulic models were applied for the simulation of the dam-break flood propagation into the areas between the dams. The flow and erosion parameters were searched via Monte–Carlo random sampling to achieve the highest peak outflow discharge below the dam.

3.1. Empirical Formulae

In order to obtain an initial idea about the peak discharges, published empirical formulae were applied. This approach is the simplest but least accurate way of predicting dam breach parameters. In dam-break simulations, direct calibration of models is usually impossible. One of the methods of verifying numerical models is to compare the modeling results with the outputs of retrospective studies of real incidents and failures. Unlike, e.g., hydrological floods, there is a lack of dam-break calibration data worldwide due to a lack of relevant incidents and reliable data from real breached dams. The empirical formulae discussed in this study (Table 2) are based on such real cases. It is true that statistical analysis is difficult due to the heterogeneity, inaccuracy, and scatter of the dam break data when subjected to regression analysis. However, the published empirical formulae and the analogy method are frequently used in dam break analysis for the comparison of real incidents with the results of numerical analysis.

Table 2. Summary of formulae used for estimating the peak breach outflow Q_b.

Author, Reference	Formulae for Q_b *) (m^3/s)
Holomek, Říha (2000) [35]	$Q_b = m\sqrt{2g}\left[bH_k^{1.5} + 0.4\frac{B-b}{h_d}H_k^{1.5}\right]$
The following breach dimensions can be used:	
USACE (1980) [36]	$b = 3h_d$ & $B = 5h_d$
ICOLD (1974) [37]	$b = 1.6h_d$ & $B = 3.6h_d$
Šimek (1988) [38]	$b = 1.2h_d$ & $B = 3.2h_d$
Froehlich (1995a, 1995b) [39,40]	$Q_b = 0.607V_w^{0.295}h_w^{1.24}$
Webby (1996) [41]	$Q_b = 0.0443 \cdot g^{0.5}V_A^{0.367}h_w^{1.40}$
Bulletin 111 (1998) [42]	$Q_b = 0.07(gh_d)^{0.5}h_d^2\left(\frac{V_b}{h_d^3}\right)^{0,5}$
Vischer, Hager (1998) [43]	$Q_b = 1.25 \cdot 10^{-2}\frac{H_a}{h_d}\sqrt{gB_a^2h_d^3}$

*) An explanation of the variables can be seen in the section "List of symbols".

One of the aims of this study was to demonstrate the difference between a single dam break and a cascade when applying empirical formulae.

3.2. The Analogy Method

A more reliable peak breach outflow may be derived via the comparison of data from real incidents at dams with similar characteristics (dam type, height, reservoir volume) to the studied dam. The data may be obtained from publications and extensive dam failure databases (both national and international) [2,3,34,42,44,45]. In Table 3, dams similar to those studied are mentioned together with basic relevant characteristics (see also Chapter 2).

Table 3. Summary of the characteristics of the real breached dams and the studied dams.

Cases	Dam Name	Dam Height (m)	Reservoir Volume (ths. m^3)	Crest Width (m)	Breach Width (m)	Peak Outflow (m^3/s)
Real cases	Metly	8.50	1930	4.0–6.9	35.0 + 15.0 *)	550
	Velký Bělčický	6.70	1063	7.4	42.0	600
	Buffalo Creek	14.00	483.5	107.0	148.0	1416
	Kelly Barnes	11.50	505	6.1	6.7	680
Studied dams	Lichnov II	14.60	400.6	5.0	-	-
	Lichnov III	10.00	117.9	4.0	-	-
	Pocheň **)	8.50	780	4.5	-	163

*) Two breach openings. **) The dam was not completely breached.

3.3. Embankment Dam Breach Modeling

A simple parametric model based on hydraulic and transport modules was used for the simulation. The tailor-made code was developed in MATLAB software to allow the dam breaching process to be modified easily according to local conditions and the failure mode.

The main output of the dam break analysis is the outflow hydrograph $Q(t)$, which results from the failure. The following steps were performed before the modeling took place.

A careful investigation was performed with regard to the dam type and structure (material, emergency spillway position, dam surface, type of pavement on the dam crest, etc.), as well as the local geological conditions (subbase material, material discontinuities).

The probable place of failure initiation was located; for an overtopping failure, the place with the minimum dam crest or auxiliary spillway height was chosen. For a piping failure, the location with the weakest subbase providing the highest seepage was chosen.

1. The shape of the dam break opening was rectangular for overtopping and circular for piping.
2. The flood hydrograph Q_{in} (t) entering the reservoir was provided by Czech Hydrometeorological Institute.

3. The bathymetry of the reservoirs was taken from a geodetic survey of the reservoir area.
4. The initial reservoir water level $H_0(t)$ was taken from the dam operation manual. It corresponds to the live storage water level.

The following relation between the reservoir inflow, outflow, and a change in water level holds:

$$\frac{dV}{dt} = Q_{in}(t) - Q_b(H) - Q_f(H) \quad (1)$$

where V is the reservoir water volume (m^3), t is time (s), Q_{in} is the reservoir inflow (m^3/s), Q_b is the breach outflow (m^3/s) (in the following text, Q_{bo} denotes overtopping discharge and Q_{bp} piping discharge), Q_f is the outflow through appurtenant works (outlets, spillways, etc.) in (m^3/s), and H is the elevation of the water level (m a. s. l.).

The outflow through the appurtenant works governed by the reservoir water level was calculated using equations taken from standard dam hydraulics [46]. The equations used and related coefficients were determined according to the arrangement of the hydraulic structures.

3.3.1. Overtopping

In the case of dam overtopping, the breach was approximated using a 1D model. The breach started at the lowest point of the dam crest, where the first overtopping occurred. The overtopping width along the dam crest was suggested as an initial breach width, b_0. The shape of the breach opening was approximated by a rectangle, and the erosion progress was applied in both the downward and the lateral direction. The water moving along the downstream slope was assumed to exhibit quasi-steady uniform flow [3,13,45] with no submergence from the downstream side. The resistance of the dam body against the surface erosion was evaluated with respect to the water flow velocity at the downstream slope, with non-scouring velocity being used for the evaluation. Parallel uniform gradual backward erosion of the downstream slope was assumed [13]. The angle of the downstream slope and the width of the dam crest were set as input parameters, along with hydraulic parameters such as the discharge coefficient m, Manning roughness n, and the non-scouring velocity v_{ns} and erosion parameters of the soil α_1 and α_2, together with initial conditions (see below).

The breach outflow during overtopping Q_{bO} was calculated using an overflow formula for a rectangular broad crested weir [47]:

$$Q_{bO} = mb\sqrt{2g}(H - Z)^{1.5} \quad (2)$$

where m is the weir coefficient, b is the width of the breach opening (m), g is the acceleration due to gravity (m/s^2), and Z is the breach bottom elevation (m a. s. l.).

Using the overtopping discharge obtained from Equation (2), the water depth h_f and velocity v_f along the slope was derived from the Chezy equation [47]:

$$h_f = \left(\frac{Q_{bO}n}{b\sqrt{sin\beta}}\right)^{0.6} \quad (3)$$

$$v_f = \frac{\sqrt{sin\beta}}{n}h_f^{0.67} \quad (4)$$

where n is the roughness of the downstream slope, and β is the downstream slope angle.

At the moment, when the flow velocity along the slope exceeds the non-scouring velocity v_{ns} (the value is an input parameter), erosion occurs and the breach starts to increase in width and depth [34]:

$$\frac{dZ}{dt} = -\alpha_1 v_f, \text{ for } v_f > v_{ns} \quad (5)$$

$$\frac{\mathrm{d}b}{\mathrm{d}t} = \alpha_2 v_f, \text{ for } v_f > v_{ns} \tag{6}$$

where α_1 and α_2 are erosion parameters.

Initial conditions specify the reservoir water level H_0, the lowest dam crest elevation Z_0, and initial overtopping width b_0 at the beginning of the solution (t = t_0). The set of Equations (1) to (6) was solved numerically using the Newton method with a small-time step Δt = 1 s.

3.3.2. Piping

For the piping model, a circular opening with a constant leakage pipe diameter and constant shear stress and hydraulic gradient along the pipe were assumed [19–22,28,48]. The input parameters were the length of the pipe L, the initial reservoir water level H_0, the elevation of the pipe outflow $H_{outflow}$, the critical shear stress along the walls of the pipe τ_c, the soil erodibility C_e, the initial pipe diameter D_0, and absolute roughness Δ.

The outflow through the pipe Q_{bP} and average water velocity v_{pip} in the pipe were determined using pipe hydraulics [47]:

$$Q_{bP} = \frac{1}{\sqrt{\alpha + \lambda \frac{L}{D}}} S \sqrt{2g\left(H - H_{outflow}\right)} \tag{7}$$

$$v_{pip} = \frac{Q_{bP}}{S} \tag{8}$$

where α is the kinetic energy (Coriolis) coefficient, L is the length of the pipe (m), D is the pipe diameter (m), λ is the friction loss coefficient, S is the pipe cross-sectional area (m^2), and $H_{outflow}$ is the pipe outflow elevation (m a. s. l.).

Using the water velocity in the pipe, the average hydraulic gradient along the pipe I_E and shear stress τ acting on the pipe walls were obtained:

$$I_E = \lambda \frac{1}{D} \frac{v_{pip}^2}{2g} \tag{9}$$

$$\tau = \rho g \frac{D}{4} I_E \tag{10}$$

where ρ is the water density. When the critical shear stress τ_c on the pipe wall was exceeded, erosion and an increase in the pipe diameter initiated. The rate of erosion $\dot{\varepsilon}$ and the change in pipe radius dr were determined according to equations suggested by Wan and Fell [48]:

$$\dot{\varepsilon} = C_e(\tau - \tau_c) \tag{11}$$

$$\mathrm{d}r = \frac{\dot{\varepsilon}}{\rho_b} \mathrm{d}t \tag{12}$$

where $\dot{\varepsilon}$ is the rate of erosion (kg/s/m^2), C_e is the coefficient of soil erosion (s/m), ρ_b is the bulk density of the eroded soil (kg/m^3), and $r = D/2$ is the pipe radius.

The initial reservoir water level H_0 and initial pipe diameter D_0 are the initial conditions. As in the case of overtopping, the solution of Equations (1) and (7) to (12) was carried out using the Newton method with a very small-time step of Δt = 1 s.

During the solution, the relation between the pipe diameter and the thickness of the roof of the conduit (overburden) was checked. After the pipe diameter exceeded the overburden thickness, it was assumed that the soil layer above the opening collapsed and the process continued similarly as in the case of overtopping. The width of the breach opening corresponded to the pipe diameter at the instant of the overburden collapse, while the breach bottom was assumed to be at the top of the seepage pipe at the instant of the collapse.

3.3.3. Optimization Procedure

Most of the input parameters in both scenarios are rather uncertain, and realistic intervals of their values were estimated from literature sources and real cases [3,41]. The combination of parameter values from their intervals may result in significant differences in outflow hydrographs and peak outflow discharge. In order to obtain the least favorable result with the highest peak outflow, the "optimization" of the parameters mentioned in Table 4 was performed via the Monte–Carlo sampling procedure [49].

Table 4. Intervals of parameters for the Monte–Carlo sampling procedure.

Overtopping.			Piping		
Parameter (units)	**Minimum**	**Maximum**	**Parameter (units)**	**Minimum**	**Maximum**
n (s/m$^{1/3}$)	0.025	0.035	d_0 (mm)	5	100
α_1 (-)	0.0002	0.0030	Δ (m)	4.56×10^{-4}	9.58×10^{-4}
α_2 (-)	0.0002	0.0015	τ_c (Pa)	1.3	64
v_{ns} (m/s)	1.8	3.0	C_e (s/m)	1.38×10^{-3}	4.24×10^{-3}

The most influential parameters for the optimization were determined based on the equations governing the breach outflow discharge. The shape of the dam break hydrograph and namely the peak outflow discharge is generally governed dominantly by the size of the breach opening and the water level in the reservoir (Equations (2) and (7)). Crucial variables influencing the shape and size of the breach opening are the erodibility parameters (Equations (4)–(6) for overtopping and Equations (8) and (12) for piping). The reservoir water level is determined via level pool routing through the reservoir, Equation (1), and depends on the reservoir bathymetry $V(H)$, the inflow flood hydrograph Q_{in}, the capacity of the appurtenant works Q_f, and the breach outflow Q_b. The importance of the mentioned variables was proved by sensitivity and statistical analyses carried out in a previous study [50]. In this study, the members in Equation (1) were not subject to optimization as they were "prescribed" by the Czech Hydrometeorological Institute (inflow hydrograph), by the reservoir bathymetry, and by the capacity of the spillway.

For each dam, about 10 thousand simulations were carried out using parameter values obtained from random sampling. A uniform probability distribution was used for each parameter. As a result, a "sub-optimal" scenario was found after a reasonable amount of computational time with sufficient accuracy when compared to other simplifications adopted (the mathematical model, for example). The scenario with the highest peak outflow was used for further processing.

3.4. *Dam-Break Flood Routing*

The dam-break flood routing between reservoirs was simulated using a two-dimensional (2D) model based on shallow-water flow equations [8]. The solution was performed using the commercial code HEC-RAS v. 5.0.7 [5]. As the model used is standard and the procedures are routine, the model is not described here in detail, and the authors refer those interested to the available literature [5,46]).

The inputs for the calculation were the topology and geometry of the area between the dams taken from a digital model of the terrain (DTM). Civil structures were assumed to be fixed, with no damage suffered due to the flow of water.

The upstream boundary condition was represented by the dam breach flood hydrograph and hydrological flood hydrographs at contributing local streams. The initial condition was set as a stream discharge and water level corresponding to a 20-year return period, which corresponds approximately to the bank-full capacity of the streams.

4. Simulations and Results

First, the peak outflow from the dams was estimated using empirical formulae (Table 2) based on the dam and reservoir characteristics (Table 3). A comparison with real dam failures was also performed.

The numerical simulation started from the uppermost dam, Lichnov II. From an analysis of hydrological and empirical data, a check flood inflow hydrograph from the area upstream of Lichnov II was derived by the Czech Hydrometeorological Institute (Figure 7), providing a peak discharge of 50 m^3/s. This approximately corresponds to the outflow of the largest experienced rainfalls at small river basins in the Czech Republic (more than 200 mm/day) during the floods in the years 1996, 1997, and 2010 at relatively small catchments with areas of 5 to 10 km^2.

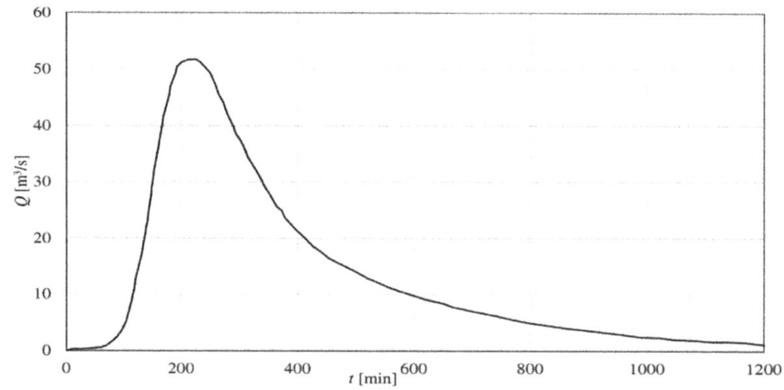

Figure 7. Hydrograph of the check flood at the inflow to the Lichnov II reservoir.

The runoff from the larger catchment area between the reservoirs at the time of the dam break events corresponds to a flood with a 20-year return period. This assumption is also based on past experience from real floods at catchments with areas of 50 to 60 km^2. This also corresponds to the capacity of related streams in smaller urban areas.

The dam break study proceeded from the upstream reservoir in the downstream direction in five steps containing level pool routing:

- through the Lichnov II reservoir (simulation of the dam breaching due to both overtopping and piping),
- between Lichnov II and Lichnov III,
- through the Lichnov III reservoir (simulation of the dam breaching due to overtopping),
- between Lichnov III and Pocheň,
- through the Pocheň reservoir (simulation of the dam breaching due to overtopping).

In order to achieve the worst-case scenario of dam breaching, in which the maximum peak outflow occurs from the collapsed dams, parameter optimization was carried out. The optimization was conducted using the parameters shown in Table 4, with their realistic intervals adopted from the available literature [2,3,37,40,45,50].

A summary of the final input values, including the final parameters obtained from the optimization procedure, is listed in Table 5. The input values mentioned in Table 5 result in the maximum outflow discharge from the reservoirs during the breach event.

Table 5. Summary of inputs for the numerical model.

Inputs for the Numerical Model (units)	Notation	Input Overtopping and Piping Values		
		Lichnov II	**Lichnov III**	**Pocheň**
Initial water level in the reservoir (m a. s. l.)	H_0	444.30	424.00	315.90
Dam crest elevation (m a. s. l.)	C_L	453.50	425.30	317.50
Dam bottom elevation (m a. s. l.)	B_L	442.00	414.50	308.00
Bottom outlet axis elevation (m a. s. l.)	H_{SV}	440.00	414.24	-
Auxiliary spillway crest elevation (m a. s. l.)	H_{aux}	452.80	424.80	-
Initial elevation of the lowest point on the dam crest (m a. s. l.)	Z_0	-	-	317.40
Discharge coefficient (-)	m	0.33	0.33	0.33
Initial overflow width at the auxiliary spillway/lowest point on the dam crest (m)	b_{spill}	25	12	15
Non-scouring velocity (m/s)	v_{ns}	2.2	2	3.1 **)
Manning roughness coefficient of the downstream slope (s/m$^{1/3}$)	n	0.031	0.032	0.030
Erosion parameter of the breach bottom (-)	α_1	0.0007	0.0008	0.0002
Erosion parameter of the breach banks (-)	α_2	0.0004 *)	0.0003 *)	0.00021
Initial pipe diameter (mm)	D_0	5	-	-
Critical shear stress (Pa)	τ_c	1.3	-	-
Soil erodibility (s/m)	C_e	1.3×10^{-3}	-	-
Elevation of the erosion pipe axis (m a. s. l.)	$H_{outflow}$	438.00	-	-
Length of the pipe (m)	L	74.5	-	-
Absolute roughness (m)	Δ	9.32×10^{-4}	-	-

*) Lateral erosion in one direction. **) Determined from real incidents.

At the Lichnov II dam, both overtopping and piping failure were modeled. Figure 8 shows that the piping initiated approximately at the time of the peak inflow while overtopping occurred about 25 min after the inflow peak. The initialization phase at overtopping was followed by the gradual enlargement of the breach opening up to a final width of 68 m, which was reached about 125 min after the first overtopping as a result of intensive erosion only at the left side towards the central axis of the dam. A comparison of the modeled breach outflow hydrographs (Figure 8) shows that the peak discharge caused by the overtopping failure (802 m^3/s) was more than 2.5 times higher than that caused by piping (316 m^3/s). Therefore, the hydrograph from the overtopping failure was adopted for further analysis (Figure 8) of the cascade. This hydrograph provides a quite realistic duration for the breach event (approimately 30 min), which corresponded to that experienced during the breaching of small dams in the Czech Republic [34].

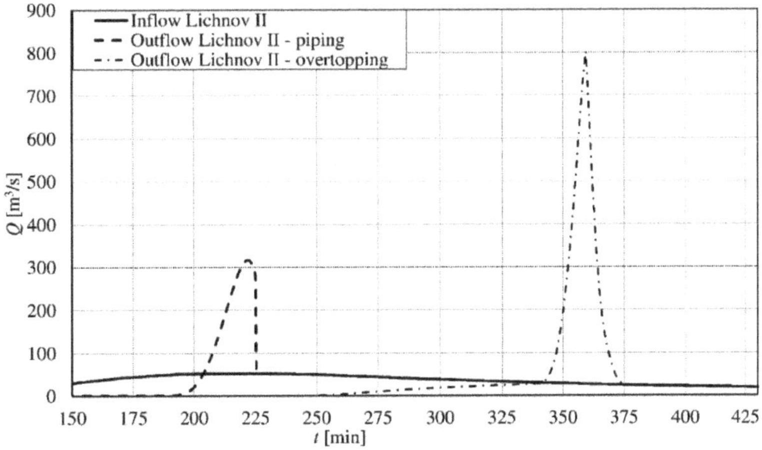

Figure 8. Comparison of the maximum breach outflow from Lichnov II due to overtopping and piping.

Most of the studied real dam failures exhibited peak discharges of 550 to 680 m^3/s, except for in the case of the Buffalo Creek dam, which had 1416 m^3/s (Table 6). Empirical formulae provide values, which are closer to the higher modeled discharge (485 to 1763 m^3/s).

Table 6. Summary results—peak breach outflows from Lichnov II, Lichnov III, and Pocheň.

Method of Determination		Maximum Breach Outflow Q_{bmax} (m^3/s)		
		Lichnov II	**Lichnov III**	**Pocheň**
Comparison with real incidents at dams	Metly	550	550	550
	Velký Bělčický	600	600	600
	Buffalo Creek	1416	-	-
	Kelly Barnes	680	-	680
Empirical formulae	U.S. Army Corps of Engineers (1980) [36]	1763	1225	975
	ICOLD (1974) [37]	1078	754	595
	Šimek (1988) [38]	980	686	541
	Froehlich (1995a, 1995b) [39,40]	565	307	506
	Webby (1996) [41]	485	231	438
	Bulletin 111 (1998) [42]	578	224	628
	Vischer and Hager (1998) [43]	795	637	259
	Overtopping	802	841	557
	Piping	316	-	-

The flood routing between the Lichnov II and Lichnov III reservoirs along the 500 m long valley reach provided relatively small attenuation of the peak discharge from 802 m^3/s to approximately 756 m^3/s with a shift in the peak of about 3 min (Figure 9).

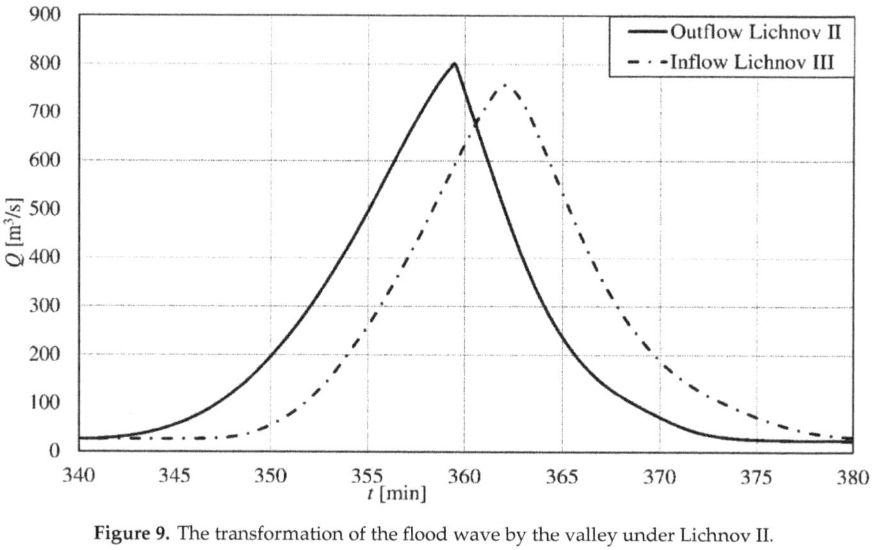

Figure 9. The transformation of the flood wave by the valley under Lichnov II.

As mentioned above, the Lichnov III dam would be breached primarily by overtopping. The flood arrival was quite rapid (10 min), so the breaching started shortly after the rise in the reservoir water level. About 1 h was needed for the initiation of piping (Figure 8), and, therefore, the failure mode was primarily overtopping.

The modeled peak outflow of 841 m^3/s from the dam occurred a few minutes after the arrival of the flood from the upper Lichnov II reservoir. The duration of the breach event was about 25 min, and the time at which peak outflow occurred is slightly after the time that peak inflow to the reservoir took place (Figure 10). The maximum breach width was 25 m. From Table 6 it can be seen that the range of values for the peak outflow derived from individual formulae varies from 231 to 1225 m^3/s;

the modeled peak outflow exceeds practically all calculated values (except for those obtained from USACE methodology). This could be attributed to the domino cascade effect that occurs when two dams are in very close proximity. Comparisons with empirical relationships and real dam failures may not be entirely relevant, as the difference between single and cascade dam breach peak discharge is about 50–55%. It should be noted that empirical equations correspond to single dam failure and do not take into account the domino effect.

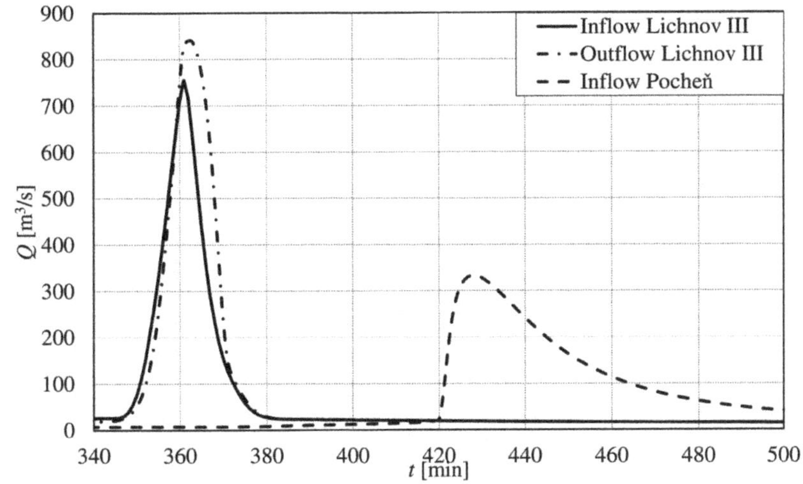

Figure 10. The breach inflow and dam break outflow of Lichnov III during an overtopping failure, and the transformation of the flood wave by the valley under Lichnov III.

If the Lichnov III dam were breached, the dam-break flood would arrive at the village of Lichnov fast, flooding most of the built-up area within 100 to 150 m of both banks of the Čižina River. Attenuation would be more significant below the village, where the floodplain of the Čižina River would be inundated over a 300 to 500 m wide area.

Flood routing along the remaining reach of Tetřeví Stream, and further on along the Čižina River resulted in a transformed hydrograph entering the Pocheň reservoir. From Figure 10, it could be seen that the peak discharge dropped from 841 m^3/s to about 330 m^3/s at the entrance to the Pocheň reservoir due to the attenuation along the 10 km long river valley between the Lichnov III and Pocheň dams (Figure 2). The time shift in the flood arrival time between both reservoirs is about 1 h.

The input values of the erosion resistance characteristics for the simulation of the Pocheň dam breaching were verified using data from real overtopping events that took place during floods in 1977 and 1996 (Figure 11). They correspond to a non-scouring velocity of $v_n \approx 3.1$ m/s, which was set in the calculation (it was not subject to optimization).

The resulting reservoir inflow and outflow hydrographs obtained by the simulation are shown in Figure 12. It can be seen that the peak outflow was delayed for approximately 1 h and 20 min after the culmination at the inflow. This time corresponds to the gradual filling of the flood attenuation reservoir storage, the overtopping of the relatively horizontal crest of the dam, and the slow development of the breach opening in the relatively resistant dam material. It can be seen that the peak inflow of 330 m^3/s into the reservoir was increased by the breaching to a peak outflow of 557 m^3/s, which is slightly higher than the average peak discharge obtained via comparative analysis and empirical formulae (Table 6). The minor difference between the simulated and empirical values was due to the characteristics of the cascade and other influences, including significant flood attenuation between the dams, the improved spillway capacity, and the rather higher dam body resistance against erosion.

Figure 11. Overtopping of the Pocheň dam in 1996.

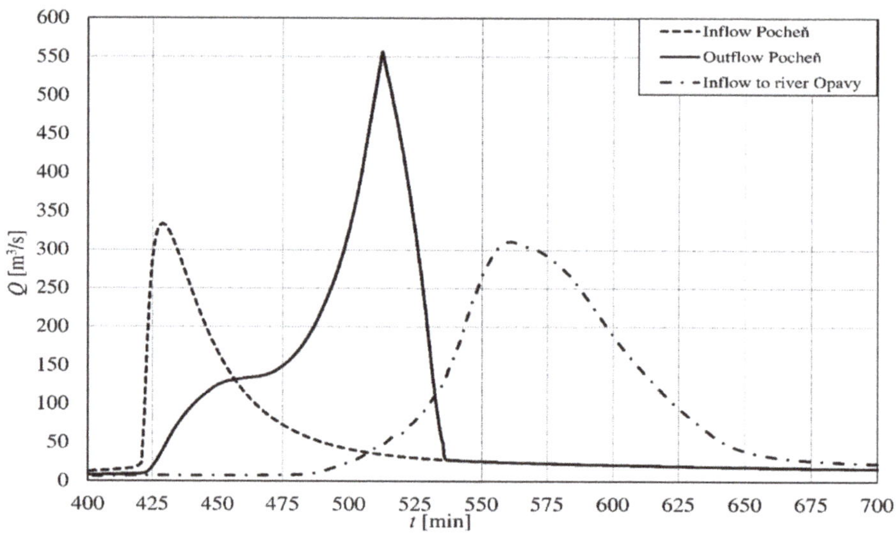

Figure 12. The breach inflow and dam break outflow at Pocheň due to overtopping failure, and the transformation of the flood wave by the valley under Pocheň.

The flood routing simulation continued down to the Opava River, which is located in a wide floodplain with road and railway embankments that create large "reservoirs" when flooded by the dam-break flood. This resulted in a relatively significant reduction of the 557 m^3/s peak outflow from the Pocheň dam to approximately 310 m^3/s at the junction with the Opava River (Figure 12). This peak discharge is lower than the 100-year return period discharge Q_{100} = 352 m^3/s of the hydrological flood in the Opava River downstream of the junction with the Čižina. As recommended in Guideline 14 [51], the dam-break flood simulation was terminated at the profile where the peak discharge induced by the dam break dropped below the 100-year hydrological flood discharge.

A summary of the simulations is shown in Figure 13, where the flood hydrographs are plotted starting with the initial hydrological flood that entered the Lichnov II reservoir (on the left) and induced the dam-break flood progression through the system of three reservoirs along the Tetřeví Stream and the Čižina River. In Table 6, the dam breach peak discharges obtained from simulations are compared with the results of the comparisons with real incidents and with data from empirical equations. The comparison indicates that the use of comparisons and empirical equations for a dam break in a cascade may be confusing, especially if the breached dams are very close to one another. These methods may be used for the comparison of the breach peak discharge failures at single dams, but with some caution due to the significant uncertainty involved.

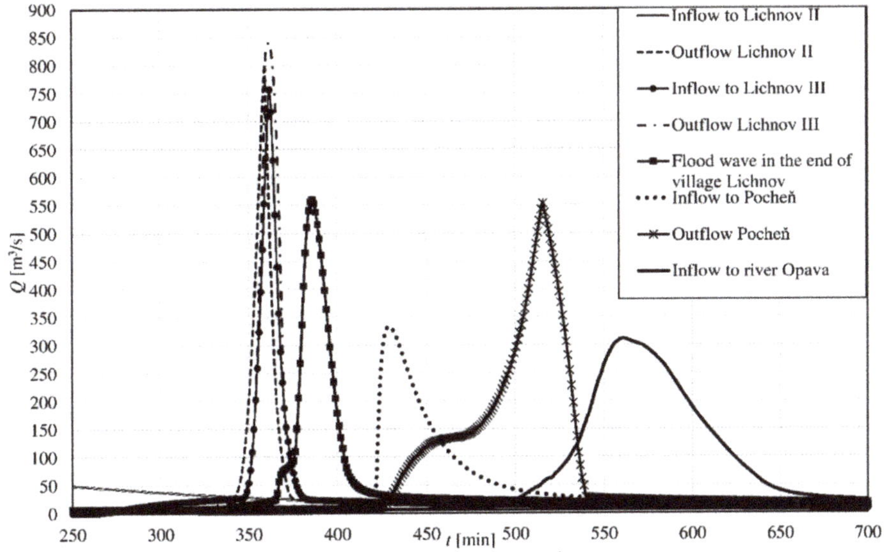

Figure 13. Progression of the dam-break flood in selected critical profiles in the area.

Figure 14 shows a map depicting the zone inundated by a dam-break flood corresponding to the worst-case scenario identified by Monte–Carlo optimization. The map displays the maximum water depth reached during the event.

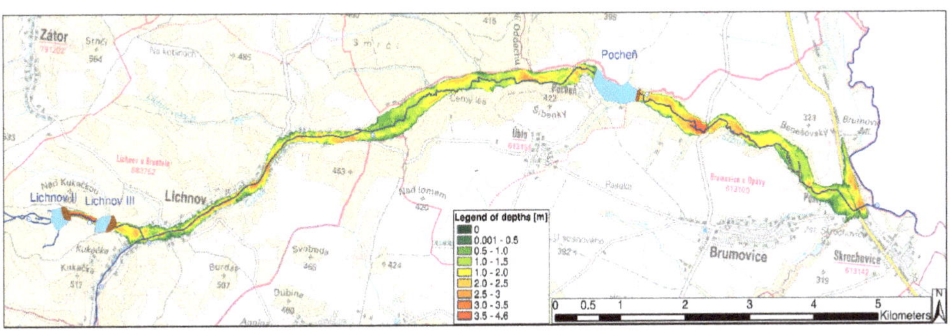

Figure 14. Map of the maximum water depths on the map base, depth in meters.

5. Discussion

The simulation of a dam break in a cascade consists of two basic procedures, namely single dam-break analysis (Sections 3.1–3.3), and the analysis of the dam-break-induced flood between the dams (Section 3.4).

Basically, two failure modes—overtopping and piping—should be simulated at the uppermost dam. These modes are completely independent; they occur for different reasons at different places in the dam body, and at different times (Figures 3 and 8). The initiation of piping is governed by the threshold hydraulic gradient when critical shear stress is exceeded. Therefore, piping usually starts at the rising limb of the flood hydrograph and progresses during the arrival of the flood peak at the maximum water level in the reservoir. This may happen before dam crest overtopping occurs. At the moment when the "roof" of the pipe becomes thin (here this is assumed to be when it is equal to the "pipe's" diameter), the overburden collapses, and the simulation switches to the overtopping mode at the location of the original piping. This may result in a temporary drop in outflow during piping. The duration of the piping failure until the maximum breach opening developed was more than 1 h (Figure 8). The worst-case scenario of an overtopping failure occurred at the receding limb of the inflow hydrograph, at the moment when the reservoir was completely full, and critical velocity along the downstream slope was reached.

The optimization procedure showed that the outflow from a reservoir is predominantly governed by the size of the breach opening, and the reservoir water level in relation to the base of the breach opening. Thus, the peak outflow is limited by the location of the first overtopping, the morphology of the dam profile, and the erodibility of the dam material.

With large dams, the impact of the breaching of an upper dam on the failure risk of a downstream one is usually governed by the ratio between the released volume from the upper reservoir and the flood attenuation volume of the lower one. In the case of small dams, the flood attenuation volume is relatively small when compared to the volume of the arriving extreme flood, with the ratio varying between 1:10 and 1:25. The attenuation of peak discharge usually varies between 5–10%. Therefore, the attenuation effect of small reservoirs is not so important for dam break analysis in the case of a cascade of small dams. The capacity of the bottom outlets at small dams usually has a negligible effect on the flood passing through the reservoir and may be neglected.

At lower-lying small dams in a cascade where the reservoir only provides a limited attenuation effect, the rising limb is steep (Figure 9), and the increase in the reservoir water level is quite rapid (5–10 min). In the case of the scheme investigated here, the overtopping and breaching started shortly after the rapid increase in the reservoir water level. About 1 h was needed for the initiation of the piping, and, therefore, the expected failure mode was primarily overtopping. Therefore, the piping failure mode was not taken into account for the downstream dams.

The results of the presented study prove that flood attenuation between reservoirs located in close proximity to one another is practically negligible and that simplifications can be applied (Figure 9) [52]. Flood attenuation increases with the distance from the breached dam and follows an approximately exponential trend. Our results (Figure 13) correspond well with the results of parametric simulations published in [53], where the longitudinal slope, the dam-break flood volume, the width, and the "roughness" of the channel between the reservoirs are taken into account.

The empirical formulae tested in this study were derived using real collapses of single dams. When used for peak discharge estimation, empirical formulae derived for a single dam break should be applied carefully, as they may underestimate the peak outflow by up to 10% in the case of a dam cascade.

6. Conclusions

The paper describes the analysis of a dam break process in a cascade of small dams, including the simulation of flood routing along the adjacent streams. The aim is to fill a gap in the existing research concerning the breaching of small embankment dams in a cascade. This is a specific type of event which may differ from a dam break in a cascade of large dams, where the reservoir attenuation volume may play a significant role. The case study concerns an area along Tetřeví Stream and the Čižina River in the Moravian-Silesian Region of the Czech Republic. The area contains three small dams: Lichnov II, Lichnov III, and Pocheň. A simplified parametric model for piping and overtopping erosion with parameter optimization was used for the dam break simulations, while a standard 2D unsteady shallow-water flow model was applied for the flood in the areas downstream of each reservoir. As no calibration data were available, the comparison of the simulation results was carried out with the use of empirical formulae and by means of comparison with real small dam failures.

It is recommended that the solution for a dam cascade be divided into separate sequences corresponding to the dam breaching process and flood routing along the streams downstream of the dams. The location of the first overtopping and piping event should be identified based on the dam geometry, the arrangement of appurtenant works (namely the auxiliary spillway), observed leaks, etc. It is desirable to carry out at least a simplified optimization procedure, which guarantees that a combination of parameters providing the worst possible realistic scenario for the simulated event is obtained. When applying empirical formulae or comparison with real dam failures to a dam cascade, one should be aware that these methods may underestimate the peak discharge by up to 10%.

All flood hydrographs (Figure 13) showing arrival times and flood zones with water depth (Figure 14) provide important information for warning and emergency plans, and for evacuation procedures.

Further research should focus on the comprehensive probabilistic analysis of dam failures and flood routing between reservoirs in order to create a basis for the development of risk maps for endangered areas.

Author Contributions: Conceptualization, S.K. and L.P.; methodology, J.Ř., L.P.; software, S.K., L.P. and J.Ř.; validation, S.K.; formal analysis, S.K.; investigation, S.K., L.P. and J.Ř.; resources, J.Ř.; data curation, J.Ř.; writing—original draft preparation, L.P. and S.K.; writing—review and editing, S.K. and J.Ř.; visualization, S.K.; supervision, J.Ř.; project administration, J.Ř.; funding acquisition, S.K., L.P. and J.Ř. All authors have read and agreed to the published version of the manuscript.

Funding: "This research was funded by Determination of uncertainties of selected hydraulic problems, grant number FAST-J-20-6315" and Probability assessment of internal instability in earth structures and in hydraulic structures' foundations, grant number FAST-S-19-5714.

Conflicts of Interest: The authors declare no conflict of interest.

List of Symbols

B	width of the breach opening in the crest (m)
b	width of the breach opening (m)
b_0	initial overtopping width (m)
b_{spill}	overflow width at the auxiliary spillway/lowest point of the dam crest (m)
B_a	average breach width (m)
B_L	dam bottom elevation (m a. s. l.)
C_e	coefficient of soil erosion (s/m)
C_L	dam crest elevation
D	pipe diameter (m)
D_0	initial pipe diameter (m)
g	acceleration due to gravity (m/s^2)
H	elevation of water level (m a. s. l.)
H_0	initial reservoir water level (m a. s. l.)
H_a	characteristic dimension of the reservoir (-)
H_k	water depth in the reservoir at the time of peak inflow (m)
H_{aux}	auxiliary spillway crest elevation (m a. s. l.)
h_d	dam height (m)
h_f	water depth along the slope (m)

$H_{outflow}$	elevation of the pipe outflow (m a. s. l.)
H_{SV}	bottom outlet axis elevation (m a. s. l.)
h_w	hydraulic depth in the breach (m)
I_E	average energy gradient along the pipe (-)
K_w	elevation of water level (m a. s. l.)
L	length of the pipe (m)
m	weir coefficient (-)
n	Manning roughness coefficient of the downstream slope (s/m$^{1/3}$)
Q_b	breach outflow (m^3/s)
Q_{bO}	breach outflow during overtopping (m^3/s)
Q_{bP}	outflow via the pipe (m^3/s)
Q_f	outflow through appurtenant works (outlets, spillways, etc.) (m^3/s)
Q_{in}	reservoir inflow (m^3/s)
r	pipe radius (m)
S	pipe cross-sectional area (m^2)
t	time (s)
V	reservoir water volume (m^3)
V_A	volume in the reservoir under the lowest point of the breach at the time failure starts (m^3)
V_b	reservoir water volume (m^3)
v_f	velocity along the slope (m/s)
v_{ns}	non-scouring velocity (m/s)
v_{pip}	average water velocity in the pipe (m/s)
V_w	volume in the reservoir for the elevation of water level K_w (m^3)
Z	breach bottom elevation (m a. s. l.)
Z_0	initial lowest dam crest elevation (m a. s. l.)
α	kinetic energy (Coriolis) coefficient (-)
α_1, α_2	erosion parameters (m/s)
β	downstream slope angle (°)
Δ	absolute roughness (m)
Δt	time step (s)
$\dot{\varepsilon}$	rate of erosion (kg/s/m^2)
λ	friction loss coefficient (-)
ρ	water density (kg/m^3)
ρ_b	bulk density of the eroded soil (kg/m^3)
τ	shear stress acting on the pipe walls (Pa)
τ_c	critical shear stress (Pa)

References

1. Cristofano, E.A. *Method of Computing Erosion Rate for Failure of Earth Fill Dams;* United States Department of the Interior, Engineering and Research Center: Denver, CO, USA, 1973; No. 727.
2. Jandora, J.; Říha, J. *The Failure of Embankment Dams due to Overtopping;* VUTIUM: Brno, Czec Republic, 2008; 168p.
3. Wahl, T. *Prediction of Embankment Dam Breach Parameters;* Report DSO-98-004; Dam Safety Office: Washington, DC, USA, 1998; 67p.
4. Jandora, J.; Říha, J. Incidents and failures affecting small dams in the Czech Republic. In Proceedings of the 8th ICOLD European Symposium, Dam safety. Sustainability in a Changing Environment, Innsbruck, Austria, 22–23 September 2010; pp. 419–424.
5. Brunner, G.W. *HEC-RAS River Analysis System: Hydraulic Reference Manual;* Version 5.0; US Army Corps of Engineers Hydrologic Engineering Center (HEC): Davis, CA, USA, 2017; 538p.
6. Jain, S.C. *Open Channel Flow;* John Wiley & Sons: New York, NY, USA, 2001; 328p.
7. Smith, R.H. Development of Flood Routing Model for Small Meandering Rivers. Ph.D. Thesis, University of Missouri at Rolla, Rolla, MO, USA, 1978; 159p.

8. Vreugdenhil, C.B. *Numerical Methods for Shallow Water Flow*; Kluwer Academic Press: Dodrecht, The Netherlands, 1998; 261p.
9. Brown, R.J.; Rogers, D.C. *BRDAM Users' Manual*; U.S. Department of the Interior: Denver, CO, USA, 1981; 67p.
10. Ponce, V.M.; Tsivoglou, A.J. Modelling gradual dam breaches. *J. Hydraul. Div.* **1981**, *107*, 829–838.
11. Nogueira, V.D.Q. A mathematical model of progressive earth dam failure. Ph.D. Thesis, Colorado State University, Fort Collins, CO, USA, 1984.
12. Fread, D.L. *DAMBRK: The NWS Dam Break Flood Forecasting Model. National Oceanic and Atmospheric Administration*; National Weather Service: Silver Spring, MD, USA, 1984; 27p.
13. Fread, D.L. *BREACH: An Erosion Model for Earthen Dam Failures (Model Description and User Manual)*; National Oceanic and Atmospheric Administration, National Weather Service: Silver Spring, MD, USA, 1988; 35p.
14. Fread, D.L. *The NWS DAMBRK Model: Theoretical Background/User Documentation*; National Weather Service: Silver Spring, MD, USA, 1988; 315p.
15. Singh, V.P.; Scarlatos, C.A. Breach erosion of earthfill dams and flood routing: BEED model. *J. Nat. Hazards* **1988**, *1*, 161–180. [CrossRef]
16. Říha, J.; Daněček, J. Mathematical modelling of earth dam breach due to overtopping. *J. Hydrol. Hydromech.* **2000**, *48*, 165–179.
17. Tingsanchali, T.; Chinnarasri, C. Numerical modelling of dam failure due to flow overtopping. *Hydrol. Sci. J.* **2001**, *46*, 113–130. [CrossRef]
18. Davison, M.; Hassan, M.; Gimeno, O.; Van Damme, M. A Benchmark study on dam breach and consequence estimation using EMBREA and Life Safety Model. In *ICOLD Theme C: Computational challenges in Consequence Estimation for Risk Management*; HR Wallingford: Wallingford, UK, 2013; pp. 241–254.
19. Mohamed, A.A.A.; Samuels, P.G.; Morris, M.W.; Ghataora, G.S. Improving the accuracy of prediction of breach formation through embankment dams and flood embankments. In Proceedings of the International Conference on Fluvial Hydraulics, Louvain-la-Neuve, Belgium, 4–6 September 2002.
20. Morris, M.W. Breaching of Earth Embankments and Dams. Ph.D. Thesis, The Open University, London, UK, 2011; 338p.
21. Hanson, G.J.; Cook, K.R.; Hunt, S.L. Physical modeling of overtopping erosion and breach formation of cohesive embankments. *Trans. ASAE* **2005**, *48*, 1783–1794. [CrossRef]
22. Van Damme, M.; Morris, M.W.; Hassan, M. A New Approach to Rapid Assessment of Breach Driven Embankment Failures. FRMRC Research Report SWP4.4. Available online: http://fcerm.net/sites/default/files/resources/frmrc2wp44sciencereport.pdf (accessed on 29 May 2020).
23. Macchione, F. Model for predicting floods due to earthen dam breaching. I: Formulation and Evaluation. *J. Hydraul. Eng.* **2008**, *134*, 1688–1696. [CrossRef]
24. Volz, C.; Rousselot, P.; Vetsch, D.F.; Faeh, R. Numerical modelling of non-cohesive embankment breach with the dual-mesh approach. *J. Hydraul. Res.* **2012**, *50*, 587–598. [CrossRef]
25. Wu, W. Simplified Physically Based Model of Earthen Embankment Breaching. *J. Hydraul. Eng.* **2013**, *139*, 837–851. [CrossRef]
26. Dewals, B.; Archambeau, P.; Erpicum, S.; Mouzelard, T.; Pirotton, M. Coupled computations of highly erosive flows with WOLF software. In Proceedings of the 5th International Conference on Hydro-Science & Engineering, Warsaw, Poland, 18–21 September 2002.
27. DHI. *MIKE-11: A Modelling System for Rivers and Channels (Manual)*; DHI: Horsholm, Denmark, 2003; 510p.
28. Paquier, A. Rupro, Breach Model Used by CEMAGREF during Impact Project. In Proceedings of the 1st Project Workshop (Wallingford Meeting), Wallingford, England, 16–17 May 2002.
29. King, L.M.; Simonovic, S. A Deterministic Monte Carlo Simulation Framework for Dam Safety Flow Control Assessment. *Water* **2020**, *12*, 505. [CrossRef]
30. Cai, W.; Zhu, X.; Peng, A.; Wang, X.; Fan, Z.; Cai; Zhu, P.; Wang, Z.; Fan, P. Flood Risk Analysis for Cascade Dam Systems: A Case Study in the Dadu River Basin in China. *Water* **2019**, *11*, 1365. [CrossRef]
31. Liu, Z.; Guo, X.; Zhou, X.; Fu, H.; Xia, Q.; Li, S. Cascading dam breach process simulation using a coupled modeling platform. *Sci. China Ser. E Technol. Sci.* **2018**, *62*, 1455–1466. [CrossRef]
32. Stoyanova, V.; Coombs, R. Predictive breach analyses for reservoir cascades. In *Sustainable and Safe Dams around the World*; Canadian Dam Association: Ottawa, Canada, 2019; pp. 220–234.

33. Hu, L.; Yang, X.; Li, Q.; Li, S. Numerical simulation and Risk Assessment of Cascade Reservoir Dam-Break. *Water* **2020**, *12*, 1730. [CrossRef]
34. Alhasan, Z.; Jandora, J.; Říha, J. Study of Dam-break Due to Overtopping of Four Small Dams in the Czech Republic. *Acta Univ. Agric. Silvic. Mendel. Brun.* **2015**, *63*, 717–729. [CrossRef]
35. Holomek, P.; Říha, J. A comparison of breach modelling methods applied to the Slusovice earth dam. *Dam Eng.* **2000**, *9*, 171–202.
36. Owen, H.J. *Flood Emergency Plans: Guidelines for Corps Dams*; The Hydrologic Engineering Center: Davis, CA, USA, 1980; 52p.
37. ICOLD. *Lessons from Dams Incidents (Complete Edition)*; International Commission on Large Dams: Paris, France, 1974; 1069p.
38. Šimek, M. *Verification of the Methods of Dam Categorisation in Czechoslovakia based on the Risk Factor*; Final Report TR-805; VRV Prague: Prague, Czech Republic, 1988; 82p. (In Czech)
39. Froehlich, D.C. Peak Outflow from Breached Embankment Dam. *J. Water Resour. Plan. Manag.* **1995**, *121*, 90–97. [CrossRef]
40. Froehlich, D.C. Embankment Dam Breach Parameters Revisited. In Proceedings of the 1995 ASCE Conference on Water Resources Engineering, San Antonio, TX, USA, 14–18 August 1995; pp. 887–891.
41. Webby, M.G. Discussion on "Peak Outflow from Embankment Dam" by Froehlich, 1995a. *J. Water Resour. Plan. Manag.* **1996**, *122*, 316–317. [CrossRef]
42. Bulletin 111. *Dam Break Flood Analysis*; ICOLD Bulletin No. 111: Paris, France, 1998; 301p.
43. Vischer, D.L.; Hager, W.H. *Dam Hydraulics*; Wiley: Hoboken, NJ, USA, 1998; 316p.
44. Bulletin 99. *Dam Failures Statistical Analysis*; ICOLD Bulletin No. 99: Paris, France, 1995; 73p.
45. Singh, V.P. *Dam Breach Modeling Technology*; Louisiana State University: Baton Rouge, LA, USA; Kluwer Academic Publishers: Dodrecht, The Netherlands, 1996; 242p.
46. Sentűrk, F. *Hydraulics of Dams and Reservoirs. Highland Ranch*; Water Resources Publications: Ankara, Turkey, 1994; 778p.
47. Munson, R.B.; Young, D.F.; Okiishi, T.H.; Huebsch, W.W. *Fundamentals of Fluid Mechanics*; Wiley & Sons: Hoboken, NJ, USA, 2006; 724p.
48. Wan, C.F.; Fell, R. *Investigation of Internal Erosion and Piping of Soils in Embankment Dams by the Slot Erosion Test and the Hole Erosion Test—Interpretative Report*; University of New South Wales: Sydney, Australia, 2002; 142p.
49. Raychaudhuri, S. Introduction to Monte Carlo simulation. In Proceedings of the 2008 Winter Simulation Conference, Miami, FL, USA, 7–10 December 2008; pp. 91–100.
50. Alhasan, Z.; Duchan, D.; Říha, J. Influence of surface lining type on the probability of dike breaching due to overtopping. *J. Flood Risk Manag.* **2019**, *12*, e12534. [CrossRef]
51. Guidelines 14. *Guidelines for Elaboration of Plans for Protection of Areas below Dams against Dam Break Floods*; Bulletin of the Ministry of the Environment of the Czech Republic: Prague, Czech Republic, 2005; pp. 77–90.
52. DEFRA. *Risk Assessment for Reservoir Safety Management: Volume 2—Methodology*; Environment Agency: Bristol, UK, 2013; 310p.
53. DSP. *Estimating Dam Break Downstream Inundation. Dam Safety Program*; Ministry of Forests, Lands and Natural Resource Operations: Vancouver, BC, Canada, 2016; 15p.

water

MDPI

Article

Vertically Averaged and Moment Equations for Dam-Break Wave Modeling: Shallow Water Hypotheses

Francisco N. Cantero-Chinchilla [1], Rafael J. Bergillos [2], Pedro Gamero [2], Oscar Castro-Orgaz [2,*], Luis Cea [3] and Willi H. Hager [4]

1 Soil and Water Division, IAS-CSIC, Spanish National Research Council, Alameda del Obispo s/n, 14004 Córdoba, Spain; fncantero@ias.csic.es

2 Hydraulic Engineering Area, Rabanales Campus, Leonardo da Vinci Building, University of Cordoba, 14071 Córdoba, Spain; rafael.bergillos@uco.es (R.J.B.); ag2gaojp@uco.es (P.G.)

3 Water and Environmental Engineering Research Team (GEAMA), Civil Engineering School, Universidade da Coruña, Elviña, 15071 Coruña, Spain; luis.cea@udc.es

4 Laboratory of Hydraulics, Hydrology and Glaciology, Swiss Federal Institute of Technology, 8093 Zürich, Switzerland; hager@vaw.baug.ethz.ch

* Correspondence: ag2caoro@uco.es

Received: 14 October 2020; Accepted: 16 November 2020; Published: 18 November 2020

check for updates

Abstract: The dam-break wave modeling technology relies upon the so-called shallow water equations (SWE), i.e., mass and momentum vertically averaged equations by implementing the shallow water hypotheses, namely (i) horizontal velocity component independent of the vertical coordinate, (ii) vertical velocity component is null, (iii) pressure distribution is hydrostatic, (iv) turbulence is neglected. While this model often yields a satisfactory answer from an engineering standpoint, flows with vertical length scales not negligible cannot be modeled with accuracy, including the undular surge generated after a dam break for relatively high tailwater levels. These flows are modeled by the Serre–Green–Naghdi equations (SGNE), which fail to mimic wave breaking for low tailwater levels, however. Neither SWE nor SGNE produce a fully satisfactory answer for modeling dam break waves, therefore. A higher-order model using vertically averaged and moment equations (VAM) is used in this work to simulate dam break waves, thereby showing good results for arbitrary values of the tailwater level. The model contains four perturbation parameters implemented to overcome the shallow water hypotheses; two for the velocity components and two for fluid pressure. The role of each parameter in relaxing the limitations of the SWE is systematically investigated, depicting a complex and necessary interplay between the dynamic component of fluid pressure and the modeling of the velocity profile in producing accurate solutions for both non-hydrostatic and broken waves in dam break flows. The results highlight how the shallow water hypotheses can be relaxed in the vertically averaged modeling of dam break waves, producing an outcome of both theoretical and practical interest in the field. The results generated are tested with available experimental data, resulting in acceptable agreement.

Keywords: dam break; non-hydrostatic flow; shallow water hypotheses; vertically averaged model; weighted residual method

1. Introduction

Dam-break waves are rapidly varied, unsteady free surface phenomena occurring frequently in hydraulic and environmental engineering. For example, the fast operation of gates in man-made canals may produce these waves with the consequent risk of overtopping, whereas the failure of a dam would release a large body of water moving down a valley with the consequences in terms of inundation hazard [1–3]. Depending on the initial conditions, i.e. flow depths and velocities before wave release, the resulting dam-break flows can lead to different phenomena, e.g. the front surge may be undular or broken, in addition to a rarefaction wave propagating into the upstream direction [4]. The prediction of dam-break waves is, therefore, of paramount importance in the river environment [5] as well as canals and their hydraulic structures.

In this work, a novel application of vertically averaged and moment equations [3] is presented for dam-break wave modeling, where the different degrees of freedom available to model the field variables, e.g., velocity components and pressure, are analyzed in depth, including the impact on the wave-breaking phenomena. The most general method to attack the problem computationally is to use three-dimensional (3D) models, albeit they are still costly in terms of simulation time at large scales of the computational domain. Therefore, a feasible approximation widely used in practice is to simulate dam-break waves resorting to vertically integrated models. In those models, the basic idea is to remove the vertical coordinate from the model equations by undertaking a vertical integration and adopting suitable hypotheses on the field variables, e.g., the velocity and pressure fields, and, sometimes, on the turbulence. The most applied vertically integrated system of equations in this modeling technology appears to be the so-called de Saint Venant equations or shallow water equations (SWE) [3,6]. These equations are a system of two hyperbolic conservation laws for mass and momentum (1D case) derived by implementing the so-called *shallow water hypotheses* in the vertical-integration of the RANS equations, namely

(i) horizontal velocity component independent of the vertical coordinate,
(ii) vertical velocity component is null,
(iii) pressure distribution is hydrostatic,
(iv) turbulence is neglected.

It can be demonstrated that these hypotheses are in reality linked through the general vertically integrated form of the RANS equations [7]. For example, if the vertical velocity is zero, and turbulence is neglected, of necessity the pressure distribution is hydrostatic [3]. Or in other words, those models accounting for non-hydrostatic pressure of necessity assume that the vertical velocity is not zero. However, it is frequent in the literature to state them separately, as introduced here. Despite these limitations, the ensuing model and its predictions are good from the engineering standpoint [3]. It should be kept in mind that flows with vertical scales not null cannot be expected to be predicted with any accuracy. Many efforts are conducted, therefore, to increase the range of validity of the SWE by relaxing one or some of the *shallow water hypotheses*. In the river environment, these studies are rather recent [8–10]. However, there is a long tradition in maritime hydraulics working with higher order models accounting for a non-hydrostatic pressure distribution ([11–22], among others).

Among the vertically averaged non-hydrostatic models available, the most frequently used in practice are the so-called Boussinesq-type models. These incorporate dynamic pressures into the governing equations through additional terms depending on high-order velocity derivatives, which usually require special numerical discretization [8,14,18,23–29]. Another family of vertically averaged non-hydrostatic models is the layer-averaged equations, where the flow is discretized in a multilayer fashion [16,30–34]. Castro-Orgaz et al. [7] demonstrated that the single-layer case is a special example of the Boussinesq-type model assuming a linear non-hydrostatic pressure distribution. The Serre-Green-Naghdi Equations (SGNE) [4,35–37] is a Boussinesq type model obtained by assuming that

(i) the horizontal velocity is independent of the vertical coordinate,
(ii) vertical velocity is linear,
(iii) pressure distribution is non-hydrostatic, following a parabolic law.

It is evident under which conditions the SGNE improves as compared with the SWE. However, these improvements on the shallow water hypotheses inherent of the SGNE are not independent, e.g., assuming (i), improvement (ii) follows from continuity considerations, and (iii) by resorting to the vertical momentum balance. The SGNE are widely used for nearshore flow modelling [38–40], and more recently in dam break wave modeling [4,37]. The so-called hybrid shock-capturing Boussinesq models are able to simulate the wave propagation from deep water to the coastline, including wave breaking, by switching off the dispersive terms when and where the ratio between the wave height and the water depth is equal to a certain threshold value, usually taken in practice equal to 0.8. Effective applications of this kind of Boussinesq-type models to complex engineering case studies include the numerical simulation of coastal sediment transport [41] or the tsunami propagation over complex bathymetry [42].

At this stage it is clear that the shallow water hypotheses, on which the SWE rely, are in a way relaxed by the SGNE. However, the SGNE have their own limitations, much of them because of the nature of the improvements implemented. A first issue of pure numerical nature is that the non-hydrostatic effects are modeled through high-order derivatives of the velocity field, demanding extreme care in the discretization. Second, the assumption of horizontal velocity independent of the vertical coordinate produces a linear dispersion relation in the SGNE limited to long wave conditions, e.g., only shallow water waves can be modeled. The secondary waves generated in dam break flows are rather short, however, and thus behave like deep water waves. It is well-known that the SGNE are unable to mimic wave breaking. This is due to the interaction between non-linear effects and frequency dispersion of the model, a balance kept behind realistic conditions. A consequence is that wave breaking shall be mimicked in SGNE by introducing different types of breaking-sub-models, like an eddy viscosity term, a roller flow model, or a hybrid treatment switching to the SWE at breaking nodes. It was recently revealed that the lack of the wave-breaking ability in the SGNE is due to the lack of modeling of differential advection of momentum, e.g., the horizontal velocity profile is not modeled in the vertically integrated equations [4].

A third family of vertically averaged models to simulate non-hydrostatic flows is based on the application of the weighted residual method to derive vertically averaged and moment equations from the RANS equations [43–45]. This third family of non-hydrostatic models is known as vertically averaged and moment (VAM) equations model [10]. The VAM model is a physical system of equations that is not yet really familiar to the scientific community. The rationale behind this technique is as follows. It is possible to prescribe the laws of variation with elevation of the field variables, e.g., velocity and pressure fields, as polynomials containing undetermined coefficients acting as perturbation parameters [43]. These parameters are used to deviate the actual flow from that modeled with the SWE, namely hydrostatic with uniform velocity in the streamwise direction. The perturbation parameters introduce additional degrees of freedom to increase the vertical resolution of the vertically averaged model. Each perturbation parameter introduced needs an equation for mathematical closure. The idea is to produce weighted-averaged RANS equations in addition to the vertically averaged flow equations, which are regarded as weighted-averaged equations using the unit weighting function. The selection of the weighting function is not unique; in principle, an unbounded number of perturbation parameters could be used simply by employing different weighting functions. In [43,44] a weighting function is used which consists in taking moments around the centroid of a flow sections, and therefore, their equations are called Moment equations. At the current state of knowledge, the VAM equations are still applied in isolated works, despite the excellent results [10,43–47]. In addition of producing good results, there are unique features rendering the VAM model attractive to simulate vertically averaged flows. First, the highest order derivative in the VAM model is one, permitting a simple and robust numerical discretization of the system. Second, the VAM model is accurate in predicting flows from shallow to intermediate water depths without any empirical tuning of the linear dispersion relation [10,46]. Third, in contrast to Boussinesq-type models, the VAM model possesses the ability to tackle wave breaking without introducing a sub-model to confer this feature or using an empirical condition to define the onset of wave breaking [46]. Despite the good performance of the VAM model, the role of each perturbation parameter in producing the solutions is unknown. The propagation of dam break waves is therefore an ideal scenario to investigate how each perturbation in the field variables contributes to improve the limitations posed by the shallow water hypotheses.

The VAM model uses four perturbation parameters; two for the velocity components (u_1, w_2) and two for fluid pressure (p_1, p_2). The main objective of this work is to investigate the influence of each perturbation parameter in the solution of the VAM model while producing simulations of dam-break flows, thereby revealing how the shallow water hypotheses are improved by the more general VAM model as compared with the SWE. The results are further compared with laboratory observations of dam-break flows over horizontal fixed beds collected by [5,48].

The manuscript is structured as follows. The governing VAM equations and reduced versions are presented in Section 2; the numerical scheme implemented is described in Section 3. Results are contained in Section 4, including the comparative analysis. Finally, the conclusions drawn based on the results are summarized in Section 5.

2. Vertically Averaged Models for Dam-Break Flows

Consider a dam-break flow in a straight and horizontal flume of rectangular cross-section. After dam release, often simulated with a gate of negligible opening time, a rarefaction wave propagates in the upstream direction while a bore advances toward the tailwater (Figure 1a). The features of this surge depend on the ratio r of downstream (h_d) to upstream (h_u) flow depths. If r is equal to or greater than the transition domain 0.4–0.55 [4,5,48], non-hydrostatic effects are important, producing an advancing train of waves, essentially formed by a leading solitary-like wave followed by a packet of shorter waves (Figure 1a). This type of front surge is called undular surge and reproduced by the SGNE [4]. For smaller depth ratios, wave breaking occurs at the leading wave of the bore, reducing the dispersive effect over the breaker extension (Figure 1b). These cases are reasonably well modeled by the SWE based on the hydrostatic pressure distribution, which thus are dispersion-less. The physical surge is in good agreement with the shock wave modeled mathematically by the SWE [3]. Dry bed conditions are a limiting case where the bore does not exist [37,49].

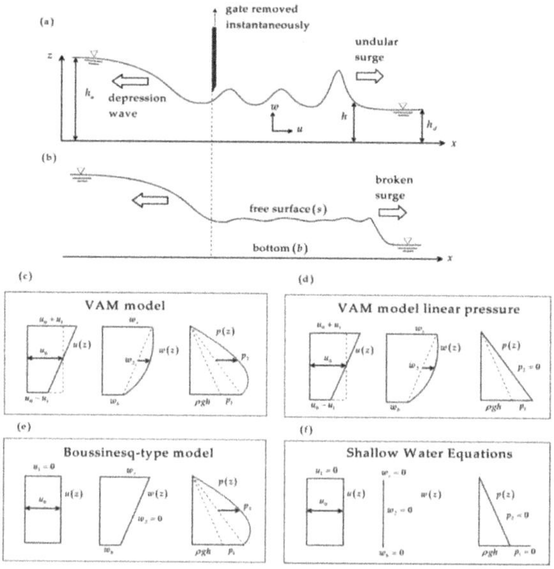

Figure 1. Surges formed in a 1D dam-break flow (**a**) undular, (**b**) broken; and hypotheses considered for field variables in the models implemented: (**c**) vertically averaged and moment (VAM) model, (**d**) VAM model with linear non-hydrostatic pressure distribution, (**e**) fully non-linear Boussinesq-type model (similar to Serre–Green–Naghdi), (**f**) de Saint-Venant model.

Pertinent questions are: What is a suitable model to produce good simulations regardless of r? Are the shallow water hypotheses relaxed by these good simulations? Modeling of dam break waves using the VAM model, to be presented below, reveal that they produce good results for an arbitrary value of r, whereas neither the SWE nor the SGNE are able to do so. The full VAM model [10,43,44] as well as several reduced VAM models (sub-models) and the classical SWE models are detailed below, to highlight the contribution of the perturbation parameters to overcome the shallow water hypotheses.

2.1. Vertically Averaged and Moment (VAM) Equations Model

Prior to the vertical integration of the Reynolds Averaged Navier–Stokes (RANS) equations, the field variables are expanded using polynomials in the VAM model as [43] (Figure 1c):

$$u(x,z,t) = u_0 + u_1(x,t)(2\chi - 1), \tag{1}$$

$$w(x,z,t) = [w_b + w_2(x,t)4\chi](1-\chi) + w_s\chi, \tag{2}$$

$$p(x,z,t) = [\rho g h + p_1(x,t) + p_2(x,t)4\chi](1-\chi), \tag{3}$$

where $u(x,z,t)$ and $w(x,z,t)$ are the horizontal and vertical velocity components, $p(x,z,t)$ is the fluid pressure, x is the horizontal Cartesian coordinate, z is the vertical Cartesian coordinate, t is time, $u_0(x,t)$ is the depth-averaged horizontal velocity, $u_1(x,t)$ is the x-velocity at the free surface in excess of u_0, $\chi = (z - z_b)/h$, z_b is the bed level, h is the flow depth, $w_b(x,t)$ is the vertical velocity at the bed, $w_2(x,t)$ is the mid-depth z-velocity in excess of the average of the vertical velocities at the bed and surface, $w_s(x,t)$ is the vertical velocity at the free surface. Further, ρ is fluid density, g is the gravity acceleration, $p_1(x,t)$ is the bed pressure in excess of hydrostatic, and $p_2(x,t)$ is the mid-depth deviation from the linear non-hydrostatic law (Figure 1c). In this work, the analysis focuses on dam-break waves over horizontal beds, thus $z_b = 0$, although the equations are presented in general form.

Four perturbation parameters (u_1, w_2, p_1, and p_2) as functions of (x,t) are introduced in Equations (1)–(3) to deviate the actual field variables from those used in dam-break wave modeling based on the SWE, namely $u = u_0$, $w = 0$, $p = \rho g h(1-\chi)$. Using Equations (1) and (2) in the vertical integration process of the continuity, x-momentum, z-momentum RANS equations yields, respectively [10,44]:

$$\frac{\partial h}{\partial t} + \frac{\partial q}{\partial x} = 0, \tag{4}$$

$$\frac{\partial q}{\partial t} + \frac{\partial}{\partial x}\left(g\frac{h^2}{2} + \frac{q^2}{h}\right) = -\frac{\partial}{\partial x}\left(\frac{{u_1}^2 h}{3} + \frac{hp_1}{2\rho} + \frac{2hp_2}{3\rho} - \frac{h\overline{\sigma_x}}{\rho}\right) - gh\frac{\partial z_b}{\partial x} - \frac{p_1}{\rho}\frac{\partial z_b}{\partial x} - \frac{\tau_b}{\rho}, \tag{5}$$

$$\frac{\partial h\overline{w}}{\partial t} + \frac{\partial q\overline{w}}{\partial x} = \frac{1}{6}\frac{\partial (hu_1 w_*)}{\partial x} + \frac{p_1}{\rho} - \frac{\tau_b}{\rho}\frac{\partial z_b}{\partial x} + \frac{1}{\rho}\frac{\partial h\overline{\tau_{xz}}}{\partial x}, \tag{6}$$

where $q(x,t)$ is the unit discharge, $\overline{\sigma}_x$ the depth-averaged normal stress term in the x-direction, τ_b the bed shear stress, $\overline{w} = w_b/2 + 2w_2/3 + w_s/2$ the mean vertical velocity, $w_* = w_b - w_s$, and $\overline{\tau}_{xz}$ is the depth-averaged shear stress term. Assuming a slip velocity condition at the bed level, the kinematic boundary conditions used for the bed and free surface levels in the derivation process of Equations (4)–(6) are $w_b = u_b\partial z_b/\partial x$ and $w_s = \partial h/\partial t + u_s\partial z_s/\partial x$, where u_b and u_s are the horizontal velocity component evaluated at the bed and free surface levels, respectively. The resulting system of equations [Equations (4)–(6)] is, however, not closed mathematically, given that only three PDEs are available to solve six independent variables (h, u_0, u_1, w_2, p_1, and p_2). Thus, the weighted residual method is used in this work to derive vertically integrated moment equations [43] from the continuity, x-momentum, and z-momentum RANS equations. A weighting function involving moments around the centroid of a section is selected to apply the method, resulting in [44]:

$$\frac{h}{2}\frac{\partial h}{\partial t} + q\frac{\partial \overline{z}}{\partial x} + \frac{1}{6}\frac{\partial (h^2 u_1)}{\partial x} = h\overline{w}, \tag{7}$$

$$\frac{\partial u_1}{\partial t} + \frac{\partial}{\partial x}\left(\frac{qu_1}{h}\right) = \frac{1}{2\rho}\left(\frac{\partial p_1}{\partial x} - \frac{p_1}{h}\frac{\partial h}{\partial x}\right) - \frac{4p_2}{h\rho}\frac{\partial \overline{z}}{\partial x} + \frac{6}{h\rho}\left(\frac{\tau_b}{2} + \overline{\sigma_x}\frac{\partial \overline{z}}{\partial x} - \overline{\tau_{xz}}\right), \tag{8}$$

$$\begin{aligned}\frac{\partial}{\partial t}\left[\frac{h^2 w_*}{12}\right] + \frac{\partial}{\partial x}\left[\frac{hqw_*}{12}\right] = \frac{h\overline{w}}{2}\frac{\partial h}{\partial t} + \left[q\overline{w} - \frac{hu_1 w_*}{6}\right]\frac{\partial \overline{z}}{\partial x} + \frac{\partial}{\partial x}\left[\frac{h^2 u_1}{10}\left(\overline{w} + \frac{w_b}{3} + \frac{w_s}{3}\right)\right] \\ -h\overline{w^2} - \frac{h}{\rho}\left(\overline{\tau_{xz}}\frac{\partial \overline{z}}{\partial x} - \overline{\sigma_z}\right) - \frac{h\tau_b}{2\rho}\frac{\partial z_b}{\partial x} - \frac{2hp_2}{3\rho}\end{aligned} \tag{9}$$

Here $\overline{z} = z_b + h/2$ is the height of the section centroid, $\overline{\sigma}_z$ the depth-averaged normal stress term in z-direction, and $\overline{w^2} = \overline{w}^2 + w_b^2/12 + w_s^2/12 - w_b w_s/6 + (2\overline{w} - w_b - w_s)^2/20$. Note that Equations (7)–(9) stand for the moment of continuity, x-momentum, and z-momentum RANS equations, respectively. The moment of continuity [Equation (7)] is a reactive equation, as the mathematical structure of transport equations is not conserved. The variable w_* is defined by

$$w_* = \frac{\partial q}{\partial x} + \frac{q}{h}\left(\frac{\partial z_s}{\partial x} + \frac{\partial z_b}{\partial x}\right) + u_1\frac{\partial h}{\partial x}. \tag{10}$$

In matrix form, the transport equations in the VAM model read:

$$\frac{\partial \mathbf{U}}{\partial t} + \frac{\partial \mathbf{F}}{\partial x} = \mathbf{S}_o + \mathbf{S}_\tau, \tag{11}$$

$$\mathbf{U} = \begin{bmatrix} h \\ q \\ h\overline{w} \\ u_1 \\ \frac{h^2 w_*}{12} \end{bmatrix}, \mathbf{F} = \begin{bmatrix} q \\ \frac{q^2}{h} + g\frac{h^2}{2} \\ q\overline{w} \\ \frac{qu_1}{h} \\ \frac{hqw_*}{12} \end{bmatrix}, \mathbf{S}_o = \begin{bmatrix} 0 \\ -\frac{\partial}{\partial x}\left(\frac{u_1{}^2 h}{3} + \frac{hp_1}{2\rho} + \frac{2hp_2}{3\rho}\right) - gh\frac{\partial z_b}{\partial x} - \frac{p_1}{\rho}\frac{\partial z_b}{\partial x} \\ \frac{1}{6}\frac{\partial (hu_1 w_*)}{\partial x} + \frac{p_1}{\rho} \\ \frac{1}{2\rho}\left(\frac{\partial p_1}{\partial x} - \frac{p_1}{h}\frac{\partial h}{\partial x}\right) - \frac{4p_2}{h\rho}\frac{\partial \overline{z}}{\partial x} \\ \left[-\frac{h\overline{w}}{2}\frac{\partial q}{\partial x} + \left[q\overline{w} - \frac{hu_1 w_*}{6}\right]\frac{\partial \overline{z}}{\partial x}\right. \\ \left. + \frac{\partial}{\partial x}\left[\frac{h^2 u_1}{10}\left(\overline{w} + \frac{w_b}{3} + \frac{w_s}{3}\right)\right] - h\overline{w^2} - \frac{2hp_2}{3\rho}\right] \end{bmatrix}, \mathbf{S}_\tau = \begin{bmatrix} 0 \\ -\frac{\tau_b}{\rho} \\ -\frac{\tau_b}{\rho}\frac{\partial z_b}{\partial x} + \frac{1}{2\rho}\frac{\partial h\tau_b}{\partial x} \\ 0 \\ -\frac{h\tau_b}{2\rho}\left(\frac{\partial \overline{z}}{\partial x} + \frac{\partial z_b}{\partial x}\right) \end{bmatrix} \tag{12}$$

Here **U**, **F**, and **S** are vectors for the flow conservative variables, fluxes, and source terms, respectively, and the subscripts o and τ refer to the inviscid and turbulent stress source terms. These latter stresses are modelled to the lowest order of approximation, assuming that: (i) The turbulent normal stress terms are neglected, i.e., $\overline{\sigma}_x = \overline{\sigma}_z = 0$, (ii) the turbulent shear stress is modelled by a linear law, i.e., $\tau_{xz} = \tau_b(1 - z/h)$ that yields $\overline{\tau}_{xz} = \tau_b/2$, and (iii) the bed-shear stress is modelled by Manning's equation accounting for the vertical velocity component, i.e.$\tau_b = \rho g n^2\left(u_b{}^2 + w_b{}^2\right)/h^{1/3}$. More details of the mathematical derivation of the governing equations in the VAM model and the model closures are found in [46].

2.2. VAM Model with Simplified Non-Hydrostatic Modeling ($p_2 = 0$)

A reduced VAM model is proposed to evaluate the impact of accounting for the extra degree of detail in the non-hydrostatic pressure modeling given by the perturbation parameter p_2. The model is a version of the VAM model in which p_2 is removed from the flow field hypotheses (Figure 1d). As a result, the number of independent variables in the ensuing system of equations decreases to five. To produce a compatible system of equations, the number of equations must, therefore, also be reduced by one. To that end, the moment of the z-momentum equation is removed from the system, resulting in:

$$\mathbf{U}=\begin{bmatrix} h \\ q \\ h\overline{w} \\ u_1 \end{bmatrix}, \mathbf{F}=\begin{bmatrix} q \\ \frac{q^2}{h}+g\frac{h^2}{2} \\ q\overline{w} \\ \frac{qu_1}{h} \end{bmatrix},$$
$$\mathbf{S}_0=\begin{bmatrix} 0 \\ -\frac{\partial}{\partial x}\left(\frac{u_1^2 h}{3}+\frac{hp_1}{2\rho}\right)-gh\frac{\partial z_b}{\partial x}-\frac{p_1}{\rho}\frac{\partial z_b}{\partial x} \\ \frac{1}{6}\frac{\partial(hu_1 w_*)}{\partial x}+\frac{p_1}{\rho} \\ \frac{1}{2\rho}\left(\frac{\partial p_1}{\partial x}-\frac{p_1}{h}\frac{\partial h}{\partial x}\right) \end{bmatrix}, \mathbf{S}_\tau=\begin{bmatrix} 0 \\ -\frac{\tau_b}{\rho} \\ -\frac{\tau_b}{\rho}\frac{\partial z_b}{\partial x}+\frac{1}{2\rho}\frac{\partial h\tau_b}{\partial x} \\ 0 \end{bmatrix} \tag{13}$$

2.3. VAM Model Simplified to Boussinesq-Type Model ($u_1 = w_2 = 0$)

Following the methodology exposed above, a reduced VAM model is produced to assess the impact of considering u_1, that is, the modelling of the horizontal velocity field. The model, imposing $u_1 = 0$, uses a constant-in-depth u-velocity law (Figure 1e), yielding five independent variables. As a consequence of imposing $u_1 = 0$ results in $w_2 = 0$ from the moment of continuity Equation (7). Thus, both u_1 and w_2 are set to zero simultaneously. The moment of the x-momentum equation is removed, therefore, to confer compatibility to the ensuing system of equations. In the reduced VAM model without u_1, the vectors for the flow conservative variables, fluxes, and source terms are modified, respectively, as follows:

$$\mathbf{U}=\begin{bmatrix} h \\ q \\ h\overline{w} \\ \frac{h^2 w_*}{12} \end{bmatrix}, \mathbf{F}=\begin{bmatrix} q \\ \frac{q^2}{h}+g\frac{h^2}{2} \\ q\overline{w} \\ \frac{hqw_*}{12} \end{bmatrix},$$
$$\mathbf{S}_o=\begin{bmatrix} 0 \\ -\frac{\partial}{\partial x}\left(\frac{hp_1}{2\rho}+\frac{2hp_2}{3\rho}\right)-gh\frac{\partial z_b}{\partial x}-\frac{p_1}{\rho}\frac{\partial z_b}{\partial x} \\ \frac{1}{2\rho}\left(\frac{\partial p_1}{\partial x}-\frac{p_1}{h}\frac{\partial h}{\partial x}\right)-\frac{4p_2}{h\rho}\frac{\partial \overline{z}}{\partial x} \\ -\frac{h\overline{w}}{2}\frac{\partial q}{\partial x}+q\overline{w}\frac{\partial \overline{z}}{\partial x}-h\overline{w}^2-\frac{2hp_2}{3\rho} \end{bmatrix}, \mathbf{S}_\tau=\begin{bmatrix} 0 \\ -\frac{\tau_b}{\rho} \\ -\frac{\tau_b}{\rho}\frac{\partial z_b}{\partial x}+\frac{1}{2\rho}\frac{\partial h\tau_b}{\partial x} \\ -\frac{h\tau_b}{2\rho}\left(\frac{\partial \overline{z}}{\partial x}+\frac{\partial z_b}{\partial x}\right) \end{bmatrix} \tag{14}$$

This system of equations can be demonstrated to be a fully non-linear Boussinesq-type model, e.g., similar to the SGNE, involving a uniform horizontal velocity profile, linear vertical velocity profile, and parabolic pressure profile [3] (Figure 1e).

2.4. De Saint Venant Model ($u_1 = w_2 = p_1 = p_2 = 0$)

If all the perturbation parameters (i.e., u_1, w_2, p_1 and p_2) are neglected in the VAM model, the flow field is that predicted by the SWE (Figure 1f). Neither the z-momentum equation nor the moment equations are therefore required to solve the SWE, in which h and q are the only conservative variables. In the SWE, **U**, **F**, and **S** reduce to:

$$\mathbf{U}=\begin{bmatrix} h \\ q \end{bmatrix}, \mathbf{F}=\begin{bmatrix} q \\ \frac{q^2}{h}+g\frac{h^2}{2} \end{bmatrix}, \mathbf{S}_o=\begin{bmatrix} 0 \\ -gh\frac{\partial z_b}{\partial x} \end{bmatrix}, \mathbf{S}_\tau=\begin{bmatrix} 0 \\ -\frac{\tau_b}{\rho} \end{bmatrix}, \tag{15}$$

3. Numerical Scheme

The ensuing system of PDEs for the full VAM model and the subsequent reduced VAM models described in the previous section requires accurate numerical schemes to their solution [10,50]. Herein, a semi-implicit hybrid finite volume-finite difference scheme is used for the numerical solution of the VAM models following a splitting strategy [3,6] (Figure 2). This strategy entails three steps:

- Homogeneous solution of Equation (11) in a finite volume explicit step

$$\frac{\partial \mathbf{U}}{\partial t} + \frac{\partial \mathbf{F}}{\partial x} = 0 \Rightarrow \hat{\mathbf{U}}, \tag{16}$$

- Inviscid update of the solution by Equation (16) in a finite-difference implicit step

$$\frac{\partial \mathbf{U}}{\partial t} = \mathbf{S}_o(\hat{\mathbf{U}}) \Rightarrow \widetilde{\mathbf{U}}, \tag{17}$$

- Turbulent update of the solution by Equation (17) in a finite-difference explicit step

$$\frac{\partial \mathbf{U}}{\partial t} = \mathbf{S}_\tau(\widetilde{\mathbf{U}}) \Rightarrow \mathbf{U}^{k+1} \tag{18}$$

Here $\hat{\mathbf{U}}$ is the solution of the homogeneous part, $\widetilde{\mathbf{U}}$ the solution updated by the inviscid terms, and $\mathbf{U}^{k+1}$ the solution of the flow variables at the time level k+1, where k is the time level index.

First, a finite volume, Godunov-type scheme is applied to Equation (16) where fluxes at the interfaces need to be determined. To that purpose, the flow variables are reconstructed at the intercell faces by means of a fourth-order total variation diminishing Monotone UpStream centered scheme for Conservation Laws (MUSCL-TVD-4th) [51]. The high-order reconstruction is needed when non-hydrostatic pressure is modeled [46]. During the reconstruction, to grant the C-property, the weighted surface-depth gradient method (WSDGM) [52] is applied, by which the water depth (depth gradient method (DGM)) and the free surface level (surface gradient method (SGM)) are weighted to reconstruct the water depth. Should the SGM be used, the water depth at the interfaces is later obtained by using the bed level evaluated at the cell faces. For the current application of dam break waves over horizontal beds the SGM and DGM reconstructions coincide, but for the sake of generality the WSDGM method is implemented in our solver for alternative applications to uneven beds. Once the variables are reconstructed, the numerical fluxes are calculated using the HLLC approximate Riemann solver [6,53], designed to tackle contact waves in the solution, such as those produced by the additional conservative flow variables considered in the VAM model.

In a second step, $\hat{\mathbf{U}}$ is updated, including the effects of the inviscid source terms as outlined by Equation (17). A Newton–Raphson iteration scheme is applied to solve the implicit system of equations, where a Jacobian matrix of $6N \times 6N$ dimensions (with N as the number of cells) is computed calculating 108 analytical partial derivatives per cell for the VAM model. In the solution of the reduced VAM models (VAM without p_2 and VAM without u_1 and w_2), the Jacobians are reduced. The equations are ordered to construct Jacobians without zeros in the diagonal. To save computational time, the Jacobians are frozen during the iterations [10]. The updated solution by the inviscid terms $\widetilde{\mathbf{U}}$ is accepted once a prescribed tolerance is reached in the mean relative error.

In the final step, the intermediate $\widetilde{\mathbf{U}}$ solution is updated using the turbulent terms as prescribed by Equation (18). To that purpose, a one-step forward Euler scheme is applied to Equation (18). The Courant-Friedrich-Lewy condition (CFL) is used to ensure stability of the computations. The VAM model is stable for CFL < 1, but to avoid secondary instabilities in the form of wiggles detected for some specific runs, CFL < 0.5 is recommended based on trial-and-error numerical experiments. The value used in some computations is smaller to get mesh size-independent results, not because of stability constrains. Further details of the present hybrid numerical scheme are provided in [46].

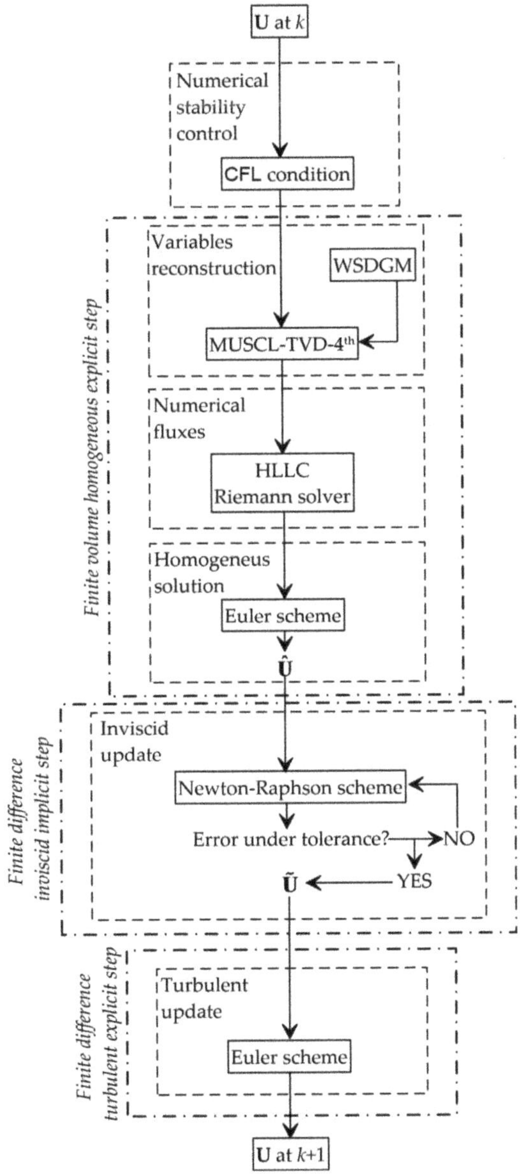

Figure 2. Flow chart of numerical scheme used in this work.

4. Results and Discussion

To conduct the comparison between the VAM model [10,43,44], the reduced VAM models and the SWE, three data sets are selected from [5,48] for dam-break flows over horizontal fixed bed with different up- and downstream flow conditions. The dam-break tests by [5] were conducted in a 15.24-m-long, 0.4-m-wide, and 0.4-m-high flume for two different upstream flow depths: $h_u = 0.1$ m for small scale experiments and $h_u = 0.36$ m for large scale experiments, implementing three depth ratios r each, namely 0, 0.1, and 0.45. The dam-break experiments by [48] were conducted in a 9-m-long, 0.3-m-wide, and 0.34-m-high flume with $h_u = 0.25$ m for three different r values: 0, 0.1, and 0.4.

The three data sets, hence, correspond to each of the h_u selected. To assess the impact of considering the extra perturbation parameters in the dam-break wave modeling, the four models are compared two-by-two. The full VAM model is used as the reference in each plot.

4.1. Hydrostatic versus Non-Hydrostatic Modelling

Figure 3 compares the VAM and SWE models based on [48] for $r = 0$ and $h_u = 0.25$ m, and based on [5] for $r = 0.45$ and $h_u = 0.36$ m at different times. The latter data describe an undular non-broken bore after the dam release, which is both highly non-linear and non-hydrostatic, so that these data are ideal for analyzing the importance of the non-hydrostatic effects on dam-break flow modelling. Given that no experimental data are available to describe the rarefaction wave for $r = 0.45$, the data for the dry downstream bed ($r = 0$) by [48] were also selected here for comparison purposes. The simulations were conducted by maintaining a grid step of $\Delta x = 0.01$ m and $\texttt{CFL} = 0.1$, while the bed roughness was accounted for by using $n = 0.01$ $\text{m}^{-1/3}\text{s}$ in the turbulence closure. The ability of the VAM model to tackle the non-hydrostatic dam-break flows in both dry and wet beds with undular non-broken surges downstream is remarkable, as shown in Figure 3e–h. The dry bed front is accurately simulated by the VAM model at 0.18 s, 0.44 s, and 0.8 s (Figure 3a–c). It is somewhat delayed at 1.06 s (Figure 3d) showing an accumulative delay in the predictions that may be due to inaccuracies in the Manning's parameter modelling. However, both the front of the undular non-broken bore and the secondary wave trains are accurately reproduced by the VAM model for $r = 0.45$ (Figure 3e–h). Small overestimations are observed at 0.4 s and 0.64 s (Figure 3e,f) for the peak of the advancing solitary-like wave, for which the non-linear effects are prominent. On the contrary, the SWE model fails to reproduce these highly non-hydrostatic flows. The front of the bore is overestimated for $r = 0.45$ (Figure 3e–h) in the SWE model, while the dry front estimated in line with the results by the VAM model for $r = 0$ (Figure 3a–d). For the wave train simulation for $r = 0.45$, only the trend of the waves is predicted (Figure 3e–h). In addition, the rarefaction wave is wrongly predicted by the SWE model, as flow curvature is not accounted for (Figure 3a–d). The non-hydrostatic modelling of dam-break flows while using depth-averaged shallow water models is, therefore, important for an accurate description of these flows. Note that non-hydrostatic effects are not important only at initiation of motion, as sometimes argued to discard non-hydrostatic modeling in practice for long simulation times. The advancing undular bore keeps its non-hydrostaticity for long simulation times, producing an evolution of the surge wave train, that ultimately tends to split in a sequence of cnoidal waves [25]. Therefore, inclusion of non-hydrostatic effects in the dam break wave modeling technology is of both theoretical and practical interest.

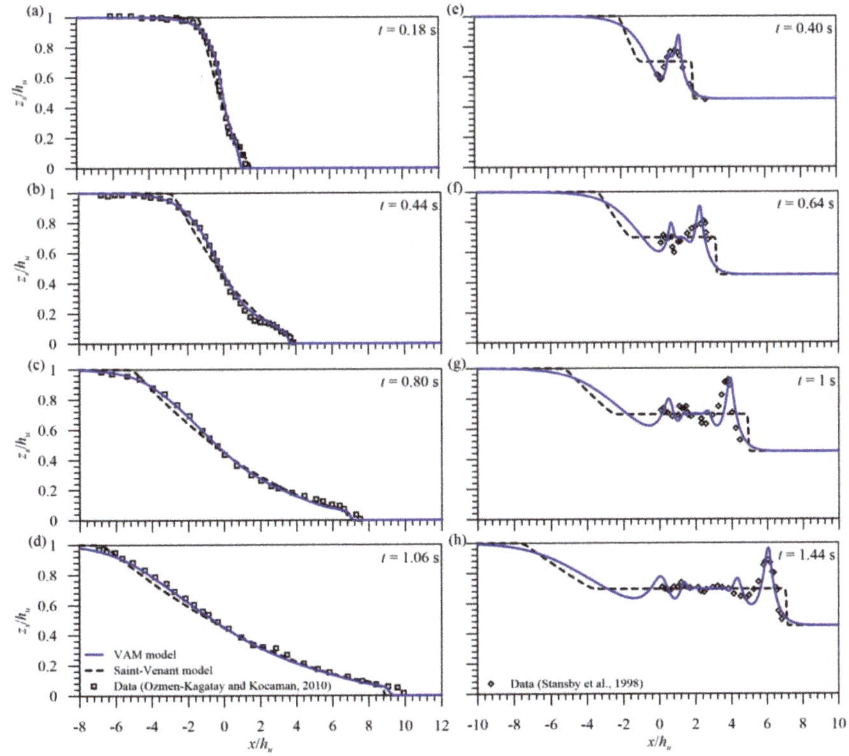

Figure 3. Comparison between VAM and shallow water equation (SWE) models to evaluate the importance of non-hydrostatic dam-break flow using the data by [48] for dry bed conditions, and the data by [5] (large scale) for wet bed conditions with $r = 0.45$ at different times t [s]: (**a**) 0.18, (**b**) 0.44, (**c**) 0.80, (**d**) 1.06, (**e**) 0.40, (**f**) 0.64, (**g**) 1, and (**h**) 1.44.

4.2. Simplified Non-Hydrostatic Flow Modeling

In this section, the impact of the non-hydrostatic perturbation parameters p_1 and p_2 is analyzed. To that purpose, Figure 4 shows the results of the VAM model (which accounts for both p_1 and p_2) and those of the VAM sub-model without p_2 (only accounting for p_1). They are compared to experimental data of dam-break flow over horizontal fixed bed by [48] with $r = 0.4$ and $h_u = 0.25$ m, and those by [5] with $r = 0.45$ and $h_u = 0.1$ m at different times. While the data set by [5] with $r = 0.45$ describes undular non-broken bores after the dam release with significant non-hydrostatic and non-linear effects, the data by [48] with $r = 0.4$ show the breaking of the leading wave, yielding diminished non-linear effects at the front. The simulations were conducted using $\Delta x = 0.01$ m, CFL = 0.1, and $n = 0.01$ m$^{-1/3}$s. The VAM models, however, fail to predict wave breaking for $r = 0.4$, possibly because of the oversimplified turbulence closure used, yielding, instead, solitary-like wave profile (Figure 4c–d). In line with the previous results, the VAM models accurately tackle the front of the bore and the wave train behind (Figure 4a,b,d–h) as well as the rarefaction wave (Figure 4c,d,g,h). Based on the comparison between both VAM models, it can be concluded that the impact of p_2 is minimum while modelling the non-hydrostaticity effect of dam-break flows over horizontal beds, for the selected tests. It is worth noting that, in contrast, if the bed bathymetry is uneven, p_2 is essential to tackle the free surface position [54].

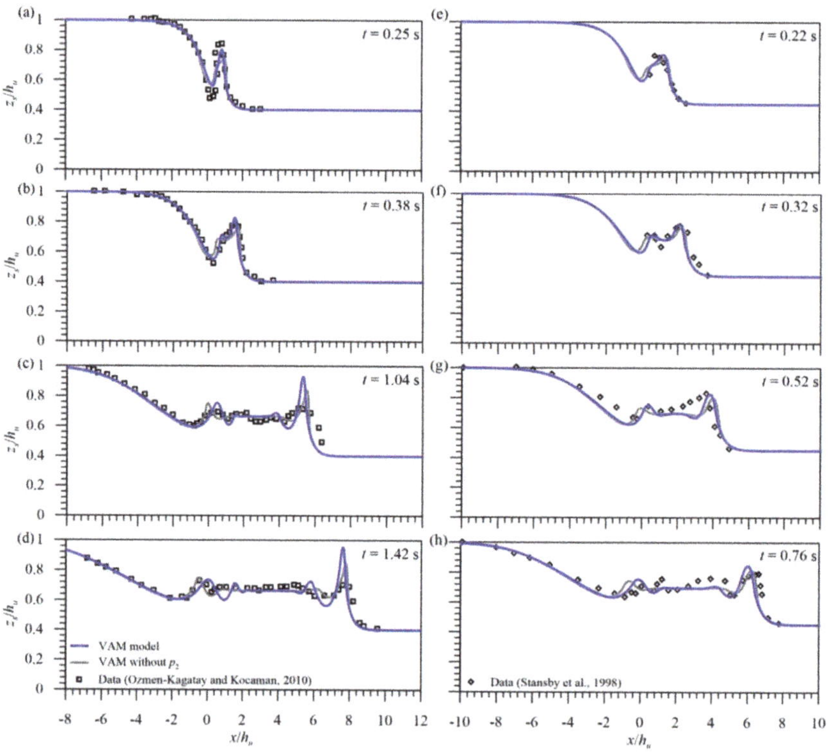

Figure 4. Comparison between VAM model and VAM sub-model without p_2 to evaluate the effect of non-hydrostaticity modelling of dam-break flows using the data by [48] and [5] (small scale) for wet bed conditions with $r = 0.4$ and $r = 0.45$, respectively, at times t [s]: (**a**) 0.25, (**b**) 0.38, (**c**) 1.04, (**d**) 1.42, (**e**) 0.22, (**f**) 0.32, (**g**) 0.52, and (**h**) 0.76.

On the other hand, the exclusion of p_2 alters the frequency dispersion relation of the original VAM model. After linearization of the VAM equations, the frequency dispersion relation of the VAM model is [10,43,46]

$$\frac{c^2}{gh_m} = \frac{1 + \frac{1}{12}(kh_m)^2}{1 + \frac{5}{12}(kh_m)^2 + \frac{1}{144}(kh_m)^4}, \tag{19}$$

where c is the celerity, k the wave number, and h_m is the mean flow depth. A detailed derivation of Equation (19) is available in [46]. Eliminating p_2 from the VAM model, the corresponding dispersion relation is as follows:

$$\frac{c^2}{gh_m} = \frac{1 + \frac{1}{12}(kh_m)^2}{1 + \frac{4}{12}(kh_m)^2}. \tag{20}$$

Figure 5 illustrates, in relative errors, the performance of the frequency dispersion relations of the VAM model [Equation (19)] and the VAM reduced without p_2 [Equation (20)] as compared to the exact frequency dispersion relation [25], i.e., $c^2/gh_m = \tanh(kh_m)/(kh_m)$. The VAM model shows a remarkable accuracy for the spectra of shallow water waves (or long waves, having $kh_m << 1$) and intermediate water waves (with a relative error bounded by ±4% for $kh_m < 6$). The VAM sub-model without p_2 provides important errors for $kh_m > 0.8$. Therefore, it is generally important to retain p_2, given its role in controlling the frequency dispersion at intermediate water depths.

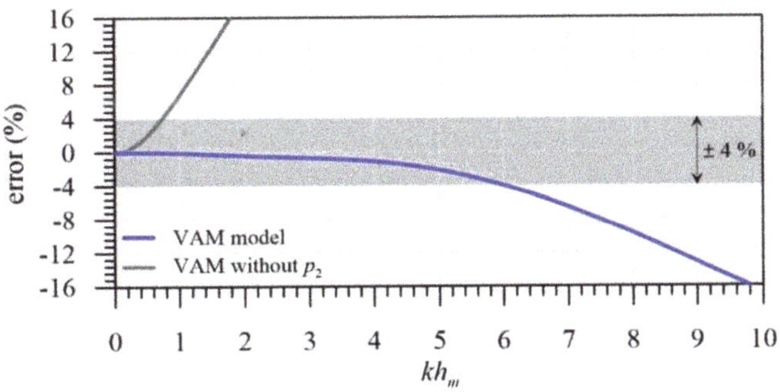

Figure 5. Frequency dispersion relations of VAM model and VAM sub-model without p_2, compared to the exact solution.

A penalty of the VAM model as compared to the SWE relates to the computational time. In the hyperbolic step of the solver computational time of the VAM model increases as compared to that of the SWE, given that the numerical flux with the HLLC solver is computed for five PDEs instead of only two. In the elliptic step of the solver six algebraic equations are solved per cell, producing a large system of implicit equations for the whole computational domain. Computational times of VAM and SWE equations for the test by [48] with $h_u = 0.25$ m, $h_d = 0.1$ m ($r = 0.4$), at simulation time $t = 1.04$ s ($T = 6.51$) are compared in Table 1 for different time and space discretizations. For a mesh of $\Delta x = 0.05$ m and CFL $= 0.8$, computational time is only a few seconds both for VAM and the SWE. Note that the MUSCL reconstruction is fourth-order accurate for both models. A reduction in CFL increases the computational time in the VAM model, but it is the reduction in Δx which really affects the computational time. A large mesh produces stable results in the VAM model, but the accuracy in the prediction of some portions of the waves, especially the undular fronts, was inadequate, given the numerical diffusion. Therefore, Δx was reduced and thus the number of cells increased, thereby increasing the total number of assembled equations to solve by the elliptic component of the numerical solver. This research focuses on testing the physical accuracy of the VAM model and its simplifications for dam break wave modeling rather than producing an optimized numerical solver, e.g., by resorting to a frozen Jacobian or using parallel Fortran codes. Therefore, we settled $\Delta x = 0.01$ m and CFL $= 0.1$ common to all simulations, although the same accuracy in the output could be obtained with the VAM model with larger time-space discretizations in some runs. In other words, we selected discretizations to ensure accuracy for all runs, and kept this choice constant during all this research. The resulting system of the elliptic step is quite large, with typically 700 cells and 6 equations per cell, that is, 4200 unknown variables to solve in each iteration cycle of the Newton–Raphson solver. By using a frozen Jacobian matrix the computational time reported in Table 1 is easily reduced by 30%.

Table 1. Computational times of VAM and SWE equations (seconds) using MATLAB for the test by [48] with $h_u = 0.25$ m, $h_d = 0.1$ m ($r = 0.4$), simulation time $t = 1.04$ s ($T = 6.51$)

Δx (m)	CFL	Full VAM Model	SWE Model
0.05	0.8	8.06	1.78
0.01	0.8	455.56	16.92
0.05	0.1	51.54	6.84
0.01	0.1	1749.98	125.58

To the authors knowledge, this is so far the only finite-volume solver available for the VAM model. Therefore, tools for optimizing its solution are not yet available. Improvements and further developments in this regard are advisable to produce more efficient solvers.

4.3. Impact of Horizontal Velocity Modelling

In this section, the impact of modelling of the horizontal velocity field is evaluated for dam-break flow situations in which the velocity field is significantly non-uniform, such as at the breaking portion of the leading wave of the bore. To that purpose, the VAM model is compared with a VAM sub-model without u_1 using the experimental data by [48] with $r = 0.1$ and $h_u = 0.25$ m, and the data by [5] with $r = 0.1$ and $h_u = 0.36$ m. In these tests, wave breaking occurs in front of the bore after the dam release and during the subsequent instants, as shown in Figure 6. In line with Figures 3 and 4, the simulations were conducted using $\Delta x = 0.01$ m, CFL = 0.1, and $n = 0.01$ m$^{-1/3}$s. The consideration of the linear, non-constant u-velocity profile in the VAM model by inclusion of the u_1 perturbation parameter is crucial to simulate the wave-breaking phenomenon. If u_1 were neglected, and consequently w_2 too, as already explained, the computational results of the ensuing model display highly non-linear non-breaking waves (Figure 6), similar to those obtained with Boussinesq-type models [4], in full disagreement with the experimental observations. The perturbation parameter u_1 controls the differential advection of momentum in the vertically averaged model, and thus, the dissipation needed at the wave front via velocity profile modeling to mimic wave breaking (Figure 6). In other words, the VAM model simulates the momentum velocity correction coefficient as $\beta = 1 + 1/3(u_1/u_0)^2$. It has been highlighted in the literature that the hydraulic approximation $\beta = 1$ is inaccurate for modeling flows with recirculation and/or significant velocity gradients [55–57], such as in (steady) hydraulic jumps. The broken waves simulated herein can be looked at, therefore, from the perspective of moving hydraulic jumps: the momentum flux term due to velocity profile modeling $\partial/\partial x[(\beta{-}1)u_0{}^2h]$ partially balances the dynamic pressure terms of the x-momentum balance, namely $\partial/\partial x[(1/2p_1 + 2/3p_2)h/\rho]$, at breakers. These dynamic terms are responsible for the undulations of a surge in Boussinesq-type models, and therefore, these effects are partially suppressed at broken waves by the velocity profile modeling through the inclusion of the coefficient β. Therefore, the broken waves are significantly influenced by the velocity profile modeling, with a complex interaction between u_1, w_2, p_1, and p_2 at breakers.

This comparison shows the advantages of the VAM model over a Boussinesq-type model, namely a precise determination of non-hydrostatic pressure effects and simultaneously wave breaking, by the former. Note that wave breaking is automatically accounted for; neither a wave-breaking index is used to define its onset, as in Boussinesq-type models, nor a breaking sub-module to produce the broken fronts, as in Boussinesq-type models. Thus, relaxing the shallow water hypothesis only introducing the modeling of w and p by linear and parabolic profiles, respectively, produces a model similar to the SGNE that is still highly constrained by the unimproved assumption, namely the modeling of $u(z)$.

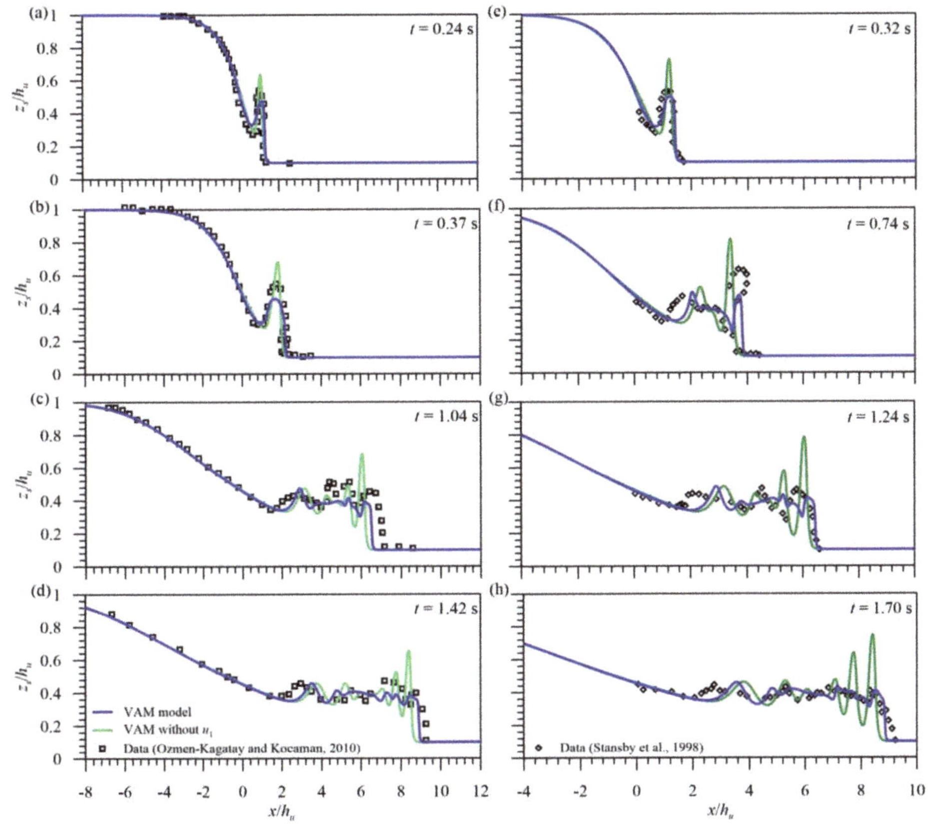

Figure 6. Comparison between VAM model and VAM sub-model without u_1 to explore the effects of the horizontal velocity field modelling of dam-break flows using the data by [48] and [5] for wet bed conditions with $r = 0.1$ at different times t [s]: (**a**) 0.24, (**b**) 0.37, (**c**) 1.04, (**d**) 1.42, (**e**) 0.32, (**f**) 0.74, (**g**) 1.24, and (**h**) 1.70.

The exclusion of u_1 from the VAM model also alters the frequency dispersion relation. For the VAM sub-model without u_1 (and thus also without w_2), it reads:

$$\frac{c^2}{gh_m} = \frac{1}{1 + \frac{1}{3}(kh_m)^2}. \tag{21}$$

Note that the dispersion relation in Equation (20) is identical to that for the SGN equations [25] indicating that the VAM sub-model without u_1 and w_2, and the SGN equations are physically equivalent for modeling linear dispersive effects. Figure 7 shows that, in contrast with the full VAM model, excluding u_1 from the VAM model leads to significant errors, where only the shallow water waves spectrum are approximated with errors bounded by ±4%.

A complete comparison of the three data sets selected follows in the Appendix A, where, in addition, a VAM with only p_1 model is considered [8] for illustrative purposes. This reduced VAM model is also of Boussinesq-type, but based on a linear pressure profile instead of a parabolic, as with the SGNE. Since no additional insights are revealed by this model, it was excluded from the main core of the discussion presented here.

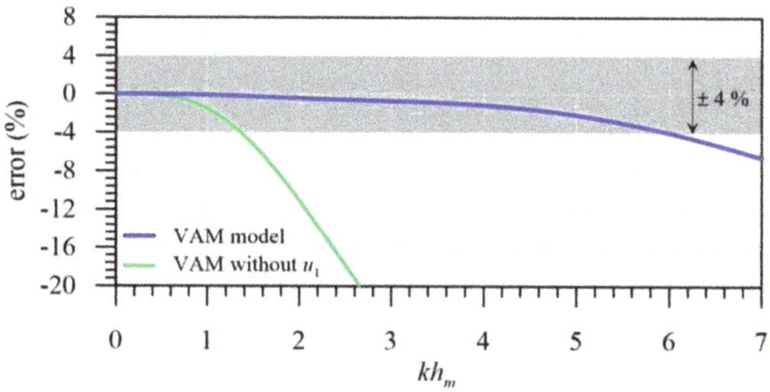

Figure 7. Frequency dispersion relations of VAM model and VAM sub-model without u_1 compared to the exact solution.

5. Conclusions

The *shallow water hypotheses* in dam-break flow modeling are analyzed within a vertically averaged framework using a vertically averaged and moment (VAM) equations model containing perturbation parameters introduced to overcome the limitations of the SWE.

First, the importance of considering non-hydrostatic effects was assessed by comparing the VAM model, the SWE model, and the experimental data for wet and dry bed conditions, respectively. The results reveal that the predictions by the VAM model are more accurate than those obtained with the SWE model, concluding that the inclusion of non-hydrostatic pressure effects is important to model dam-break flows.

The impact of a simplified non-hydrostatic pressure modeling was then analyzed by comparing the results of the full VAM model with those obtained by a sub-model of the VAM removing the perturbation parameter p_2. The accuracy of the sub-model for the simulations conducted was similar to that with the full VAM model, but the prediction of the frequency dispersion is inacceptable, concluding that the pressure must be modeled *non*-hydrostatic with a quadratic predictor.

Finally, the impact of the horizontal velocity profile modeling was quantified by comparing the full VAM model, a reduced VAM removing the variable u_1 and thus equivalent to a Boussinesq-type model, and experimental data for wet bed conditions with depth ratios producing breaking bores. The VAM sub-model without u_1 leads to non-breaking waves, as expected from Boussinesq's theory, in full disagreement with experiments, whereas the full VAM model automatically produces broken waves. As a conclusion, it is essential to consider the horizontal velocity modeling in dam-break flows simultaneously to the non-hydrostatic pressure modeling.

SGNE and SWE correspond to extremes in the modeling of dam break waves, each having advantages and disadvantages. Only a higher-order model, such as the VAM system, has the ability of coupling simultaneously the best of both modeling technologies, by implementing non-hydrostatic pressure *and* velocity profile modeling. These results demonstrate that relaxing the *shallow water hypotheses* is important to increase the accuracy range of vertically averaged models. The VAM model via its perturbation parameters is therefore a possible tool for improving the dam-break wave modeling technology, with numerous applications in hydraulic engineering, such as the design strategies to mitigate flood hazards.

Author Contributions: Conceptualization, F.N.C.-C., O.C.-O. and W.H.H.; methodology, F.N.C.-C., R.J.B., P.G. and O.C.-O.; software, F.N.C.-C. and O.C.-O.; validation, F.N.C.-C., R.J.B. and O.C.-O.; formal analysis, F.N.C.-C., O.C.-O. and W.H.H.; investigation, F.N.C.-C., R.J.B., O.C.-O. and W.H.H.; resources, F.N.C.-C., R.J.B. and P.G.; data curation, F.N.C.-C., R.J.B. and P.G.; writing—original draft preparation, F.N.C.-C., R.J.B. and O.C.-O.; writing—review and editing, O.C.-O, L.C. and W.H.H.; visualization, O.C.-O, L.C. and W.H.H.; supervision, O.C.-O, L.C. and W.H.H.; project administration, O.C.-O. and L.C.; funding acquisition, O.C.-O. and L.C. All authors have read and agreed to the published version of the manuscript.

Funding: This research was funded by the Spanish State Research Agency; project number CTM2017-85171-C2-1-R. FNCC and RB were partly funded by the Spanish Ministry of Science, Innovation and Universities through the research contracts FJCI-2016-28009 and FJCI-2017-31781, respectively.

Conflicts of Interest: The authors declare no conflict of interest.

Appendix A. Complete Comparison of All the Models for the Three Data Sets Selected

This appendix compares the results obtained with all the models used in this work. An additional VAM sub-model only considering p_1 as perturbation parameter, is included for illustrative purposes. It is mathematically equivalent to a Boussinesq-type model with linear vertical pressure distribution, which is also the multilayer model for the single-layer case [7]. The sub-model with only p_1 is obtained from the VAM model by neglecting the perturbation parameters u_1, w_2, and p_2 and, hence, the moment Equations (7), (8), and (9). The ensuing system of equations to be solved in the implicit step of the numerical scheme, Equation (17), yields a $3N \times 3N$ sized Jacobian.

Figure A1 compares all models using the data by [48] for all the depth ratios considered in their experiments. Figures A2 and A3 show the comparison of all models using the data by [5] for all the depth ratio conditions considered in the small and large-scale experiments, respectively. The results of the sub-model of the VAM with only p_1 (equivalent to a Boussinesq model) are very similar to those obtained by the sub-model without u_1 model, i.e., equivalent to the SGNE (fully non-linear Boussinesq-type model), highlighting that the non-hydrostaticity of flow is mainly accounted for by the perturbation parameter p_1; while the crucial parameter in the depth-averaged dam-break modelling is u_1.

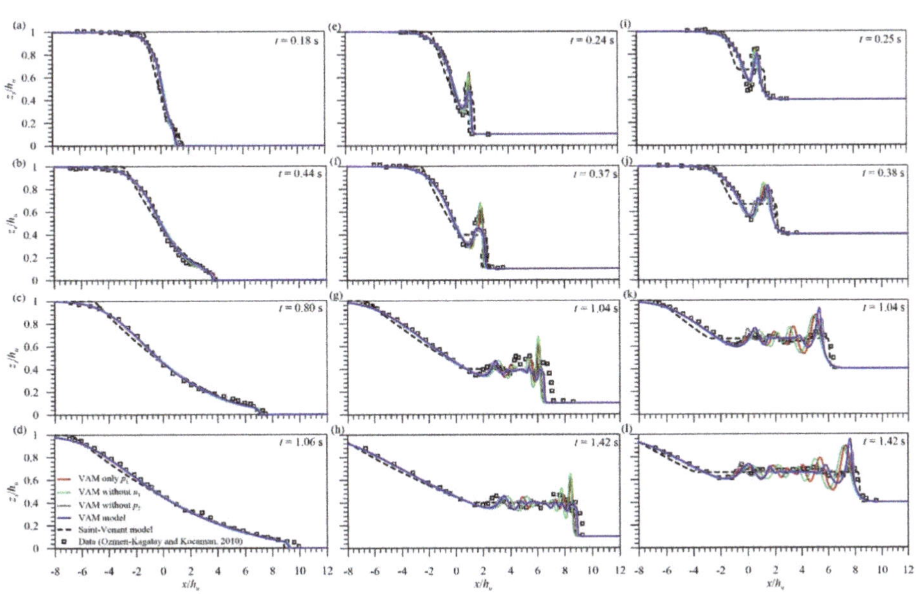

Figure A1. Comparison between all models using the experimental data by [48]. Dry bed conditions at times: (**a**) 0.18 s, (**b**) 0.44 s, (**c**) 0.80 s, (**d**) 1.06 s; wet bed conditions with $r = 0.1$ at times t [s]: (**e**) 0.24, (**f**) 0.37, (**g**) 1.04, (**h**) 1.42; and wet bed conditions with $r = 0.4$ at times t [s]: (**i**) 0.25, (**j**) 0.38, (**k**) 1.04, (**l**) 1.42.

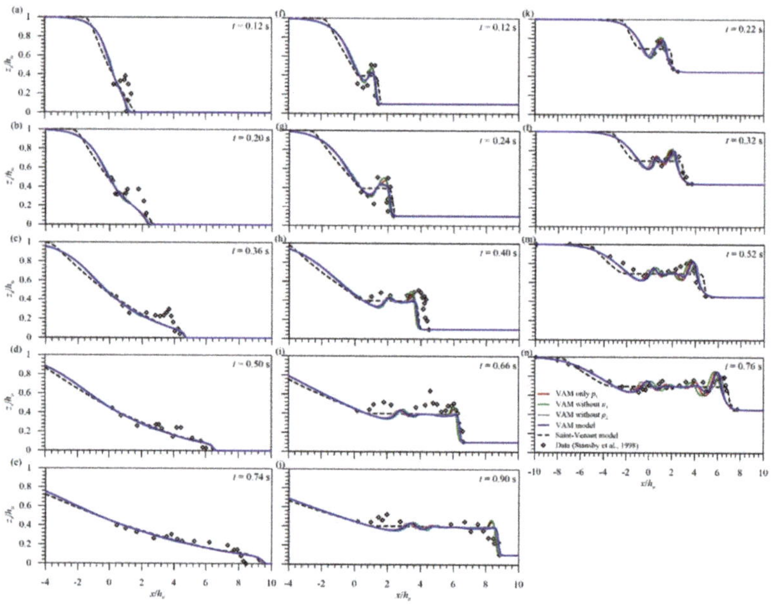

Figure A2. Comparison between all models using the experimental data of small scale dam-break flows by [5]. Dry bed conditions at times t [s]: (**a**) 0.12, (**b**) 0.20, (**c**) 0.36, (**d**) 0.50, (**e**) 0.74; wet bed conditions with $r = 0.1$ at times t [s]: (**f**) 0.12, (**g**) 0.24, (**h**) 0.40, (**i**) 0.66, (**j**) 0.90; and wet bed conditions with $r = 0.45$ at times t [s]: (**k**) 0.22, (**l**) 0.32, (**m**) 0.52, (**n**) 0.76.

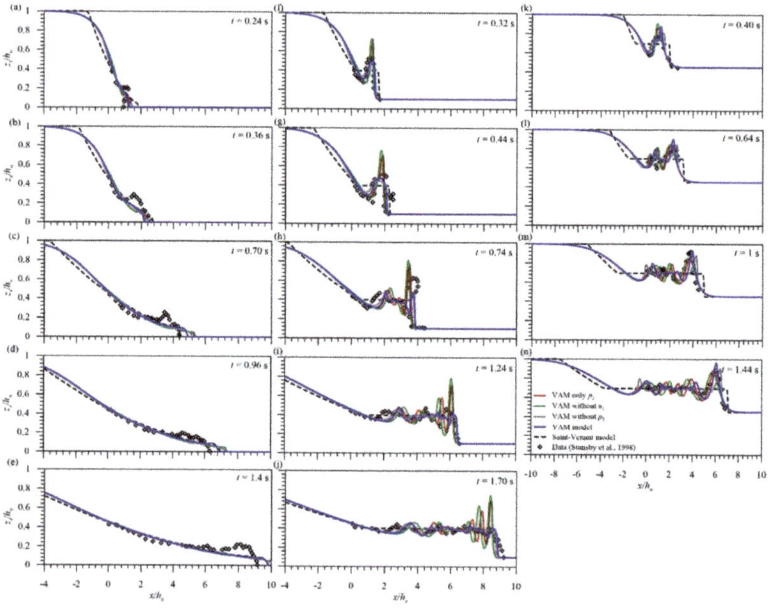

Figure A3. Comparison between all models using the experimental data of large scale dam-break flows by [5]. Dry bed conditions at times t [s]: (**a**) 0.24, (**b**) 0.36, (**c**) 0.70, (**d**) 0.96, (**e**) 1.4; wet bed conditions with $r = 0.1$ at times t [s]: (**f**) 0.32, (**g**) 0.44, (**h**) 0.74, (**i**) 1.24, (**j**) 1.70; and wet bed conditions with $r = 0.45$ at times t [s]: (**k**) 0.40, (**l**) 0.64, (**m**) 1, (**n**) 1.44.

The outflow discharge at the dam section obtained with SWE model and the VAM model, along with their different approximations, is plotted in Figure A4 for the dry bed case in [48]. For $t > 0.6$ s all models tend to predict the discharge given by Ritter's analytical solution [3], namely $q(0,t) = (8/27)g^{1/2} h_u^{3/2} = 0.116$ m. For $t < 0.6$ s predictions of the VAM model differ from the SWE, with a peak discharge above $q(0,t)$. Differences among VAM approximations are visible, but small.

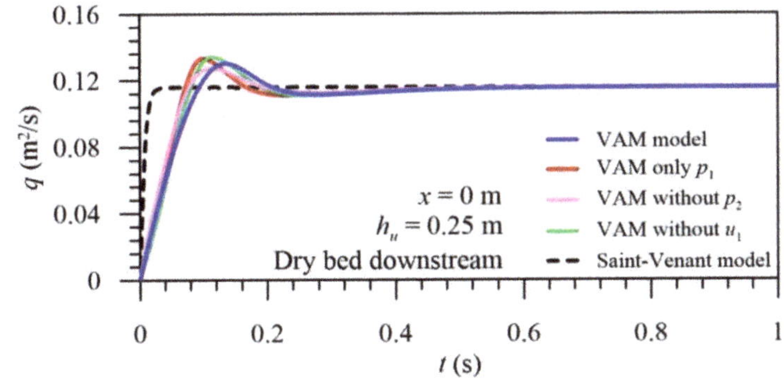

Figure A4. Comparison of discharge hydrograph at dam break section ($x = 0$ m) computed with all models for dry bed conditions [48].

References

1. De Almeida, A.B.; Franco, A.B. Modeling of dam-break flow. In *Computer Modeling of Free Surface and Pressurized Flows*; Springer: Berlin, Germany, 1994; pp. 343–373.
2. Chaudhry, M.H. *Open-Channel Flow*; Springer: Berlin, Germany, 2008.
3. Castro-Orgaz, O.; Hager, W.H. *Shallow Water Hydraulics*; Springer: Berlin, Germany, 2019.
4. Castro-Orgaz, O.; Chanson, H. Undular and broken surges in dam-break flows: A review of wave breaking strategies in a Boussinesq-type framework. *Environ. Fluid Mech.* **2020**, *20*, 1383–1416. [CrossRef]
5. Stansby, P.K.; Chegini, A.; Barnes, T.C.D. The initial stages of dam-break flow. *J. Fluid Mech.* **1998**, *374*, 407–424. [CrossRef]
6. Toro, E.F. *Shock-Capturing Methods for Free-Surface Shallow Flows*; John Wiley & Sons: Singapore, 2001.
7. Castro-Orgaz, O.; Hutter, K.; Giraldez, J.V.; Hager, W.H. Non-hydrostatic granular flow over 3D terrain: New Boussinesq-type gravity waves? *J. Geophys. Res. Earth Surf.* **2015**, *120*, 1–28. [CrossRef]
8. Cantero-Chinchilla, F.N.; Castro-Orgaz, O.; Dey, S.; Ayuso, J.L. Non-hydrostatic dam break flows. I: Physical equations and numerical schemes. *J. Hydraul. Eng.* **2016**, *142*, 04016068. [CrossRef]
9. Cantero-Chinchilla, F.N.; Castro-Orgaz, O.; Dey, S.; Ayuso, J.L. Non-hydrostatic dam break flows. II: One-dimensional depth-averaged modeling for movable bed flows. *J. Hydraul. Eng.* **2016**, *142*, 04016069. [CrossRef]
10. Cantero-Chinchilla, F.N.; Castro-Orgaz, O.; Khan, A.A. Depth-integrated non-hydrostatic free-surface flow modeling using weighted-averaged equations. *Int. J. Numer. Methods Fluid* **2018**, *87*, 27–50. [CrossRef]
11. Hibberd, S.; Peregrine, D.H. Surf and run-up on a beach: A uniform bore. *J. Fluid Mech.* **1979**, *95*, 323–345. [CrossRef]
12. Kobayashi, N.; Otta, A.K.; Roy, I. Wave reflection and run-up on rough slopes. *J. Waterw. Port Coast. Ocean Eng.* **1987**, *113*, 282–298. [CrossRef]
13. Titov, V.V.; Synolakis, C.E. Modeling of breaking and nonbreaking long-wave evolution and runup using VTCS-2. *J. Waterw. Port Coast. Ocean Eng.* **1995**, *121*, 308–316. [CrossRef]
14. Brocchini, M.; Peregrine, D.H. Integral flow properties of the swash zone and averaging. *J. Fluid Mech.* **1996**, *317*, 241–273. [CrossRef]
15. Hu, K.; Mingham, C.G.; Causon, D.M. Numerical simulation of wave overtopping of coastal structures using the non-linear shallow water equations. *Coast. Eng.* **2000**, *41*, 433–465. [CrossRef]

16. Zijlema, M.; Stelling, G.S. Further experiences with computing non hydrostatic free-surface flows involving water waves. *Int. J. Numer. Methods Fluid* **2005**, *48*, 169–197. [CrossRef]
17. Zijlema, M.; Stelling, G.S. Efficient computation of surf zone waves using the nonlinear shallow water equations with non-hydrostatic pressure. *Coast. Eng.* **2008**, *55*, 780–790. [CrossRef]
18. Yamazaki, Y.; Kowalik, Z.; Cheung, K.F. Depth-integrated, non-hydrostatic model for wave breaking and run-up. *Int. J. Numer. Methods Fluid* **2009**, *61*, 473–497. [CrossRef]
19. Bai, Y.; Cheung, K.F. Dispersion and nonlinearity of multi-layer non-hydrostatic free-surface flow. *J. Fluid Mech.* **2013**, *726*, 226–260. [CrossRef]
20. Wei, Z.; Jia, Y. A depth-integrated non-hydrostatic finite element model for wave propagation. *Int. J. Numer. Methods Fluid* **2013**, *73*, 976–1000. [CrossRef]
21. Aissiouene, N.; Bristeau, M.O.; Godlewski, E.; Sainte-Marie, J. A combined finite volume-finite element scheme for a dispersive shallow water system. *Netw. Heterog. Media* **2015**, *11*, 1–27. [CrossRef]
22. Escalante, C.; Morales de Luna, T.; Castro, M.J. Non-hydrostatic pressure shallow flows: GPU implementation using finite volume and finite difference scheme. *Appl. Math. Comput.* **2018**, *338*, 631–659. [CrossRef]
23. Castro-Orgaz, O. Weakly undular hydraulic jump: Effects of friction. *J. Hydraul. Res.* **2010**, *48*, 453–465. [CrossRef]
24. Castro-Orgaz, O.; Hager, W.H. 1D modelling of curvilinear free surface flow: Generalized matthew theory. *J. Hydraul. Res.* **2014**, *52*, 14–23. [CrossRef]
25. Castro-Orgaz, O.; Hager, W.H. *Non-Hydrostatic Free Surface Flows*; Springer: Berlin, Germany, 2017.
26. Peregrine, D.H. Calculations of the development of an undular bore. *J. Fluid Mech.* **1966**, *25*, 321–330. [CrossRef]
27. Madsen, P.A.; Murray, R.; Sørensen, O.R. A new form of the Boussinesq equations with improved linear dispersion characteristics. *Coast. Eng.* **1991**, *15*, 371–388. [CrossRef]
28. Madsen, P.A.; Sørensen, O.R. A new form of the Boussinesq equations with improved linear dispersion characteristics. Part 2. A slowly-varying bathymetry. *Coast. Eng.* **1992**, *18*, 183–204. [CrossRef]
29. Brocchini, M. A reasoned overview on Boussinesq-type models: The interplay between physics, mathematics and numerics. *Proc. Math. Phys. Eng. Sci.* **2013**, *469*, 20130496. [CrossRef]
30. Stansby, P.K.; Zhou, J.G. Shallow-water flow solver with non-hydrostatic pressure: 2D vertical plane problems. *Int. J. Numer. Methods Fluid* **1998**, *28*, 541–563. [CrossRef]
31. Casulli, V.A. A semi-implicit finite difference method for non-hydrostatic, free-surface flows. *Int. J. Numer. Methods Fluid* **1999**, *30*, 425–440. [CrossRef]
32. Cea, L.; Stelling, G.; Zijlema, M. Non-hydrostatic 3D free surface layer-structured finite volume model for short wave propagation. *Int. J. Numer. Methods Fluid* **2009**, *61*, 382–410. [CrossRef]
33. Fernandez-Nieto, E.D.; Parisot, M.; Penel, Y.; Sainte-Marie, J. A hierarchy of dispersive layer-averaged approximations of Euler equations for free surface flows. *Commun. Math. Sci.* **2018**, *16*, 1169–1202. [CrossRef]
34. Escalante, C.; Fernández-Nieto, E.D.; Morales de Luna, T.; Castro, M.J. An efficient two-layer non-hydrostatic approach for dispersive water waves. *J. Sci. Comput.* **2019**, *79*, 273–320. [CrossRef]
35. Serre, F. Contribution à l'étude des écoulements permanents et variables dans les canaux (contribution to the study of steady and unsteady channel flows). *La Houille Blanche* **1953**, *8*, 374–388. (In French) [CrossRef]
36. Green, A.E.; Naghdi, P.M. A derivation of equations for wave propagation in water of variable depth. *J. Fluid Mech.* **1976**, *78*, 237–246. [CrossRef]
37. Castro-Orgaz, O.; Chanson, H. Ritter's dry-bed dam-break flows: Positive and negative wave dynamics. *Environ. Fluid Mech.* **2017**, *17*, 665–694. [CrossRef]
38. Do Carmo, J.S.A. Boussinesq and Serre type models with improved linear dispersion characteristics: Applications. *J. Hydraul. Res.* **2013**, *51*, 719–727. [CrossRef]
39. Popinet, S. A quadtree-adaptive multigrid solver for the Serre–Green–Naghdi equations. *J. Comput. Phys.* **2015**, *302*, 336–358. [CrossRef]
40. Kirby, J.T. Recent advances in nearshore wave, circulation, and sediment transport modeling. *J. Mar. Res.* **2017**, *75*, 263–300. [CrossRef]
41. Gallerano, F.; Cannata, G.; Scarpone, S. Bottom changes in coastal areas with complex shorelines. *Eng. Appl. Comput. Fluid Mech.* **2017**, *11*, 396–416. [CrossRef]
42. Roeber, V.; Cheung, K.F. Boussinesq-type model for energetic breaking waves in fringing reef environment. *Coast. Eng.* **2012**, *70*, 1–20. [CrossRef]

43. Steffler, P.M.; Jin, Y.C. Depth averaged and moment equations for moderately shallow free surface flow. *J. Hydraul. Res.* **1993**, *31*, 5–17. [CrossRef]
44. Khan, A.A.; Steffler, P.M. Modeling overfalls using vertically averaged and moment equations. *J. Hydraul. Eng.* **1996**, *122*, 397–402. [CrossRef]
45. Khan, A.A.; Steffler, P.M. Vertically averaged and moment equations model for flow over curved beds. *J. Hydraul. Eng.* **1996**, *122*, 3–9. [CrossRef]
46. Cantero-Chinchilla, F.N.; Bergillos, R.J.; Castro-Orgaz, O. Nearshore coastal flow processes using weighted-averaged equations. *Ocean. Eng.* **2020**, *211*, 107480. [CrossRef]
47. Gamero, P.; Bergillos, R.J.; Cantero-Chinchilla, F.N.; Castro-Orgaz, O. A MATLAB software platform for modelling vertically-integrated non-hydrostatic flows with moment equations. *Environ. Model. Softw.* **2020**, *127*, 104674. [CrossRef]
48. Ozmen-Cagatay, H.; Kocaman, S. Dam-break flows during initial stage using SWE and RANS approaches. *J. Hydraul. Res.* **2010**, *48*, 603–611. [CrossRef]
49. Montes, J.S. *Hydraulics of Open Channel Flow*; ASCE Press: Reston, VA, USA, 1998.
50. Hicks, F.E.; Steffler, P.M. Characteristic dissipative Galerkin scheme for open-channel flow. *J. Hydraul. Eng.* **1992**, *118*, 337–352. [CrossRef]
51. Yamamoto, S.; Daiguji, H. Higher-order-accurate upwind schemes for solving the compressible Euler and Navier-Stokes equations. *Comput. Fluids* **1993**, *22*, 259–270. [CrossRef]
52. Aureli, F.; Maranzoni, A.; Mignosa, P.; Ziveri, C. A weighted surface-depth gradient method for the numerical integration of the 2D shallow water equations with topography. *Adv. Water Resour.* **2008**, *31*, 962–974. [CrossRef]
53. Toro, E.F. *Riemann Solvers and Numerical Methods for Fluid Dynamics: A Practical Introduction*; Springer: Berlin, Germany, 2009.
54. Cantero-Chinchilla, F.N.; Castro-Orgaz, O.; Khan, A.A. Vertically-averaged and moment equations for flow and sediment transport. *Adv. Water Resour.* **2019**, *132*, 103387. [CrossRef]
55. Castro-Orgaz, O.; Hager, W.H. Classical hydraulic jump: Basic flow features. *J. Hydraul. Res.* **2009**, *47*, 744–754. [CrossRef]
56. Castro-Orgaz, O.; Hager, W.H. Subcritical side-weir flow at high lateral discharge. *J. Hydraul. Eng.* **2012**, *138*, 777–787. [CrossRef]
57. Castro-Orgaz, O.; Lozano, D.; Mateos, L. Energy and momentum velocity coefficients for calibration of submerged sluice gates in irrigation canals. *J. Irrg. Drain. Eng.* **2010**, *136*, 610–616. [CrossRef]

water

MDPI

Article

A New Dam-Break Outflow-Rate Concept and Its Installation to a Hydro-Morphodynamics Simulation Model Based on FDM (An Example on Amagase Dam of Japan)

Sayed Masihullah Ahmadi [1,*] and Yoshimichi Yamamoto [2]

1 Graduate School of Science and Technology, Tokai University, Kanagawa 259-1292, Japan
2 Department of Civil Engineering, Tokai University, Kanagawa 259-1292, Japan; yo-yamamo@tokai.ac.jp
* Correspondence: 9ltad003@mail.u-tokai.ac.jp; Tel.: +81-463-58-1211; Fax: +81-463-50-2045

Abstract: Dams are constructed to benefit humans; however, dam-break disasters are unpredictable and inevitable leading to economic and human life losses. The sequential catastrophe of a dam break directly depends on its outflow hydrograph and the extent of population centers that are located downstream of an affected dam. The population density of the cities located in the vicinity of dams has increased in recent times and since a dam break hydrograph relies on many uncertainties and complexities in devising a dam-break outflow hydrograph, more researches for the accurate estimation of a dam-break flood propagation, extent and topography change becomes valuable; therefore, in this paper, the authors propose a novel and simplified dam-break outflow rate equation that is applicable for sudden-partial dam breaks. The proposed equation is extensively affected by a dam-break shape. Therefore, the inference of a dam-break shape on a dam-break outflow rate is investigated in the current study by executing hydraulic experiments in a long, dry bed, frictionless and rectangular water channel connected to a finite water tank to acquire a mean break-shape factor. The proposed equation is further validated by regenerating the Malpasset dam-break hydrograph and comparing it to the existing methods and also by installing it on an existing 2D hydro-morphodynamics flood simulation model. Finally, Amagase Dam's (arch-reaction dam in Japan) break simulation is executed as a case study. The results of the simulations revealed that the greater the height of a dam-break section, the more devastating its flood consequences would be.

Keywords: dam-break simulation; hydro-morphodynamics; dam out-flow rate; amagase dam; shallow water equations; FDM

Citation: Ahmadi, S.M.; Yamamoto, Y. A New Dam-Break Outflow-Rate Concept and Its Installation to a Hydro-Morphodynamics Simulation Model Based on FDM (An Example on Amagase Dam of Japan). *Water* **2021**, *13*, 1759. https://doi.org/10.3390/w13131759

Academic Editors: Francesca Aureli, Andrea Maranzoni and Gabriella Petaccia

Received: 4 May 2021
Accepted: 21 June 2021
Published: 25 June 2021

1. Introduction

As water is stored behind a dam, enormous potential energy is formed that in the case of a dam-break, devastating catastrophe may follow [1] especially, when densely populated cities are located downstream of a dam [2], in the last two decades, floods resulting from dam breaks are responsible for some of the most devastating man-made disasters. Therefore, every effort to further reduce the severe effects of a dam break and generating a better forecast of its flood extent in the tailwater areas is of immense value [3].

The development of effective emergency action plans and the design of an early warning system that might reduce or eliminate the consequences of a dam break requires its inundation information downstream [4]. The development of the inundation information from a dam break consists of three steps: the routing of the inflow flood through a reservoir, estimating the dam breach characteristics and the downstream flood routing [5]. Out of these three steps, estimating the characteristics and parameters of the dam breach contains the greatest uncertainties of all aspects of dam-break [6]. Therefore, in this paper, the authors present a simplified novel approach relevant to sudden-partial dam-break outflow rate for non-erodible type dams.

Generating precise dam-break hydrographs has been a major concern since the end of the 10th century [3]. The first documented experimental dam-break experiment in a channel might be the one proposed by Bazin [7], following which, Ritter [8], derived an analytical solution of the famous one-dimensional De-Saint Venart equation for sudden dam-breaks. Later on, Su and Barnes [9], extended Ritter's solution considering the effect of different channel cross-sections and proposed a useful power type equation for this phenomenon. However, until recent years, very few advances have been reported to address the dam-break problem [3].

In most dam-break studies, the reservoir is usually assumed as a long infinite water supply until Aureli et al. [10], presented useful approaches by numerically solving the Ritter, and Su and Barnes [9] methods in combination, as well as considering dam-break bathymetry effects on the produced outflow hydrograph. Fundamentally, an outflow hydrograph is affected by the bathymetry of a dam, valley shape, dam height and shape and extension of the breach. In the papers [3,11–13], the effects of dam bathymetry on the dam-break hydrograph have been evaluated and many more researchers have concentrated on the breach formation shape and breach formation time [14].

In all studies predating [3], all researchers had assumed a total dam-break scenario which is not realistic and does not take into account the types of dams that can be partially breached. Concerning this fact, Piloti et al. [3] and Aureli et al. [10] proposed methods to address sudden partial dam breaks. However, based on the equation (23) of Aureli et al. [10] and equation (19) of Pilotti et al. [3] the details of the storage-depth curve of the reservoir are necessary to precisely calculate the outflow hydrograph.

In the practical scenario, the storage-depth curves might either not be available or difficult to acquire for some existing dams. Hence to overcome this issue, the authors propose a new dam-break outflow rate equation that requires only the height of a dam for calculating a dam break outflow rate. The proposed equation is applicable to many concrete dam types. However, because the proposed equation is extensively affected by the dam-break cross-section a flow rate coefficient (break shape factor) must be initially decided. In the present work, for the arch dams (dams with thin width) a mean shape factor is acquired after executing many hydraulic experiments in a channel.

To show the validity of the proposed equation, the hydrograph of the famous Malpasset dam-break that happened in 1959 in Fréjus city of France, is generated and compared to that of [10]. The results reveal that the proposed equation comparatively produces a higher peak discharge and a faster emptying time. Finally, the inference of a dam-break height and width on the extent in inundation area and initial flood wave velocity is investigated by executing many dam-break simulations on the Amagase dam which is a concrete arch-reaction dam in Japan. The results suggest that, as the break section's height increases, the inundation extent and initial flood velocity increase even if the break cross-sectional area tends to remain the same.

2. Calculation Methods

2.1. New Model of Dam-Break Out-Flow Rate

For the development of the dam-break hydrograph, a finite volume reservoir with a surface area of $A(h)$ is assumed with a break section of width B and heigh H. shown in Figure 1a,b. The shape of a break section can vary from a rectangle to a triangle depending on the shape of a proposed dam valley. Since in this paper, it is intended to develop a dam-break outflow rate equation that is independent of a dam's stage-discharge curve, an imaginary coefficient D_d is assumed for the dam-break section. D_d is measured from the water surface level of a dam's reservoir and it is many times the height of a dam. If D_d is set to a very high value compared to the respected dam height, it replicates an almost rectangular dam-break cross-section, and vice versa, if this value is set to a relatively small value, it replicates a trapezoidal and triangle reservoir cross-section. Therefore, the proposed equation can produce the outflow rate of a dam break with different cross-sections.

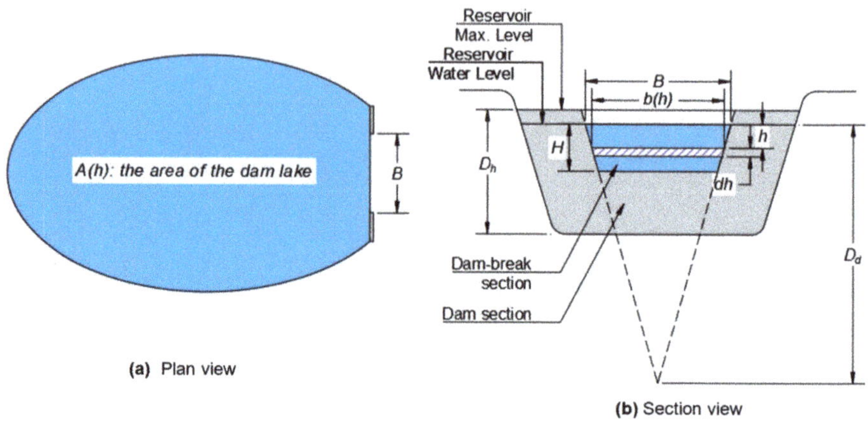

Figure 1. Dam-break plan (**a**) and section view (**b**).

The run-off rate from the dam lake in the case that the overflow height H is assumed constant (water supply is infinite) can be calculated using the below equations:

$$Q_r = \int_0^H b(h) \times c\sqrt{2gh}dh = \int_0^H \frac{B(D_d - h)}{D_d} c\sqrt{2gh}dh \tag{1}$$

$$Q_r = \frac{B}{D_d} c\sqrt{2g}\left[D_d \frac{2}{3} h^{\frac{3}{2}} - \frac{2}{5} h^{\frac{5}{2}}\right]_0^H \tag{2}$$

$$Q_r = \frac{B}{D_d} c\sqrt{2g}\left(\frac{2}{3} D_d H^{\frac{3}{2}} - \frac{2}{5} H^{\frac{5}{2}}\right) \tag{3}$$

Here, Q_r is the discharge rate; H is the final depth of a break section; g is the earth's gravitational force; $b(h)$ and dh are the width and minute height of the break portion at depth h; h is falling depth of water surface in a dam lake after a break; c is the flow rate coefficient; D_d is the imaginary depth from the dam crown level as shown in Figure 1b.

Since the case of finite water supply from a dam lake is more realistic in the actual dam-break cases, in Equation (3) H is substituted with $\left(H - \frac{dh}{dt}T\right)$. Here, T is the run-off elapse time (sec). In the following, equations for dam-break outflow rate (Q_r), and the total volume of discharge (V_r), are developed with considering the case of finite water supply from a dam reservoir;

$$Q_r = \frac{B}{D_d} c\sqrt{2g}\left(\frac{2}{3} D_d \left(H - \frac{dh}{dt}T\right)^{3/2} - \frac{2}{5}\left(H - \frac{dh}{dt}T\right)^{5/2}\right) \tag{4}$$

$$V_r = \int_0^H A(h)dh = \int_0^{\frac{H}{\frac{dh}{dt}}} QdT \tag{5}$$

$$V_r = \int_0^{\frac{H}{\frac{dh}{dt}}} \frac{B}{D_d} c\sqrt{2g}\left(\frac{2}{3} D_d \left(H - \frac{dh}{dt}T\right)^{3/2} - \frac{2}{5}\left(H - \frac{dh}{dt}T\right)^{5/2}\right) dT \tag{6}$$

$$\text{Here, when } H - \frac{dh}{dt}T \text{ is replaced by X, } \frac{dX}{dT} = -\frac{dh}{dt}, \text{ and} \tag{7}$$

$$T = 0 \to T = \frac{H}{\frac{dh}{dt}} \text{ is changed to } X = H \to X = 0, \text{ therefore;} \tag{8}$$

$$V_r = \frac{B}{D_d} c\sqrt{2g} \frac{-1}{\frac{dh}{dt}} \int_H^0 \left(\frac{2}{3} D_d X^{3/2} - \frac{2}{5} X^{5/2}\right) dX \tag{9}$$

$$= \frac{B}{D_d} c\sqrt{2g} \frac{-1}{\frac{dh}{dt}} \left[\frac{4}{15} D_d X^{5/2} - \frac{4}{35} X^{7/2} \right]_H^0 \quad (10)$$

Now, substituting X with H gives the below equation for V_r;

$$\therefore V_r = \frac{B}{D_d} c\sqrt{2g} \frac{+1}{\frac{dh}{dt}} \left(\frac{4}{15} D_d H^{5/2} - \frac{4}{35} H^{7/2} \right) \quad (11)$$

$$\therefore \frac{dh}{dt} = \frac{Bc\sqrt{2g}}{D_d V_r} \left(\frac{4}{15} D_d H^{5/2} - \frac{4}{35} H^{7/2} \right) \quad (12)$$

$$\therefore T_{total} = \frac{H}{\frac{dh}{dt}} = \frac{D_d V_r}{Bc\sqrt{2g}} \frac{1}{\frac{4}{15} D_d H^{5/2} - \frac{4}{35} H^{7/2}} \quad (13)$$

Here, T_{total} is the cumulative time from the start of the calculations.

2.2. *Existing Numerical Simulation Model*

For the development of the inundation information and topographical change due to a long-wave flood such as a tsunami or dam-break, the numerical simulation model of Ca et al. [15] is used. The model was developed using the following detailed formulas.

2.2.1. Numerical Model for Fluid Motion

The numerical model used for long-wave flood simulations in this paper is based on a continuity equation of fluid (Equation (14)) and two-dimensional nonlinear long-wave equations (Equations (15) and (16)), and these governing equations are solved by finite difference method using the Crank-Nicholson scheme.

$$\frac{\partial f_y q_x}{\partial x} + \frac{\partial f_x q_y}{\partial y} + \frac{\partial S\eta}{\partial t} = 0 \quad (14)$$

$$\begin{aligned} \frac{\partial q_x}{\partial t} + \frac{1}{S}\frac{\partial}{\partial x}\left(\frac{S{q_x}^2}{d_w}\right) + \frac{1}{S}\frac{\partial}{\partial y}\left(\frac{Sq_x q_y}{d_w}\right) + g d_w \frac{\partial \eta}{\partial x} - \frac{1}{S}\frac{\partial}{\partial x}\left[d_w v_t S \frac{\partial}{\partial x}\left(\frac{q_x}{d_w}\right)\right] \\ - \frac{1}{S}\frac{\partial}{\partial y}\left[d_w v_t S \frac{\partial}{\partial y}\left(\frac{q_x}{d_w}\right)\right] + \frac{f_c}{{d_w}^2} Q q_x = 0 \end{aligned} \quad (15)$$

$$\begin{aligned} \frac{\partial q_y}{\partial t} + \frac{1}{S}\frac{\partial}{\partial x}\left(\frac{Sq_x q_y}{d_w}\right) + \frac{1}{S}\frac{\partial}{\partial y}\left(\frac{S{q_y}^2}{d_w}\right) + g d_w \frac{\partial \eta}{\partial y} - \frac{1}{S}\frac{\partial}{\partial x}\left[d_w v_t S \frac{\partial}{\partial x}\left(\frac{q_y}{d_w}\right)\right] \\ - \frac{1}{S}\frac{\partial}{\partial y}\left[d_w v_t S \frac{\partial}{\partial y}\left(\frac{q_y}{d_w}\right)\right] + \frac{f_c}{{d_w}^2} Q q_y = 0 \end{aligned} \quad (16)$$

$$q_x = \int_{\eta}^{-d_w} u_x dz \quad (17)$$

$$q_y = \int_{\eta}^{-d_w} u_y dz \quad (18)$$

Here, q_x and q_y are the horizontal fluid fluxes in the x and y directions respectively; η is the water surface elevation. f_x and f_y are the x and y direction ratios of the wet portion in a calculation mesh; S is the area ratio of the wet portion in a calculation mesh see Figure 2; d_w is the water depth (from the static water surface + η); g is the gravitational acceleration; v_t is the eddy viscosity coefficient; f_c is the ground surface friction coefficient; and $Q\left(= \sqrt{{q_x}^2 + {q_y}^2}\right)$ is the compound value of q_x and q_y.

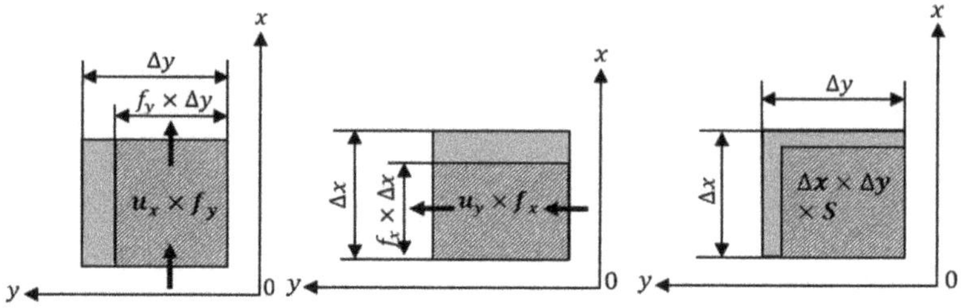

Figure 2. Explanation of f_x, f_y and S in a control volume. The shaded area shows wet portion. The gray area shows dry portion.

To calculate the eddy viscosity coefficient and the ground surface friction coefficient, the following equations are used;

$$\nu_t = \varepsilon \left[\left(\frac{\partial U}{\partial y} \right)^2 + \left(\frac{\partial V}{\partial x} \right)^2 \right]^{1/2} \cdot d_w{}^2 \tag{19}$$

$$f_c = \frac{gn^2}{d_w{}^{1/3}},\ n^2 = n_0{}^2 + 0.020 \frac{B_r}{100 - B_r} d_w{}^{4/3},\ n_0{}^2 = \frac{n_1{}^2 A_1 + n_2{}^2 A_2 + n_3{}^2 A_3}{A_1 + A_2 + A_3} \tag{20}$$

Here, U and V are the flow velocity in y and x directions; ε equals (0.1); n is Manning's roughness coefficient; B_r is the building ratio (= the ratio of the area of all vertical objects such as houses and trees to the mesh area). n_0 is the weighted average roughness coefficient of areas such as farms, roads and waste and wetlands, $A_{1,\ 2,\ \text{and}\ 3}$ respectively, with relative roughness coefficients of $n_{1,\ 2,\ \text{and}\ 3}$ [16].

2.2.2. Numerical Model for Topographical Change

The topographical change based on sediment transport by the flow can be expressed by using the continuity equation, Equation (21);

$$\frac{\partial \zeta}{\partial t} = -\frac{1}{1 - \varepsilon_s} \left(\frac{\partial q_{bx}}{\partial x} + \frac{\partial q_{by}}{\partial y} - C_s + C_{ut} \right) \tag{21}$$

Here, ζ is the ground surface elevation; q_{bx} and q_{by} are respectively the bed-load rate per unit width in x and y directions; C_s is the deposition rate of the suspended load; C_{ut} is the entrainment rate of the suspended load from the bed; and ε_s is the porosity of the sediment.

(1) Modeling of q_x and q_y on Bed-load Transport

For evaluation of the bed-load rate, Ribberink's formula [17], shown in Equation (22) is used. Yokoyama et al. [18] performed many calculations of scouring by flow and wave using indoor and outdoor data on sand and gravel with a diameter range of 0.2~10 mm and found that accurate results can be obtained using this formula.

$$q_{bi} = \begin{cases} \text{and} C_b [|\theta_s(t)| - \theta_{sc}]^{1.65} \frac{\theta_s(t)}{|\theta_s(t)|} \sqrt{\Delta g D_{50}} & (\theta_s(t) \geq \theta_{sc}) \\ \text{and} 0 & (\theta_s(t) < \theta_{sc}) \end{cases} \tag{22}$$

Here, q_{bi} is the bed-load transport rate per unit width in the i direction; C_b is the bed-load transport coefficient determined by verification simulations; however, the authors by performing many hydraulic experiments, developed useful diagrams to acquire the value of this coefficient, refer to Figures 3–5 [19]; $\theta_s(t)$ is the Shields parameter in the i direction; θ_{sc} is the critical Shields number, calculated by using the equation of van Rijn, [20]; Δ is the relative density of the sand; g is the gravitational acceleration; D_{50} is the median diameter of the sediment.

(2) Modeling of C_s and C_{ut} on Suspended Load Transport

During a long wave flood, it is necessary to consider the influence of suspended load transport. The deposition rate of the suspended load C_s and the entrainment rate from the bed C_{ut} was evaluated using Equation (23) based on the vertical distribution of suspended load concentration:

$$C_s = w_s C(z);\ z = \left(\frac{w_s}{2}\right),\ C_{ut} = -\nu_t \frac{\partial C(z)}{\partial z}|z = z_a \quad (23)$$

Here, w_s is the settling velocity of suspended particles which can be calculated using Equation (24) [21–23]; C(z) is the suspended load concentration and can be calculated using Equation (25) [24], under the assumption of the sheet flow condition for the whole area; v_t is the eddy viscosity.

$$w_s = \begin{cases} \text{and} \frac{\nu D_*^3}{18 D_{50}} & D_*^3 \le 16.187 \\ \text{and} \frac{10\nu}{D_{50}} \left[\left(1 + 0.01 D_*^3\right)^{\frac{1}{2}} - 1 \right] & 16.187 < D_*^3 \le 16187 \\ \text{and} \frac{1.1 \nu D_*^{1.5}}{D_{50}} & 16187 < D_*^3 \end{cases} \quad (24)$$

$$C(z) = C_a \left(\frac{z}{z_a}\right)^b,\ b = -\frac{w_s}{\kappa u_*} \quad (25)$$

Here, ν is the kinematic viscosity of water; D_* is van Rijin's dimensionless particle parameter; κ is Karman's constant $(= 0.4)$; u_* is the friction velocity; C_a and z_a are estimated using Equation (26) [25].

$$\left. \begin{array}{c} \text{and } C_a = \frac{0.331[\theta_s(t) - 0.045]^{1.75}}{1 + 0.720[\theta_s(t) - 0.045]^{1.75}} \\ \text{and } z_a = 2D_{50} \end{array} \right\} \quad (26)$$

3. Methods for Evaluation of Empirical Coefficients

3.1. Evaluation of the Break-Shape Coefficient of the Outflow Rate Equation

For the newly proposed outflow rate equation, (Equation (4)) the coefficient c has a great impact on a calculated hydrograph. This coefficient depends on many variables, i.e., dam bathymetry, type of break, break dimensions and dam width and height. Initially, the dam-break section's impact on this coefficient is evaluated in this paper, by performing many hydraulic experiments. Figure 3a,b shows the apparatus and experimental arrangements respectively. On the left-hand side of the apparatus, there is a water tank reserve acting as a dam reservoir, at the downstream, there is a dry-bed water channel of 0.5 m in width, 0.4 m in height and 12 m in length. At the one-meter section from the reservoir, two layers of acryl plates of one centimeter in thickness are placed and sealed using silicon glue. The two plates are arranged as both of them are cut to the desired section, listed in Table 1. One other plate placed in between these two acts as a gate where grease is used as a lubricant in between the plates to let one open the gate abruptly. This complies with the objective of sudden partial dam-break. From these experiments, it is intended to develop the hydrographs, therefore, using a digital camera which is focused on the scale bar located about 50 cm upstream of the acryl plates; the whole process of the experiments is recorded. Using this arrangement, one can measure the change in water height behind the dam, and because previously, the bathymetry measurements (reservoir surface area) are taken, the discharge rate is calculated. Figure 4 shows the calculated and measured hydrographs, where the standard deviation and correlation value between the calculated hydrograph and experimental hydrograph are also presented for each of the experimented cases. For calculated hydrographs, the c value is set to make the best fit curve with respect to the measured hydrograph. As a result, the best value of c is acquired which is tabulated in Table 1 along with their relative experimental data.

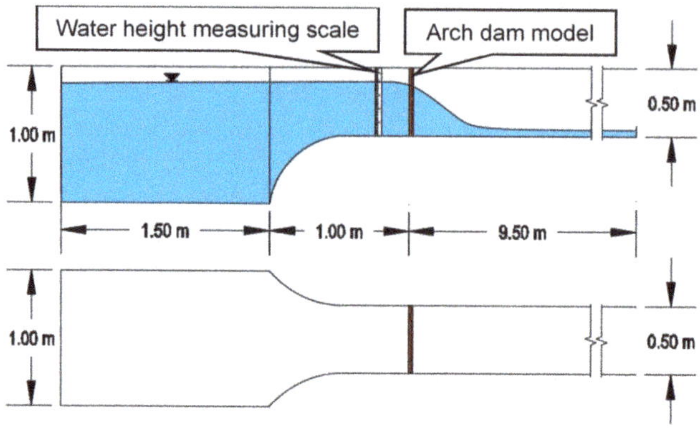

(**a**)

(**b**)

Figure 3. (**a**) Dam-break experiment apparatus. (**b**) Dam-break experiment in a rectangular channel.

Table 1. Dam-break experiment data and results.

Case	Description	Break Dimensions				c
		B (m)	H (m)	D_d (m)	Break Section Thickness (cm)	
I.1	Trapezoidal	0.30	0.25	6	3.0	0.30
1.2	Trapezoidal	0.30	0.25	6	3.0	0.30
2.1	Trapezoidal	0.30	0.25	4	3.0	0.20
2.2	Trapezoidal	0.30	0.25	4	3.0	0.20
3.1	Trapezoidal	0.20	0.30	10	3.0	0.35
3.2	Trapezoidal	0.20	0.30	10	3.0	0.35
4	Triangle	0.20	0.30	1	0.2	0.50

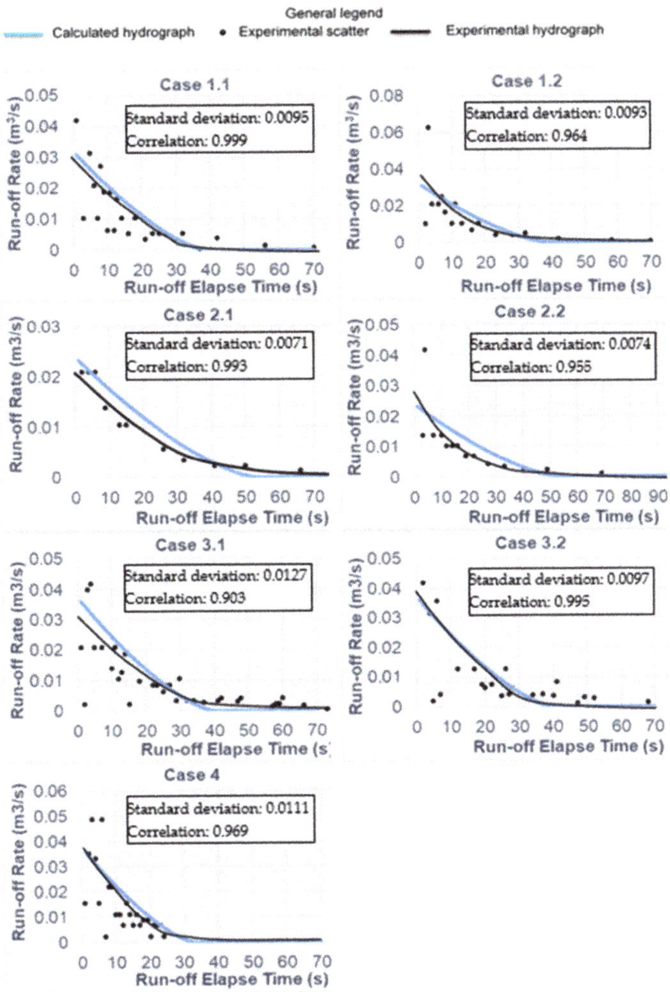

Figure 4. (Case 1–4) Dam-break hydrographs from experimental cases and calculated using Equation (4).

The flow rate coefficient (c) values presented in this paper, can be used for arch reaction dams that have relatively thin dam thickness. In the case of dam-break simulations of the Amagase concrete arch-reaction dam, illustrated in the subsequent sections of this paper, the flow rate coefficient $c = 0.28$. For Amagase dam-break simulations, the ratio of B to H is similar to that of cases (1.1–2.2. The ratio of $\frac{D_h}{D_d} = \frac{1}{15}$ for cases 1.1 and 1.2 of the experiments, D_h being the dam height as shown in Figure 1b, ($c = 0.3$), for Cases 2.1 and 2.2, $\frac{D_h}{D_d} = \frac{1}{10}$, ($\mathrm{c} = 0.2$) and for Amagase dam $\frac{D_h}{D_d} = \frac{73}{1000} \cong \frac{1}{13.8}$, therefore, the flow rate coefficient ($c = 0.28$) seems logical for Amagase dam-break simulations. In cases 3.1, 3.2 and 4 the height of the break section B is larger than the width of the break section B, therefore, they are not considered for acquiring the Amagase dam flow rate coefficient.

3.2. Rational Evaluation of the Coefficient of Bed-Load Transport Rate

To calculate the bed-load transportation, the model of Ca et al. [15] uses Equation (22) [17]. To use this equation, one needs to calculate the bedload transport coefficient C_b value using the graphs provided in Figures 5–7 as well as Equation (27). The presented graphs and illustrations of their production and use are detailed in [19].

$$C_b = C_{bo} \times C_1 \times C_2 \tag{27}$$

here, C_{b0} is bedload coefficient. It can be acquired from Figure 5 by knowing the median grain size, type of gravels (natural or crushed gravel) for a proposed area and assuming uniformity coefficient of 1.5~3, and dry density of around 1.5 g/cm^3. C_2 and C_1 are reduction coefficients, can be acquired from Figures 6 and 7 respectively.

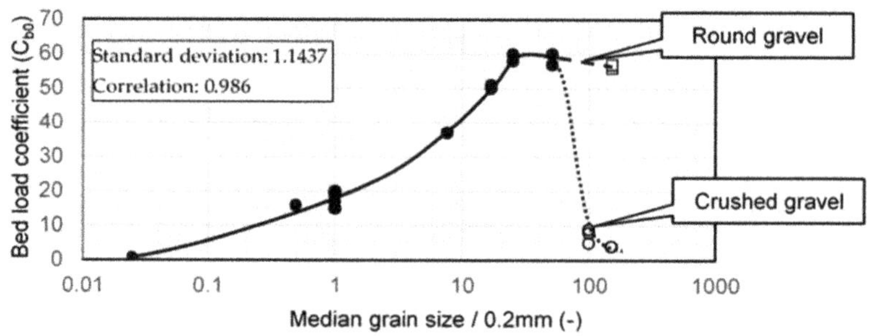

Figure 5. Influence of the median grain size to the bed-load coefficient (Uniformity coefficients are 1.5~3, dry densities are around 1.5 g/cm^3).

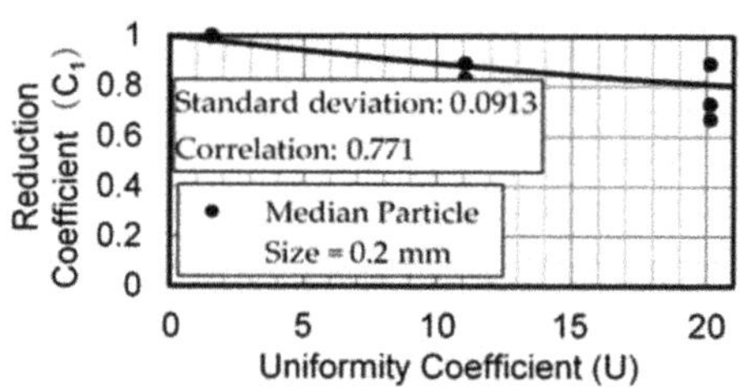

Figure 6. Influence of the uniformity coefficient to the bed-load reduction coefficient (the median grain size is 0.2 mm; dry densities are around 1.5 g/cm^3).

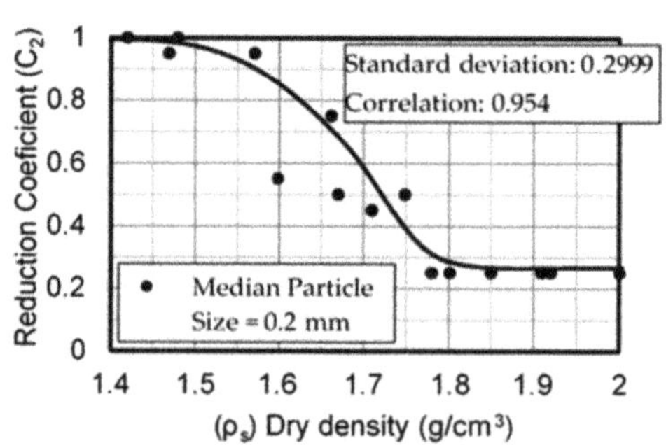

Figure 7. Influence of dry density to the bed-load reduction coefficient (the median grain size and the uniformity coefficient are 0.2 mm and 20.1).

4. Verification and Application of the Proposed Concept

For verification purposes, the famous Malpasset Dam's hydrograph is generated using the proposed dam-break outflow rate Equation (4), and it is compared to the model of Aureli et al. [10]. For applicability purposes, the proposed outflow equation is integrated into an existing 2D long-wave flood simulation model of Ca et al. [15]. Even though there are several widely used flood simulation models such as MIKE11, ISIS, ONDA, FLU-COMP and HEC-RAS [26], or the recent and more advanced SPH approach models such as SPHERA [27], the authors have used the model of Ca et al. [15] because of having enough experience of using this model which has given consistent results. This model uses the two-dimensional non-linear shallow water equations, the continuity equation for hydrodynamics calculations and Rebberink's equation [17] for modeling sediment transport. Since the data of a real dam-break are very difficult to achieve [28], and dam-break flood and a tsunami flood are almost similar [29], the validity of this model for a long wave flood routing was previously investigated by performing many tsunami simulations on the Sendai-Natori coast of Japan, which was hit by the Great East Japan Tsunami in 2011 [19].

4.1. Malpasset Dam-Break Hydrograph

Malpasset Dam was a double-curvature arch-shaped structure located in the Frejus town of France. It had a crest length of approximately 223 m, a height of 66.5 m and a total reservoir capacity of 55 million cubic meters [30]. Following a flash flood on 2nd December 1959 at 21:14, the dam broke and a flood wave as tall as 40 m gushed through the valley towards the town of Frejus. Approximately 48 million cubics of water was released from the dam which caused the devastating catastrophe resulting in 433 casualties [30]. Due to the availability of measured flood propagation data from this dam break, it is quite popular in the literature and is often used as a case study for simulation verification [31–33]. Therefore, this paper intended to compare the Maspasset dam-break hydrograph generated using the newly proposed concept and the methods proposed by Aureli et al. [10]. For comparison purposes, the authors have set the break dimensions assuming a partial dam break. 85% of the crest length = 189.6 m as well as the same percentage of the dam height = 56.53 m is assumed broken which is more realistic than a full section dam-breaking assumption assumed by Aureli et al. [10]. Furthermore, the peak discharge flow rate from the proposed method is higher than the Aureli et al. [10] method and lower than the LRA method as shown in Figure 8. Moreover, the emptying time of the proposed method is the fastest among other methods. The reason behind these differences lies extensively in the different initial assumptions for developing these methods. For example, in the proposed method, the inference of reservoir bed slope, as well as water surface gradient during the emptying process, is neglected where this part can be improved in the future.

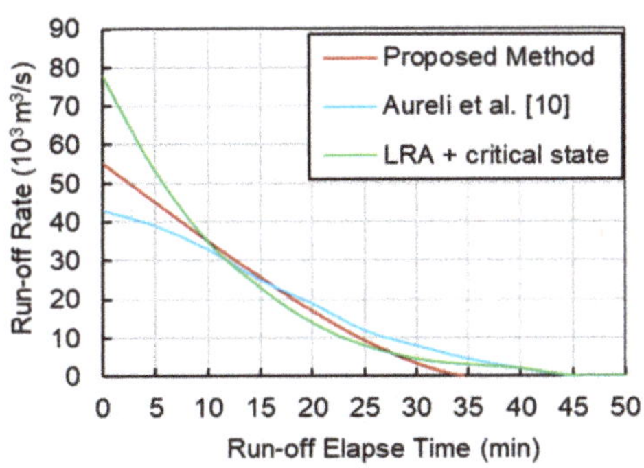

Figure 8. The Malpasset dam hydrograph comparison between the proposed method of this paper, method of Aureli et al. [10] and Level Reservoir Approximation together with the imposition of the critical state at the dam site during the emptying process, Aureli et al. [10].

4.2. Hydro-Morphodynamics Simulation in Sendai-Natori Coast

Sendai-Natori coast is located in the Miyagi prefecture of Japan. This coast experienced a magnitude 9.0 earthquake and a consequent catastrophic great tsunami in 2011. The authors executed many hydro-morphodynamics simulations to replicate the actual flooding conditions and presented the results on [19]. The authors also showed in their paper that the results of the simulations are in good agreement with the actual measured data from the area as shown in Figure 9b. Random spot elevation-change points are selected along the line (A) shown in Figure 9a, and their relative data are plotted from both measured data and the reproduction data. For the locations where measured data is available, the calculated data matches perfectly with the measured data. However, for some locations such as the seaside or underwater parts, because there is no digital elevation model data available, one cannot decide whether it differs from the actual topography change situation.

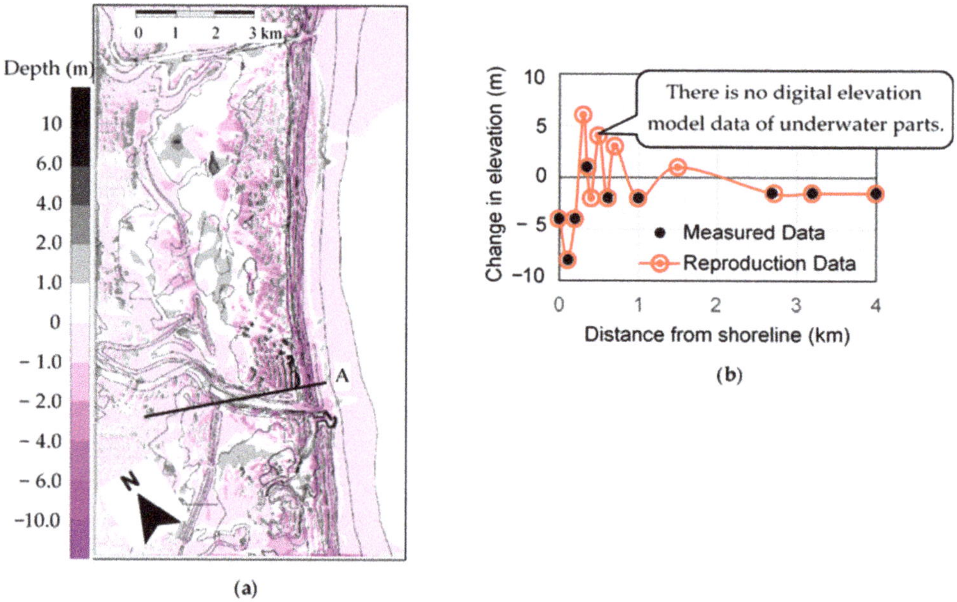

Figure 9. (**a**) Topography change results (Total bed—load), 48 min from start of the tsunami calculations [19]. (**b**) Relative accuracy of the topography-change along line (A)—shown on Figure 9. (**a**)—on the measured and reproduced simulation results [19].

4.3. Dam-Break Hydro-Morphodynamics Simulation (Amagase Dam)

Amagase Dam is located in Uji City in Kyoto Prefecture of Japan. (Figure 10). It is an arch dam made of concrete, built on the Uji-Gawa river (Figure 11). The Amagase dam's construction started in 1955 and finished in 1964. This dam is mainly constructed for flood control. It also serves as a great water supply reservoir for the Uji City residents. The catchment area of the dam is 4200 km^2 and its design volume is 122 thousand cubic meters. The height of the dam is 73 m with a crest length of 254 m. The reason behind why this dam is chosen for simulation is, about 200 thousand people live downstream of the dam. If this dam breaks due to a mega earthquake or any other possible cause, it would be catastrophic. Therefore, in this paper, it is intended to prepare a hazard map to depict the inundation height, velocity, scouring and deposition depths for the possible affected area.

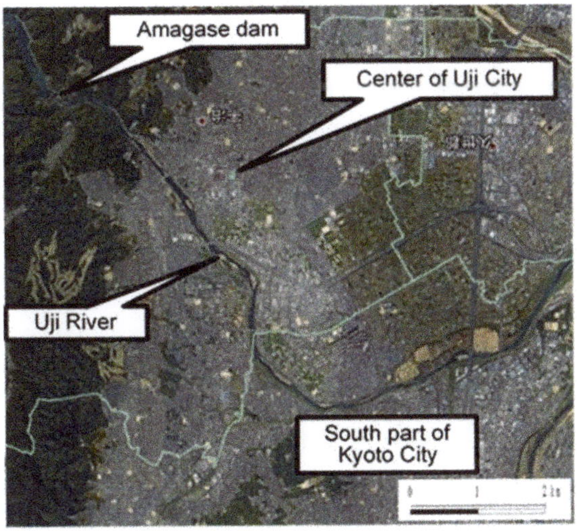

Figure 10. Locations of Amagase Dam and the high-density zone of the population of Uji City (the gray zone is the circumference of "Center of Uji City").

Figure 11. A photo of Amagase Dam (the dam height is 73 m, the dam width is 254 m).

For the proposed dam-break simulation, the following information is required as input data sets; (1) assumed broken width and height, as listed in Table 2. (2) Proposed dam's reservoir area, as listed in Table 2. (3) Water head behind the dam, as listed in Table 2. (4) Calculation meshes size, as listed in Table 2. Since the proposed model of Ca et al. [15] is based on the finite difference method (FDM), the mesh size limitation is always a concern which mainly depends on the capacity of the available computer, size of the proposed dam and required accuracy. (5) The existing topography contour map is shown in Figure 12. in this research, a 10-m accuracy digital elevation map (DEM) is used, (6) The building and trees ratio in a mesh (as shown in Figure 13, the light green color (the building ratio is 1 % in a mesh area) is referred to a soil and sand area, the dark green color (5 % in a mesh area) shows a wood area and a forest area, the light brown (30 % in a mesh area) shows a house with a garden, the gray color area (70 % in a mesh area) shows a building area, (7) Grain size distribution in the area and (8) Calculation limit boundary. (9) Acquiring the bedload roughness coefficient c_b by using Figures 5–7 and Equation (27). In the downstream of Amagase dam, from the position of topography change, there are three types of areas. (a) The high-altitude areas where the flood water cannot inundate. (b) The wetland area (Agricultura area) and (c) the city area.

Table 2. Input data for dam-break simulations.

	Case 1	Case 2	Case 3	Case 4
Broken width and height (m)	100 × 50	50 × 50	50 × 25	100 × 25
Broken area (sq. m)	2500	1250	625	1250
Dam reservoir area (sq. m)	1,880,000	1,880,000	1,880,000	1,880,000
Water height behind the dam (m)	68	68	68	68
Mesh size (m)	20 × 20	20 × 20	20 × 20	20 × 20

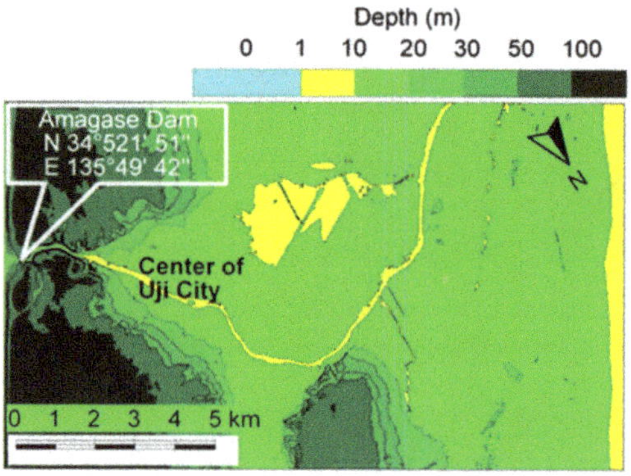

Figure 12. Existing topography—Downstream of Amagase dam (Uji City).

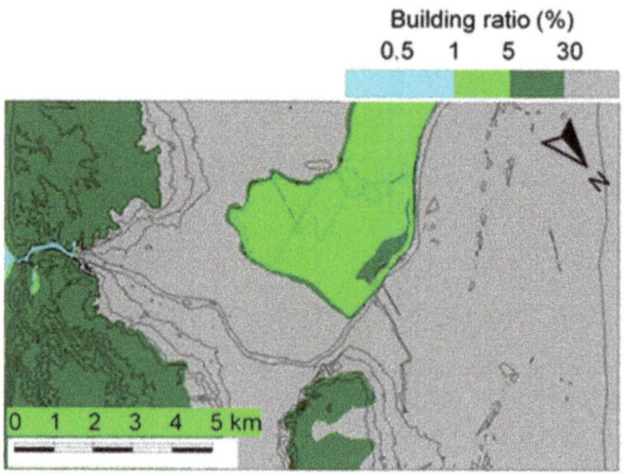

Figure 13. Building ratio in a calculation mesh (=the area ratio of buildings and trees in a mesh).

For type (a) area, there is no need for bedload coefficient. For types (b and c), the calculations are detailed in the following.

For type (b) area:

$d_{50} = 0.2$mm, using Figure 5, $\frac{d_{50}}{0.2} = 1.0 \rightarrow C_{b0} = 18$
Uniformity coefficient, $U = 10$. Using Figure 6, $C_1 = 0.9$
Dry density $\rho_d = 1.53\frac{gr}{cm^3}$. Using Figure 7, $C_2 = 0.93$
From Equation (27), $C_b = 18 \times 0.9 \times 0.93 = 15$.

For type (c) area:

$d_{50} = 0.2$mm, using Figure 5, $\frac{d_{50}}{0.2} = 1.0 \rightarrow C_{b0} = 18$

Uniformity coefficient, $U = 20$. Using Figure 6, $C_1 = 0.8$
Dry density $\rho_d = 1.65\frac{gr}{cm^3}$. Using Figure 7, $C_2 = 0.75$
From Equation (27), $C_b = 18 \times 0.8 \times 0.75 = 11$.

The value of D_d is set to 1000 m and the shape value c is set to 0.28. the method and logic behind deciding the c value is previously described in Section 3.1 of this paper.

The results of the Amagase dam-break simulation are shown in Figures 14–31. Scouring depths and deposition heights are depicted in Figures 14–16. The maximum scouring depth is about 10 m close to the dam and it decreases to about 5 m as the bore reaches further locations from the dam. Forward and return velocity distributions are depicted in Figures 17–19, the maximum forward velocity is calculated to be about 6 m per second and the maximum return velocity is found to be 3 m per second. Inundation height and extent are depicted in Figures 20–22. Since the accurate dam-break section's dimensions are difficult to anticipate, one needs to perform a dam-break with different possible break sections to find out the maximum inundation extent and velocity. Since a dam-break simulation takes time and effort, the authors in this paper executed many dam-break simulations with different heights and widths to evaluate and generalize the effect of a dam-break section's height and width on the inundation area and velocity, as listed in Table 2. Four cases of break sections are assumed. The first case's break section height is twice the height of the second case with the same width between the second case and the third case, the height is kept the same and the height of the third case is half of the second case. Between case four and the first case, again the height is kept the same however the width is half size. A brief comparison of the dam-break simulation results is depicted in Table 3.

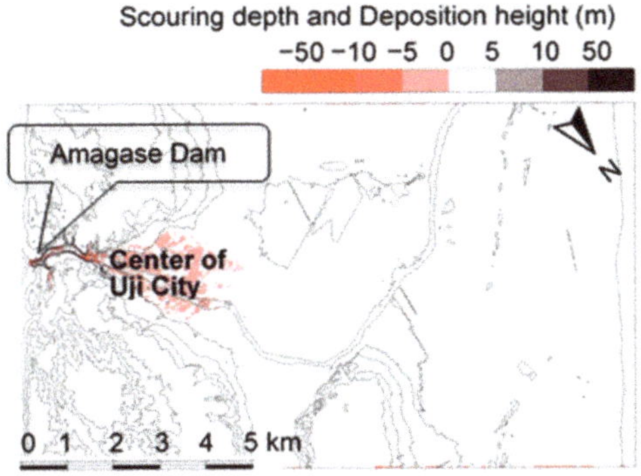

Figure 14. (Case–1). Topography change, 10 min after the start of the calculation.

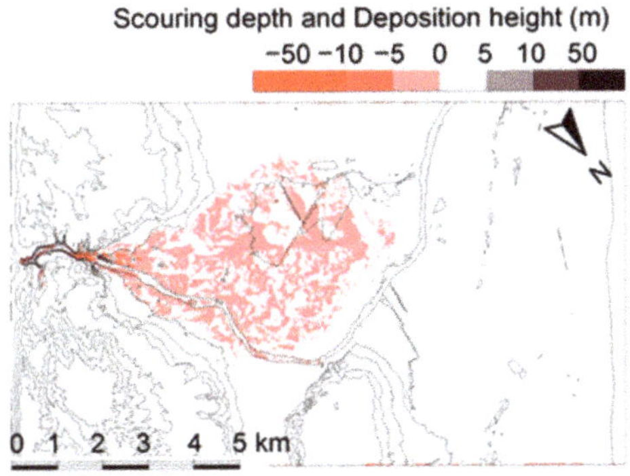

Figure 15. (Case–1). Topography change, 20 min after the start of the calculation.

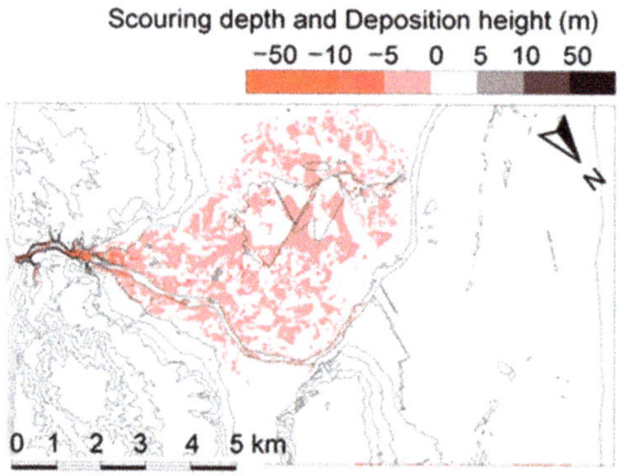

Figure 16. (Case–1). Topography change, 30 min after the start of the calculation.

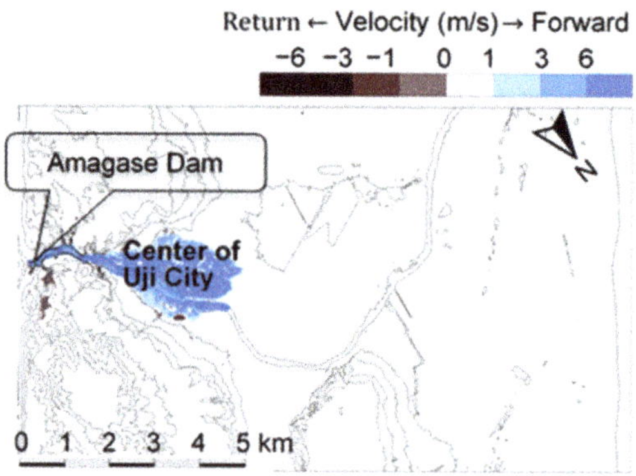

Figure 17. (Case–1). Inundation velocity, 10 min after the start of the calculation.

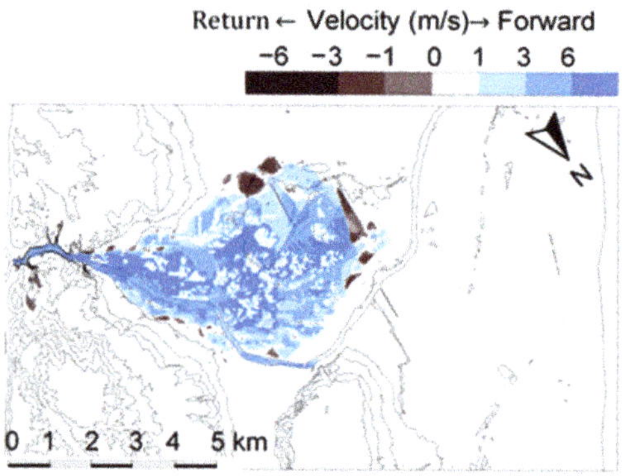

Figure 18. (Case–1). Inundation velocity 20 min after the start of the calculation.

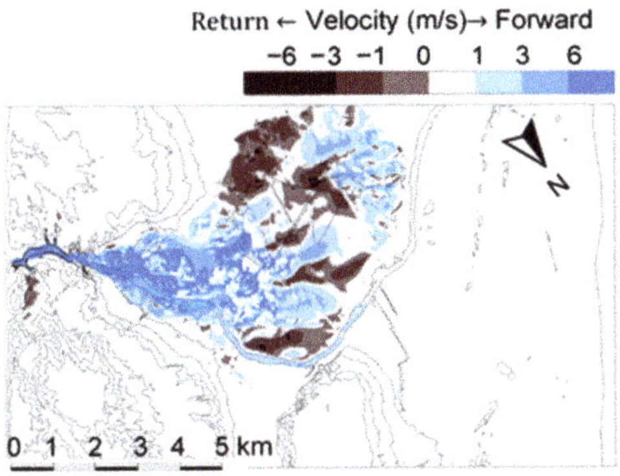

Figure 19. (Case–1). Inundation velocity 30 min after the start of the calculation.

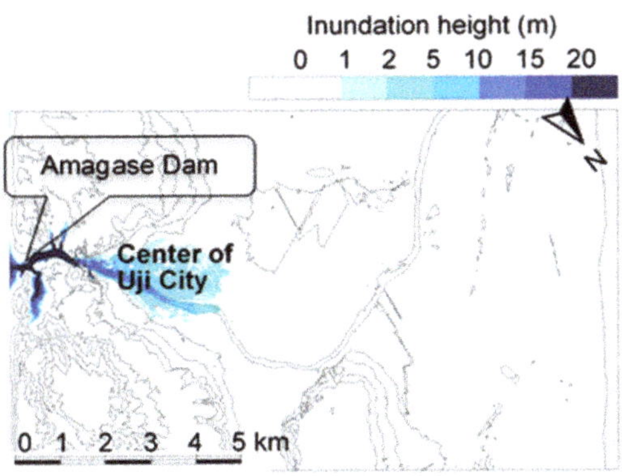

Figure 20. (Case–1). Inundation height, 10 min after start of the calculation.

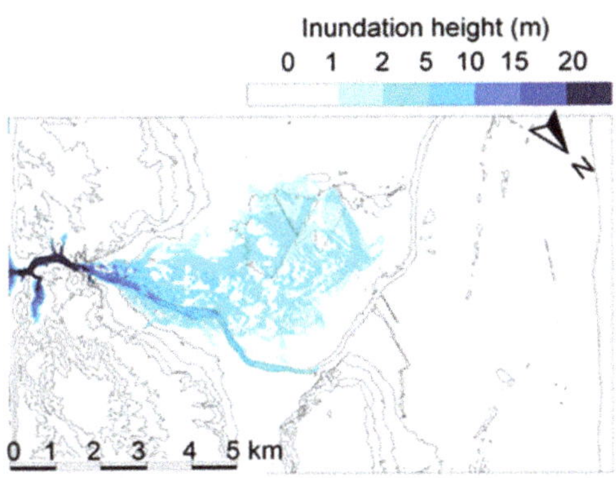

Figure 21. (Case–1). Inundation height, 20 min after start of the calculation.

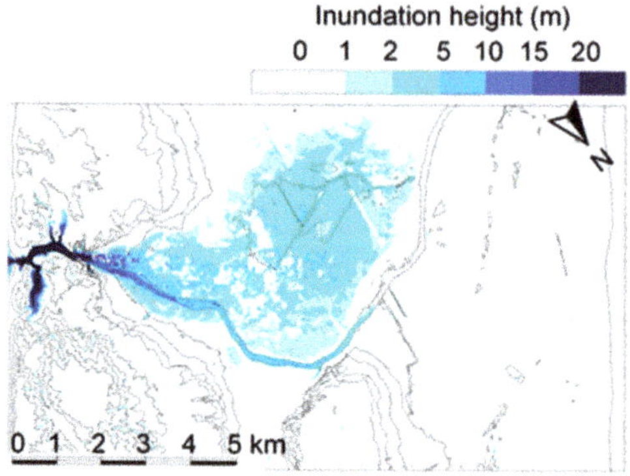

Figure 22. (Case–1). Inundation height, 30 min after start of the calculation.

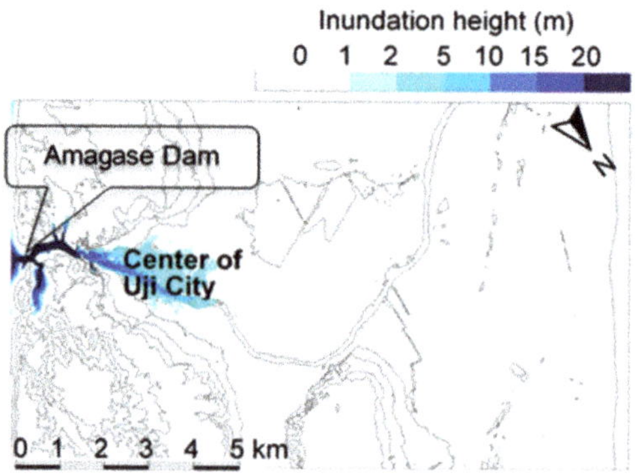

Figure 23. (Case–2). Inundation height, 10 min after start of the calculation.

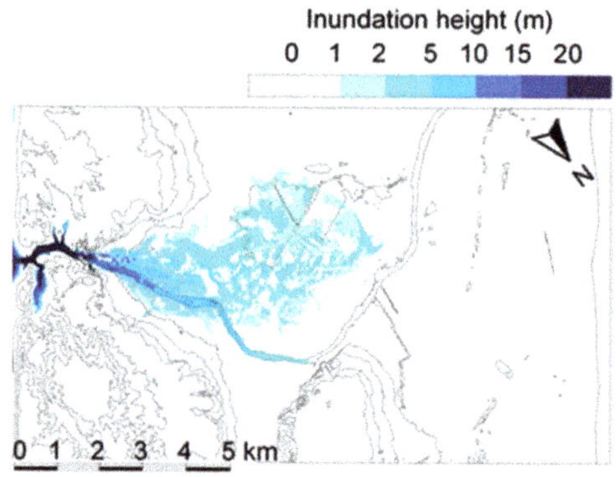

Figure 24. (Case–2). Inundation height, 20 min after start of the calculation.

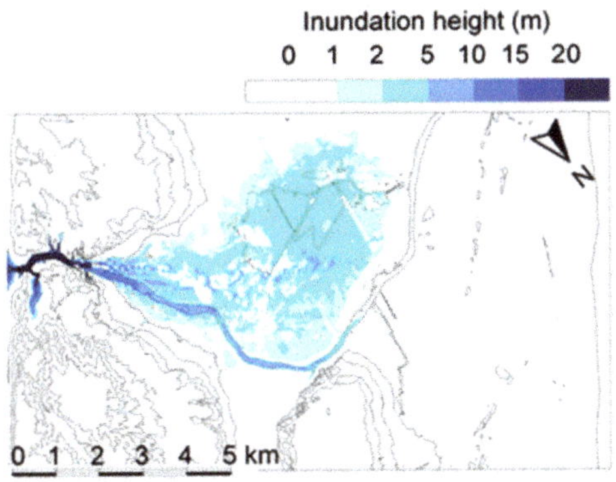

Figure 25. (Case–2). Inundation height, 30 min after start of the calculation.

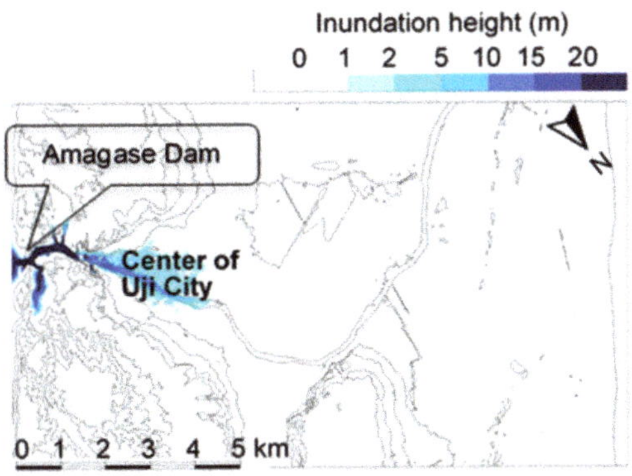

Figure 26. (Case–3). Inundation height, 10 min after start of the calculation.

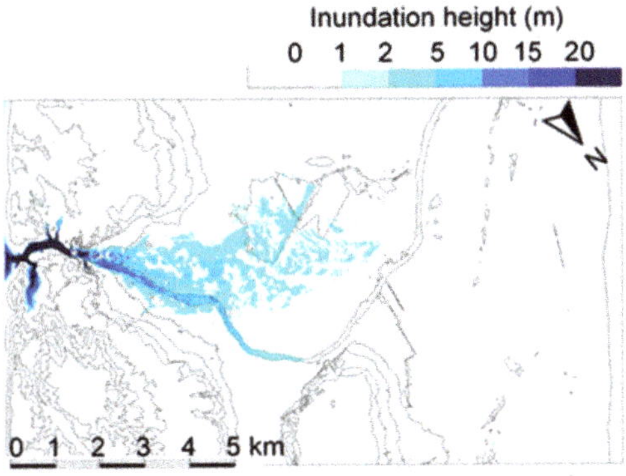

Figure 27. (Case–3). Inundation height, 20 min after start of the calculation.

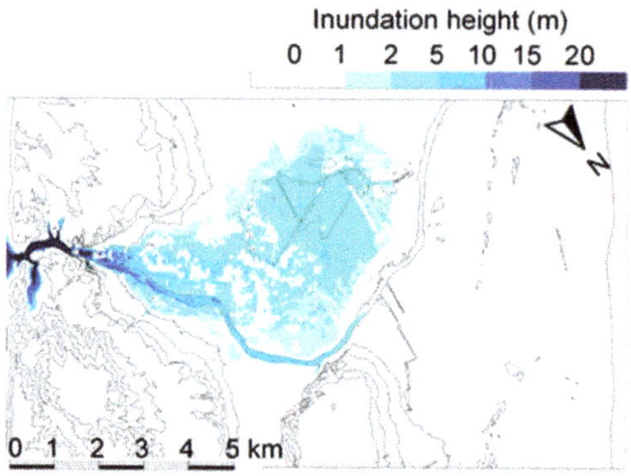

Figure 28. (Case–3). Inundation height, 30 min after start of the calculation.

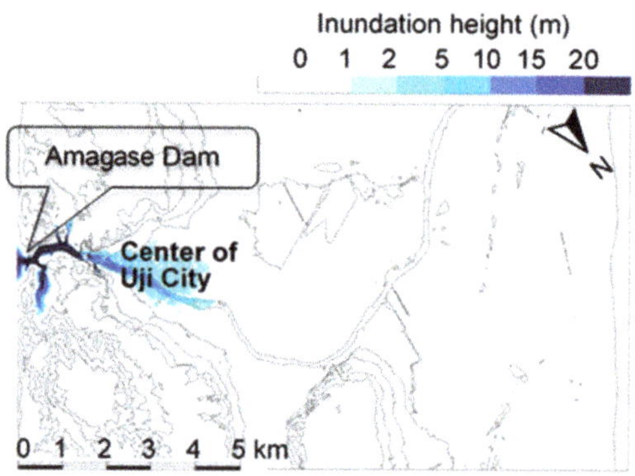

Figure 29. (Case–4). Inundation height, 10 min after start of the calculation.

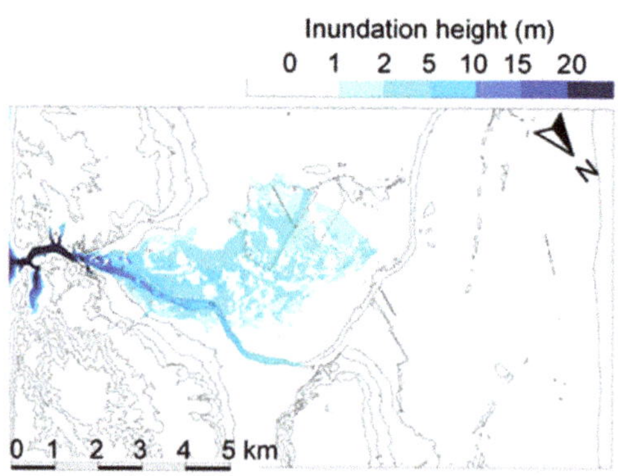

Figure 30. (Case–4). Inundation height, 20 min after start of the calculation.

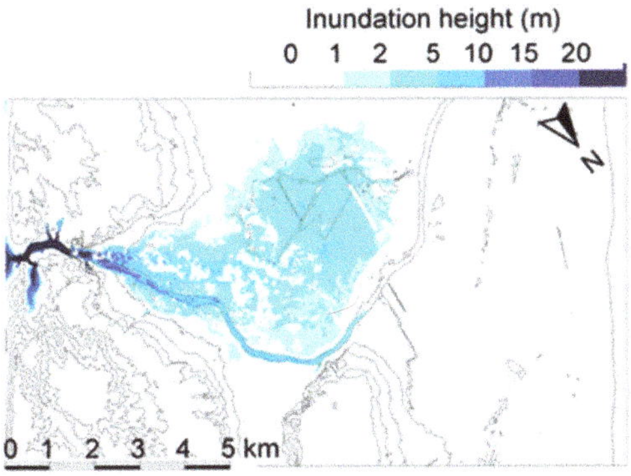

Figure 31. (Case–4). Inundation height, 30 min after start of the calculation.

Table 3. Inundation area under different dam-break sections.

Descriptions	Case 1	Case 2	Case 3	Case 4
Inundation area 10 min from the start of the calculation (m^2)	4.4×10^6	3.6×10^6	2.9×10^6	3.0×10^6
Inundation area 20 min from the start of the calculation (m^2)	1.6×10^7	1.4×10^7	1.3×10^7	1.3×10^7
Inundation area 30 min from the start of the calculation (m^2)	2.2×10^7	2.0×10^7	2.0×10^7	2.0×10^7

Considering the width and height of different break sections (cases 1 through 4), the results of the simulations show that, as the height of the break section increases, the flood velocity and flood range becomes larger.

5. Conclusions

This study presented a novel, useful and simple approach for sudden partial dam break outflow rate. Using the proposed approach, one can assume different dam-break shapes and observe the relative generated hydrographs and decide the best condition to match a proposed dam for break simulation. The merits of the proposed approach are being simple to set up, minimum input values from a dam, such as dam height, width, reservoir surface area and an assumed dam-break section.

Naturally, in the case of non-erodible dams, the dam failure is usually a partial break. Therefore, the proposed equation was developed so that one can set a percentage failure of the dam and observe its hydrograph. The proposed equation is affected by the break section's shape. Therefore, by performing many hydraulic experiments, a mean shape factor is acquired that can be set for different dam-break shapes. While executing the hydraulic experiments, it is observed that the water surface gradient is not level during the dam reservoir emptying process, therefore, the inference of water surface gradient on the proposed dam-break outflow rate concept, during a dam reservoir emptying process can be a potential future research objective.

Furthermore, to show the validity of the proposed equation, the Malpasset dam-break hydrograph is produced and compared to that of the existing literature. In this comparison, the proposed equation resulted in a relatively higher peak discharge and faster emptying time and it is concluded that the reason behind this is the different initial conditions for developing the different equations. The reliability of the proposed equation and the existing equations in the literature is the same. However, the proposed equation is easy to set up and one can simply evaluate the inference of different dam-break sections on a dam-break outflow rate.

Consequently, by integrating the proposed outflow rate equation on a two-dimensional flood simulation model, and simulating the Amagase dam-break, it is perceived that this equation can be potentially useful for the preparation of dam-break flood routing. According to the simulation results of the Amagase Dam presented in this paper, thousands of lives, as well as a countless amount of economy/businesses, are in a high-risk zone in the case of a dam break. Therefore, the authors suggest that when the dam becomes old, because it will be less durable against a great earthquake, the households living downstream of the dam and in the high-risk zone should be informed of the potential danger; i.e., in the case of superannuated dams in the area where the frequency of Mega Earthquakes is high, the preparation of hazard maps, emergency action plans and early warning systems will be very useful and will lead to the saving of lives and the economy. Especially in the case of the Amagase dam, which a densely populated city is located right at its downstream, widening the dam would result in lowering its risk of calamity in case of a dam-break.

Author Contributions: Conceptualization, S.M.A. and Y.Y.; methodology, S.M.A. and Y.Y.; software, S.M.A.; validation, S.M.A.; formal analysis, S.M.A. and Y.Y.; investigation, S.M.A. and Y.Y.; resources, Y.Y.; data curation, S.M.A.; writing—original draft preparation, S.M.A.; writing—review and editing, S.M.A. and Y.Y.; visualization, S.M.A.; supervision, Y.Y.; project administration, Y.Y.; funding acquisition, S.M.A. and Y.Y. All authors have read and agreed to the published version of the manuscript.

Funding: This research was partially funded by Japan Society for the Promotion of Science grant number 18k04667. And the APC was funded by Ministry of Education, Culture, Sports, Science and Technolog—Japan.

Institutional Review Board Statement: Not applicable.

Informed Consent Statement: Not applicable.

Data Availability Statement: The data presented in this study are available on request from the corresponding author. The data are not publicly available due to privacy reasons.

Acknowledgments: This research was executed under the partial support of Grants-in-Aid for Scientific Research (c; 18k04667) of Japan Society for the Promotion of Science (JSPS). Here, we express gratitude deeply.

Conflicts of Interest: The authors declare no conflict of interest.

References

1. Psomiadis, E.; Tomanis, L.; Kavvadias, A.; Soulis, K.X.; Charizopoulos, N.; Michas, S. Potential Dam Breach Analysis and Flood Wave Risk Assessment Using HEC-RAS and Remote Sensing Data: A Multicriteria Approach. *Water* **2021**, *13*, 364. [CrossRef]
2. Wahl, T.L. Predicting Embankment Dam Breach Parameters—A Needs Assessment. In Proceedings of the XXVIIth IAHR Congress, San Francisco, CA, USA, 10–15 August 1997.
3. Pilotti, M.; Tomirotti, M.; Valerio, G.; Bacchi, B. Simplified Method for the Characterization of the Hydrograph following a Sudden Partial Dam Break. *J. Hydraul. Eng.* **2010**, *136*, 693–704. [CrossRef]
4. Costa, J.E. *Floods from Dam Failures*; Open-File Report; U.S. Geological Survey: Denver, CO, USA, 1985.
5. Gary, G. *Using HEC-RAS for Dam Break Studies*; US Army Corps of Engineers: Washington, DC, USA, 2014.
6. Wurbs, R.A. Dam-Breach Flood Wave Models. *J. Hydraul. Eng.* **1987**, *113*, 29–46. [CrossRef]
7. Bazin, H. Recherches Expérimentales relatives aux remous et à la propagation des ondes [Experimental research on the hydraulic jump and on wave propagation]. *Deux. Partie Rech. Hydraul. Darcy Baz. Dunod Paris* **1865**, *2*, 148.
8. Ritter, A. Die Fortpflanzung der Wasserwellen. [The propagation of water waves]. *Z. Ver. Dtsch. Ing.* **1892**, *3633*, 947–954.
9. Su, S.-T.; Barnes, A.H. Geometric and Frictional Effects on Sudden Releases. *J. Hydraul. Div.* **1970**, *96*, 2185–2200. [CrossRef]
10. Aureli, F.; Maranzoni, A.; Mignosa, P. A semi-analytical method for predicting the outflow hydrograph due to dam-break in natural valleys. *Adv. Water Resour.* **2013**, *63*, 38–44. [CrossRef]
11. Saberi, O.; Zenz, G. Empirical Relationship for Calculate Outflow Hydrograph of Embankment Dam Failure due to Overtopping Flow. *Int. J. Hydraul. Eng.* **2015**, *4*, 45–53.
12. Basheer, T.A.; Wayayok, A.; Yusuf, B.; Kamal, R. Dam breach parameters and their influence on flood hydrographs for mosul dam. *J. Eng. Sci. Technol.* **2017**, *12*, 14.
13. Hakimzadeh, H.; Nourani, V.; Amini, A.B. Genetic Programming Simulation of Dam Breach Hydrograph and Peak Outflow Discharge. *J. Hydrol. Eng.* **2014**, *19*, 757–768. [CrossRef]

14. Wahl, T.L. *Prediction of Embankment Dam Breach Parameters—A Literature Review and Needs Assessment*; Dam Safety Office: Denver, CO, USA, 1998.
15. Ca, V.T.; Yamamoto, Y.; Charusrojthanadech, N. Improvement of Prediction Methods of Coastal Scour and Erosion due to Tsunami Back-flow. In Proceedings of the International Society of Offshore and Polar Engineers (ISOPE), Beijing, China, 20 June 2010. ISOPE-I-10-510.
16. Suetsugi, T.; Kuriki, M. Research on Application for Flood Disaster Prevention and Simulation of Flooding Flow by Means of New Flood Simulation Model. *Doboku Gakkai Ronbunshu* **1998**, *593*, 41–50. [CrossRef]
17. Ribberink, J.S. Bed-load transport for steady flows and unsteady oscillatory flows. *Coast. Eng.* **1998**, *34*, 59–82. [CrossRef]
18. Yokoyama, Y.; Tanabe, H.; Torii, K.; Kato, F.; Yamamoto, Y.; Vu, T.C.; Arimura, J. A quantitative evaluation of scour depth near coastal structures. In Proceedings of the Coastal Engineering 2002; World Scientific Publishing Company: Cardiff, UK, 2003; pp. 1830–1841.
19. Ahmadi, S.M.; Yamamoto, Y.; Ca, V.T. Rational Evaluation Methods of Topographical Change and Building Destruction in the Inundation Area by a Huge Tsunami. *J. Mar. Sci. Eng.* **2020**, *8*, 762. [CrossRef]
20. Van Rijn, L.C. *Principles of Sediment Transport in Rivers, Estuaries and Coastal Seas. 1: Principles of Sediment Transport in Rivers, Estuaries and Coastal Seas*; Aqua Publications: Amsterdam, The Netherlands, 1993; ISBN 978-90-800356-2-1.
21. Van Rijn, L.C. Sediment Transport, Part I: Bed Load Transport. *J. Hydraul. Eng.* **1984**, *110*, 1431–1456. [CrossRef]
22. Van Rijn, L.C. Sediment Transport, Part II: Suspended Load Transport. *J. Hydraul. Eng.* **1984**, *110*, 1613–1641. [CrossRef]
23. Van Rijn, L.C. Sediment Transport, Part III: Bed forms and Alluvial Roughness. *J. Hydraul. Eng.* **1984**, *110*, 1733–1754. [CrossRef]
24. *Richard Soulsby Dynamics of Marine Sands*; Thomas Telford Ltd: London, UK, 1997; ISBN 978-0-7277-2584-4.
25. Zyserman, J.A.; Fredsøe, J. Data Analysis of Bed Concentration of Suspended Sediment. *J. Hydraul. Eng.* **1994**, *120*, 1021–1042. [CrossRef]
26. Bates, P.D.; De Roo, A.P.J. A simple raster-based model for flood inundation simulation. *J. Hydrol.* **2000**, *236*, 54–77. [CrossRef]
27. Amicarelli, A.; Manenti, S.; Albano, R.; Agate, G.; Paggi, M.; Longoni, L.; Mirauda, D.; Ziane, L.; Viccione, G.; Todeschini, S.; et al. SPHERA v.9.0.0: A Computational Fluid Dynamics research code, based on the Smoothed Particle Hydrodynamics mesh-less method. *Comput. Phys. Commun.* **2020**, *250*, 107157. [CrossRef]
28. He, G.; Chai, J.; Qin, Y.; Xu, Z.; Li, S. Evaluation of Dam Break Social Impact Assessments Based on an Improved Variable Fuzzy Set Model. *Water* **2020**, *12*, 970. [CrossRef]
29. Kocaman, S.; Dal, K. A New Experimental Study and SPH Comparison for the Sequential Dam-Break Problem. *J. Mar. Sci. Eng.* **2020**, *8*, 905. [CrossRef]
30. Goutal, N. The Malpasset dam failure. An overview and test case definition. In Proceedings of the 4th CADAM Meeting, Zaragoza, Spain, 18–19 November 1999; pp. 18–19.
31. Alcrudo, F.; Gil, F. The Malpasset dam break case study. In Proceedings of the 4th CADAM Meeting, Zaragoza, Spain, 18–19 November 1999; pp. 18–19.
32. Hervouet, J.-M.; Petitjean, A. Malpasset dam-break revisited with two-dimensional computations. *J. Hydraul. Res.* **1999**, *37*, 777–788. [CrossRef]
33. Valiani, A.; Caleffi, V.; Zanni, A. Case Study: Malpasset Dam-Break Simulation using a Two-Dimensional Finite Volume Method. *J. Hydraul. Eng.* **2002**, *128*, 460–472. [CrossRef]

water

Article

Modeling and Risk Analysis of Dam-Break Flooding in a Semi-Arid Montane Watershed: A Case Study of the Yabous Dam, Northeastern Algeria

Aissam Gaagai [1], Hani Amir Aouissi [1,*], Andrey E. Krauklis [2], Juris Burlakovs [3], Ali Athamena [4], Ivar Zekker [5], Abderrahmane Boudoukha [6], Lahcen Benaabidate [7] and Haroun Chenchouni [8,9]

1 Scientific and Technical Research Center on Arid Regions (CRSTRA), Biskra 07000, Algeria; gaagaiaissam37@gmail.com
2 Institute for Mechanics of Materials, University of Latvia, Jelgavas Street 3, LV-1004 Riga, Latvia; andykrauklis@gmail.com
3 Rural Building and Water Management, Estonian University of Life Sciences, Kreutzwaldi 5, 51014 Tartu, Estonia; juris@geo-it.lv
4 Department of Geology, Institute of Earth and Universe Sciences, University of Batna 2, Batna 05078, Algeria; a.athamena@univ-batna2.dz
5 Institute of Chemistry, University of Tartu, 14a Ravila St., 50411 Tartu, Estonia; ivar.zekker@ut.ee
6 Laboratoire de Recherche Appliquée en Hydraulique, University of Batna 2, Batna 05078, Algeria; boudoukha_abderrahmane@yahoo.fr
7 Laboratory of Functional Ecology and Environment Engineering, University of Sidi Mohammed Ben Abdellah, Fez 30500, Morocco; benaabidate@gmail.com
8 Department of Forest Management, Higher National School of Forests, Khenchela 40000, Algeria; chenchouni@gmail.com
9 Laboratory of Natural Resources and Management of Sensitive Environments 'RNAMS', University of Oum-El-Bouaghi, Oum-El-Bouaghi 04000, Algeria
* Correspondence: aouissi.amir@gmail.com; Tel.: +213-662-387-144

Abstract: The risk related to embankment dam breaches needs to be evaluated in order to prepare emergency action plans. The physical and hydrodynamic parameters of the flood wave generated from the dam failure event correspond to various breach parameters, such as width, slope, and formation time. This study aimed to simulate the dam breach failure scenario of the Yabous dam (northeast Algeria) and analyze its influence on the related areas (urban and natural environments) downstream of the dam. The simulation was completed using the sensitivity analysis method to assess the impact of breach parameters and flooding on the dam break scenario. The flood wave propagation associated with the dam break was simulated using the one-dimensional HEC-RAS hydraulic model. This study applied a sensitivity analysis of three breach parameters (slope, width, and formation time) on five sites selected downstream of the embankment dam. The simulation showed that the maximum flow of the flood wave recorded at the level of the breach was 8768 m^3/s, which gradually attenuated along the river course to reach 1972.7 m^3/s at about 8.5 km downstream the dam. This study established the map of flood risk areas that illustrated zones threatened by the flooding wave triggered by the dam failure due to extreme rainfall events. The sensitivity analysis showed that flood wave flow, height, and width revealed positive and similar changes for the increases in adjustments ($\pm$25% and $\pm$50%) of breach width and slope in the five sites. However, flood wave parameters of breach formation time showed significant trends that changed in the opposite direction compared to breach slope and width. Meanwhile, the adjustments ($\pm$25% and $\pm$50%) of the flood hydrograph did not significantly influence the flood parameters downstream of the dam. In the present study, the HEC-RAS 1-D modeling demonstrated effectiveness in simulating the propagation of flood waves downstream of the dam in the event of dam failure and highlighted the impact of the breach parameters and the flood hydrographical pattern on flood wave parameters.

Citation: Gaagai, A.; Aouissi, H.A.; Krauklis, A.E.; Burlakovs, J.; Athamena, A.; Zekker, I.; Boudoukha, A.; Benaabidate, L.; Chenchouni, H. Modeling and Risk Analysis of Dam-Break Flooding in a Semi-Arid Montane Watershed: A Case Study of the Yabous Dam, Northeastern Algeria. *Water* **2022**, *14*, 767. https://doi.org/10.3390/w14050767

Academic Editors: Francesca Aureli, Andrea Maranzoni and Gabriella Petaccia

Received: 18 January 2022
Accepted: 25 February 2022
Published: 28 February 2022

Keywords: dam breach parameters; dam failure scenario; flood hazard; flood hydrodynamics; flood simulation; simulated natural hazard

1. Introduction

Dam-failure floods are one of the most severe environmental disasters, which result in irreversible ecological, social, and economic consequences. Previous studies assessing dam failure risk revealed that breaching risk and dam failure process can be predicted based on the past dam failure events [1–4]. The reports, which describe dam breaks causing problems around the world, are multiple, for example (1) the Marib dam (Yemen) in 575, (2) an earthen dam near Grenoble (France) in 1219, and (3) in 1923, 1935, and 1985, three dam failure cases in Italy caused the loss of many human lives in the regions of Gleno, Molare, and Stava, respectively [5]. In Algeria, a dam called 'Fergoug I' killed approximately 200 individuals in 1881 in the region of Mascara [6].

In a constantly changing world, dam failure caused by climate change is considered to be an escalating worldwide phenomenon. Usually, within studies related to dams, the dam body (height, width, and slope for earthen dams), including the ancillary structures, are dimensioned and conform to the flood hydrograph. According to recent climate changes, the flood hydrograph recorded at many dams follows an ascending curve compared to previous years [7]. This, therefore, can influence the body of the dam and even its structures. For example, if the spillway does not support the flood input, this flood will systematically discharge at the dam crest to create a breach on the dike. The breach can cause a partial or a total failure of the structure, depending on the resistance of the body of the dam. Therefore, the impact of climate change on dams is reported as a major problem that requires the rigorous and permanent monitoring of dams with the assessment of flooding hazard risk [8,9]. In this sense, in 2015, during the construction of the Yabous dam, built upstream of the city of Yabous (a mountainous area near the province of Khenchela, NE Algeria), an exceptional flood was recorded after heavy rains hit this region. The first author (AG) was a hydrology engineer monitoring the construction project during that period. He clearly observed the destructive effects of this flood on all the structures, particularly the discharge of the flood on the sails of the spillway dissipation basin.

Consequently, a comprehensive management plan for flood plains and watersheds using acute scientific knowledge is necessary to reduce flood damages and prevent socio-economic losses and environmental degradation [10–12]. Most dam failures and flooding events have resulted from mismanagement practices and/or outdated plans that did not consider or project extreme meteorological events under climate change scenarios that exceed dam dimensioning and reservoir capacity [13,14]. Although the methods for investigating flood mitigation are numerous and diverse, flood mitigation remains difficult to fully address because flooding is a complex phenomenon that involves several changing and unpredictable variables [8]. In order to reduce the risks of floods caused by dam failure, scientists and decision makers should consider all of the natural and anthropogenic factors that can predict the risk. In addition, the procedure of studying dam failure must include the delimitation of the risk zone downstream of the dam and investigate all breach formation parameters that may influence this scenario using sensitivity analysis methods to validate simulation modeling outputs accurately [15]. For instance, failure mode, dynamic failure geometry, flow conditions, breach progression, time, materials, and dam type can affect the estimation of the propagation wave resulting from a dam failure. In this regard, previous studies applying sensitivity analysis have highlighted that the parameters mentioned above have various influences on the peak flow in terms of dam geographic location, flood magnitude, etc. [16,17]. In fact, case studies and/or modeling approaches are generally used to predict the parameters of dam failure, where sensitivity analysis is an essential aspect of the use of hydraulic models. The application of sensitivity analysis is vital in decision-making processes to ensure the consistency of the final decision [18].

Different simulation scenarios could be visualized through the sensitivity analysis, which is helpful in observing the impact of the parameters changing on the final alternative ranking. The sensitivity analysis allows us to observe how the final assessment is likely to change, and it also measures the magnitude of the change generated by a certain extent of the weight differences in the predicted parameters [19,20].

Several studies have applied analytical models for assessing floods caused by dam failures, including one-dimensional (MIKE11 and HEC-RAS 1-D) and two-dimensional models (FLDWAV, DELFT-FSL, TELEMAC 2-D, MIKE2, and HEC-RAS 2-D). While all these models are designed to analyze and solve the same issue, their principle is not the same. For example, the models MIKE11 and HEC-RAS require the geometrical characteristics of the land, which allow us to draw cross- and perpendicular sections of flow direction and then determine water height and flowing velocity at each point along the river [21–24]. Moreover, the FLDWAV, DELFT-FSL, TELEMAC 2D, MIKE2, and HEC-RAS 2D models require a digital elevation model (DEM), which ensures a continuous assessment of land topography [25]. Gee et al. [26] compared aspects of dam failure using two two-dimensional models (HEC-RAS 2D and FLDWAV) and concluded that the difference in the geometry of the river is the main driver of simulation differences between the two models. In recent years, the simulation of the propagation of flood waves in open channels downstream the dams in failure events was further improved with the development of high-quality DEM and cutting-edge tools for geographic information systems (GIS) [27]. Recent studies on flood inundation modeling involve the application of combined methods, including statistical modeling techniques, remote sensing, and GIS tools [7,28,29].

This study aimed to analyze the dam failure scenario and determine its influence on the urban area downstream of the dam in a semi-arid region in northeast Algeria (Yabous, Khenchela). Using the sensitivity analysis, a simulation highlights the impact of breach parameters and the flood movement on the dam failure scenario in several sites selected downstream the dam, including different structures, i.e., mainly inhabited urban areas, the axis of the dam itself, bridges, etc. In addition, this study aimed to comprehensively model the dam failure process and its consequences using a mixture of datasets (geomatic, hydrological, and hydraulic data) with the application of the one-dimensional Hydrologic Engineering Centers–River Analysis System "HEC-RAS" model. The main benefits of this model are that the floodplain maps can be generated faster; floodplain maps can include several flow scenarios and involve the rapid preparation of geometry data. The HEC-RAS model includes steady and unsteady flow analysis, GIS tools to aide engineering analysis, automated calculation of functions (e.g., the Energy Equation), and the structural validation of hydraulic control features. Specifically, in dam-break flood modeling, the model allows users to see the impact of the breach parameters, such as the width, slope, and time formation using the sensitivity analysis [6]. According to these benefits, HEC-RAS is the best model in the hydraulic domain compared to other models, which are limited in the domain of flooding and are far from having the same advantages. Using the sensitivity analysis, the study investigated how different sources of input data influence flood simulation reliability. The purposely chosen analysis was carried out by gradually changing the values of each breach parameter in order to observe the ranking order resulting from these changes. This study aimed to: (1) evaluate the characterization of physical and hydrodynamic parameters of the simulated flood wave at different high-risk flood zones downstream of the dam; (2) assess the impact of a dam-break flood on points along the Yabous river; (3) use the sensitivity analysis to assess the impact of the breach parameters on the flood wave propagation in areas downstream of the dam; (4) validate the base model and the ability of HEC-RAS 5.0.7 to simulate the extreme flooding of an area downstream of a dam in the event of failure by applying the HEC-RAS modeling approach.

2. Materials and Methods

2.1. Study Area and Dam Characteristics

2.1.1. Geographic Location and Dam Characteristics

The Yabous dam (35°23′34.5″ N, 06°38′45.8″ E, elevation = 1222.2 m a.s.l.) is located 63 km west of Khenchela City at the northern foothills of Jebel Chélia (Figure 1). The region's climate is semi-arid, with annual precipitation varying between 350 and 400 mm, and where evapotranspiration exceeds rainfall due to frequent and prolonged drought episodes, which result in a lack of aquifer recharge [30]. The region's landscapes include montane woodlands, shrubby mountains, and high plains used for rain-fed cereal crops [31].

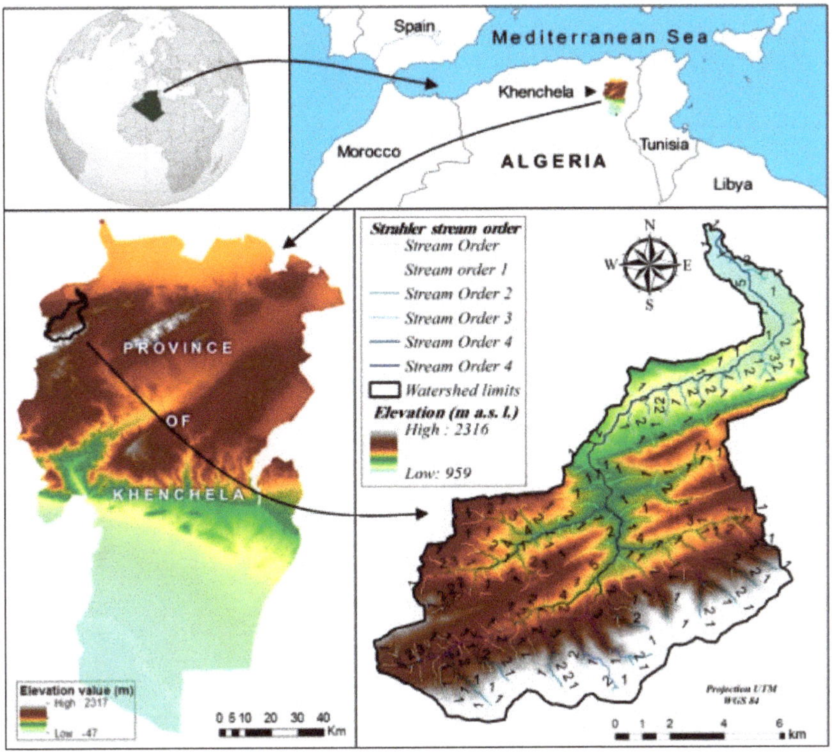

Figure 1. Geographic location and elevation maps of the study area, "Yabous watershed", located at the west of the Province of Khenchela (bottom left map) in northeastern Algeria. The map of the Yabous watershed (bottom right) includes the hydrographic network with Strahler stream orders.

At about 2 km downstream the dam, the town of Yabous (~12,000 inhabitants) is located with a position stretching for 2 km next to the Yabous river on which the dam was constructed (Figure 2b). The position of Yabous town in relation to the dam represents a real risk to the population in different ways (environmental, social, and economic), especially to the infrastructure, knowing that road # N45 stretches for more than 8 km alongside the Yabous river. However, the city of Azla, which is 15 km away from Yabous dam, is the most exposed village to exceptional floods due to the weak slope at the bed of the wadi of Yabous. Small floods have left traces on this village, so the consequences will be irreversible if the dam breaks. Moreover, the largest plain of Yabous, also known by the name "SEKOUM," with a surface of 1839 ha, can also be exposed to this flood (Figure 2a). In addition, the locals became very concerned about the hazard that the dam presents in case of failure, which urged the authorities to create an effective alarm system on the downstream zones prone to flooding associated with dam rupture events. The climate in the region is semi-arid, characterized by hot, dry summers and cold rainy winters, accompanied by snowfall on the mountains. The average annual temperature is around 14.80 °C. However, the average annual evaporation varies between 1300 and 1400 mm. The monthly evaporation of the locality of Yabous varies from 105 to 240 mm between April and September. Its monthly distribution shows steady growth from April to July. The period from October to March totals 26% of the annual average. Regarding rainfall, the average annual rainfall is between 350 and 400 mm. The Yabous station shows that the region is characterized by two rainy periods. The first extends from September to November, while the second goes from March to May.

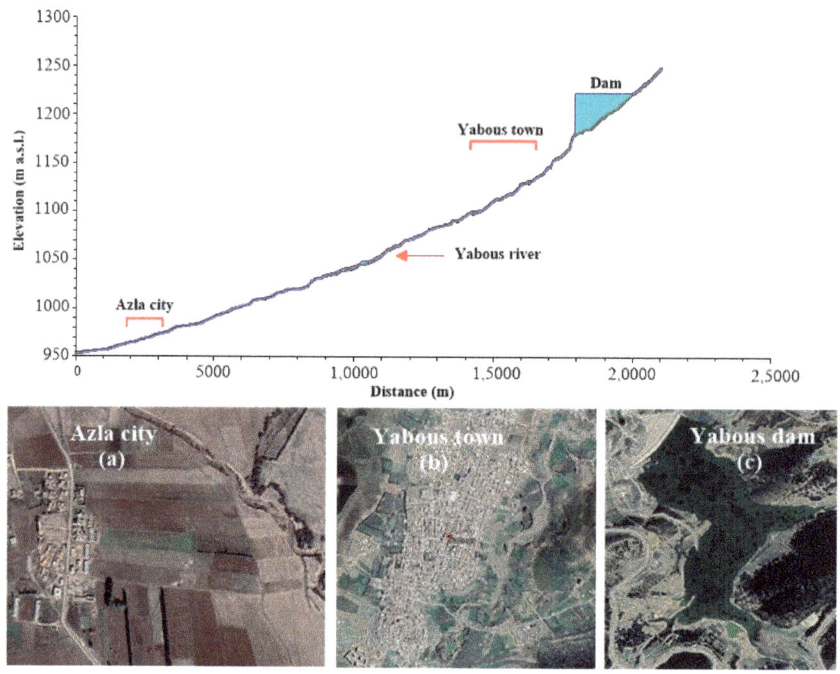

Figure 2. Longitudinal profile of Yabous river and town located downstream of the dam reservoir. (**a**) Azla city 15 Km away from Yabous dam, (**b**) Yabous town 2 Km away from Yabous dam and (**c**) Yabous dam.

Technically, the dam was built of earth with a clay core, and the dam height = 43 m, span = 460 m, and crest width = 10 m. Its storage capacity ≈ 7.5 million m^3, and the volume now ≈ 5.7 million m^3 of water. Dam water is not yet exploited, but it is projected to supply 33,000 inhabitants of surrounding regions with drinking water (~1 million m^3, with a daily allocation of 4800 m^3) and irrigation water (~2 million m^3) destined for agricultural lands (>500 ha) of mainly apple orchards.

2.1.2. Aspect and Slope Features

The slope map of the study area (Figure 3A) was established based on ASTER GDEMv2 data using the spatial analysis tool in ArcGIS-10.5 (see details in Section 2.2.1). The slope grid is identified as the maximum rate of change in the value of each cell relative to its neighboring cells. The slope degree of the Yabous watershed varies from 12° to 61°. Terrains with steep slopes result in rapid runoff and an increased erosion rate (potential loss of soil) with less groundwater recharge potential. This kind of slope is located in the southern part of the Yabous basin, which is at extremely severe risk of soil erosion. Generally, the aspect refers to the orientation of the slope. The map of aspects of the Yabous watershed (Figure 3B) displays clearly that the south-facing slopes dominate the basin. Consequently, these slopes have a lower evaporation rate and higher moisture content, even if some parts fall to the northeast, which has a lower moisture content and higher evaporation rate.

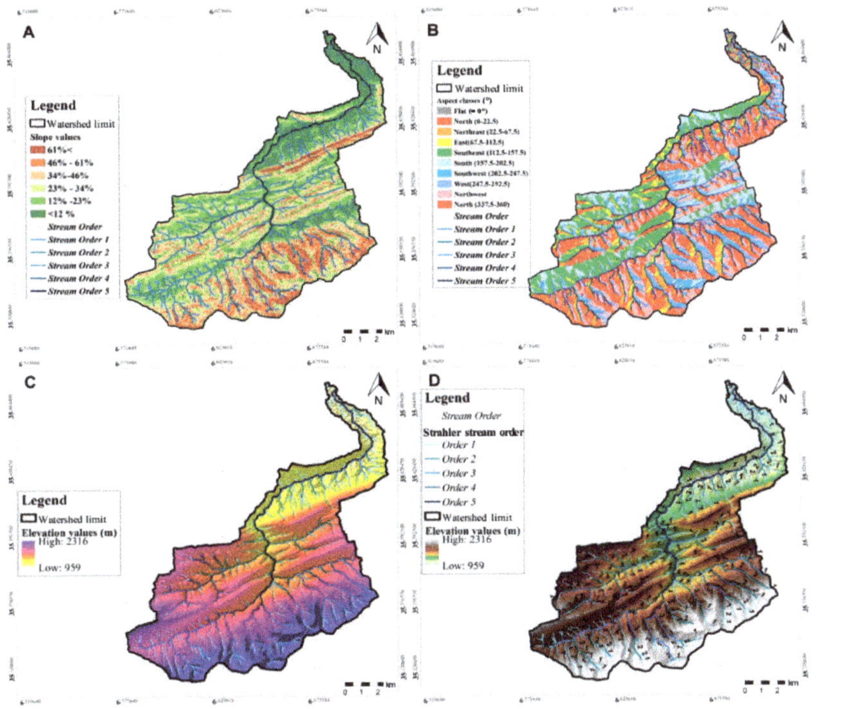

Figure 3. Maps of Yabous watershed displaying (**A**) slope gradient, (**B**) surface aspects, (**C**) altitude above sea level, and (**D**) hydrographic network with Strahler stream orders.

2.1.3. Relief and Hydrographic Network

The relief plays an important role in defining the morphometric variables used to evaluate geomorphological characteristics of any topography. The highest terrain altitude was determined to be 2316 m a.s.l., while the lowest value of elevation was recorded as 960 m a.s.l. (Figure 3C). According to the "top-down" system developed by Strahler (1957), stream orders of the Yabous basin were classified in up to five orders with a predominance of order 1 and then order 2. The number of streams of different orders and the total number of streams in the basin were counted and calculated in the GIS platform. During this procedure, it appeared that the amount of flow gradually decreases as the order of the flows increases. The variation in the size and order of streams in the basin tributaries depend primarily on the region's physiographic, geomorphological, and geological conditions. In total, 314 streamlines were recognized using five stream orders in the entire watershed of the Yabous river (Figure 3D).

2.2. Data Collection, Processing, and Modeling Approach

2.2.1. Data Processing

The satellite image of the study area was retrieved from the ASTER (Advanced Spaceborne Thermal Emission and Reflection Radiometer) satellite at 30 m spatial resolution. This image was used to build the DEM of the study area and extract the required data for the assessment of the areas downstream of the dam at risk of flooding in the event of failure.

The data were imported into ArcGIS software (version 10.5) to generate the final DEM and obtain additional mapping information, such as watershed limits, the hydrographic network used to define all the streams, altitude range, and relative position of slopes (Figure 4).

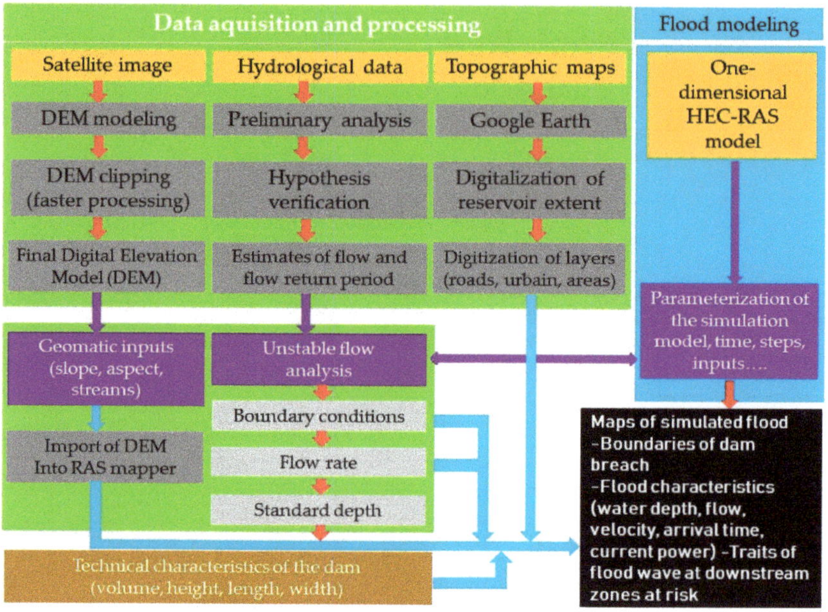

Figure 4. Methodological flowchart applied in Yabous watershed (NE Algeria) during the simulation of a dam-failure flood using 1-D HEC-RAS hydraulic model.

The flood simulation was implemented using a one-dimensional HEC-RAS model [32], which determined several physical characteristics of the flood wave, including wave height, maximum flow 'Qmax', and the velocity of the flood wave. HEC-RAS is undoubtedly one of the most used one-dimensional models because it includes other applications, such as water temperature modeling and sediment transport simulation. Since 2003, it has been developed to simulate dam failure [33].

In addition to the geomatic data described above, the simulation of the flood wave propagation downstream of the Yabous dam along the Yabous river using the HEC-RAS model was carried out using geometric, geotechnical, and hydraulic data (boundary conditions) (Figure 4).

2.2.2. Dam Dimensions

The body of the Yabous dam has the form of a zoned earth dike with a watertight clay core. Its crest length is 480 m, and its width is 7 m. The central core of the dike (zone 1) is made up of clay materials. The upstream recharge consists of argillaceous colluvium and wadi alluvium. In addition, the downstream area is made up of alluvial material from the wadi resting on a sand filter mat to protect the crumble, colluvial, and marl foundation (Figure 5).

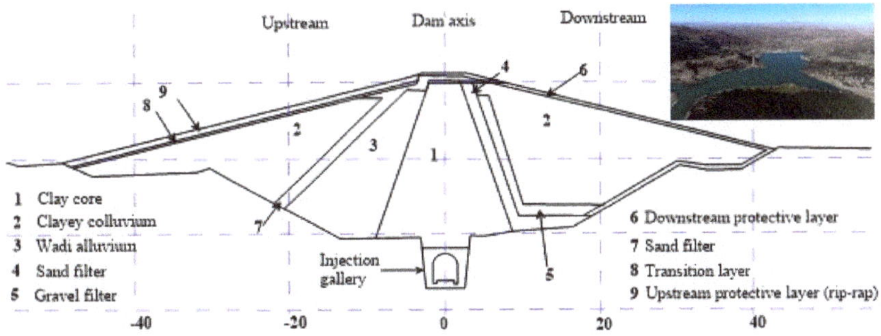

Figure 5. Dam dimension.

The embankments are protected by limestone riprap. The limestones are dense and contribute to the protection with their acceptable mechanical characteristics. The backfill's overall volume of the dam body is 1,418,810.00 m^3. There is a temporary tunnel-type diversion gallery that is 373.00 m long and 7.00 m in diameter. The tunnel has a discharging capacity of the planned (millennial) flood qualified at 485.00 m^3/s. Moreover, a tulip-type spillway, with a well connected to the temporary diversion gallery ending in a stepped dissipation basin. The spillway is capable of evacuating the designed flood. The outer diameter of the spillway's weir and that of the drum are 16.00 m and 7.00 m, respectively. Lastly, a circular water intake tower and its access walkway comprise three (03) stepped water intakes and a bottom drain.

2.2.3. Flood History

The site experienced an exceptional flood, with a significant flow, on 8 February 2015. The dam body and its annexes were unable to cope with the flow. Take the designed intake tower, the spillway, and the dissipation basin as an example. The evidence showed that the flow occupied the section of the dissipation basin and even discharged over its walls. According to Figure 6, the flooding caused the deterioration of the following structures: (i) The drainage channel to the bottom drain: the gabion mat protecting the raft from the inlet channel to the bottom drain was completely damaged (Figure 6A). (ii) The restitution channel at the outlet of the dissipation basin: the protective riprap of the dissipation basin's downstream part underwent significant degradation. Similarly, the massive support for the potable water and the irrigation pipes endured severe damage (Figure 6B).

Figure 6. Historic flood of Yabous river. (**A**) The gabion mat protecting the raft from the inlet channel to the bottom drain was completely damaged. (**B**) The protective riprap of the dissipation basin's downstream part suffered significant degradation. Similarly, the massive support for the potable water and the irrigation pipes endured severe damage.

2.2.4. Geometric Data

Cross-sections were placed along the Yabous river in order to estimate the height of the flood wave downstream of the dam at different sites, which were selected based on the change of the river's direction, width, slope, and land uses, such as agricultural lands, the presence of urban areas, and infrastructures, such as bridges and roads. Cross-sections consisted of either a short section for a narrow valley or an elongated section that functions whenever the simulated water level exceeds the height of that section. This pattern was implemented in order to ensure that the height of the flood wave resulting from a possible dam failure does not exceed the cross-sections. Using RAS Mapper (version 5.0.7), the cross-sections were converted into points supplemented with elevation values retrieved from the DEM data that were required for HEC-RAS analysis. Then, the model calculated the distance between every two cross-sections. Transversal quality control was carried out on the geometric data to ensure that no inexact information was imported. Accordingly, if required, a few cross-sections were edited in the graphic cross-section editor [34]. In total, 122 cross-sections were implemented with a total length of 21.009 km along Yabous river, with the distance between two cross-sections ranging from 50 to 150 m, and 4104 cross-sections interpolated using the model to ensure a good simulation of the flood wave propagation downstream of the dam (Figure 7a,b). These cross-sections were implemented according to: (1) the existing structures downstream of the dam; (2) the implementation of the agglomeration zones; (3) the change in the direction of the river; (4) the slope of the river; (5) the geometry of the basin. Seven sites located along an altitudinal gradient downstream of the dam were selected to assess the impact of dam-failure flood, i.e., (S1) axis of the dam, (S2) upstream urban area (Yabous town), (S3) bridge 1 (Figure 7c), (S4) bridge 2 (Figure 7d), (S5) downstream urban area, (S6) upstream Azla city, and (S7) downstream Azla city, which correspond to the cross-sections at 17,945, 17,207, 16,590, 14,546, 14,461, 2877, and 2480, respectively.

2.2.5. Geological/Geotechnical Data

The study area is part of the eastern Saharan Atlas. The area's intermediate position between the Saharan platform and the mobile Tellian domain explains the facies variations it presents (Figure 8). The Cretaceous formations characterized by the internal platform facies on the southern and southwestern margins of the basin pass to marl–limestone series to faunas and pelagic micro-faunas toward the north and northeast. The Yabous dam site is based on a tectonic accident stalling in an NS direction, which may affect its stability. The town of Yabous is located at an altitude of 900–1100 m and delimits the Aurès massif to the north. Its surface is covered by a Quaternary piedmont glacis, largely encrusted, which hides a powerful series (approximately 700 m) of sandstone sediments dated from the Tortonian Miocene. The locality of Yabous presents several types of fault structures from both pre- and post-Miocene eras. The latter appeared in Cretaceous and Miocene formations.

Due to its high precision, Manning's roughness coefficient was used in the HEC-RAS hydraulic model to compute water surface elevation. According to the classification of [35], two types of land use classes were selected for the study area. Then, the values of Manning's coefficient were adapted according to this classification. The Yabous riverbed is made of uniform gravels with rugged banks. According to Chow's classification of the river substrate, a Manning coefficient value of 0.0356 was retained for the entire riverbed in the study area. However, the Manning coefficient of agricultural land on the left and right sides of the riverbed is 0.1.

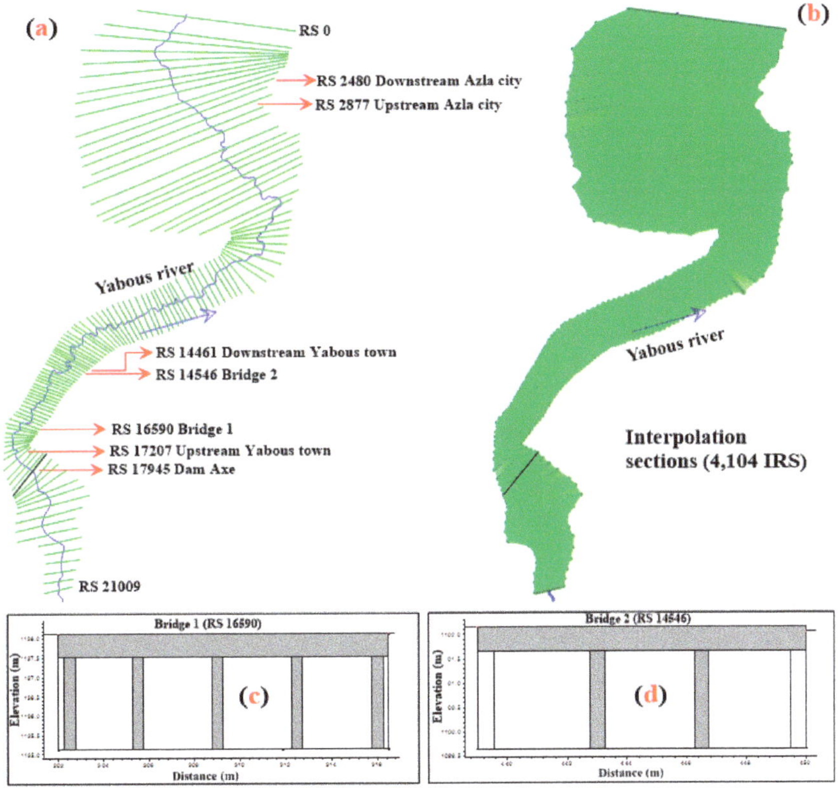

Figure 7. Location of cross-sections (n = 89, (**a**)) and interpolated sections (n = 1267, (**b**)) along the river downstream of Yabous Dam in northeastern Algeria. Numbers RS represent the cross-sections. Cross-sections Bridge 1 (**c**) and cross-sections Bridge 2 (**d**).

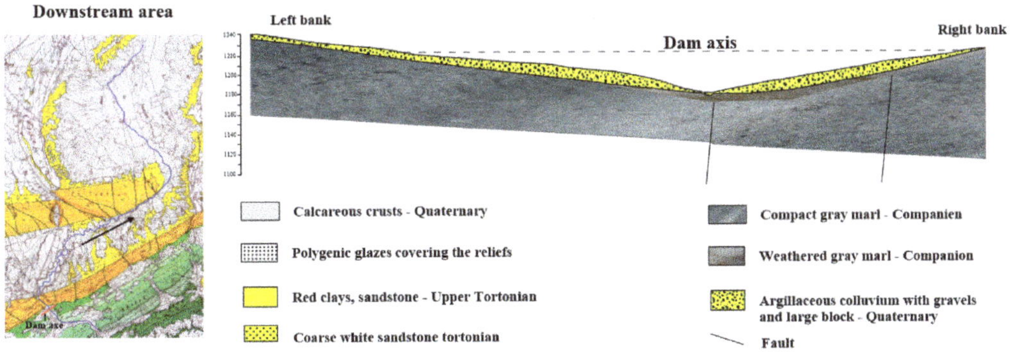

Figure 8. Geological maps of the study area, "Yabous watershed", located west the Province of Khenchela (bottom-left map) in northeastern Algeria.

2.2.6. Hydraulic Data

After processing the geometric data, the initial and boundary conditions required for the HEC-RAS model were: First, the initial condition (Q = 3 m^3/s). Second, data from the millennial series flood hydrograph with a maximum discharge of 485 m^3/s (Figure 9).

Third, the normal depth value that was obtained from the slope and calculated from the profile of the river using the HEC-RAS model. The normal depth of the Yabous River, in this study, was established at 0.002. Once the previous information was settled and data were processed, the model was launched to simulate the propagation of flood waves downstream of the dam.

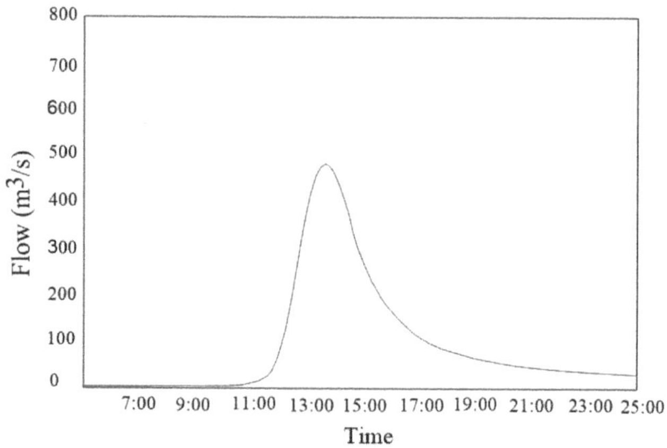

Figure 9. Flood hydrograph of Yabous river.

2.3. HEC-RAS Flood Model

As a reminder, HEC-RAS is a hydraulic model designed by the US Corporation Engineers Hydraulic Engineering Centre to model river flow [36]. It is a well-established and well-tested model globally and is sometimes used as benchmarked against the performance of other hydrodynamic model simulation software [37]. This software permits users to estimate water surface profiles along a river with steady and unsteady flow river hydraulic calculations, including sediment transport modeling [38]. Energy and momentum Equations (1) and (2) are used to derive the 1-D Saint Venant equation in generating steady and unsteady state flow water surface profile simulations within HEC-RAS using an implicit finite different method [37].

$$\frac{\delta A}{\delta t} + \frac{\delta S}{\delta t} + \frac{\delta Q}{\delta x} - q = 0 \tag{1}$$

$$\frac{\delta Q}{\delta t} + \frac{\delta VQ}{\delta x} + gA\left(\frac{\delta z}{\delta x} + S_f\right) = 0 \tag{2}$$

where A is the cross-sectional area, t is the time, S is the ineffective flow area, Q is the discharge, q is the inflow per unit area, x is the distance along the channel, V is the velocity, z is the elevation, Sf is the friction slope, and g is the gravitational acceleration.

The unsteady flow simulation component in the HEC-RAS modeling system is capable of simulating 1-D unsteady flow through a full network of open channels. In this study, 1-D HEC-RAS version 5.0.7 is used to model the flood wave in a downstream dam in the event of failure. This current version of HEC-RAS is able to perform 1-D hydrodynamic modeling and an interesting flood inundation mapping tool directly within HEC-RAS 5.0.7 through RAS Mapper.

2.3.1. One-Dimensional HEC-RAS Modeling

It is known that a good hydraulic model requires good geometry and flow data input. However, the goodness of the simulations is also affected by model type, i.e., one-dimensional (1-D), two-dimensional (2-D), or combined (1D/2D) types. A 1-D model is still the most used to simulate flow in the main river channel and, in certain cases, very effective in predicting flood extent [39]. One-dimensional modeling also has computational efficiency and simple parameterization in dealing with flows in large and complex networks and is favored even more specifically in developing countries [40]. Nevertheless, this simple model has some drawbacks when simulating a 2-D flow path in a large floodplain. In short, a 1-D model is robust, economical, provides valuable information on water profile properties, and is a popularly preferred alternative, as long as the flow paths can be identified [40,41]. The hydraulic simulation by HEC RAS allows the visualization and spatial analysis of the results, such as the limits of the flood zones, the water levels of the flood, etc. Therefore, it is considered to be a decision-making tool par excellence for flood risk management. The HEC RAS modeling is well suited to help in carrying out dam failure analyses [6]. The most significant advantage in the 1-D HEC-RAS model is that in dam-break flood modeling, the model allows users to see the impact of the breach parameters, such as the width, slope, and time formation, using the sensitivity analysis [6].

In 1-D river hydraulic modeling, all water flows are assumed to flow in a longitudinal direction. One-dimensional models represent the terrain as a sequence of cross-sections and simulate the flow in order to provide estimates of flow parameters, such as flow velocity and water depth [42]. To simplify the computation, HEC-RAS assumed a horizontal water surface at each cross-section normal to the direction of the flow such that the momentum exchange between the channel and the floodplain can be neglected [43]. The one-dimensional equations of motion then become Equations (3) and (4):

$$\frac{\partial A}{\partial t} + \frac{\partial(\varnothing Q)}{\partial x_c} + \frac{\partial[(1-\varnothing)Q]}{\partial x_f} = 0 \tag{3}$$

$$\frac{\partial Q}{\partial t} + \frac{\partial\left(\frac{\varnothing^2 Q^2}{A_c}\right)}{\partial x_c} + \frac{\partial\left[(1-\varnothing)^2 Q^2/A_f\right]}{\partial x_f} + gA_c\left[\frac{\partial Z}{\partial x_c} + S_{fc}\right] + gA_f\left[\frac{\partial Z}{\partial x_f} + S_{ff}\right] = 0 \tag{4}$$

where Q is the total flow, A is the flow area, $\varnothing$ is the quotient of channel conveyance over the total conveyance, z is the elevation of water surface, and Sf is the friction slope, in which the subscripts c and f refer to the channel and floodplain, respectively. The 1-D model unsteady flow equations were solved using a four-point implicit scheme, where the space derivatives and function values were evaluated at an interior point $(n + \theta)\Delta t$.

The performance of 1-D modeling in the hydraulic simulation is tested in a number of studies, such as Horritt and Bates [44], who concluded that the 1-D model is capable of giving a good estimation of flood level and travel time and can be used for the prediction of flood extent. In a recent study, Timbadiya et al. [45] studied the 1-D hydrodynamic model for two flood events and proposed that the flood level predicted was computationally satisfactory on the basis of the root mean square error validation. In addition, Pramanik et al. [46] simulated a 1-D model of the Brahmani River using geometry from a combination of surveyed and digital elevation models (DEM). At the same time, Patel et al. [47] executed the 1-D model for an 11 km of the lower part of Ambica River, with 359 surveyed channel cross-section inputs that highlighted the importance of accurate geometry data in 1-D hydrodynamic model. These studies confirm the cost effectiveness, reduced complexity, and reliable prediction of water level along the river of the 1-D model [38]. In this paper, the 1-D modeling approach was applied to validate the base model and ability of HEC-RAS 5.0.7 to simulate extreme events of a flood wave in a downstream dam in the event of failure.

2.3.2. Dam Break Scenario

In this study, we assumed that a historic flood could occur in the watershed of the Yabous dam and that the reservoir level of this dam is at a normal level. The flood wave created by the dam failure extended to the end of the study area. Due to the location of the lateral works, with very steep terrain and mountains downstream of the Yabous dam, the severity of the flood waves would be reduced spatially and temporally. The flood wave would change its parameters (width, height, and discharge) because of the geometry of the area downstream of the dam. In order to perform the unsteady flow analysis with the failure scenario [48,49], six steps were followed: (a) the geometry model was created in HEC-RAS 5.0. 7 using RAS Mapper, (b) dam characteristics were entered in order to create the dam, (c) failure parameters were calculated, (d) roughness data were entered, (e) the influx data were defined, and (f) the computation time was calculated [43,50].

The first step was to identify the geometry of the area, including drawing the wadi boundary and cross-sections over the entire study area and, in particular, the sections desired to see the impact of the flood wave on them. The geometric model represents a network where each section contains geometric information about the underlying terrain. In the 1-D model, each cross-section is similar to each design cell in the 2-D model, and a property table is created for each section [43,51]. More than 122 calculation sections, 4104 interpolation sections, and a 1-D flow area were created based on the extent of the flood zone with an area of 2389 ha for the present study. The flood flow is calculated based on the data obtained from the underlying terrain, such as area, elevation–volume relationship, roughness, and maximum, minimum, and average elevations. To obtain an accurate representation of the flood wave, the size of the cross-sections must be chosen [52,53]. The next step was to introduce the characteristics of the dam: dam length 480 m; dam crest width 7 m; dam height 50 m. Once the dam was created, the following parameters were introduced to breach the dam: center station, final bottom elevation, final bottom width, right- and left-side slope, and breach weir coefficient (see next Section 2.3.3).

For unsteady flow data, boundary conditions and initial conditions have been defined. As initial conditions, an initial flow is given as $Q_0 = 3\ m^3/s$. As boundary conditions, we assumed that the reservoir has a maximum water volume (100% fill) for the return interval of 0.01% (1000 years), which corresponds to a flow rate of 485 m^3/s. For this scenario, a normal depth (the energy slope of the riverbed) was defined for the 1-D flow zone [51].

The last step in the dam failure modeling process was to calculate the computation time to ensure the stability of the hydraulic model. The main effect of a bad computation time is the instability of the model [50]. Specifically, if the time step is too large, in addition to the instability of the model, an attenuation peak can occur. In addition, if the time step is too small, the calculation period will be very long. Nevertheless, in order to estimate the computation time, the current condition was used (Equation (5)) [50,51,54].

$$C = \frac{V_w \Delta T}{\Delta X} \leq 1, \qquad \Delta T = \frac{\Delta x}{V_w}, \qquad V_w = \frac{dQ}{dA'} \tag{5}$$

where C is the current, ΔT is the time step (s), Δx is the distance step in m, V_w is the flood wave speed (m/s), dQ is the change in flow rate over a short time interval (Q2 − Q1), and dA is the change in the effective section over a short time interval (A2 − A1). According to Equation (5), a time step of 3 s was used to run the model.

2.3.3. Estimation of Dam Breach Parameters

Risk assessments of embankment dam failure frequently require predicting the basic temporal and geometrical parameters of a breach to estimate peak breach flows. This flow varies depending on the breach parameters, such as the slope, width, and time of breach formation [55]. The most used models predicting these relationships have been developed from statistical analyses involving historical data of dam failures. The prediction uncertainties of these methods are broadly large, but still never concretely quantified [23]. Among many methods used to predict the parameters of dam breach, four methods were selected in this investigation [56–59] (Table 1). Regarding the adaptation of these methods for the regression analysis, a prediction analysis of these methods is important to estimate their performance. As an example, Froehlich [58,59] used parameters of 63 dam breach cases to develop their relationships (Equations (6)–(9)), whereas Von Thun and Gillette [57] used 57 dam breach cases (Equations (10) and (11)), and Macdonald and Langridge-Monopolis [56] used 42 dam breach cases (Equations (12)–(14)). The use of uncertainties of breach parameter estimates obtained from one of these methods is necessary to determine their impacts on the risk of dam failure. These uncertainties are assessed so that reasonable limits of breach parameter values can be estimated and used to determine the reliability of the flood hydrographs predicted by the hydraulic structures, as well as the flood wave peaks at the defined sites downstream of the dam. The four above-mentioned methods exist in HEC-RAS version 5.0.7. The HEC-RAS model was used to calculate each method's average width, lateral slope, and breach formation time once its requirements and input data were filled.

$$B_{ave} = 15K_0 V_w^{0.32} h_b^{0.19} \tag{6}$$

$$t_f = 3.84 V_w^{0.53} h_b^{-0.9} \tag{7}$$

$$B_{ave} = 0.27K_0 V_w^{0.32} h_b^{0.04} \tag{8}$$

$$t_f = 63.2\sqrt{\left(\frac{V_w}{g h_b^2}\right)} \tag{9}$$

$$B_{ave} = 2.5h_w + C_b \tag{10}$$

$$t_f = \frac{B_{ave}}{4h_w} \tag{11}$$

$$V_{eroded} = 0.00348(V_{out} h_w)^{0.852} \tag{12}$$

$$t_f = 0.0179(V_{eroded})^{0.364} \tag{13}$$

$$W_b = \frac{V_{eroded} - h_b^2\left(CZ_b + \frac{h_b Z_b Z_3}{3}\right)}{h_b\left(C + \frac{h_b Z_3}{2}\right)} \tag{14}$$

where: B_{ave} = average width (m); W_b is the final bottom width; h_w = depth of water above the bottom of the breach (m); h_b = height from the top of the dam to the bottom of the breach (m); Z = breach embankment slope (m); B = lower width (m); t_f = breach formation time (h); V_w = reservoir volume at time of failure (m^3); K_O = constant (1.3 for overtopping failures, 1.0 for piping); V_{eroded} = volume of material eroded from the dam embankment (m^3); V_{out} = volume of water that passes through the breach (m^3); C = crest width of the top of dam (m); $Z_3 = Z_1 + Z_2$; Z_1 = average slope (Z_1:1) of the upstream face of the dam; Z_2 = average slope (Z_2:1) of the downstream face of the dam; Z_b = side slopes of the breach (Z_b:1), which are 0.5 for the MacDonald method; C_b = coefficient, which is a function of reservoir size.

Table 1. Parameters of the flow simulation for four dam breach parameter prediction methods using the HEC-RAS in modeling dam-break flood of Yabous dam in northeastern Algeria.

Prediction Methods for Dam Breach Parameters	Flow Parameters			
	Width (m)	Slope	Flood Time (h)	Q (m^3/s)
Macdonald and Langridge-Monopolis (1984)	14	0.5	1.10	4400.2
Von Thun and Gillete (1990)	54	0.5	0.47	8767.9
Froehlich (1995)	26	1.4	0.42	7841.1
Froehlich (2008)	26	1.0	0.40	7363.0

2.4. Other Consistency Tests

As the HEC-RAS model requires a certain amount of input data, a sensitivity analysis was carried out in order to quantify the importance of the different involved quantities [18,60]. In the current study, breach formation refers to several parameters, including breach width, breach slope, and the time of breach formation. While for the propagation of the flood wave downstream, a breach involves several physical and hydrodynamic parameters, such as flow velocity, width, and height of the flood wave. For this effect, in order to highlight the influence of the breach formation parameters on the flood wave's physical and hydrodynamic parameters associated with a dam-failure flood, a sensitivity analysis method of four adjustments (−50%, −25%, +25%, and +50%) was used for each of the selected sites at risk of flooding. In this sensitivity analysis, the adjustments of the flood hydrograph also were applied to perform the hydrological modeling for four adjustments (−50%, −25%, +25%, and +50%) [60,61]. The application of this method aims to thoroughly clarify the trend of each breach parameter and flood hydrograph on each parameter of the flood wave in every downstream section selected along the flood course.

3. Results

3.1. Estimate of Dam Breach Parameters

Dam breach parameters, including average breach width, breach side slope, and failure time, were predicted. After computing these parameters using the available methods in HEC-RAS, the obtained results were reviewed, and then a single predicted value was selected. These recommended values are shown in Table 1.

3.1.1. Breach Width

The dam breach parameters analysis described earlier showed that the Froehlich methods produced comparable results for this scenario, in which the average breach width (B_{avg}) was relatively moderate (B_{avg} = 26 m). However, the Macdonald and Langridge-Monopolis (1984) method underestimated the observed breach width and showed the lowest values (B_{avg} = 14 m). The Von Thun and Gillette method recorded the highest value (B_{avg} = 54 m). The prediction intervals developed through these methods were sobering for this case study where we sought to obtain a definitive result because the results obtained varied from small values for the dam to high values for breach width. Accordingly, the results of the Macdonald and Langridge-Monopolis method were excluded in this case study due to the low values.

3.1.2. Breach Time

Failure time predictions are summarized in Table 1. All the methods indicated a short time (40–47 min) to empty the reservoir during the simulated dam failure event, except for the Macdonald and Langridge-Monopolis method, which predicted a slightly longer time = 1.1 h. According to the failure time obtained in all prediction methods, the dam break analysis showed that the flood wave associated with the simulated dam break developed very fast. Accordingly, the consequences of this flash flooding are expected to be catastrophic and irreversible if a very early warning system is not installed for alerting the citizens inhabiting this flood risk zone.

3.1.3. Flow Peak

Peak outflow estimates are shown in Table 1, sorted in order of increasing peak outflow for the dam breach scenario. The lowest peak flow predictions were obtained using the method with the least parameters. The highest peak flows were predicted using equations that incorporated a significant dependence on breach parameters. According to this analysis, a convergence was observed between results obtained using the Froehlich methods [58,59] and those of the Von Thun and Gillete method [57], whose point flow values were 7841.12 m^3/s, 7363.03 m^3/s, and 8767.87 m^3/s, respectively. However, a significant difference was detected between the results of these methods and the outputs of the Macdonald and Langridge-Monopolis method [56] (Q = 4400.18 m^3/s) (Table 1).

3.2. Analysis of Dam-Breach Flood Scenario

Hydrodynamics of the flood wave along the Yabous river during the simulated dam failure followed four steps (Figure 10). The first was upstream of the dam. The second, at the dam reservoir. The third, at the level of breach formation, and the fourth was downstream from right at the dam to the end of the river.

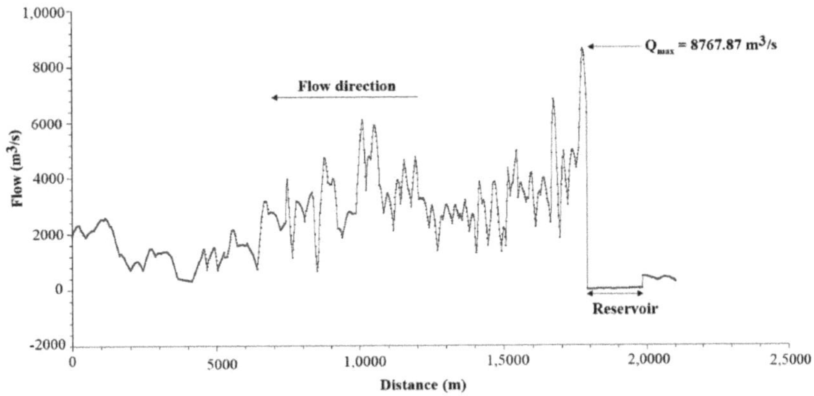

Figure 10. Flood hydrograph of dam-failure flood applied to Yabous dam using the Von Than and Gillet method.

The dynamics of the maximum flow velocity were mapped for the dam failure scenario depending on verified results of the one-dimensional hydrodynamic model (Figure 11). Regarding the flow velocity, two types of flows are distinguished during the Yabous dam-break flood. The velocity curve can be divided into three main parts: the parts (I_a, I_b) reflected a torrential flow, while part (II) indicated a fluvial flow.

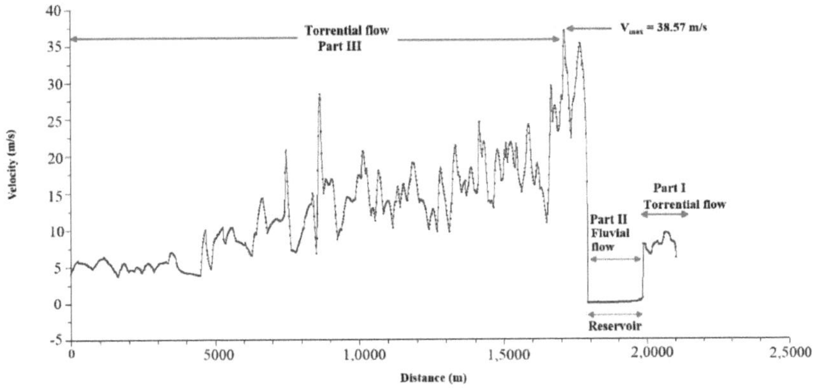

Figure 11. Dynamics of flow velocity during the dam-failure flood of Yabous dam along the river downstream.

Changes in the flood wave's width corresponded to the basin's geometry along the Yabous river (Figure 12). For steep slopes, narrow widths were recorded as the point of breach formation on the dam axis, whereas the large widths recorded at weak slopes can be related to the slowing down of water flooding at these slopes where the maximum width ($B_{reservoir}$ = 1028.18 m) was recorded at the level of the dam reservoir, just a few seconds before the dam failure. At that moment, the height of the water in the dam reservoir was equal to the dam's height. The satellite image and the cross-section show water filling the reservoir behind the Yabous dam (Figure 12). The data represented in Figure 12 indicate a hydrodynamic trend in the opposite direction compared to data shown in Figures 10 and 11, which represent the flow and velocity of the flood wave before and during the dam failure. Therefore, we can argue that Figures 10 and 11 reflect the potential energy of the flood wave. In contrast, Figure 12 represents the quantitative energy of the flood wave at each point along the Yabous river.

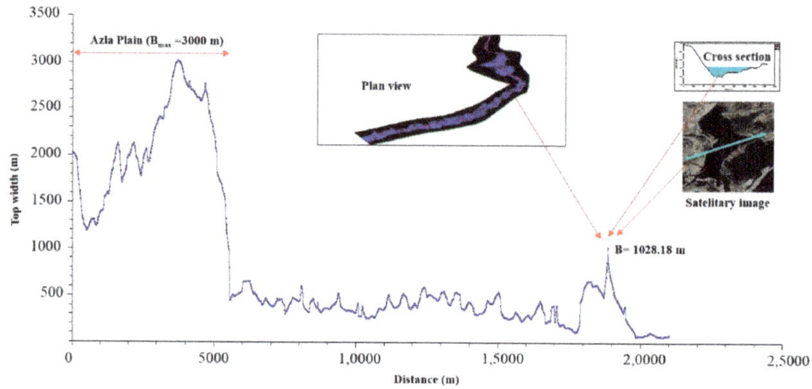

Figure 12. Hydrodynamic simulation of the width of the flood wave along Yabous river during a simulated dam-failure flood.

The profile of water current power evolved similarly with trends of water current shear stress, which represented the potential energy of the water during the flood. The two previous profiles were developing in the opposite direction to that of the surface (Figure 13). During the time of dam failure, no dynamic was recorded at the level of the dam reservoir, which is due to the stabilization of the water level in this part. However, the flow peak was observed at the level of the breach formation, where water was falling from the reservoir with the progressive formation of the breach. Meanwhile, irregularities were observed along the sections of the Yabous river for the three main factors, viz., the change in land slope (geometric), the type of soils/rock outcrops of this zone (geotechnical), and the type of flow either in load or in suspension (hydraulic).

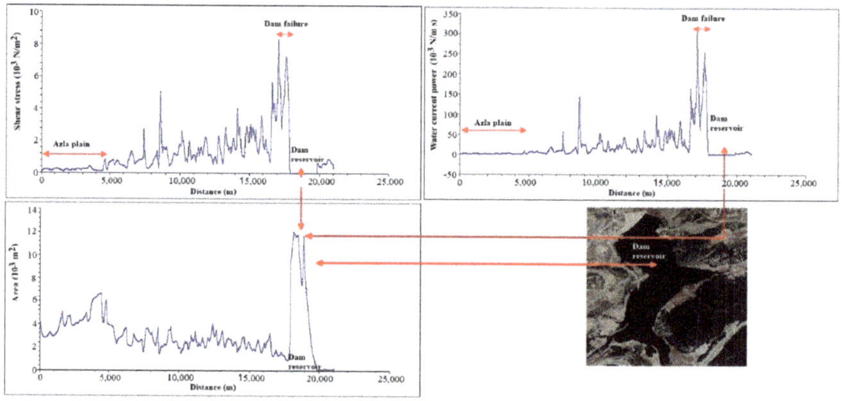

Figure 13. Power, shear stress, and water surface of water current associated with the flood wave along Yabous river during a simulated dam-break flood.

The analysis of flood wave propagation along the Yabous river during the simulated dam failure using the HEC-RAS model showed that the flood wave dynamics have four parts (Table 2, Figures 14 and 15). The first, before dam failure (time = 00:00 h), is where the reservoir is filled to the standard level. The second, after 30 min of dam failure, the flood wave arrived at the entrance of the Yabous village. The third, one hour after reaching Yabous entry, the flood wave reached the exit of the urban area. The fourth, during a time equal to 02:00 h after the dam failure, the reservoir was emptied completely.

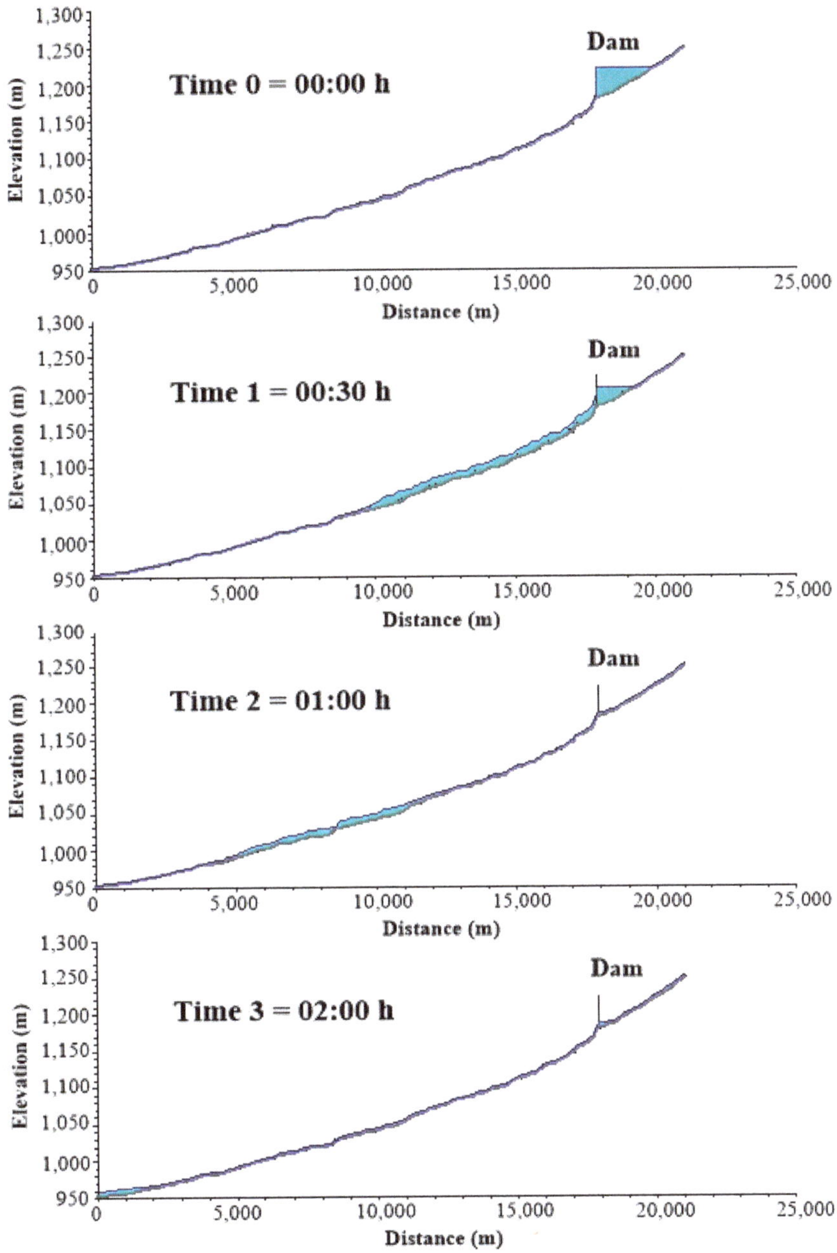

Figure 14. Longitudinal profile of the flood wave propagation before, during, and after the simulated flood of the Yabous river using 1-D HEC-RAS hydraulic model.

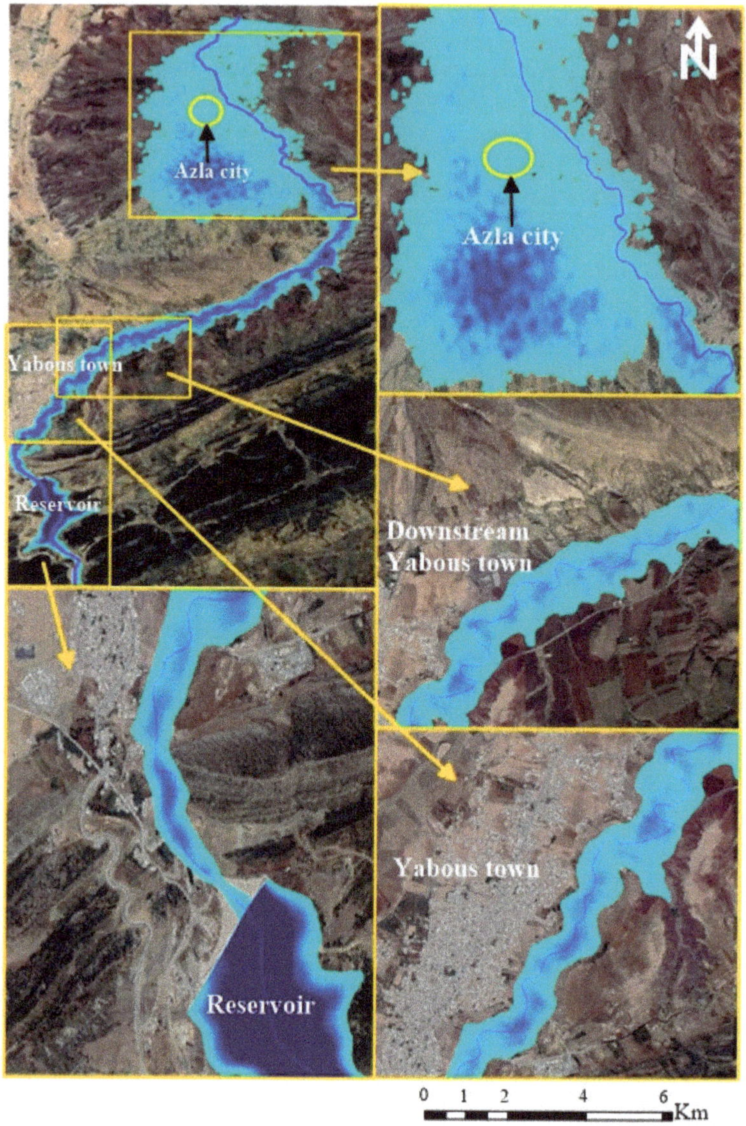

Figure 15. Satellite images showing flood risk zones downstream of the Yabous dam at high flood risk associated with simulated dam failure using 1-D HEC-RAS hydraulic model.

Eventually, the dead storage volume of the reservoir remained, and the flood wave started to fade. The frightening issue here is the high risk associated with this very quick discharge of a volume of 7 million m^3 of the reservoir in only 90 min. However, damage estimation was mostly based on water volume instead of depth [62]. It is worth mentioning that this eventuality represents a serious risk, since floods driven by dam failure can cause serious and irreversible disastrous consequences to human populations settled in downstream areas as their socio-economic and environmental livelihoods are devastated. Still, the travel time of a flood wave and the extent of flooding that would result from a dam failure should be predicted in order to prepare emergency action plans that would be used in the event of an actual dam failure [63]. Inundation maps developed in such simulation studies can serve as the core tool for decision making during the elaboration of emergency evacuation plans in the same way as in natural hazard assessment studies. These kinds of predictions have the potential to be a reference for future dam break inundation practical applications and research studies [34]. Table 2 includes the physical and hydrodynamic parameters of the flood wave for different sites selected along the Yabous river at the flood risk zone downstream of the dam (Figure 15) in order to assess the impact of the simulated dam-failure flood.

Table 2. The flood wave's physical and hydrodynamic parameters for different sites (S1–S7) were selected to assess the impact of a dam-break flood along the Yabous river in northeastern Algeria.

	Sites						
Flood Parameters	**S1**	**S2**	**S3**	**S4**	**S5**	**S6**	**S7**
Flood wave parameters	17,945	17,207	16,590	14,546	14,461	2877	2480
Maximum flow 'Q' (m^3/s)	8767.87	3603.6	3732.47	2427.49	1579.2	3265.5	3752.6
Minimum main channel elevation (m)	1180.00	1153.46	1137.28	1100.65	1100.13	972.38	968.78
Water surface elevation (m)	1194.05	1165.89	1146.97	1107.47	1106.98	974.17	970.44
Energy grade line elevation (m)	1223.87	1181.42	1155.62	1109.45	1107.72	974.33	970.58
Energy grade line slope (m/m)	0.043532	0.046506	0.036115	0.014848	0.007767	0.009170	0.007981
Velocity (m/s)	38.57	30.98	22.34	11.23	7.78	3.38	3.09
Surface (m^2)	990.72	1394.43	1757.35	1102.46	1139.24	2598.03	2560.41
Width (m)	129.96	213.85	357.17	354.88	312.54	2088.05	1376.86
Height (m)	14.05	12.43	9.69	6.82	6.85	1.79	1.66
Froude Number	2.75	2.84	2.38	1.44	1.01	0.89	0.83

S1: axis of the dam, S2: upstream Yabous town, S3: bridge 1, S4: bridge 2, S5: downstream Yabous town, S6: upstream Azla city, S7: downstream Azla city.

3.3. Sensitivity Analysis

Figure 16 illustrates the variation of flood wave parameters (flow, width, and height) for four adjustments (±25 and ±50) of each breach parameter and four adjustments (±25 and ±50) of the flood hydrograph in five sites downstream of the Yabous dam, with the initial dam failure analysis (without adjustments) given in the plot in green color as a reference value (0). An important deviation of all adjustments was observed for the breach parameters and flood hydrograph compared to the analysis of this scenario without adjustment (0). The null result was obtained in all parameters modeled in this analysis. This demonstrates the significant nature of breach parameters and flood movement on the physical parameters of the flood wave associated with dam-break flooding.

Similar trends were found between breach slope and breach width with the four adjustments (±25% and ±50%), as shown in Figure 16. Values of physical and hydrodynamic parameters of the flood wave on the five selected sites decreased with the increase in the adjustment values of breach parameters (−25%, −50%). In contrast, the flood wave physical and hydrodynamic parameters increased with the increase in adjustment values of breach parameters (i.e., +25% and +50% adjustment). Despite the similar trends obtained between breach slope and breach width, differences still remain for the adjustment values. Values of breach width for the four adjustments were higher than those of breach slope. This indicates the large impact of the breach width values on flood wave parameters (Figures 16 and 17). The underestimation of breach width or breach lateral slope can lead to unsatisfactory results, according to Singh and Scarlatos [18], particularly in earthen dams, which consist of several building materials. The results obtained for the adjustments (±25% and ±50%) of the flood hydrograph showed that the intensity of the floods upstream of the dam is generally low, and therefore the differential potential of the flood adjustments compared to adjustment (0) is also low on all of its parameters, such as flow, height and flood width (Figures 16 and 17). The adjustments (+25 and +50) make small changes that appear on the three flood parameters after the dam rupture. On the other hand, the adjustment (−50%) reveals that the hydrograph of the flood entering the dam without threat was not recorded.

In order to demonstrate how a flood wave performed for different levels of adjustments (−50%, −25, +25, and +50%) of the breach parameters (slope, width, formation time, and flood hydrography), relationships between flood wave flow, width, and height were tested using generalized linear mixed-effects models (GLMM), with Gaussian distribution error and 'identity' link. The selected sites downstream of Yabous dam were considered to be random effects, whereas flood wave variables (height and width) and breach adjustment levels were the fixed-effect factors. The GLMMs were carried out using the package {nlme} in R [64]. For each breach parameter (slope, width, and formation time), three GLMMs were created to test the effects of flood wave width on wave flow and wave height and the effect of flood wave height on wave flow. The R package {ggplot2} was employed in plotting the data of the dam breach parameters.

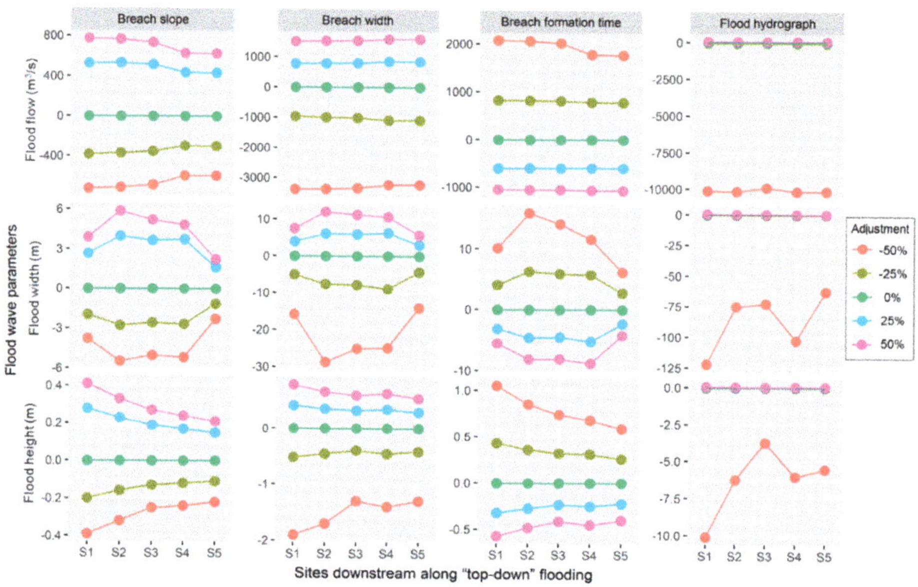

Figure 16. Results of sensitivity analysis of dam breach parameters (flood flow, width, and height) showing various adjustments of the breach slope, breach width, breach formation time, and flood hydrograph for different sites (S1–S5) selected along the Yabous river in northeastern Algeria for appraising the simulated dam-break flood.

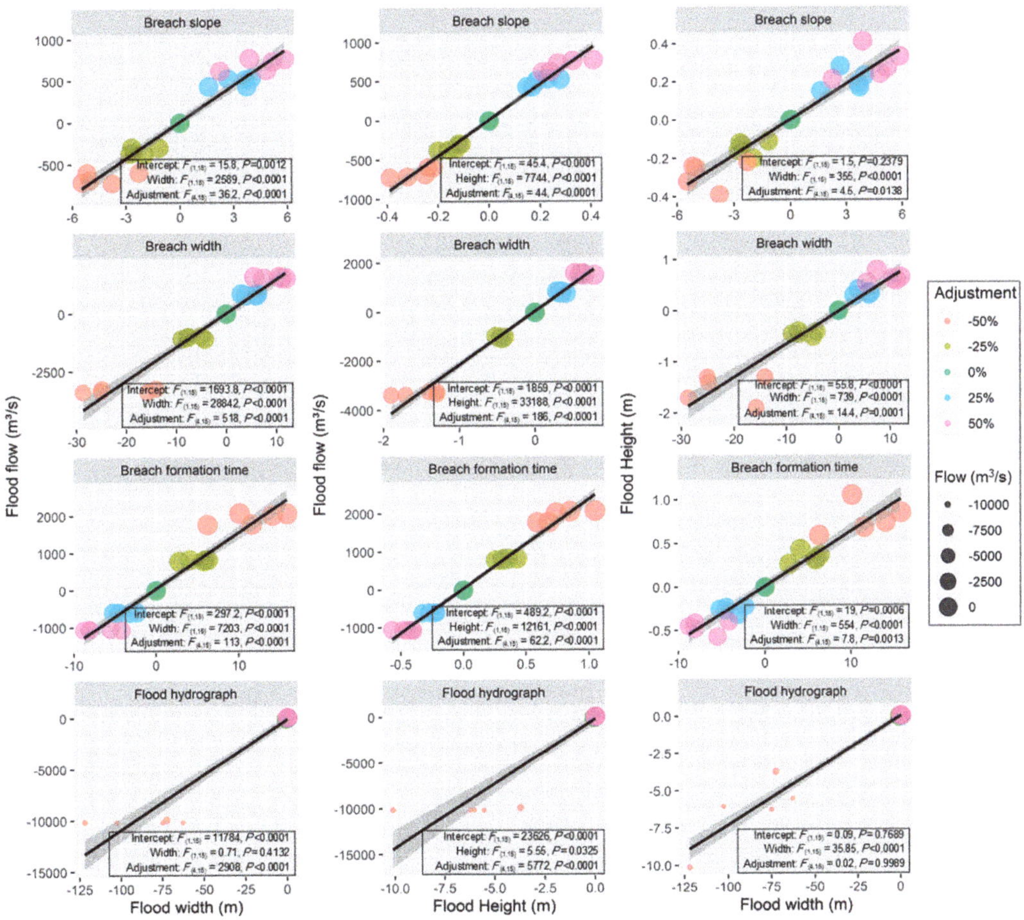

Figure 17. Relationships between simulated flood wave parameters (flood flow, width, and height) for various adjustments of the breach slope, breach width, breach formation time, and flood hydrograph simulated on the Yabous river in northeastern Algeria. Point sizes are set to water flow values and the color mapped to the levels of breach adjustments (−50%, −25, +25, and +50%). The solid black lines represent linear regressions obtained by a Gaussian GLM (generalized linear model) fit with 95% confidence regions in light gray. Statistics reported in the boxes are results of fixed-effects terms of GLMMs, with F-statistics and p-values being *type-II* tests 'sequential'. Figures between brackets represent the numerator and denominator degrees of freedom, respectively.

4. Discussion

Four methods were used in this work including MacDonald and Langridge-Monopolis [56], Von Thun and Gillete [57], two by Froehlich [58,59], to determine the dam breach parameters. The results for the four methods for all parameters are different (Table 1). This difference can be explained by the lack of necessary data for the estimation of breach parameters. On the one hand, this reflects the reliability of the results obtained using the first three methods for the estimation of the breach parameters, including width, slope, and time of breach formation, and the simulation of flood wave propagation downstream following dam failure, on the other hand. The Macdonald and Langridge-Monopolis' method can no longer be adopted for a full study because of its aberrant and unrealistic results. Overall, Froehlich's methods showed good results in the field. They remained the best methods in this field, given the high number of real dam failure cases (63) used while predicting dam breach parameters, and because they are the most suitable for earthen dams [65]. The methods offered by Froehlich (1995, 2008) distinctly had the best prediction performance in the uncertainty analysis (Table 1). These methods had the smallest mean prediction error and the narrowest prediction interval by a significant margin [23,66]. The results of Von Thun and Gillete's method highlighted the worst flood case with a high flow level (Q = 8767.87m^3/s) (Figure 10). This clearly indicates the limits of the risk zone downstream of the dam. The dam breach scenario proved to be a large event when analyzed by this method, so the peak outflow is expected to occur when the breach has reached its maximum size before the occurrence of a significant drawdown in the reservoir.

The hydrodynamics of the flood wave along the Yabous river during the simulated dam failure followed four steps. For step (1), a weak hydrodynamic simulation of the flood wave upstream of the dam was represented by a flow hydrograph given to the HEC-RAS model as a condition at the upstream limit with a flood rate (Qmax = 485 m^3/s). For step (2), no water dynamics were observed at the entire length of the dam reservoir because of the stability of the water level in the reservoir. In step (3), water discharge began at the dam crest during the dam break with a very slow flow. A significant development was recorded during the breach formation. This reflects the discharge of reservoir waters in the event of a rupture on a small breach that begins to form gradually during the dam rupture, where flow recorded a peak (Qmax = 8767.87 m^3/s). Given the large water quantity discharged over the dam crest, the propagation of the flood wave downstream of the dam is seriously hazardous and may have irreversible and disastrous consequences on urban areas, local populations, and the environment, with heavy economic losses [28]. For step (4), the dynamics of the flood wave downstream of the dam were simulated. This is most likely because the slope changes along the Yabous river, on the one hand. On the other hand, the basin section widens for each point of the river. According to the time of breach formation (0.47 h) and the geometry of the breach (flood width = 54 m, slope = 0.5), the flow peak of the flood wave propagated very quickly downstream (Figure 10).

Regarding the flow velocity, two types of flows were distinguished during the Yabous dam-break flood ($I_{a,b,}$ and II). Part I_a represents the flood hydrograph given to the model as a condition at the upstream limit (S0). Moreover, it extended from section (S0) to the beginning of the dam reservoir with an average velocity (V = 9 m/s). For part II, it represented the fluvial flow velocity over the length of the dam dike with almost zero velocity. This flow resulted from the dissipation of flow energy by the stable water of the dam reservoir. Regarding part $\mathrm{I_b}$, it indicated the torrential flow, which reflects the potential energy of the flood wave from the formation of the breach of the dam to the final section of the study area (Figure 11). The flow torrentiality is due to the change in the slope of the river on the whole part (I_b). Indeed, the flow velocity increases rapidly in the wave tip zone, where the flow resistance dominates [67,68]. High-velocity peaks represent steep slopes, where flow speed recorded the highest velocity at the level of the breach formation (Vmax = 38.57 m/s). The low peaks represent the weak slopes where small reservoirs occurred naturally and contributed to the dissipation of water flow on the river. The flood wave velocity shows a better match with the DEM model implanted in HEC-RAS. This analysis can provide a reasonable approximation of the impacts of a flood wave on areas located downstream in the event of dam failure. Thus, this information is essential for flood hazard classification and developing emergency action plans [68].

The simulation of the flood wave propagation downstream of the Yabous river was carried out using the 1-D HEC-RAS model, which delineated the zone that can be affected by this flood wave downstream of the dam in the case of the latter's failure. The flood wave related to dam failure was recorded at the breach level with a flow point of Qmax = 8767.87 m^3/s, and this reflects the impact of this flood downstream of the dam (Table 2). Subsequently, this wave arrived at site S2 (N° 17207), which represents the beginning of Yabous village with a flow and a flood-wave width of Q = 3603.60 m^3/s and B_{Flood} = 213.85 m, respectively. This width confirms that the upstream limit of the

Yabous village is far from this flood (Figure 15). The flood wave reached site S3 at the first bridge with four square openings (3m × 3m) with a flow of Q = 3732.47 m^3/s, flow width = 357.17 m, and a height = 9.69 m, which is expected to completely submerge the bridge (Table 2). This demonstrates the violent and destructive impact of the rapid flood resulting from dam failure. The wave arrived at the second three-opening bridge (3 m × 3 m) located downstream the village (S4: N° 14546) with a flow Q = 2427.49 m^3/s and a height H = 6.82 m, which may induce the total damage of this bridge and, thus, cuts the road connecting the town of Yabous to the Khenchela City. The flooding caused by dam failures in an urban area is a complex phenomenon, specifically with the appearance of structural ruptures, suspended debris, and incisal currents, especially when associated with a very unstable and turbulent flood flow [6,69]. It could be possible that the town of Yabous is a little far from the flood generated by the rupture of the dam, but the village of Azla (N° 2877, 2480), which is 15 km from the dam, will be completely submerged, and this is due to the geometry of this village being almost the same altitude compared to the wadi of Yabous bed. The consequences could be really destructive and irreversible on the urban and the social side (Figure 15). In addition to these disasters, Figure 15 can also reveal the destruction of the largest Yabous plain, as approximately 80% of this plain could be submerged (Figure 15). The consequences of this flood do not stop here; the road of this village will also be destroyed. According to Figure 15, the only way to deal with this flood is to make a good alarm system that responds to this type of risk, given the fact that the flood will reach the village in about two hours (02 h) (Figure 14).

The flood wave arrived at Azla city with a flow = 3265.5 m^3/s and width = 2088 m. These findings demonstrate the real need to protect the local population and their livelihood from possible flood events generated by dam failure. Heavy human and economic losses are expected with the rapid passage of the flood wave. This includes the partial destruction of the urban zone, agricultural lands, and crops located downstream. The analysis also demonstrated that because of the dam-break flood event, the town of Yabous may lose some houses built on the left edge of the Yabous river (Figure 15). Despite the acceptable protection status of Yabous town against this simulated flood, the flood wave is expected to seriously harm the region at various levels, such as all the flooded agricultural lands, which are expected to be severely destroyed and lose their productivity features. Indeed, crop yields are destroyed, too. Agricultural activities and livestock farming represent a large part of this area's population income. More than 80,000 different orchards trees (80% are apple) can be affected by this flood event.

Ecologically, the passage of flood waves is often associated with detritus deposits and drastic changes in soil characteristics. All this is expected to disrupt the behavior of fauna and flora and the entire ecosystem functioning of the region [70–73]. The negative effects on flora and riparian river habitats are projected to be severe and most likely persistent throughout the risk area [74]. Flood waves can influence ecosystem processes, affecting within-channel productivity, particulate organic matter retention, decomposition, nutrient transformations, and microbial processes [75]. It can also influence surface water and groundwater quality and recharge [76]; flood waves may also cause 'sand slugs,' alter channel morphology, scour substrata, remove in situ basal resources, such as macrophytes, detritus, and translocate, injure or kill biota, and affect the relative importance of autochthonous and allochthonous carbon sources [77]. In general, floods also alter water quality [78] and facilitate the spread of non-native and invasive species [79]. Floods waves vary greatly in their character and their impact, as does the vulnerability of the populations they affect. It is also known that the areas at greatest risk are located downstream from a dam [80]. The health impacts of floods waves can be direct or indirect and depend upon various factors, including the characteristics of the flood hazard, patterns of exposure, and the underlying vulnerability of the population. The health impacts of a particular flood event are context-specific and are very different between developed and developing countries, ranging from injuries to epidemiological consequences, which will have the effect of increased mortality [81].

Simulation analysis can measure the extent and the intensity of the damage caused by the deposits of solid residues (tailings, metals, plastics, woods, etc.) transported with the flood wave, of which the loss of vegetation would trigger bottom-up land degradation [82]. Consequently, the loss of soil stability and runoff control increases the risk of further geomorphological disturbances, including landslides, bank ruptures, and massive creep movements. To this end, the priority is given to mitigation measures to deal with this type of risk through policy making that ensures a strict application of integrated environmental impact assessments and preventive monitoring. This should avoid or significantly minimize dam failure impacts [74].

In the case of the Yabous dam, the damage that the flood can cause to the dam body building and the type of construction materials, especially of the core (usually clay), can increase the breach width once the flood has dumped on the dam crest. This is expected, therefore, to increase the flood flow while reducing the flood propagation time. The resistance of materials, regardless of the friction angle and coefficient of cohesion, can be considered in the modeling flood scenario. The low resistance of materials causes its easy removal during the flooding on the ridge, which causes an increase in the breach width and lateral slope. Thereby, the rate of eroded materials displaced from the dam body to downstream areas of the dam and river increases as a laden flow [23].

Regardless of breach parameters (slope, width, formation time, and flood hydrography), the GLMMs indicated that the flood wave flow increased significantly with the increase in wave height ($p < 0.0001$) and wave width ($p < 0.0001$). The effect of the flood wave width of the variation in wave height in the five selected sites were highly significant ($p < 0.0001$) for all breach parameters (Figure 17). The statistical models showed that the variables of wave flow, both the physical and hydrodynamic, differed significantly among breach adjustment levels ($p < 0.001$), with values of flood wave variables increasing along a negative-to-positive gradient of adjustment levels (−50–50%) of breach slope and width. However, flood wave flow, height, and width increased inversely to the increase in the breach adjustment levels of breach formation time (Figure 17).

Indeed, the physical and hydrodynamic parameters of the breach slope and the breach width went in the opposite direction compared to the breach formation time (Figure 16). Accordingly, the more the adjustment values of the breach formation time move in the negative direction (i.e., 25% and −50%), the higher the physical parameters of the breach formation time are. However, the physical and hydrodynamic parameters of the flood wave at the five selected sites decreased for the positive adjustments (+25% and +50%) of the breach formation times (Figure 17). However, the adjustments of the flood hydrograph (± 25% and ±50%) have shown that the flood can only increase the possible risk of rupture of the dam whose spillway does not support these flows that exceed the flow rate for which this structure was designed (Figure 16). However, this flood with the adjustments (± 25% and ±50%) does not have a great influence on the flood parameters downstream of the dam (Figures 16 and 17). This is due to the large difference in the peak flow that leaves the dam in the event of a break compared to the flow of the flood entering the dam.

As a result, the actual dam break mechanisms are not well understood. Neither the current physical models nor the empirical models have been able to fully explain the mechanisms and impacts of dam failures [83]. For practical applications, these results contribute to the simple identification of the most prominent breach parameters so that we can prioritize resources towards a more precise estimation of their values [84,85]. Otherwise, in earthen dams, several technical factors can influence the general parameters of the breach formation. The first example to mention is the volume of the dam reservoir; the larger the volume, the more the breach shape increases [86]. The second, the type of clay used for the dike core; the more plastic the clay used is (cohesion and friction nail), the more the dam dike will resist the various floods on the dike crest. Therefore, the resulting breach shape is weak [87,88]. The third, the thickness of the filters upstream and downstream of the dike core; a wide filter thickness will induce the phenomenon of erosion in the event of a flood spill on the crest. This will systematically give a large breach shape [89]. The fourth is the amount of compaction applied on the layers of the dike material during its construction [90]. The fifth, the thickness of the layer of the material used for the realization of the dike and, in particular, the clay material; the lower the thickness of the layer, the more the dike body can resist the flood spill on the dam crest [91]. The sixth, the number of the layers watering the dam material (water content), especially the core material (clay) [90]. Additionally, the seventh is the protection of upstream and downstream embankments [92]. All these factors have an influence on the general dimensions of the breach (slope, width, and height); therefore, the latter can directly influence positively or negatively physical parameters of the flood wave. In general, the physical parameters of the flood wave are greatly influenced by the width of the breach and the slope of the breach, while it remains essentially insensitive to the time of breach formation [93,94]. In return for smaller reservoirs (a larger breach), the formation time has more influence on the outflow than the width of the breach. To this end, successful proof has been established concerning the approval and significance of these breach parameters on the development of the physical parameters of the flood in the event of a dam failure.

5. Conclusions

Dam-break floods cause truly catastrophic and irreparable environmental, social, and economic consequences. In order to mitigate flood risks, prior knowledge of flooding hazards is required. The present study demonstrated the effectiveness of the HEC-RAS 1-D model in simulating the propagation of flood waves downstream of the dam in the event of dam failure. Our findings revealed that the maximum flow of the Yabous dam-break flood at the level of the breach was 8768 m^3/s, which attenuated gradually along the Yabous river, where it recorded a flow of 1972.7 m^3/s at 3.3 km from the dam. According to Von Thun and Gillette's method, the dam will run out of water in a very short time (~47 min), which stresses the need to install a fast and efficient alarm system. According to the risk map, it turned out that, despite the fact that the village of Yabous is predominantly far from flood waves, severe environmental and economic consequences could potentially occur, in particular the damage of two bridges and the destruction of the agricultural land.

The interrelationships and contribution rates of the model input parameters are studied by using the sensitivity analysis, performed on three breach parameters, i.e., width, slope, and training time, for four adjustments (−25%, −50%, +25%, and +50%) over five sites located at an altitudinal gradient downstream of the dam. Sensitivity analysis highlighted the significant impact of breach parameters on the dynamic of the flood wave, either in the negative or positive direction following the adjustment. The analysis revealed that the influence of breach width and slope on the flood wave is in the same direction, on the one hand. On the other, it is in the opposite direction with respect to the time of breach formation. Nonetheless, the adjustments of the flood hydrograph (−25%, −50%, +25%, and +50%) have shown that the flood can only increase the possible risk of rupture of the dam whose spillway does not support these flows. That exceeds the flow rate by which this structure was designed. In addition, it provided valuable insights into the uncertainty of the input parameters selection for any dam design and dam feature of any work. The application of the sensitivity analysis method on dam-break flooding scenarios proved the significant association of the physical parameters of the flood wave with the parameters of breach formation. In the end, such flexibility is vital for climate change and measures that deal with the long-term consequences of water resource management. This study provides opportunities to monitor performance indicators and enhance effectiveness in our quest to manage water effectively and sustainably.

Author Contributions: Conceptualization, A.G. and H.A.A.; methodology, A.G. and H.A.A.; software, A.G. and H.C.; validation, A.B., L.B., and H.A.A.; formal analysis, J.B., A.E.K., and I.Z.; investigation, A.G. and A.A.; resources, A.E.K., J.B., and I.Z.; data curation, A.B., L.B., and H.C.; writing—original draft preparation, A.G., H.A.A., A.B., and L.B.; writing—review and editing, H.A.A., A.E.K., J.B., I.Z., and H.C.; visualization, A.B. and H.C.; supervision, L.B. and H.C.; project administration, A.B.; funding acquisition, H.A.A. All authors have read and agreed to the published version of the manuscript.

Funding: A. E. Krauklis' contribution was supported by the European Regional Development Fund within the Activity 1.1.1.2 "Post-doctoral Research Aid" of the Specific Aid Objective 1.1.1 of the Operational Programme "Growth and Employment" (Nr.1.1.1.2/VIAA/4/20/606).

Institutional Review Board Statement: Not Applicable.

Informed Consent Statement: Not Applicable.

Data Availability Statement: The data presented in this study are available on request from the corresponding author.

Acknowledgments: Many thanks are addressed to the DGRSDT (Directorate General of Scientific Research in Algeria) and the MESRS (Ministry of Higher Education and Scientific Research of Algeria) for their support. Many thanks are also addressed to the European Regional Development Fund within the Activity 1.1.1.2 "Post-doctoral Research Aid" of the Specific Aid Objective 1.1.1 of the Operational Programme "Growth and Employment" (Nr.1.1.1.2/VIAA/4/20/606) for their support.

Conflicts of Interest: The authors declare no conflict of interest.

References

1. Derdous, O.; Djemili, L.; Bouchehed, H.; Tachi, S.E. A GIS based approach for the prediction of the dam break flood hazard—A case study of Zardezas reservoir—Skikda, Algeria. *J. Water Land Dev.* **2015**, *27*, 15–20. [CrossRef]
2. Boussekine, M.; Djemili, L. Modelling approach for gravity dam break analysis. *J. Water Land Dev.* **2016**, *30*, 29–34. [CrossRef]
3. Berghout, A.; Meddi, M. Sediment transport modelling in Wadi Chemora during flood flow events. *J. Water Land Dev.* **2016**, *31*, 23–31. [CrossRef]
4. Wang, B.; Chen, Y.; Wu, C.; Dong, J.; Ma, X.; Song, J. Semi-analytical model for predicting peak discharge of dam-break flood. *Hydrol. Processes* **2016**, *30*, 3682–3691. [CrossRef]
5. Luino, F.; Tosatti, G.; Bonaria, V. Dam Failures in the 20th Century: Nearly 1000 Avoidable victims in Italy Alone. *J. Environ. Sci. Eng.* **2014**, *3*, 19–31.
6. Gaagai, A.; Boudoukha, A.; Benaabidate, L. Failure simulation of Babar dam—Algeria and its impact on the valley downstream section. *J. Water Land Dev.* **2020**, *44*, 75–89. [CrossRef]
7. Negm, A.; Bouderbala, A.; Chenchouni, H.; Barcelo, D. *Water Resources in Algeria—Part I: Assessment of Surface and Groundwater*; The Handbook of Environmental Chemistry Series; Springer: Cham, Germany, 2020. [CrossRef]
8. Schreider, S.Y.; Smith, D.I.; Jakeman, A.J. Climate Change Impacts on Urban Flooding. *Clim. Change* **2000**, *47*, 91–115. [CrossRef]
9. Nassopoulos, H.; Dumas, P.; Hallegatte, S. Adaptation to an uncertain climate change: Cost benefit analysis and robust decision making for dam dimensioning. *Clim. Change* **2012**, *114*, 497–508. [CrossRef]
10. Arar, A.; Chenchouni, H. How could geomatics promote our knowledge for environmental management in Eastern Algeria. *J. Environ. Sci. Technol.* **2012**, *5*, 291–305. [CrossRef]
11. ShahiriParsa, A.; Noori, M.; Heydari, M.; Rashidi, M. Floodplain zoning simulation by using HEC-RAS and CCHE2D models in the Sungai Maka river. *Air Soil Water Res.* **2016**, *9*, 55–62. [CrossRef]
12. Gaagai, A. Etude de L'évolution de la Qualité des Eaux du Barrage de Babar (Sud-Est Algérien) et L'impact de la Rupture de la Digue sur L'environnement. Ph.D. Thesis, University of Batna 2, Batna, Algeria, 2017. [CrossRef]
13. Fluixá-Sanmartín, J.; Morales-Torres, A.; Escuder-Bueno, I.; Paredes-Arquiola, J. Quantification of climate change impact on dam failure risk under hydrological scenarios: A case study from a Spanish dam. *Nat. Hazards Earth Syst. Sci.* **2019**, *19*, 2117–2139. [CrossRef]
14. Lee, B.S.; You, G.J.Y. An assessment of long-term overtopping risk and optimal termination time of dam under climate change. *J. Environ. Manag.* **2013**, *121*, 57–71. [CrossRef] [PubMed]
15. Kundzewicz, Z.W.; Kanae, S.; Seneviratne, S.I.; Handmer, J.W.; Nicholls, N.; Peduzzi, P.; Mechler, R.; Bouwer, L.M.; Arnell, N.; Mach, K.J.; et al. Flood risk and climate change: Global and regional perspectives. *Hydrol. Sci. J.* **2014**, *59*, 1–28. [CrossRef]
16. Delenne, C.; Cappelaere, B.; Guinot, V. Uncertainty analysis of river flooding and dam failure risks using local sensitivity computations, Reliab. *Eng. Syst. Saf.* **2012**, *107*, 171–183. [CrossRef]
17. Guo, X.; Dias, D.; Pan, Q. Probabilistic stability analysis of an embankment dam considering soil spatial variability. *Comput. Geotech.* **2019**, *113*, 93–103. [CrossRef]
18. Singh, V.P.; Scarlatos, P.D. Analysis of Gradual Earth-Dam Failure. *J. Hydraul. Eng.* **1988**, *114*, 21–42. [CrossRef]
19. Israa, D.A.; Qassem, H.J.; Jasim, M.A. Sensitivity Analysis for Dam Breach Parameters Using Different Approaches for Hamrin Dam. *Diyala J. Eng. Sci.* **2021**, *14*, 90–97. [CrossRef]
20. Pektas, A.O.; Erdik, T. Peak discharge prediction due to embankment dam break by using sensitivity analysis-based ANN. *KSCE J. Civ. Eng.* **2014**, *18*, 1868–1876. [CrossRef]

21. Xanthopoulos, T.; Koutitas, C. Numerical simulation of two-dimensional flood wave propagation due to dam failure. *J. Hydraul. Res.* **1976**, *14*, 321–331. [CrossRef]
22. Hervouet, J.M. A high-resolution 2-D dam-break model using parallelization. *Hydrol. Processes* **2000**, *14*, 2211–2230. [CrossRef]
23. Wahl, T.L. Uncertainty of predictions of embankment dam breach parameters. *J. Hydraul. Eng.* **2004**, *130*, 389. [CrossRef]
24. Sadrolashrafi, S.S.; Thamer, A.M.; Ahmad, R.B.M.; MajidK, K.; Amir, S. Integrated modeling for flood hazard mapping using watershed modeling system. *Am. J. Eng. Appl. Sci.* **2008**, *1*, 149–156. [CrossRef]
25. Mao, J.; Wang, S.; Ni, J.; Xi, C.; Wang, J. Management System for Dam-Break Hazard Mapping in a Complex Basin Environment. *ISPRS Int. J. Geo-Inf.* **2017**, *6*, 162. [CrossRef]
26. Gee, M.; Brunner, G. Dam Break Flood Routing Using HEC-RAS and NWS-FLDWAV Conference. *World Water Environ. Resour. Congr.* **2005**, *3*, 6. [CrossRef]
27. Albu, L.M.; Enea, A.; Iosub, M.; Breabăn, L.G. Dam breach size comparison for flood simulations. A HEC-RAS based, gis approach for Drăcșani Lake, Sitna River, Romania. *Water* **2020**, *12*, 1090. [CrossRef]
28. Hafnaoui, M.A.; Madi, M.; Hachemi, A.; Farhi, Y. El Bayadh city against flash floods: Case study. *Urban Water J.* **2020**, *17*, 390–395. [CrossRef]
29. Tamm, O.; Tamm, T. Verification of a robust method for sizing and siting the small hydropower run-of-river plant potential by using GIS. *Renew. Energy* **2020**, *155*, 153–159. [CrossRef]
30. Oudni, A.; Dinar, H. Caractérisation Géologique et Géotechnique de la Cuvette du Barrage Tagharist. Master's Thesis, University of Oum El Bouaghi, Oum El Bouaghi, Algeria, 2016.
31. Chenchouni, H. Statuts de protection et de conservation des oiseaux recensés dans les Aurès et ses alentours (nord-est algérien). In *Actes du Séminaire International sur la Biodiversité Faunistique en Zones Arides et Semi-Arides*; Ouargla University: Ouargla, Algeria, 2010; pp. 56–75.
32. USACE. *HEC-RAS Version 4.1 [Hydraulic Reference Manual]*; Army Corps of Engineers Hydrologic Engineering Center, U.S. Army Corps of Engineers: Davis, CA, USA, 2008; Available online: https://www.hec.usace.army.mil/software/hec-hms/documentation/HEC-HMS_Applications_Guide_March2008.pdf (accessed on 13 January 2022).
33. Hu, H.H.; Walton, R. Advanced guidance on use of steady HEC-RAS. *World Environ. Water Resour. Congr.* **2008**, *4*, 12–16. [CrossRef]
34. Haltas, I.; Tayfur, G.; Elci, S. Two-dimensional numerical modeling of flood wave propagation in an urban area due to Urkmez dam-break, Izmir, Turkey. *Nat. Hazards* **2016**, *81*, 2103–2119. [CrossRef]
35. Chow, V.T. *Open-Channel Hydraulics*; McGraw-Hill: New York, NY, USA, 1959.
36. Brunner, G.W. HEC-RAS: River Analysis System User's Manual Version 5.0. 2016. Available online: https://www.hec.usace.army.mil/software/hec-ras/documentation/HEC-RAS%205.0%20Reference%20Manual.pdf (accessed on 12 January 2022).
37. Zellou, B.; Rahali, H. Assessment of reduced-complexity landscape evolution model suitability to adequately simulate flood events in complex flow conditions. *Nat. Hazards* **2016**, *86*, 1–29. [CrossRef]
38. Zainalfikry, M.K.; Ab Ghani, A.; Zakaria, N.A.; Chan, N.W. HEC-RAS One-Dimensional Hydrodynamic Modelling for Recent Major Flood Events in Pahang River. In *Proceedings of AICCE'19*; Mohamed Nazri, F., Ed.; Springer: Cham, Switzerland; Berlin/Heidelberg, Germany, 2020; Volume 53. [CrossRef]
39. Vozinaki, A.E.K.; Morianou, G.G.; Alexakis, D.D.; Tsanis, I.K. Comparing 1D and combined 1D/2D hydraulic simulations using high resolution topographic data a case study of the Koiliaris basin. *Greece Hydrol. Sci. J.* **2016**, *62*, 642–646. [CrossRef]
40. Ahmad, B.; Hassan, Z.A. Flood map of Tupai River using combined 1D and 2D modelling. In Proceedings of the 3rd International Conference on Managing Rivers in the 21st Century: Sustainable Solutions for Global Crisis of Flooding, Pollution and Water Scarcity, Penang, Malaysia, 6–9 December 2011.
41. Leow, C.S.; Abdullah, R.; Zakaria, N.A.; Ghani, A.; Chang, C.K. Modelling urban river catchment: A case study in Malaysia. *Proc. Inst. Civ. Eng. Water Manag.* **2008**, *162*, 25–34. [CrossRef]
42. Dasallas, L.; Kim, Y.; An, H. Case Study of HEC-RAS 1D–2D Coupling Simulation: 2002 Baeksan Flood Event in Korea. *Water* **2019**, *11*, 2048. [CrossRef]
43. Brunner, G. *HEC-RAS River Analysis System Hydraulic Reference Manual Version 5.0*; USACE CEC: Davis, CA, USA, 2016.
44. Horritt, M.; Bates, P. Evaluation of 1D and 2D numerical models for predicting river flood inundation. *J. Hydrol.* **2002**, *268*, 87–99. [CrossRef]
45. Timbadiya, P.V.; Patel, P.L.; Porey, P.D. One-dimensional hydrodynamic modelling of flooding and stage hydrographs in the lower Tapi River in India. *Curr. Sci.* **2014**, *106*, 708–716.
46. Pramanik, N.; Panda, R.K.; Sen, D. One dimensional hydrodynamic modeling of river flow using DEM extracted river cross-sections. *J. Water Resour. Manag.* **2010**, *24*, 835–852. [CrossRef]
47. Patel, S.B.; Mehta, D.J.; Yadav, S.M. One dimensional hydrodynamic flood modeling for Ambica River, South Gujarat. *J. Emerg. Technol. Innov. Res.* **2018**, *5*, 595–601.
48. Kostecki, S.; Banasiak, R. The Catastrophe of the Niedów Dam—The Causes of the Dam's Breach, Its Development, and Consequences. *Water* **2021**, *13*, 3254. [CrossRef]
49. Yerramilli, S. Potential Impact of Climate Changes on the Inundation Risk Levels in a Dam Break Scenario. *ISPRS Int. J. Geo-Inf.* **2013**, *2*, 110–134. [CrossRef]

50. Brunner, G.W. Using HEC-RAS for Dam Break Studies. 2014. Available online: https://www.hec.usace.army.mil/publications (accessed on 2 February 2020).
51. Mihu-Pintilie, A.; Cîmpianu, C.I.; Stoleriu, C.C.; Pérez, M.N.; Paveluc, L.E. Using High-Density LiDAR Data and 2D Streamflow Hydraulic Modeling to Improve Urban Flood Hazard Maps: A HEC-RAS Multi-Scenario Approach. *Water* **2019**, *11*, 1832. [CrossRef]
52. Menéndez-Pidal, I.; Martín, J.A.H.; Alonso-Muñoyerro, J.M.; Sanz, E. Real-Time Data and Flood Forecasting in Tagus Basin. A Case Study: Rosarito and El Burguillo Reservoirs from 8–12 March 2018. *Water* **2020**, *12*, 1004. [CrossRef]
53. Arseni, M.; Rosu, A.; Calmuc, M.; Calmuc, V.; Iticescu, C.; Georgescu, L.P. Development of Flood Risk and Hazard Maps for the Lower Course of the Siret River, Romania. *Sustainability* **2020**, *12*, 6588. [CrossRef]
54. Urzică, A.; Mihu-Pintilie, A.; Stoleriu, C.C.; Cîmpianu, C.I.; Hu¸tanu, E.; Pricop, C.I.; Grozavu, A. Using 2D HEC-RAS Modeling and Embankment Dam Break Scenario for Assessing the Flood Control Capacity of a MultiReservoir System (NE Romania). *Water* **2021**, *13*, 57. [CrossRef]
55. Sammen, S.S.; Mohamed, T.A.; Ghazali, A.H.; Sidek, L.M.; El-Shafie, A. An evaluation of existent methods for estimation of embankment dam breach parameters. *Nat. Hazards* **2017**, *87*, 545–566. [CrossRef]
56. MacDonald, T.C.; Langridge-Monopolis, J. Breaching characteristics of dam failures. *J. Hydraul. Eng.* **1984**, *110*, 567–586. [CrossRef]
57. Von Thun, J.L.; Gillette, D.R. *Guidance on Breach Parameters*; Unpublished internal document; U.S. Bureau of Reclamation: Denver, CO, USA, 1990.
58. Froehlich, D.C. Embankment dam breach parameters and their uncertainties. *J. Hydraul. Eng.* **2008**, *134*, 1708–1721. [CrossRef]
59. Froehlich, D.C. Peak Outflow from Breached Embankment Dam. *J. Water Resour. Plan. Manag.* **1995**, *121*, 90–97. [CrossRef]
60. Andreadakis, E.; Diakakis, M.; Vassilakis, E.; Deligiannakis, G.; Antoniadis, A.; Andriopoulos, P.; Spyrou, N.I.; Nikolopoulos, E.I. Unmanned Aerial Systems-Aided Post-Flood Peak Discharge Estimation in Ephemeral Streams. *Remote Sens.* **2020**, *12*, 4183. [CrossRef]
61. Kastridis, A.; Theodosiou, G.; Fotiadis, G. Investigation of Flood Management and Mitigation Measures in Ungauged NATURA Protected Watersheds. *Hydrology* **2021**, *8*, 170. [CrossRef]
62. Prettenthaler, F.; Amrusch, P.; Habsburg-Lothringen, C. Estimation of an absolute flood damage curve based on an Austrian case study under a dam breach Scenario. *Nat. Hazards Earth Syst. Sci.* **2010**, *10*, 881–894. [CrossRef]
63. McMahon, G.F. Developing dam-break flood zone ordinance. *J. Water Resour. Plan. Manag. Div.* **1981**, *1072*, 461–476. [CrossRef]
64. Core Team, R. *R: A Language and Environment for Statistical Computing*; R Foundation for Statistical Computing: Vienna, Austria, 2021; Available online: https://www.R-project.org. (accessed on 16 January 2022).
65. Bellos, V.; Tsakiris, V.K.; Kopsiaftis, G.; Tsakiris, G. Propagating Dam Breach Parametric Uncertainty in a River Reach Using the HEC-RAS Software. *Hydrology* **2020**, *7*, 72. [CrossRef]
66. Shahrim, M.F.; Ros, F.C. Estimation of breach outflow hydrograph using selected regression breach equations. *IOP Conf. Ser. Earth Environ. Sci.* **2020**, *476*, 012129. [CrossRef]
67. Chanson, H. Application of the method of characteristics to the dam break wave problem. *J. Hydraul. Res.* **2009**, *47*, 41–49. [CrossRef]
68. Wang, B.; Chen, Y.; Wu, C.; Peng, Y.; Ma, X.; Song, J. Analytical solution of dam-break flood wave propagation in a dry sloped channel with an irregular-shaped cross-section. *J. Hydro-Environ. Res.* **2017**, *14*, 93–104. [CrossRef]
69. Gallegos, H.A.; Schubert, J.E.; Sanders, B.F. Structural damage prediction in a high-velocity urban dam-break flood: Field-scale assessment of predictive skill. *J. Eng. Mech.* **2012**, *138*, 1249–1262. [CrossRef]
70. Gaagai, A.; Boudoukha, A.; Boumezbeur, A.; Benaabidate, L. Hydrochemical characterization of surface water in the Babar watershed (Algeria) using environmetric techniques and time series analysis. *Int. J. River Basin Manag.* **2017**, *15*, 361–372. [CrossRef]
71. Tamm, T.; Noges, T.; Jarvet, A.; Bouraoui, F. Contributions of DOC from Surface and Groundflow into Lake Vortsjarv (Estonia). *Hydrobiologia* **2008**, *599*, 213–220. [CrossRef]
72. Aouissi, H.A.; Ababsa, M.; Gaagai, A.; Bouslama, Z.; Farhi, Y.; Chenchouni, H. Does melanin-based plumage coloration reflect health status of free-living birds in urban environments? *Avian Res.* **2021**, *12*, 45. [CrossRef]
73. Aouissi, H.A.; Petrişor, A.I.; Ababsa, M.; Boştenaru-Dan, M.; Tourki, M.; Bouslama, Z. Influence of Land Use on Avian Diversity in North African Urban Environments. *Land* **2021**, *10*, 434. [CrossRef]
74. Fernandes, G.W.; Goulart, F.F.; Ranieri, B.D.; Coelho, M.S.; Dales, K.; Boesche, N.; Bustamante, M.; Carvalho, F.A.; Carvalho, D.C.; Dirzo, R.; et al. Ecological and socio-economic impacts of the dam breach in Mariana, Brazil. *Nat. Conserv.* **2016**, *14*, 35–45. [CrossRef]
75. Leigh, C.; Bush, A.; Harrison, E.T.; Ho, S.S.; Luke, L.; Rolls, R.J.; Ledger, M.E. Ecological effects of extreme climatic events on riverine ecosystems: Insights from A ustralia. *Freshw. Biol.* **2015**, *60*, 2620–2638. [CrossRef]
76. Zhu, M.; Wang, S.; Kong, X.; Zheng, W.; Feng, W.; Zhang, X.; Yuan, R.; Song, X.; Sprenger, M. Interaction of Surface Water and Groundwater Influenced by Groundwater Over-Extraction, Waste Water Discharge and Water Transfer in Xiong'an New Area, China. *Water* **2019**, *11*, 539. [CrossRef]

77. Bastos, R.F.; Lippi, D.L.; Gaspar, A.L.B.; Yogui, G.T.; Frédou, T.; Garcia, A.M.; Ferreira, B.P. Ontogeny drives allochthonous trophic support of snappers: Seascape connectivity along the mangrove-seagrass-coral reef continuum of a tropical marine protected area. *Estuar. Coast. Shelf Sci.* **2022**, *264*, 107591. [CrossRef]
78. Gholizadeh, M. Effects of floods on macroinvertebrate communities in the Zarin Gol River of northern Iran: Implications for water quality monitoring and biological assessment. *Ecol. Processes* **2021**, *10*, 46. [CrossRef]
79. Čuda, J.; Rumlerová, Z.; Brůna, J.; Skálová, H.; Pyšek, P. Floods affect the abundance of invasive Impatiens glandulifera and its spread from river corridors. *Divers. Distrib.* **2017**, *23*, 342–354. [CrossRef]
80. Du, W.; FitzGerald, G.J.; Clark, M.; Hou, X.Y. Health impacts of floods. *Prehospital Disaster Med.* **2010**, *25*, 265–272. [CrossRef]
81. Ahern, M.; Kovats, R.S.; Wilkinson, P.; Few, R.; Matthies, F. Global health impacts of floods: Epidemiologic evidence. *Epidemiol. Rev.* **2005**, *27*, 36–46. [CrossRef]
82. Hatje, V.; Pedreira, R.M.; de Rezende, C.E.; Schettini, C.A.F.; de Souza, G.C.; Marin, D.C.; Hackspacher, P.C. The environmental impacts of one of the largest tailing dam failures worldwide. *Sci. Rep.* **2017**, *7*, 10706. [CrossRef]
83. Xiong, Y. A Dam Break Analysis Using HEC-RAS. *J. Water Resour. Prot.* **2011**, *3*, 370–379. [CrossRef]
84. Dewals, B.; Erpicum, S.; Detrembleur, S.; Archambeau, P.; Pirotton, M. Failure of dams arranged in series or in complex. *Nat. Hazards* **2011**, *56*, 917–939. [CrossRef]
85. Bencedira, S.; Bechiri, O. Degradation of fuchsine acid using the HP2W15Mo3Co2.5O62, $20H_2O/H_2O_2$ system: Effect of organic and inorganic additives. *Euro-Mediterr. J. Environ. Integr.* **2021**, *6*, 60. [CrossRef]
86. Clague, J.J.; Evans, S.G. A Review of catastrophic drainage of moraine-dammed lakes in British Columbia. *Quat. Sci. Rev.* **2000**, *19*, 1763–1783. [CrossRef]
87. Al-Riffai, M.; Nistor, I. Impact and analysis of geotechnical processes on earth fill dam breaching. *Nat. Hazards* **2010**, *55*, 15–27. [CrossRef]
88. Al-Riffai, M.; Nistor, I.; Vanapalli, S.K.; Orendorff, B. Overtopping of earth embankments: Sensitivity analysis on dam breach parameters using two numerical models. In Proceedings of the 60th Canadian Geotechnical Conference, Ottawa, ON, Canada, 21–24 October 2007; pp. 1213–1220.
89. Minguez, R.; Delgado, F.; Escuder, I.; de Membrillera, M.G. Reliability assessment of granular filters in embankment dams. *Int. J. Numer. Anal. Methods Geomech.* **2006**, *30*, 1019–1037. [CrossRef]
90. Zhang, L.; Gelet, R.; Marot, D.; Smith, M.; Konrad, J.M. A method to assess the suffusion susceptibility of low permeability core soils in compacted dams based on construction data. *Eur. J. Environ. Civ. Eng.* **2018**, *23*, 626–644. [CrossRef]
91. Zhong, Q.M.; Chen, I.S.S.; Mei, I.S.A.; Cao, I.W. Numerical simulation of landslide dam breaching due to overtopping. *Landslides* **2017**, *8*, 321–332. [CrossRef]
92. Neilsen, M.L. Global sensitivity analysis of dam erosion models. In Proceedings of the 10th International Conference on Scientific Computing, Las Vegas, NV, USA, 22–25 July 2013.
93. Singh, K.P.; Snorrason, A. Sensitivity of outflow peaks and flood stages to the selection of dam breach parameters and simulation models. *J. Hydrol.* **1984**, *68*, 295–310. [CrossRef]
94. Dewals, B.J.; Archambeau, P.; Erpicum, S.; Detrembleur, S.; Pirotton, M. Sensitivity analysis of the peak outflow induced by the breaching of embankment dams. In Proceedings of the 14th German Dam Symposium & 7th ICOLD European Club Dam Symposium, Munich, Germany, 17–19 September 2007; Rutschmann, P., Ed.; Technische Universität München: Munich, Germany, 2007; pp. 86–92.

water

MDPI

Article

Influence of Dam Breach Parameter Statistical Definition on Resulting Rupture Maximum Discharge

Diego Bello [1], Hernán Alcayaga [1,*], Diego Caamaño [2] and Alonso Pizarro [1]

[1] Escuela de Ingeniería en Obras Civiles, Universidad Diego Portales, Santiago 8370109, Chile; diego.bello@mail.udp.cl (D.B.); alonso.pizarro@mail.udp.cl (A.P.)
[2] Departamento de Ingeniería Civil, Universidad Católica de la Santísima Concepción, Concepción 4030000, Chile; dcaamano@ucsc.cl
* Correspondence: hernan.alcayaga@udp.cl

Abstract: However rare, dam breach occurrences are recently reported and associated with significant damage to life and property. The rupture of the structural dam wall generates severe flow rates that exceed spillway capacity consequently generating unprecedented flooding scenarios. The present research aims to assess the influence of the dam breach statistical configuration on the most relevant parameters to predict the rupture maximum discharge (RMD). McBreach© software was used to provide the necessary inputs for the operation of the HEC-RAS dam breach module. McBreach© automates the process of batch mode simulations providing a Monte Carlo approach to characterize the breach parameters stochastically. Thus, a sensitivity analysis was performed to identify the most influential breach parameters, followed by an uncertainty assessment regarding their statistical definition of the resultant RMD. Analysis showed that the overtopping failure mode discharges are most sensitive to the breach formation time (tf) parameter, followed by the final height breach (Inv) and the final width of the breach (B), which combined are responsible for 85% of the rupture's maximum discharge. Further results indicated highly variable RMD magnitudes (up to 300%) depending on the breach parameter's statistical definition (i.e., probability density function and associated statistical parameters). The latter significantly impacts the estimated flood risk associated with the breach, the flood zone delimitation, preparation of emergency action plans (EAP) and scaling of future dam projects. Consequently, there is a plausible need for additional investigations to reduce this uncertainty and, therefore, the risk associated with it.

Keywords: dam breach; rupture maximum discharge; statistical definition

Citation: Bello, D.; Alcayaga, H.; Caamaño, D.; Pizarro, A. Influence of Dam Breach Parameter Statistical Definition on Resulting Rupture Maximum Discharge. *Water* **2022**, *14*, 1776. https://doi.org/10.3390/w14111776

Academic Editors: Francesca Aureli, Andrea Maranzoni and Gabriella Petaccia

Received: 5 May 2022
Accepted: 28 May 2022
Published: 1 June 2022

1. Introduction

Dams are used worldwide owing to their extensive contribution to the management of water demand in cities, agriculture, industry, power generation, and flood management. However, it is known that dams have a potential risk of failure [1]. A dam breach scenario involves the rapid evacuation of enormous volumes of water and sediment, which are associated with aggressive floods, economic losses, environmental damage, and the possible loss of human life. An example of this occurred in 1975 in Henan, China, where a five-day storm generated 1631 mm of rainfall that led to the overtopping of two large dams (Banqiao and Shimandan), two medium dams (Tiangang and Zhugou), and 58 small dams. For the Banqiao reservoir, a peak breach flow of 78,100 m^3/s was reached, resulting in inundations totaling 12,000 km^2, a death toll of more than 26,000, and economic losses exceeding USD 1.6 billion [2].

Based on the evidence, it is crucial to know the combination of possible hydrographs generated in a failure scenario. In these terms, the traditional way of conducting a dam breach analysis consists of three phases: (i) Hydraulic simulations of the dam breach; (ii) Deterministic analysis of breach parameters to model the hydrograph that would be generated from the rupture; and, (iii) A hydraulic model to route the flood downstream. As established in the FEMA Guide [3], there are a series of mathematical models and methods for analyzing each of these three phases. It is important to indicate that dam breach simulations are designed to predict the resulting flood hydrograph and not the moment of failure [4]. This prediction would be highly sensitive to the in situ conditions (dam and field conditions) because a wide range of possible alternatives are physically valid in classifying the required modeling parameters. Therefore, determining breach parameters usually involves reviewing and evaluating parametric regression equations, revising probable failure mode reports, and expert judgment [5]. The latter is generally associated with a significant level of uncertainty [6,7].

Faced with this problem, many studies have investigated different methods for reducing this uncertainty [8–14]. For example, Pierce [13] used a series of existing equations and empirical methods to assign values to the breach parameters (e.g., breach width and breach formation time). This type of approach is usually based on the analysis of observed data from existing dam failures, such as the analyzes performed by Wahl [7,15]. Despite these efforts, deterministic models fail to adequately incorporate the process of dam breaches because they do not consider critical factors such as the randomness of a natural system, the limitations of the modeling approach, and possible errors in the observed data [16]. The challenge for engineers and scientists is to use statistical approaches to make more accurate estimates from limited data sets [17]. In this context, probabilistic methodologies have been proposed to quantify the associated uncertainty. Among them, Goodel [18] states the importance of knowing the uncertainty related to hydraulic models for flood risk studies and mentions the approaches to risk analysis established in different continents such as Africa [14], Europe [19], Asia [20], and North America [3]. The latter quantified the effect on the RMD results given several models fed by different magnitudes of the breach parameters, making it possible to assign probabilities of occurrence to the results and thus determine flood risks [16].

Despite significant efforts in this regard, some questions remain unanswered [21]. One of the most important gaps in the literature is related to the sensitivity of the solution with respect to the required parametrizations. The latter can be posed as a function of questions that are briefly mentioned as follows: (i) Which breach parameters have the most significant weight for any given case? (ii) How does the parameter's statistical definition affect the rupture's maximum discharge (RMD)? (iii) Does this characterization affect the peak flow in the same way that it affects the shape of the hydrograph? Given the above open questions, the main goal of this paper is to determine the sensitivity of the RMD with respect to the most relevant parameters required for the dam breach simulation.

The rest of the paper is organized as follows: Section 2 presents the methodology used for the hydraulic simulations (using McBreach and HEC-RAS); the Chacrillas case study; and the different configurations for the sensitivity analysis. Section 3 shows the results, highlighting the breach parameters that have the most significant influence on the RMD and the influence of statistical parameters, respectively. Conclusions are provided at the end.

2. Methods

2.1. Workflow Adopted

Hydraulic simulations were performed using the United States Army Corp of Engineers software HEC-RAS v5.0.7 (Hydrologic Engineering Center's, CEIWR-HEC, Davis, CA, USA) [22]. In contrast, a Monte Carlo approach provides the required breach parameter characterization using the software package McBreach© v5.07 (Pittsfield, ME, USA) by Kleinschmidt [23].

The dam failure mode addressed in the present study corresponds to overflow or overtopping. This type of failure occurs when the water surface elevation in the reservoir exceeds the dam's height, with water flowing through the upper ridge until it causes the wall to break, culminating in the evacuation downstream of a volume of water of great magnitude. The overflow implies that the weir system of the dam is not working properly, or there is more water than the designed spillway capacity. The overtopping failure mode is the one that occurs most frequently in the different types of dams [24] and the Concrete-face Rockfill Dam (CFRD) is the type of dam widely used around the world [3]

HEC-RAS is equipped with a dam break module that uses nine breach parameters. These parameters can be grouped according to their characteristics: (i) three geometric parameters (width of the breach, slopes of the breach and final height of the breach); (ii) one to define the breach formation time; and (iii) five parameters to characterize the breach flow (discharge coefficient, height of the breach, tolerable height of the water before a failure occurs, weir coefficient and height where internal erosion begins) [25]. HEC-RAS simulates the overflow discharge over the dam as a flow over a weir. Then, to measure the magnitude of the overflow discharge, three variables are used: (i) the weir discharge coefficient; (ii) the hydraulic head; and (iii) the crest length, with the last two representing functions of time [22]. Seven of the above-mentioned breach parameters describe the overtopping failure mode in HEC-RAS (see Table 1 for definitions and abbreviations of variables).

Table 1. Dam breach parameters used by HEC-RAS when overtopping is considered.

Overtopping Breach Parameters in HEC-RAS	Abbreviation
Final breach height	Inv [m.s.n.m]
Final breach width	B [m]
Left side slope of the breach	LSS [H:V]
Right side slope of the breach	RSS [H:V]
Breach formation time	tf [h]
Height of the water for failure to occur	Init [m.s.n.m]
Discharge coefficient	Cd

Many combinations of these variables can be met in field conditions causing different scenarios. The randomness of the parameter combinations is conditioned by probability density functions, which in turn are characterized by statistical parameters (i.e., mean, standard deviation and mode). Four different distribution functions are considered in McBreach, and the statistical parameters that define them are presented in Table 2. The physical range of operation of each breach parameter is identified based on empirical equations such as those proposed by [7,13] combined with the in situ conditions of the studied dam (e.g., the final breach width cannot exceed the width of the wall crest, which defines the maximum magnitude). The given magnitude for each simulation is provided by a Monte Carlo approach, which through many iterations of complex calculations, obtains a statistically valid result [26]. The minimum number of simulations that allows the statistical convergence of the data (i.e., an optimal sample size) corresponds to 10,000 breach parameter combinations. Thus, McBreach© calculates 10,000 simulations of the HEC-RAS dam breach model for each statistical scenario.

Table 2. Statistical indicators and distribution functions available in McBreach.

Statistical Distribution Functions	Statistical Parameter
Uniform	NA
Triangular	Mode
Normal	Mean and standard deviation
Log-Normal	Mean and mode

The probabilistic analysis results were used to generate a series of scenarios to quantify the uncertainty associated with the combination of the breach parameters and the RMD. This allowed the variability of the generated peak breach flows to be analyzed following the hydraulic and probabilistic model simulations. Figure 1 illustrates the stages used in the uncertainty and sensitivity analyzes, which are treated in detail below.

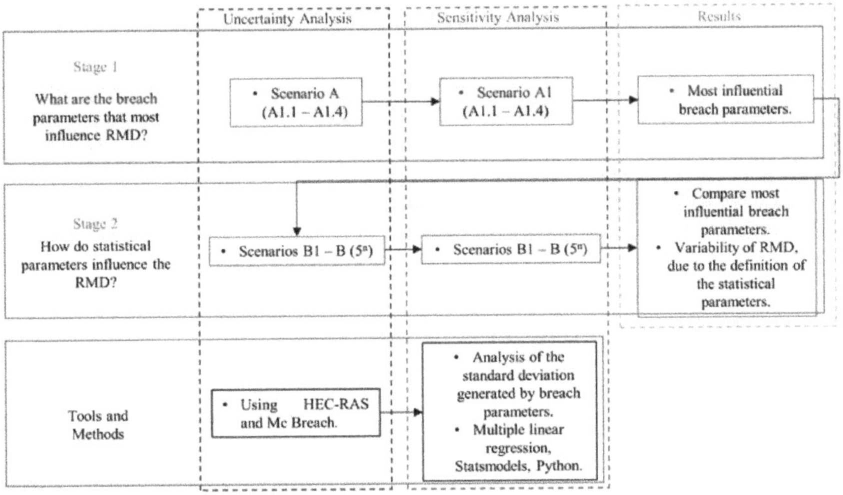

Figure 1. Illustration of the workflow adopted in this paper.

Although research has been conducted on the selection of breach parameters [8–14], the probability distributions that best fit these parameters and the associated uncertainty are generally unknown. Despite this matter, the normal distribution is usually adopted for uncertainty and sensibility analyses [18,27,28]. This work aligns with the former assumption, and Table 3 shows the statistical parameters for the base case.

Table 3. Statistical parameter values for the base case.

PDF	Statistical Parameter	Inv [m.s.n.m]	B [m]	LSS	RSS	Tf [h]	Init [m.s.n.m.]	Cd
Normal	Mean	1272	31.50	0.80	0.80	1.55	1338.85	2.80
	Standard deviation	9.30	7.80	0.26	0.26	0.48	0.11	0.06

Furthermore, six sub-scenarios were established with the base case distribution but with different statistical parameters to analyze the mean and standard deviation sensibility. These sub-scenarios change the mean and the standard deviation by 50% (i.e., +50% of the mean value; −50% of the mean value; +50% of the standard deviation; −50% of the standard deviation; +50% of both the mean value and standard deviation; and −50% of both the mean value and standard deviation). The variability of the RMD against the magnitude of statistical parameters and influential breach variables was also determined. In this case, the number of scenarios was determined by the expression 5^n, where the number 5 represents the sets of statistical parameters varying from a base value and ranging −50% to +50% with 25% intervals (i.e., −50%, −25%, base value, +25%, and 50%). Whereas "n" is the number of breach variables established as probabilistic in McBreach. Therefore, B1 to B(5n) scenarios are generated.

Worthy of mentioning is that the breach parameters—having less influence on the RMD—were assumed with a constant value in McBreach. Consequently, their influence was considered deterministic (their values were fixed by the combination of breach parameters associated with an RMD with an exceedance probability of 50%). An exceedance probability of 50% was selected because the magnitude of the RMD generated by the combination of the breach parameters was found at the middle rather than the upper or lower extremes.

2.2. Case Study: The Chacrillas Reservoir

The defined methodology was applied to the Chacrillas Reservoir in the Andes Mountains of central Chile (70°30′ W, 32°30′ S). Its wall is located 2.5 km upstream of the confluence of the Rocín River and Chalaco Estuary (Figure 2). The delimited reservoir basin has a surface area of 630 km^2. Downstream from the Chacrillas Reservoir wall (Figure 2b), the Rocín River has a 1.8% longitudinal slope and a gravel bed (d_{50} = 14 mm). The most important urban settlement is the town of Putaendo (around 17,000 inhabitants), located 13 km upstream from the confluence of the Putaendo and the Aconcagua Rivers. The average annual precipitation in the basin is 307.7 mm (according to the Resguardo de los Patos station, maintained by the Chilean organization in charge). The hydrological regime is nivo-pluvial with an average annual flow of 1.19 m^3/s. The dam wall is of Concrete Face Rockfill Dam type, and its construction began in 2011 and ended in 2017 [29]. The construction characteristics of the Chacrillas Reservoir wall are presented in Table 4.

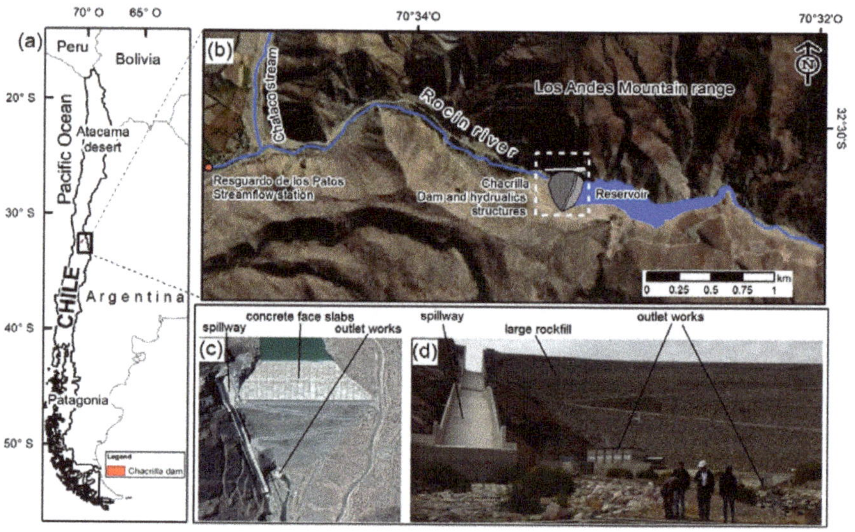

Figure 2. (**a**,**b**) Location of the Chacrillas Reservoir; (**c**) Zoom of the Chacrillas Reservoir with outlet works and concrete face slabs; (**d**) Zoom of the spillway.

Table 4. Characteristics of the Chacrillas dam.

Dam Element	Value	Unit
Crest height	1338.5	m.s.n.m
Crest length	319	m
Crest width	10	m
Dam height	102.5	m
Reservoir capacity	31	Hm^3
Regulation volume	27	Hm^3
Maximum water level	1337.5	m.s.n.m
Minimum water level	1264	m.s.n.m
Upstream slope	1.5:1	H:V
Downstream slope	1.6:1	H:V

A one-dimensional (1D) hydraulic model was created for a non-steady flow, where the hydrograph influencing the reservoir—with a Probable Maximum Flood of 2770 m^3/s—was considered as the upstream boundary condition [21]. Downstream, the geometric characteristics of the Rocín River channel (i.e., the shape of the cross-sections without abrupt changes, stable longitudinal slope, and homogeneous roughness) suggest a downstream boundary condition established by the normal depth. The topography used for the modeling was the same as that used to design the reservoir, contemplating 4.5 km upstream of the dam and up to 1 km downstream (Figure 2b). The spatial discretization adopted was 10 m between cross-sections, which allowed the definition of the calculation time step to satisfy the Courant numerical stability criteria [22].

Table 5 shows the range of magnitudes defined for each breach parameter based on empirical equations proposed by [7–13], and adjusted to the characteristics of the Chacrillas in situ conditions. This configuration was used in each of the simulations conducted with McBreach.

Table 5. Range of breach parameters for the probabilistic model.

Inv [m.s.n.m]	B [m]	LSS	RSS	tf [h]	Init [m.s.n.m]	Cd
1244–1300	8–55	0–1.6	0–1.6	0.1–3.0	1338.82–1339.82	2.6–3.0

3. Results

The uncertainty that exists when determining the magnitude of the RMD was quantified for each combination of breach parameters in each of the sub-scenarios. Figure 3 shows the variability in the magnitude of the peak breach flows with respect to the variation of the statistical parameters that define the distribution (mean and standard deviation). The curve with the largest flows corresponds to scenario A2, where a 50% decrease in the mean value was considered. The curve with the smallest magnitudes in its peak breach flows is scenario A1, corresponding to a 50% increase from the mean. As shown in the A sub-scenarios, the mean has a greater weight than the standard deviation on the results of the RMD calculation.

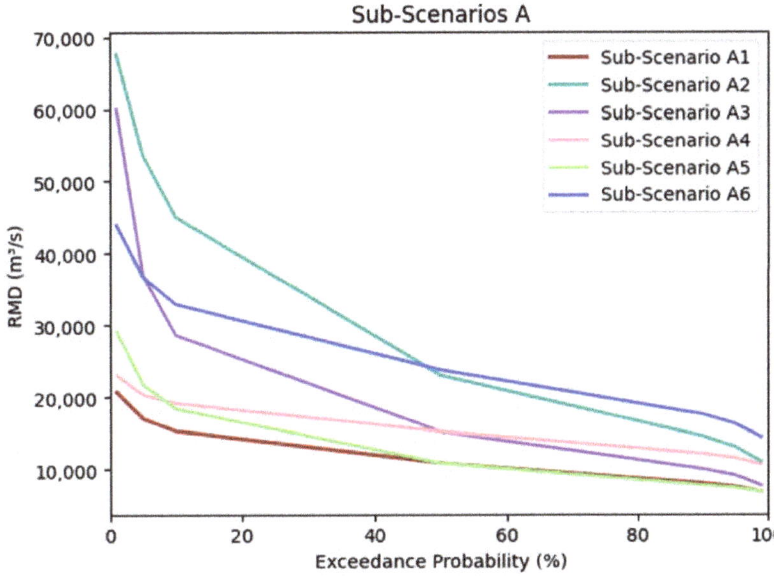

Figure 3. RMD exceedance probability curves for A2 scenarios.

Despite being able to visualize certain variations in the magnitude of the RMD, it is not possible to relate this variability to a specific breach parameter. To identify the parameter(s) that were most relevant to the calculation of the RMD, the sensitivity analysis described below was performed.

3.1. Sensitivity Analysis

Following the scatter plot analysis procedure described in [30], the standard deviations of the peak flows were calculated. In Figure 4, the results of these standard deviations are presented as a function of the breach parameters. It is evident that the breach parameter with the highest standard deviation value and, therefore, the greatest influence on the RMD calculation is the breach formation time (tf). Given that the order of breach parameter importance is different according to the scenario, we could not define at this point which parameters were the most influential with respect to the RMD after tf. For this reason, Multiple Linear Regressions (MLR) were performed for each sub-scenario to understand the variability of the relationship that exists between breach parameters (predictors) and the RMDs (response variable).

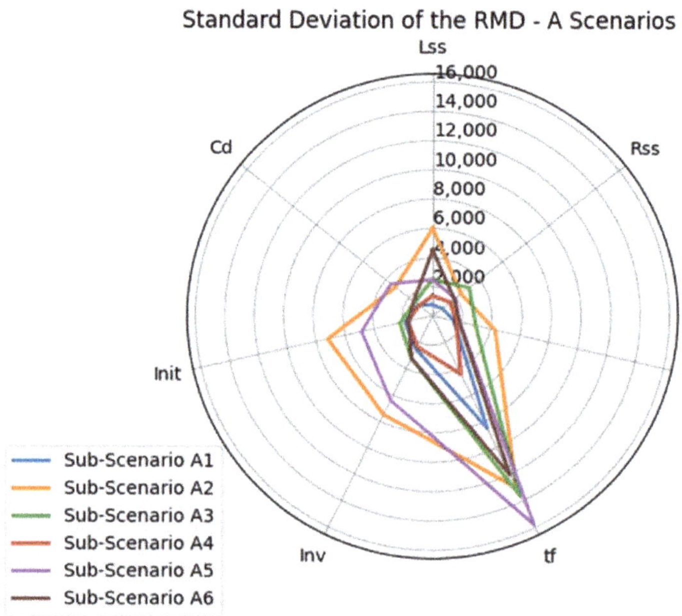

Figure 4. Standard deviation of the RMD produced by each breach parameter for A sub-scenarios.

MLR were performed between 10,000 combinations of simulated breach parameters and their respective RMDs for each scenario using Python's Statsmodels module [31]. The MLR allowed the obtention of several coefficients (Bj), which represent the weight of each parameter compared to the RMD. In Equation (1), the relationship between the coefficients, the breach parameters and the RMD is shown.

$$\begin{aligned} \mathrm{RMD} = \mathrm{Bj}_{\mathrm{Inv}} \times \mathrm{Inv}^* + \mathrm{Bj}_{\mathrm{B}} \times \mathrm{B}^* + \mathrm{B}_{\mathrm{jLSS}} \times \mathrm{LSS}^* + \mathrm{B}_{\mathrm{jRSS}} \times \mathrm{RSS}^* + \mathrm{Bj}_{\mathrm{tf}} \times \mathrm{tf}^* \\ \times \mathrm{Init}^* + \mathrm{Bj}_{\mathrm{Cd}} \times \mathrm{Cd}^* + \mathrm{e} \end{aligned} \tag{1}$$

where the coefficients (Bj) represent each breach parameter, the RMD is measured in m^3/s and "e" represents the residual of the MLR. Worthy of mentioning is that all the breach variables were normalized before fitting Equation (1) (e.g., $\mathrm{Inv}^* = \mathrm{Inv}/\mathrm{Inv}_{\mathrm{max}}$).

The results of the MLR analysis are shown in Figure 5. The values correspond to the coefficients of the predictors of the MLR of each sub-scenario, representing the average change expected for the response variable (RMD) by increasing the predictive variable (breach parameters) by one unit. The graphs confirm that the parameter to which the RMD is most sensitive is tf. In addition, it clearly shows that the most influential breach parameters after tf are Inv and B, respectively, and that the remaining four parameters (LSS, RSS, Init and Cd) have little relevance in determining the RMD.

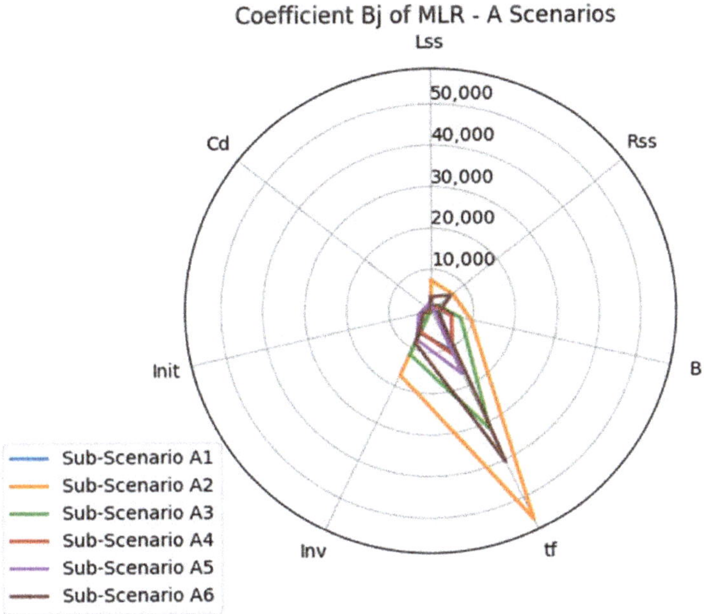

Figure 5. Coefficients of the predictors for the sub-scenarios.

Following these results, a distinction was made between the breach parameters to minimize computational costs while capturing the main dynamics of the process. As only three variables were the most influential ones, they were defined as probabilistic parameters in McBreach. Therefore, 125 scenarios (5^3) were generated. For these three parameters defined as probabilistic, five sets of statistical parameters (mean and standard deviation) were established, as seen in Table 6. The percentage variations were applied to the base condition of the normal distribution presented in Table 3.

Table 6. Stage 2 statistical parameter set.

Most Influential Breach Parameter with Respect to RMD	Mean [%]	Std. Deviation [%]
tf, Inv, B	+50 +25 Base condition −25 −50	+50 +25 Base condition −25 −50

The combinations of breach parameters and statistical parameters (mean and standard deviation) that characterize the normal distributions associated with the 125 scenarios are presented in the supplementary material (Table S1: Combinations of statistical parameters. Table S2: Combinations of statistical and deterministic parameters. Table S3: Combinations of breach parameters).

Consequently, since the LSS, RSS, Init and Cd parameters were found to have a low influence on the RMD, they were considered deterministic and were defined based on the combination of breach parameters which generated an RMD with an exceedance probability of 50% (Table 7).

Table 7. Combination of breach parameters associated with an RMD with an exceedance probability of 50% for sub-scenario A1.3.

Exc. P 50% Sub-Scenario A1.3	Inv [m.s.n.m]	B [m]	Tf [h]	LSS	RSS	Init [m.s.n.m]	Cd	Qmax [m^3/s]
	1272.39	32.06	1.91	1.07	0.66	1338.82	2.83	11,575

3.2. Uncertainty Analysis

The exceedance probability curves for the RMD were obtained from the results of the 125 simulations in McBreach (Figure 6). This allowed the uncertainty associated with using the probabilistic parameters that generated these flows to be quantified. The results showed that for the exceedance probability of 99%, the highest and lowest flows obtained correspond to approximately 17,000 m^3/s and 6000 m^3/s, respectively. Furthermore, the variation is almost three times greater due to the value defined for the mean and standard deviation, demonstrating the enormous importance of selecting these probabilistic parameters. For the exceedance probability of 1%, the highest and lowest flows reached an approximate magnitude of 60,000 m^3/s and 22,000 m^3/s, respectively, maintaining the same variation indicated above (i.e., almost three times). The curve that represents the most extraordinary magnitude flows (i.e., Max in Figure 6) corresponds to sub-scenario B121 (variation of −50% of mean and standard deviation to the base case for tf and Inv, and +50% of mean and standard deviation to the base case). Sub-scenario B5 had the lowest mean and standard deviation value of tf and Inv of the entire set of sub-scenarios and the highest mean and standard deviation of B for the entire simulated set. This is coherent since the RMD is inversely proportional to tf and Inv, and directly proportional to B. The curve representing the lowest magnitudes of RMD (i.e., Min in Figure 6) corresponds to sub-scenario B5, which is the opposite sub-scenario of B121. Sub-scenario B5 has a variation of +50% of mean and standard deviation to the base case for tf and Inv, and −50% of mean and standard deviation to the base case for B; that is, it has the highest mean and standard deviation values for tf and Inv, and the lowest mean and standard deviation values for B.

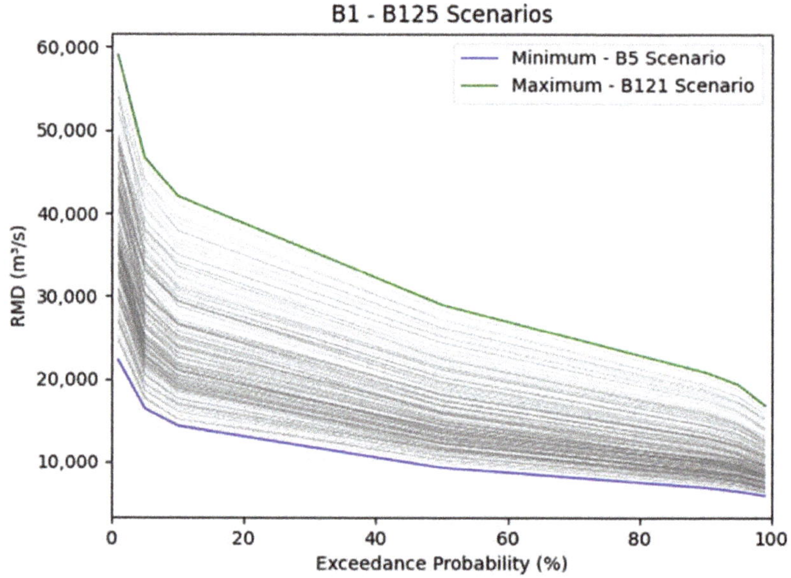

Figure 6. Exceedance probability curves and RMD for scenarios B1 to B125.

Figure 7 shows the calculated standard deviations of the breach parameters tf, Inv and B (Figure 7a), and the coefficient values of the predictors for the MLR (Figure 7b) for the 125 sub-scenarios of Stage 2. Regardless of the simulated scenario, the highest standard deviation of the RMD always corresponded to the parameter tf, and its magnitude was also much higher than that of the breach parameters Inv and B. The results of the MLR, which are expressed through the predictive coefficients Bj, clearly show that tf is the breach parameter that has the most significant influence on the RMD calculation. The parameter that has the second most influence in determining the RMD corresponds to Inv, while the third most influential parameter is B.

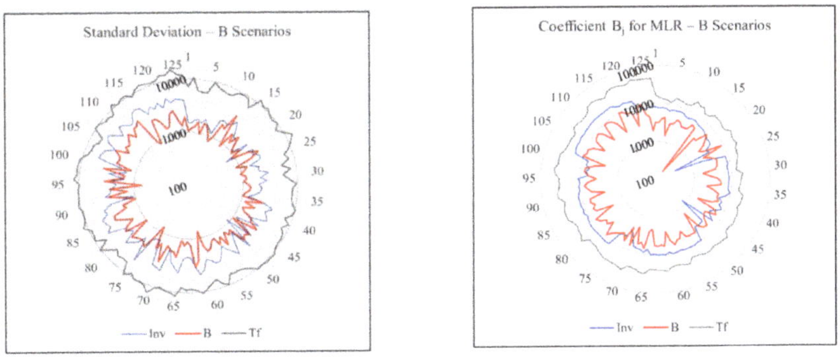

(**a**) (**b**)

Figure 7. (**a**) Standard deviation of the RMD produced by each breach parameter for B sub-scenarios; (**b**) Coefficients of the predictors for the B sub-scenarios.

After determining the order of the breach parameters with the most significant weight in the RMD calculation, the relationship between the magnitude of the statistical parameters (mean and standard deviation) with respect to the RMD was analyzed. For the analysis, the extreme values of the RMD were used (exceedance probabilities of 99% and 1%) so that all possible variations of the RMD were captured. Figure 8 shows the results of this analysis for the exceedance probabilities of 99% and 1%, as well as the means of the parameters tf, Inv and B (Figure 8a,b, respectively).

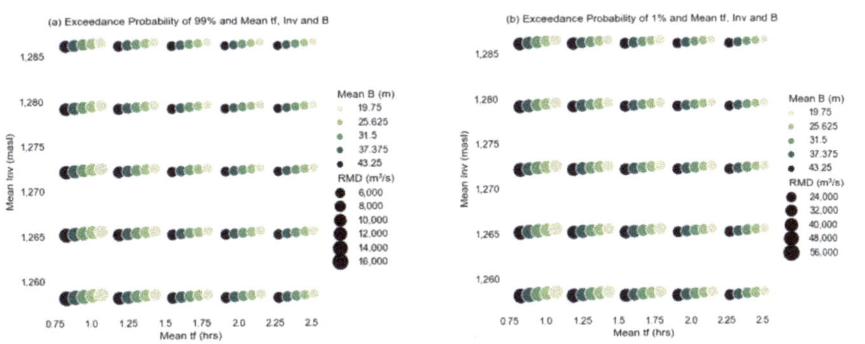

Figure 8. (**a**) Exceedance probability of 99% and mean tf, Inv and B; (**b**) Exceedance probability of 1% and mean tf, Inv and B.

Figure 8a,b show the relationship between the proportionality of the breach parameters (tf, Inv and B) indicated above and the definition of the statistical parameters (the mean, in this case). The set of results in the lower-left corner represents the RMDs with the highest magnitudes of the 125 scenarios. In contrast, the set of results in the upper right corner of the graphs represents the opposite, corroborate what was expected.

As a function of the variation in the magnitude of the RMD, we can see that the Inv parameter is more influential than B, since the proportion by which the diameter of the circumference decreases as Inv increases is greater than compared to B. Continuing with the same analysis, the mean of the parameter tf changes along the horizontal axis of the graph, reaching the greatest magnitude flows with the lowest value of tf, with the magnitude decreasing as the mean increases. The radius of the circumference, which represents the proportion by which the magnitude of each flow decreases, is much greater than that displayed for Inv and B, corroborating that tf is the most influential of the three parameters.

For the graphs of each standard deviation associated with the flow magnitudes, the same results described above were obtained since the variation of the set of parameters considered the mean and the standard deviation as the same data set. Therefore, the variation results in the same percentage since the same flow magnitudes would be plotted with the same distribution as in Figure 8 but with the standard deviation values instead of the mean.

4. Discussion and Conclusions

The uncertainty associated with the magnitude of the breach hydrograph resulting rupture maximum discharge (RMD) was clearly shown in the results obtained in Stage 2, where for the same distribution analysis, a difference of three times was obtained for the exceedance probabilities of 99% and 1%. These significant differences between RMD values are mainly due to the selection of the statistical parameters that in turn define the breach parameters when a Monte Carlo approach is used.

This means that studying and defining the breach parameters of the hydraulic model in HEC-RAS in the best possible way is not enough, and the definition of the statistical parameters should also be a focus of analysis. These findings have significant implications in the operational dimension, regardless of the sophistication of the hydraulic model used. Among these implications are the estimated risks associated with the breach, flood zone determination, scaling of projects and the preparation of emergency action plans (EAP). In the case of this study site, for the same distribution function and an exceedance probability of 99%, the RMD values ranged from 6000 m^3/s to 17,000 m^3/s. The variations in the flow magnitudes indicated above directly affect the analysis of the breach hydrograph, which conditions the flood downstream, affecting the wave travel time, the depths and velocities, and thus the flooded areas.

For the sensitivity analyzes of all the simulated scenarios, it was possible to establish that the most influential breach parameter in the generation of the RMD corresponds to the breach formation time (tf), which showed magnitudes much higher than the rest of the parameters in both its standard deviation and the MLR coefficient. Therefore, since tf is the most heavily weighted breach parameter in the RMD calculation, its accurate determination is essential. Secondly, the breach parameters Inv and B have a significant degree of sensitivity. The parameter Inv is more influential on average than B, so the numerical definition should also be considered an essential focus of analysis. The variations produced by the rest of the parameters (LSS, RSS, Init and Cd) are not significant as they have minimal impact on the RMD.

After understanding the influence of the parameters tf, Inv and B, it was possible to corroborate what was expected for the proportionality of these breach parameters with respect to the breach flow generated, where the parameters tf and Inv behave in an inversely proportional manner, and parameter B is directly proportional to the variation of the RMD. Therefore, based on the above, the highest flow for any exceedance probability is associated with the scenario with the lowest mean and standard deviation for tf and Inv, the highest mean and standard deviation for B, and vice versa.

Finally, it should be noted that the conclusions established in this investigation are specific to the study site analyzed, corresponding to the Chacrillas Reservoir. However, the methodology could be adapted and replicated in other reservoirs of interest, and thus determine whether the behavior is similar to that presented in this study. In the future, it would also be advantageous to compile sensitivity analyses of breach parameters for different dams to relate the variability of the influence of these breach parameters with respect to the characteristics of each study site (e.g., valley slope, wall height, available reservoir volume, roughness, among others). Finally, with respect to the probability density functions and the statistical parameters that are used for probabilistic analysis, it is important to add and work with other density functions than those already available in McBreach.

Supplementary Materials: The following supporting information can be downloaded at https://www.mdpi.com/article/10.3390/w14111776/s1. Table S1: Combinations of statistical parameters. Table S2: Combinations of statistical and deterministic parameters. Table S3: Combinations of breach parameters.

Author Contributions: Conceptualization, H.A. and D.C.; software, D.B.; formal analysis, D.B.; writing—original draft preparation, review and editing, D.B., D.C., H.A. and A.P. All authors have read and agreed to the published version of the manuscript.

Funding: This research was partially funded by Universidad Diego Portales through the internal project "Bridge Erosion Critical Modelling" (grant No. 1100327026), and the project ANID InES Ciencia Abierta INCA210005.

Institutional Review Board Statement: Not applicable.

Informed Consent Statement: Not applicable.

Acknowledgments: Hydraulic Works Department of Chile (Dirección de Obras Hidráulicas MOP) and the project ANID InES Ciencia Abierta INCA210005.

Conflicts of Interest: The authors declare no conflict of interest.

References

1. Álvarez, M.; Puertas, J.; Peña, E.; Bermúdez, M. Two-Dimensional Dam-Break Flood Analysis in Data-Scarce Regions: The Case Study of Chipembe Dam, Mozambique. *Water* **2017**, *9*, 432. [CrossRef]
2. Xu, Y.; Zhang, L.; Jia, J. Lessons from catastrophic dam failures in August 1975 in Zhumadian, China. In Proceedings of the Geosustainability and Geohazard Mitigation, New Orleans, LA, USA, 9–12 March 2008; pp. 162–169. [CrossRef]
3. FEMA. *Fedearl Guidelines for Inundation Mapping of Flood Risks Associated with Dam Incidents and Failures*, 5th ed.; FEMA P-946/July 2013; Technical Report Federal Emergency Management Agency: Washington, DC, USA, 2013.
4. Dhiman, S.; Patra, K.C. Experimental study of embankment breach based on its construction parameters. *Nat. Hazards Earth Syst. Sci. Discuss.* **2017**, *2017*, 1–16. [CrossRef]
5. Godell, C.; Johnson, D.; Raeburn, R.; Monk, S.; Karki, A.; Lee, A. Probabilistic dam breach modelling using HEC-RAS and McBreach; United States Society on Dams. In Proceedings of the 38th Annual Conference and Exhibition, Miami, FL, USA, 4 April 2018.
6. Basheer, T.A.; Wayayok, A.; Yusuf, B.; Kamal, R. Dam breach parameters and their influence on flood hydrographs for mosul dam. *J. Eng. Sci. Technol.* **2017**, *12*, 2896–2908.
7. Wahl, T.L. Uncertainty of Predictions of Embankment Dam Breach Parameters. *J. Hydraul. Eng.* **2004**, *130*, 389–397. [CrossRef]
8. Kirkpatrick, G.W. Evaluation guidelines for spillway adequacy. The evaluation of dam safety. In Proceedings of the Engineering Foundation Conference ASCE, New York, NY, USA, 2–5 May 1977; pp. 395–414.
9. U.S. Bureau of Reclamation. *Guidelines for Defining Inundated Areas Downstream from Bureau of Reclamation Dams*; Reclamation Planning Instruction Report No. 82-11; U.S. Bureau of Reclamation: Washington, DC, USA, 1982.
10. Singh, K.P.; Snorrason, A. Sensitivity of outflow peaks and flood stages to the selection of dam breach parameters and simulation models. *J. Hydrol.* **1984**, *68*, 295–310. [CrossRef]
11. Froehlich, D.C. Peak Outflow from Breached Embankment Dam. *J. Water Resour. Plan. Manag.* **1995**, *121*, 90–97. [CrossRef]
12. Macdonald, T.C.; Langridge-Monopolis, J. Breaching Characteristics of Dam Failures. *J. Hydraul. Eng.* **1984**, *110*, 567–586. [CrossRef]
13. Pierce, M.W.; Thornton, C.I.; Asce, M.; Abt, S.R.; Asce, F. Predicting Peak Outflow from Breached Embankment Dams. *J. Hydrol. Eng.* **2010**, *15*, 338–349. [CrossRef]
14. Gaagai, A.; Aouissi, H.A.; Krauklis, A.E.; Burlakovs, J.; Athamena, A.; Zekker, I.; Boudoukha, A.; Benaabidate, L.; Chenchouni, H. Modeling and Risk Analysis of Dam-Break Flooding in a Semi-Arid Montane Watershed: A Case Study of the Yabous Dam, Northeastern Algeria. *Water* **2022**, *14*, 767. [CrossRef]

15. Wahl, T.L. *Prediction of Embankment Dam Breach Parameters a Literature Review and Needs Assessment*; Dam Safety Research Report; Bureau of Reclamation Dam Safety Office, U.S. Department of Interior: Washington, DC, USA, 1998. Available online: https://www.nrc.gov/docs/ML0901/ML090150051.pdf (accessed on 6 May 2021).
16. Tsai, C.W.; Yeh, J.-J.; Huang, C.-H. Development of probabilistic inundation mapping for dam failure induced floods. *Stoch. Hydrol. Hydraul.* **2019**, *33*, 91–110. [CrossRef]
17. Moglen, G.E.; Hood, K.; Hromadka, T.V. Examination of Multiple Predictive Approaches for Estimating Dam Breach Peak Discharges. *J. Hydrol. Eng.* **2019**, *24*, 04018065. [CrossRef]
18. Goodell, C. An Approach to Dam Breach Modeling. In Proceedings of the FloodRisk 2012 Conference, Rotterdam, The Netherlands, 20–22 November 2012. [CrossRef]
19. Dirección General de Obras Hidráulicas y Calidad de Aguas. *Guía Técnica para la Elaboración de los Planes de Emergencia de Presas*; Ministerio del Medio Ambiente España: Madrid, Spanish, 2021. Available online: https://www.miteco.gob.es/es/agua/publicaciones/ (accessed on 26 May 2021). (In Spanish)
20. Zhou, X.; Zhou, J.; Du, X.; Li, S. Study on dam risk classification in China. *Water Supply* **2014**, *15*, 483–489. [CrossRef]
21. Bello, D.; Alcayaga, H.; Caamaño, D. *Modelación Hidráulica de Rotura del Embalse Chacrillas y Evaluación de la Amenaza Asociada*; Repositorio UDP; Universidad Diego Portales: Santiago, Chile, 2020.
22. USACE. *HEC-RAS Hydraulic Reference Manual*; Version 6.0; U.S. Army Corps of Engineers, Hydrologic Engineering Center: Davis, CA, USA, 2020. Available online: https://www.hec.usace.army.mil/publications/ (accessed on 26 May 2021).
23. Kleinschmidt. Probabilistic Dam Breach Modeling User's Manual—McBreach. 2019. Available online: https://www.kleinschmidtgroup.com/mcbreach/ (accessed on 26 May 2021).
24. International Commission on Large Dams. World Register of Dams. 2011. Available online: http://www.icold-cigb.org (accessed on 22 February 2020).
25. Bellos, V.; Tsakiris, V.K.; Kopsiaftis, G.; Tsakiris, G. Propagating Dam Breach Parametric Uncertainty in a River Reach Using the HEC-RAS Software. *Hydrology* **2020**, *7*, 72. [CrossRef]
26. Yanmaz, A.M.; Gunindi, M.E. Assessment of overtopping reliability and benefits of a flood detention dam. *Can. J. Civ. Eng.* **2008**, *35*, 1177–1182. [CrossRef]
27. Peter, S.J.; Siviglia, A.; Nagel, J.; Marelli, S.; Boes, R.M.; Vetsch, D.; Sudret, B. Development of Probabilistic Dam Breach Model Using Bayesian Inference. *Water Resour. Res.* **2018**, *54*, 4376–4400. [CrossRef]
28. Ahmadisharaf, E.; Kalyanapu, A.J.; Thames, B.A.; Lillywhite, J. A probabilistic framework for comparison of dam breach parameters and outflow hydrograph generated by different empirical prediction methods. *Environ. Model. Softw.* **2016**, *86*, 248–263. [CrossRef]
29. Janke, J.R.; Ng, S.; Bellisario, A. An inventory and estimate of water stored in firn fields, glaciers, debris-covered glaciers, and rock glaciers in the Aconcagua River Basin, Chile, Geomorphology. *Geomorphology* **2017**, *296*, 142–152. [CrossRef]
30. Saltelli, A.; Ratto, M.; Campolongo, F.; Cariboni, J.; Gatelli, D.; Saisana, M.; Tarantola, S. (Eds.) Methods and Settings for Sensitivity Analysis—An Introduction. In *Global Sensitivity Analysis. The Primer*; John Wiley & Sons, Ltda: Hoboken, NJ, USA, 2008; pp. 16–23.
31. Seabold, S.; Perktold, J. Statsmodels: Econometric and Statistical Modeling with Python. In Proceedings of the 9th Python in Science Conference, Austin, TX, USA, 28 June–3 July 2010; pp. 92–96. [CrossRef]

water

MDPI

Article

Probabilistic Analysis of Floods from Tailings Dam Failures: A Method to Analyze the Impact of Rheological Parameters on the HEC-RAS Bingham and Herschel-Bulkley Models

Malena Melo [1] and Julian Eleutério [2,*]

[1] Post-Graduation in Sanitation, Environment and Water Resources, Federal University of Minas Gerais, Belo Horizonte 31270-901, Brazil; malena_melo@outlook.com
[2] Department of Hydraulic and Water Resources Engineering, Federal University of Minas Gerais, Belo Horizonte 31270-901, Brazil
* Correspondence: julian.eleuterio@ehr.ufmg.br

Abstract: The difficulty in determining the rheological characteristics of tailings inside reservoirs as well as their intrinsic variability adds uncertainty to tailings dam failures in flood studies. Uncertainty propagation in non-Newtonian hydrodynamic models stands as a great scientific challenge. This article explores the sensibility of tailings dam breach flood mapping to rheological parameters in Bingham and Herschel-Bulkley (H-B) models. The developed approach was based on the probabilistic Latin Hypercube Sampling of rheological parameters. It was automated to propagate uncertainty throughout multiple hydrodynamic simulations using the HEC-RAS v.6.1 software. Rheological parameter ranges and distributions were based on a broad bibliographic review. Bingham models were revealed to be more sensitive than H-B in terms of simulated min-max area values: for Bingham, flood areas, maximum depths, and arrival times varied by 17.9%, 9.3%, and 8.2%, respectively; for H-B, variations were 25.7%, 5.1%, and 3.9%. However, Bingham was less sensitive in terms of hydrodynamically associated probability: high probability ratios were related to a small range of simulated areas in Bingham, while H-B presented great variability. Finally, for each model, the parameters that affect uncertainty the most were identified, reinforcing the importance of determining them properly. Furthermore, the identified parameter ranges for both models should be valuable for defining variable value boundaries for flood sensitivity tests on specific tailings materials for other case studies. The automated algorithm can be used or adapted for specific tests with other hydrodynamic simulations.

Keywords: non-Newtonian fluid; hyperconcentrated flow; mudflow; dam-break study; flood-risk assessment; uncertainty boundaries; automated computational routine

Citation: Melo, M.; Eleutério, J. Probabilistic Analysis of Floods from Tailings Dam Failures: A Method to Analyze the Impact of Rheological Parameters on the HEC-RAS Bingham and Herschel-Bulkley Models. *Water* **2023**, *15*, 2866. https://doi.org/10.3390/w15162866

Academic Editors: Francesca Aureli, Andrea Maranzoni and Gabriella Petaccia

Received: 1 July 2023
Revised: 31 July 2023
Accepted: 3 August 2023
Published: 8 August 2023

1. Introduction

Failures of tailings dams occur at a significantly higher frequency compared to those of water storage dams [1]. Recent incidents have highlighted the immense destructive potential associated with tailings inundations [2,3]. Forecasting and mapping flood hydrodynamics are crucial for implementing effective risk mitigation measures. Different uncertainties may impact flood risk assessments. During the last decade, the simulation of floods resulting from tailings dam failures has commonly been performed using Newtonian models. However, the tailings comprise a combination of fine materials and water, and depending on the proportions in terms of solid concentration, their properties can resemble non-Newtonian fluids (hyper-concentrated), which brings uncertainty to dam-break studies.

Rheology investigated the flow and deformation behavior of fluids in different contexts [4,5]. Rheological parameters are obtained and analyzed through rheological tests conducted on various materials over time [6–8]. Several factors can influence the variability of rheological parameters, including shear rate, solid concentration, pH, and temperature [9]. The effects of physical and geotechnical parameters on rheological properties were also evaluated [10]. Different models have been employed for simulating hyper-concentrated flows for decades [11–13].

The term "hyper-concentrated flow" commonly refers to natural mudflows or debris flows [14]. Numerous investigations have delved into the influence of rheological parameters on the natural flow of mud, particularly debris flow [15–22]. These studies yielded good results when compared to recorded events. However, for predictive applications, it is widely recognized that the difficulties in determining representative rheological parameters, along with their considerable spatial variability, remain significant sources of uncertainty in this type of model [17,23,24]. Notably, the authors in [24] conducted a sensitivity analysis of non-Newtonian hydrodynamic modeling of debris flow with regards to rheological parameters, highlighting the relevance of these parameters in hydrodynamics.

Over the course of time, numerous in-depth investigations have been undertaken to explore various dimensions of this uncertainty. Specifically, the author in [25] investigated the uncertainties inherent in physical and rheological parameters during the numerical simulation of non-Newtonian flow phenomena. The authors in [26] focused on the examination of uncertainties pertaining to input parameters employed in the modeling of natural soil landslides within steep terrains. Moreover, the authors in [27] evaluated the parameter variability concerning the uncertainties associated with rheological characterization. These studies highlighted the relevance of rheological characterization for hydrodynamic analyses.

Sensitivity analyses have also been applied in the study of rheology, with methods such as the Monte Carlo Method (MCM) being used to investigate various scenarios, including mudflow and natural debris events [26,28] and pipeline and pump flows associated with tailings transportation [29]. The studies encompassed the utilization of Bingham models [26,28,30] and the Voellmy model [26]. In contrast, the authors in [24] uniquely conducted a global sensitivity analysis of the Quadratic model in FLO-2D software, employing Latin hypercube sampling (LHS), to evaluate the impact of rheological parameters on natural mudflows.

Despite the numerous studies that have focused on these parameters in the context of debris flows, limited attention has been given to their investigation in the context of tailings dam failures. Studies that performed sensitivity assessments of the rheological parameters linked to tailings dam failure, the specific subject of this article, have not been found. Despite the potential similarity in behavior between flood wave propagation in tailings dam failures and debris flows, there is currently a lack of studies specifically focusing on the propagation of uncertainties in the hydrodynamic modeling of tailings dam failure floods. In this context, the significance of the pronounced variability of rheological parameters is underscored among the challenges associated with the modeling of tailings dam failure floods. With the purpose of filling this scientific gap, this article explores rheology uncertainty in the context of dam-break studies.

Currently, several hydrodynamic models allow for the simulation of hyper-concentrated flow and are already being used for dam-break studies, such as FLO-2D [15,24,31,32], RIVERFLOW2D [33], and FLWAV [34]. Recently, the widely used and freely available program HEC-RAS has also incorporated non-Newtonian flow modeling. However, sensitivity studies of HEC-RAS to rheological parameters in the modeling of hyper-concentrated flow, particularly for tailings dam failures, have not been performed.

In order to assess the effects of rheology uncertainty and to explore the applicability of the HEC-RAS model for practical purposes related to mapping floods with hyper-concentrated flow, this study seeks to address the following question: How does the rheological variability of stored materials impact the results of modeling floods caused by mining tailings dam failures? To address this question, this article presents a probabilistic method to analyze the sensitivity of flood maps to input parameters of the HEC-RAS Bingham [35] and Herschel-Bulkley [36] rheological models, which are the most adapted rheological models for hyper-concentrated flow simulations [9,27,37]. Within this context, the developed research compares the influence of rheological characterization in both models, raising the question of the suitability of these models for dam-break analysis regarding uncertainty boundaries. For this purpose, ranges of variation for rheological parameters were explored based on an extensive literature review and through a case study probabilistic analysis. Thus, this study demonstrates the importance of rheological characterization in dam-break studies and introduces a probabilistic method to assist in the analysis of rheological parameter uncertainties due to the heterogeneity of tailings materials.

2. Materials and Methods

The sensitivity analysis proposed in this research is based on four main steps: (1) definition of test parameters; (2) literature review of parameter ranges and probabilistic sampling; (3) automated downstream propagation simulations of the rupture wave; (4) compilation and analysis of results.

The Bingham and Herschel-Bulkley models were tested, and the Latin Hypercube Sampling (LHS) technique was used to randomly sample the rheological parameters under investigation based on a comprehensive literature review. The sensitivity analysis is applicable to any deterministic hydrodynamic model developed using the HEC-RAS framework, that encompasses the propagation of wave rupture in tailings dam scenarios. Similar to the existing approach [24], the impact on flood extent and maximum depth was also assessed. Additionally, the analysis encompassed the examination of flood wave arrival times, considering the significant importance of this parameter in studies concerning impacts and loss of life [38,39]. The automated repetition of these steps is determined by the parameters being tested and the sampling approach. The developed sensitivity analysis was applied to the conceptual case study conceived and analyzed in [40], proposed by the International Commission on Large Dams (ICOLD).

2.1. Mathematical Rheological Models and Parameterization

Given the nonlinear relationships between shear stress and strain rate, several mathematical models were developed to represent the behavior of non-Newtonian fluids [14,35,36,41,42]. Laboratory experimental results [43] cited in the study [4] confirmed that under deformation rates observed in the field, fluids with high concentrations of fine particles exhibit plastic behavior resembling Bingham fluids. In this context, the rheological models commonly used to represent the flow resulting from the rupture of ore tailings dams are the Bingham and Herschel-Bulkley models [9,24,27,37].

The Bingham model, applied to plastic fluids, was developed in 1922 based on its conceptualization of the yield limit phenomenon, where a residual shear stress value must be exceeded for the material to flow. The Bingham plastic model tends to remain undeformed under the application of small shear stresses [44], represented by the sum of yield and viscous stresses, as given by Equations (1) and (2) [35].

$$\tau = \tau_y + \mu \frac{dv_x}{dz}, \; \tau > \tau_y \tag{1}$$

$$\frac{dv_x}{dz} = 0, \; \tau < \tau_y \tag{2}$$

where τ is the shear stress; τ_y is the yield stress or limit stress; μ is the dynamic viscosity of the mixture or dynamic viscosity; and dv_x/dz is the strain rate or velocity gradient.

Pseudoplastic fluids with yield stress can be represented by the Herschel-Bulkley model. This model, also known as the "generalized Bingham model", was developed based on the Bingham and Power-Law models, aiming to incorporate the non-linear flow curve for stresses higher than the yield stress τ_y [45]. In this model, the shear stress and strain rate exhibit a non-linear relationship, serving as a generalized proposition of the Bingham plastic fluid. This relationship is described by Equations (3) and (4), indicating the sum of yield and viscous/turbulent stresses [46].

$$\tau = \tau_y + K\left(\frac{dv_x}{dz}\right)^n, \ \tau > \tau_y \tag{3}$$

$$\frac{dv_x}{dz} = 0, \ \tau < \tau_y \tag{4}$$

The relationship between shear stress and strain rate depends on the dimensionless exponent n, which is specific for each fluid and determined experimentally [44]. This coefficient is the same as the behavior index in the Power Law model. In addition to the flow behavior index n, the consistency index K in the Herschel-Bulkley model is also an empirical parameter [6].

HEC-RAS has four rheological models in its structure [46]: the Bingham model, which is standard for hyperconcentrated flows; the O'Brien (Quadratic) and Generalized Herschel-Bulkley models for mud and debris flows; and the Clastic-Grain Flow model (Mohr-Coloumb geotechnical model) for clastic flows.

In this research, the test parameters were defined according to the input data of the non-Newtonian fluid modeling in HEC-RAS for the Bingham and Herschel-Bulkley rheological models. The following parameters were defined for the Bingham model: volumetric concentration of solids (C_v); maximum volumetric concentration of solids (Max C_v); and the coefficients *a* and *b* of the exponential yield stress curve. These parameters were determined in view of the predefined HEC-RAS approaches. For the Herschel-Bulkley model, the following parameters were defined: the yield stress (τ_y), the consistency index (K), and the behavior index (n). In this case, the predefined approach of the exponential curve, i.e., the calibration parameters *a* and *b*, was not used since these parameters are usually adjusted based on the yield stress (τ_y) of the Bingham model. Thus, Herschel-Bulkley yield stress (τ_y) values obtained from previous studies were used.

2.2. Parameter Intervals and Sampling

To conduct the sensitivity tests in this study, parameter value ranges for the equations were determined based on an extensive literature review related to mining tailings [2,7–10,14,27,37,38,47–53]. The tailings studied by these authors included coal, copper, iron, lithium, base metal, nickel, gold, pyrophyllite, uranium, and zinc, as well as synthetic and theoretical materials, and the obtained values are listed in Appendix A. Although there are limited studies on tailings rheology analysis, the identified intervals served the purpose of constituting a sample space that allowed for the variation of values and the evaluation of their influence.

The establishment of physically plausible value ranges for the rheological parameters of interest was performed (Table 1) based on the provided literature review (Appendix A).

Table 1. Value ranges of rheological parameters defined for the sensitivity assessment of HEC-RAS using Bingham and Herschel-Bulkley models.

Rheological Model	Varied Parameters	Literature Value Ranges	Adopted Interval
Bingham	C_v (%)	9.3–65.0	20.0–60.0
	Max C_v (%)	49.9–76.5	49.0–61.5
	a (Pa)	0.0000001–3.45	0.067–3.450
	b	1.2–40.0	1.2–10.359
Herschel-Bulkley (H-B)	C_v (%)	20.0–60.0	
	τ_y (Pa)	0.6–251.9	
	K (Pa·s^n)	0.008–130.0	
	n	0.192–1.5	

In the case of the Bingham model, the parameters C_V, Max C_V, *a*, and *b* were subject to variation. The decision to employ independent random sampling for these parameters is justified by the inherent limitation of HEC-RAS, which utilizes a fixed C_V value. Consequently, the resulting values of yield stress (τ_y) and dynamic viscosity (μ) remain constant when C_V is employed in their calculation. Additionally, to ensure physically meaningful parameter combinations, the value ranges and relationships between the parameters were adjusted to yield τ_y and μ values within the ranges documented in the literature (Table 1), following a similar approach [24].

To encompass the classification of flow types proposed by the authors [54], the interval for solid volume concentration (C_V) in the context of hyperconcentrated flow was defined as 20% to 60%. This interval was chosen to cover a range of flow conditions, including floods, mudflows, and hyperconcentrated flows. Based on the literature, the study [54] suggested a C_V range of 20% to 50% for floods and mudflows, while [54] proposed a C_V range of 30% to 60% specifically for hyperconcentrated flows.

In relation to the maximum volumetric concentration parameter (Max C_V), the chosen interval considered a maximum value of 61.5%, which corresponds to the value reported in [53]. This value is pre-defined in HEC-RAS [46], and since it exceeds the upper limit of the adopted interval for solid volume concentration (C_V of 60%), it was established as the maximum value for the Max C_V interval.

The values of C_V and Max C_V were independently sampled, along with the parameters *a* and *b* in the Bingham model. Subsequently, the sampled values of Max C_V were reorganized to ensure that within each parameter set, the value of Max C_V was always greater than C_V. Another criterion in combining these parameters was to adhere to the limits of dynamic viscosity (μ) reported in the literature, with a minimum μ of 0.002 Pa·s [9] and a maximum μ of 100.0 Pa·s [38]. In brief, the Max C_V samples were rearranged, based on the C_V samples, in such a way that the relationship between the parameters satisfied two criteria: Max $C_V > C_V$, and the relationship between Max C_V and C_V in the calculation of μ using the Maron and Pierce method [46] yielded values within the range of 0.002–100.0 Pa·s.

The values of *a* and *b* in the yield stress curve were adjusted to ensure that the combination of parameters yielded values within the range of 0.085 Pa [49] to 1726.33 Pa [48]. The value of 2396.53 Pa presented in Appendix A as the maximum value from the study [48] is related to a C_V of 65% in the slump test. In accordance with the C_V range of 20% to 60% adopted in this study, the value of 1726.33 Pa was considered the maximum value for τ_y, also obtained in the slump test for a C_V of 60%.

The parameters that varied in the Herschel-Bulkley (H-B) model were C_V, τ_y, *K*, and *n*. For this test, the ranges defined in the literature review (Table 2) were maintained, as the parameters *K* and *n* are calibration parameters of the H-B model and do not have a direct relationship that allows for evaluating or limiting their ranges. The range of solid volume concentration (C_V) was kept from 20% to 60% [14,54].

Similarly, the range of τ_y values in the H-B model remained the same as observed in previous studies. Although they share the same name, the yield stresses of the Bingham and H-B models have different values for the same fluid. Thus, the adopted yield stress range considers only the values adjusted according to the H-B model.

The random sampling of parameters using Latin Hypercube Sampling (LHS) was performed independently with the aim of analyzing the sensitivity of HEC-RAS to the parameters within their variability intervals. A total of 1000 parameter sets were selected for the simulations of each model, as the random sampling conducted by LHS allows for a reduced number of samples compared to the Monte Carlo method [55]. To achieve convergence in the Monte Carlo simulation, other authors have employed 5000 [26,29] and 10,000 [25] realizations, respectively. LHS is a statistical method [56] that generates a quasi-random sample of parameter values from a multidimensional distribution. In LHS, each variable set is divided into equal intervals, and one value is obtained for each stratum [56]. Assuming these variables are independent, the sample sets are randomly combined with each other, forming a multidimensional sample [56].

To obtain a set of samples generated by the LHS, it is necessary to adopt a probability distribution for each parameter to be varied. Few studies have evaluated the best-fit distributions for the variations in rheological parameter values. The research in [25] and the more recent work [30] considered that the dynamic viscosity (μ) follows a normal distribution. The probability distribution of the volumetric concentration of solid particles (C_v) in tailings was compared to the normal distribution using measured data [57]. The authors in [27] fitted normal, lognormal, gamma, and beta statistical distributions for rheological parameter sets. Due to the scarcity of studies and data used in fitting probability distributions in the literature for these parameters, the approach in [24] was adopted in this paper, which used a uniform distribution to account for the variation of all rheological parameters tested.

2.3. Automation of Sensitivity Analysis in HEC-RAS

To conduct probabilistic analysis, it was necessary to automate the modeling processes due to the repetitive nature of sampling and simulation. The automation of simulations in HEC-RAS was based on the principles exposed in [58] and achieved using HEC-RASController v.6.1, which is part of the model's programming interface and consists of a collection of subroutines and functions in Visual Basic [59]. HEC-RASController can be used with Visual Basic for Applications (VBA), but according to [60], this programming language has limited usage, albeit highly useful. Consequently, other languages such as MATLAB and Python have been implemented in the HEC-RASController [60–62]. In this investigation, the automation of the four methodological steps: (1) sampling parameters with LHS and uniform distribution using the SCIPY.STATS library; (2) changing rheological parameters by editing the file "unsteady flow" (.u01) in text format; (3) simulation via HEC-RASController; and (4) Storing the results of flooded area, maximum depths, and arrival times using the H5PY library, was implemented using Python based on the same principles exposed by [58] to achieve a probabilistic full 2D and complete hydrodynamic model propagation algorithm.

Following the proposed methodology for sensitivity analysis, the developed computational routine (code provided in Appendix B) begins with Step 1, generating the sampling for each parameter to be varied. Contrary to [58], the developed algorithm for sampling was performed using the Latin Hypercube Sampling (LHS) method, similar to [24]. These tests with associated probability distributions (uniform) were performed using the SCYPY.STATS library in Python.

For achieving Step 2, the parameterization of the hydrodynamic model is done by modifying the "unsteady flow data" file (extension .u01), which stores input data, including the rheological parameters, as a text file.

After modifying the parameter set, simulations (Step 3) are performed using functions from HEC-RASController, opening and running the project with the altered parameters. Following each simulation, in Step 4, the results of the flooded area, maximum depths, and arrival times in each cell of the 2D model are stored using the H5PY library in HDF files, which allow for the organization and storage of large amounts of numerical data (such as matrix data).

Finally, upon completion of each simulation, the Python code generates the following outputs: the extent of the flooded area in a vector file, which is utilized to determine the flooded areas; maximum depths and arrival times derived from the "unsteady flow analysis" file (extension .p01) of HEC-RAS, encompassing the results for each cell of the model; relative errors in volume balance to assess potential instabilities in the simulations that could influence the results and consequently should be disregarded. With these results, it is possible to assess the sensitivity of the hydrodynamic model by observing the impacts caused by varying each rheological parameter. All Python codes developed are provided in Appendix B and can be used or adapted for other studies where automation is feasible.

2.4. Case Study: The Hydrodynamic Model

The sensitivity analysis methodology developed in this study was applied to the ICOLD case study, which involves a hypothetical dam presented by [40], and was adapted here for the proposed testing purpose. This fictional dam is implemented in a virtual environment and has undergone extensive testing in a study conducted by the International Commission on Large Dams (ICOLD), which was presented at the 12th International Benchmark Workshop on Numerical Analyses of Dams. Eight studies evaluating the risks associated with the dam's failure under different methodologies were published, and these studies were compiled by [40]. More recently, other studies have conducted dam failure analyses using the same case study [58,63]. None of these studies focused on rheological aspects, which is a novelty promoted by the present study.

The fictional dam has a height of 61 m (Elevation 272 m), upstream and downstream slopes of 3:1, a base width of 416 m, and a crest width of 24 m. The crest length is 360 m, forming a reservoir with a capacity of 38×10^6 m^3. It is located in a virtual mountainous terrain, with a confined V-shaped downstream valley along the first 3.6 km and a flat-floored urban area valley ending in a lake along approximately 17.6 km. It should be noted that for this study, it was assumed that the dam is an earth dam storing undefined mining tailings.

The Digital Elevation Model (DEM) used has a horizontal resolution of 9.5 m. The classification of land use and land cover, along with the assignment of Manning's roughness coefficient values, is based on 14 typologies (Figure 1): open water (0.019); low-intensity open space developed areas (0.065); medium-intensity open space developed areas (0.074); high-intensity open space developed areas (0.070); barren land (0.056); deciduous forest (0.138); evergreen forest (0.158); mixed forest (0.123); shrubs (0.057); grass (0.038); pasture (0.070); cultivated land (0.068); and wetlands with shrubs (0.155) [40].

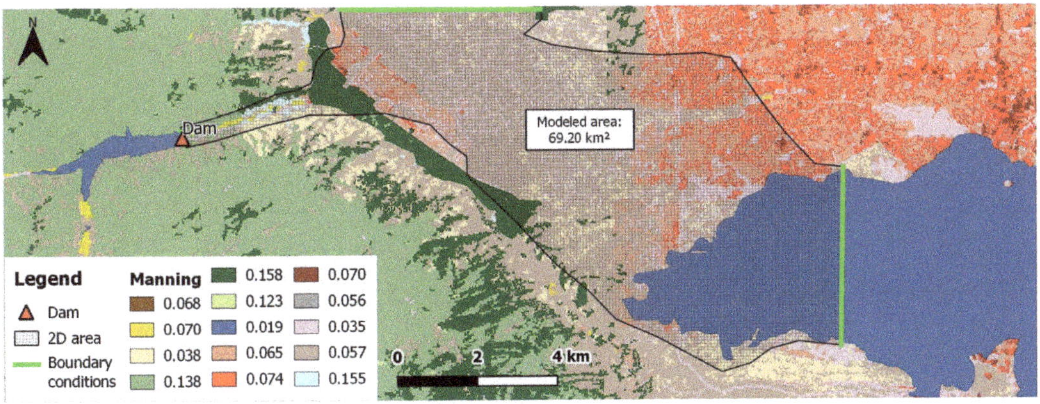

Figure 1. Two-dimensional model of the ICOLD case study used in the simulations with HEC-RAS.

The computational domain of the developed model in HEC-RAS covers an area of 69.8 km^2. The computational mesh consists of cells with dimensions of 50×50 m, with a refinement of 40×40 m in the downstream thalweg of the confined valley, and cells of 100×100 m in the lake area. The downstream boundary conditions are defined as normal depth in two sections, with a slope of 0.005 m/m for the thalweg and 0.0001 m/m for the downstream lake. The computational time step used in the simulations was one second, satisfying the Courant condition with a maximum value of unity. The shallow water equations (SWE) were employed to perform the simulations.

Table 2. Breach parameters calculated based on Froehlich.

Parameters	Data	Froehlich [64]
Failure Mode		Overflow
Total volume (V_w) (1.000 m^3)	38,276.34	
Average breach width (B_m) (m)		116.27
Minimum breach width (m)		55.27
Elevation of the dam crest (m)	272.00	
Elevation of the base of the dam (m)	211.00	
Elevation of the bottom of the breach (m)	211.00	
Dam height (m)	61.0	
Height of the breach (H_b) (m)	61.0	
Time of breach formation (h)		0.54
Left Lateral Slope (H:1V)		1.0
Right Side Slope (H:1V)		1.0
Mode of progression		Sine Curve

The flood hydrograph used in the analysis was obtained using the parametric model proposed by Froehlich [64], which is widely applied for defining dam failure debris flow. No considerations were made for breach formation and mobilized volume parameter variations. The theoretical failure mode assumed the complete opening of a breach with a height equal to the height of the dam. The calculated parameters can be observed in Table 2.

After the parameter calculation, the obtained values were integrated into the HEC-RAS software to derive the rupture hydrograph, representing the flow characteristics ensuing from breach initiation and subsequent reservoir discharge. The analysis yielded a peak flow rate of 37,826.19 m^3/s.

3. Results

The sensitivity evaluation was performed according to the results of each simulation. A total of 2000 simulations were performed, 1000 with the Bingham model and 1000 with the Herschel-Bulkley (H-B) model.

The results obtained were evaluated in terms of flooded area, maximum depths, and arrival times. To analyze the results of the 2000 simulations, percentage coefficient of variation (COV), Spearman's rank correlation coefficient, and histograms were used. The details of these results are presented in the following sections: A summary for each model with the computational specifications of interest for the probabilistic simulations is presented in Appendix C. Complementary exhaustive result sets analyzed in relation to their variability, their correlations with the input parameters, and their probability distributions with histograms are presented in Appendix D.

3.1. Probabilistic Maps Related to Flooded Areas

Probabilistic maps were generated by manipulating the maximum depth outcomes from each simulation and transforming these results into binary data indicating whether or not each cell area was flooded during each simulation: the value of one was assigned to flooded cells, and zero was assigned to non-flooded cells [58,61]. The flooding probabilities for each model cell were calculated using Equation (5) presented by [58].

$$P_{flood} = \frac{\text{NUMBER OF SCENARIOS WITH FLOODED CELL}}{\text{TOTAL NUMBER OF SIMULATED SCENARIOS}} \quad (5)$$

For both the Bingham (Figure 2) and H-B models (Figure 3), the influence of the variation of the rheological parameters on the inundation area occurs mainly in the extension of the simulated area, i.e., the downstream flood wave extent.

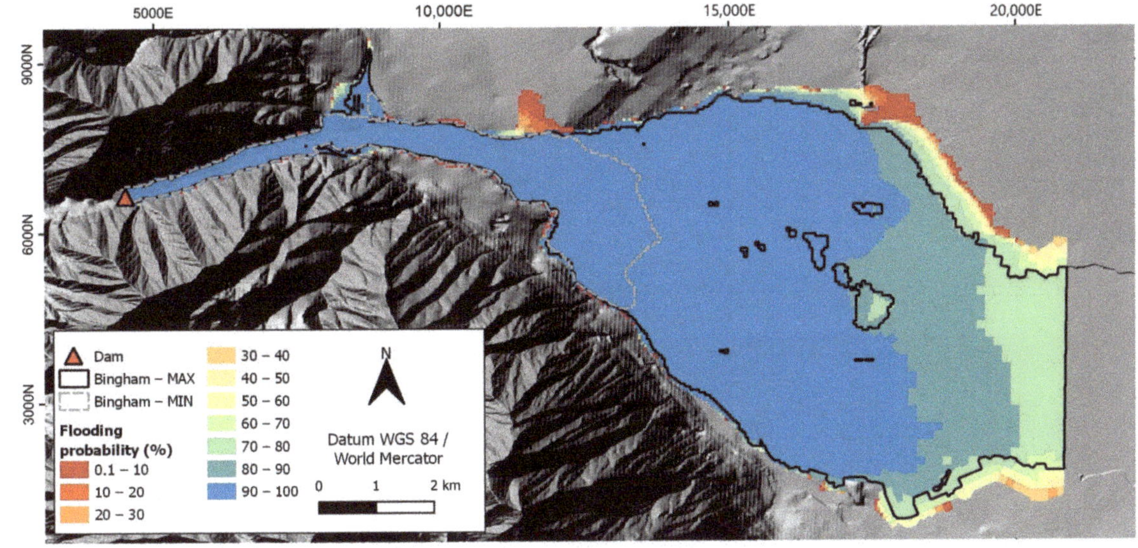

Figure 2. Probabilistic mapping of the Bingham model.

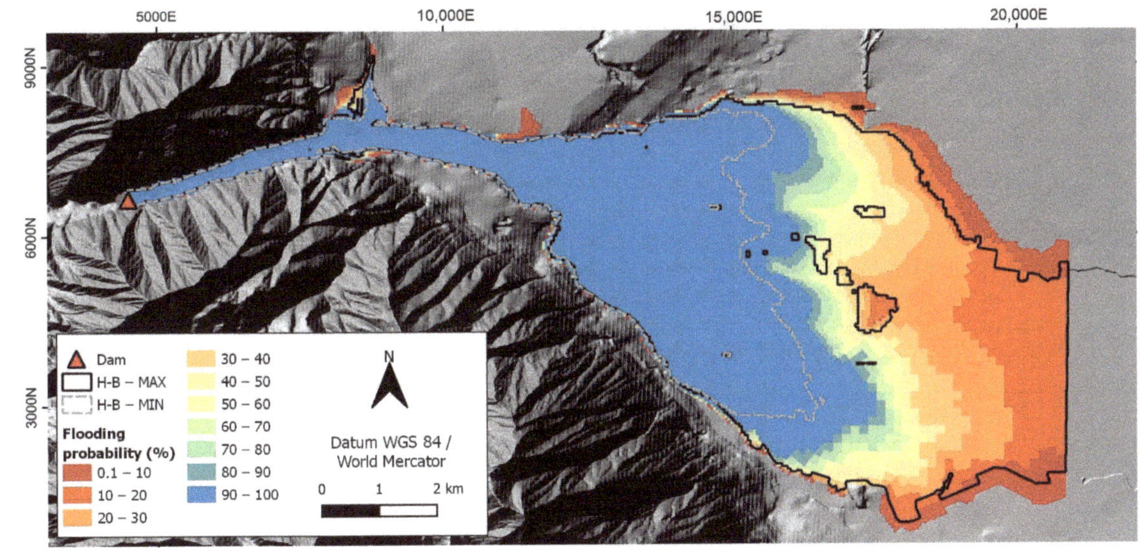

Figure 3. Probabilistic map of the Herschel-Bulkley (H-B) model.

Within the ranges of simulated parameter values for the model, the H-B (Figure 3) presented a greater discretization of simulated areas according to the flood probabilities. Regarding the flood extent, only the maps obtained from flood probabilities between 0.1 and 20% reached the model output, and the other probability range maps indicate that the material was retained upstream. In the Bingham model (Figure 2), the flooded areas with a probability of occurrence between 50 and 80% extended to the border of the model.

Taking as a reference the area relative to the cells that were flooded in all simulations, the extent of the flooded area varied between 15.5 km (100% probability flooded) and 17.6 km (border of the simulated area) in the Bingham model. In the H-B model, the difference in flood extent was greater, about 3.2 km (from 14.3 km to 17.6 km). However, the associated probabilities were significantly lower at the border of the model for the H-B model. The Bingham model showed greater variation regarding the minimum and maximum simulated areas, with flood extents varying from 10.1 km to 17.6 km (extent difference of 7.5 km). The H-B model simulations resulted in a maximum difference in extent of 3.6 km, varying from 14.0 km to 17.6 km.

3.2. Arrival Times and Maximum Depth Variation along the Valley

To understand the impact on the dynamics of the flood event along the valley, 12 cross-sections (CS) were used for extracting and analyzing the simulation results (Figure 4): CS-00 to CS-10 (1 km of distance from each section) and CS-11, located 2 km from CS-10.

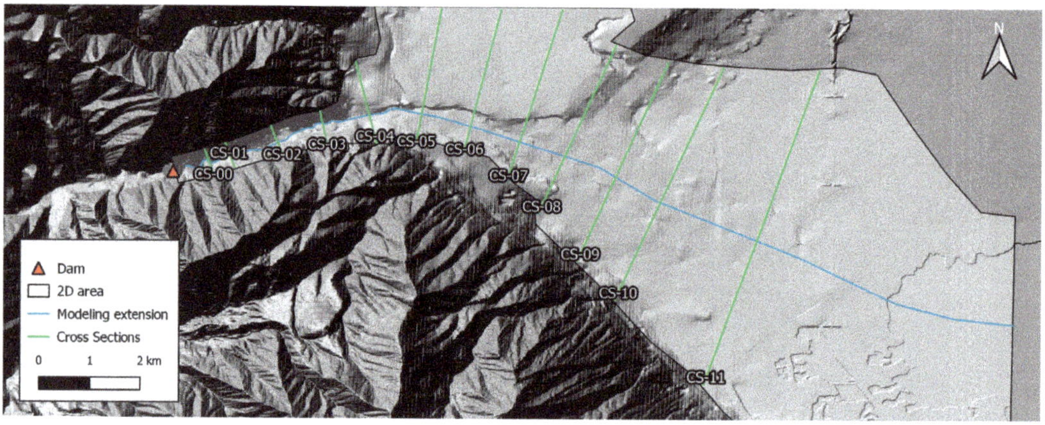

Figure 4. Results analysis sections.

The arrival time was calculated for each CS, considering the time from the beginning of the event to the moment the depth of 0.3 m (1 foot) is reached, following the standard approach in [65]. The results presented include the maximum and minimum values and the coefficient of variation (COV %) of the arrival times comparing the 1000 simulations for both rheological models at the 11 cross-sections (Figure 5).

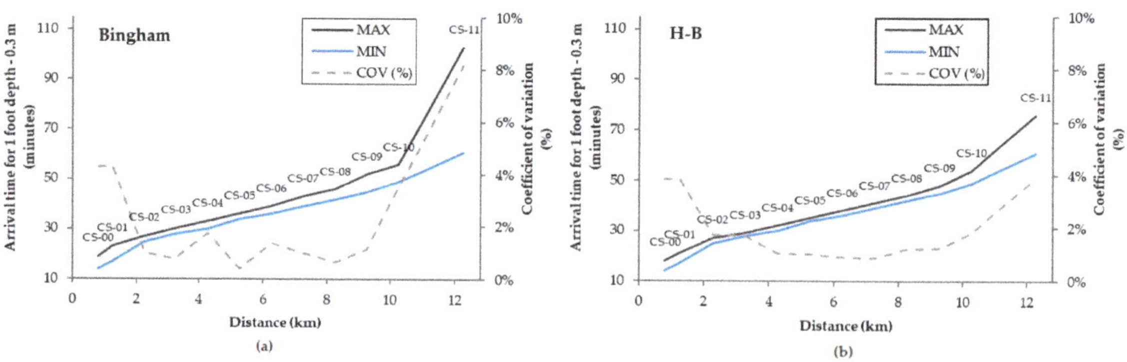

Figure 5. Variation of arrival time along the downstream talweg for simulations using the (**a**) Bingham model and (**b**) Herschel-Bulkley (H-B) model.

In general, arrival times varied between 2 and 7 min in the Bingham simulations, except for the last section, where a range of 42 min (variation from minimum to maximum values) was observed. For H-B, we observed arrival times varying from 1 to 5 min in the simulations for CS-01 to CS-10. For the last section (CS-11), a range of 15 min was observed.

For the Bingham model (Figure 5a), the greatest variation (8.2%) occurred in the last section of the model (CS-11), where the wave arrival time for the 0.3 m depth varied between 61 and 103 min; the smallest variation occurred in section CS-05 (between 34 and 36 min) with a COV equal to 0.4%. The CS-05 is located at the transition between the V-shaped valley and the flat-floored valley. In the H-B model (Figure 5b), the largest variation (3.91%) also occurred in the downstream section CS-11, where the wave arrival time for the 0.3 m depth varied between 61 and 76 min; the smallest variation occurred in section CS-07 (0.88%), which is located in the transition region between the embedded shaped valley and the flat-floored area, similar to what was observed in the Bingham model.

Regarding the depth variations simulated with the Bingham model, it can be seen that between sections CS-00 and CS-09, the maximum depths varied from a minimum of 1.8 m (CS-00) to a maximum of 4.1 m (CS-02) (Figure 6a). In the downstream sections, the variations of simulated maximum depths were higher: from 0.14 m to 6.84 m and from 0 m to 4.93 m in sections CS-10 and CS-11, respectively. These higher influences in downstream areas may be related to upstream volume retention caused by rheology parameter assumptions related to smaller flow velocities for more viscous materials. The highest COV value, 9.3%, was observed in the last downstream section, CS-11, and the lowest, 0.4%, in the most upstream section, CS-00, indicating that changes in rheological parameters were more expressive in the depths further from the dam. Although the coefficients of variation obtained are less than 10%, it is observed that the variations are relevant in absolute values of maximum depth since depths above 0.3 or 0.61 m already have significant associated risks [65].

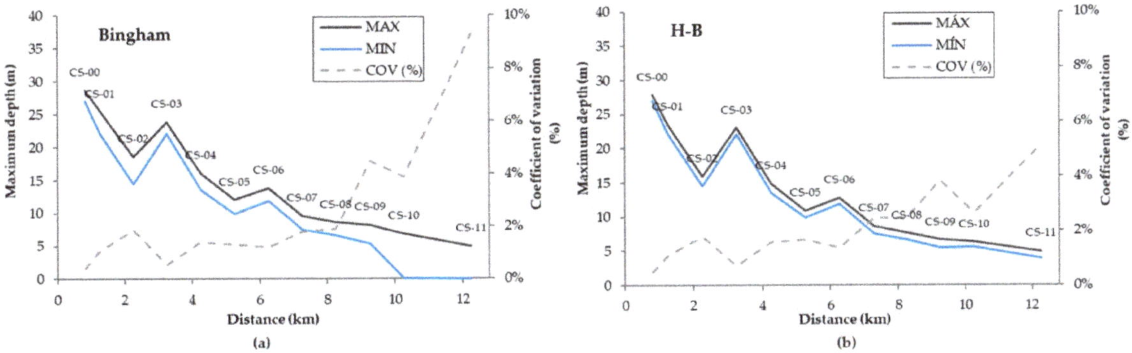

Figure 6. Variation of maximum depths along the downstream talweg for the (**a**) Bingham and (**b**) H-B models.

In the H-B model, the maximum depths obtained in the analysis sections experienced the greatest variation in section CS-11 (5.1%), ranging from 3.9 m to 4.9 m (Figure 6b). The least variation in simulated maximum depths occurred in section CS-00 (0.5%), from 27.0 m to 27.9 m. Overall, the variation in maximum depths in the sections was small, ranging between 0.8 m and 1.5 m from changes in the rheological parameters of the H-B model. The qualitative variation was similar when comparing the H-B model and the Bingham one regarding the effect of the distance along the valley, with increasing variations from upstream to downstream sections.

3.3. Rheological Parameters vs. Simulated Areas

With Bingham, the minimum and maximum simulated areas were 8.1 km^2 and 50.8 km^2 (Figure 7), respectively, and the area results showed a COV of 17.9%. It is observed that most simulations returned values close to the maximum one: about 68% of the simulations resulted in flooded areas within a 5% variation from the maximum simulated area (from 48.2 km^2 to 50.8 km^2). It revealed that most sets of rheological parameters resulted in higher flooded area values. The relationships of the four parameters (C_V, Max C_V, *a* and *b*) with the flooded area indicated a tendency toward reduction in the flooded area with the increase of these input parameters, as indicated by the Spearman correlation coefficient (negative). This influence was most noticeable between the flooded area and the values of *b*, the exponent of the yield stress curve, which indicated the strongest correlation ($\rho = -0.7301$) (Figure 7). The parameters C_V and *a* showed moderate correlation with the flooded area, with Spearman's correlation coefficients equal to -0.4116 and 0.4606, respectively. The parameter Max C_V showed a weak correlation with the flooded area ($\rho = -0.0529$). Thus, the simulations that returned smaller flooded areas are mainly related to the increase in the value of the parameter *b*.

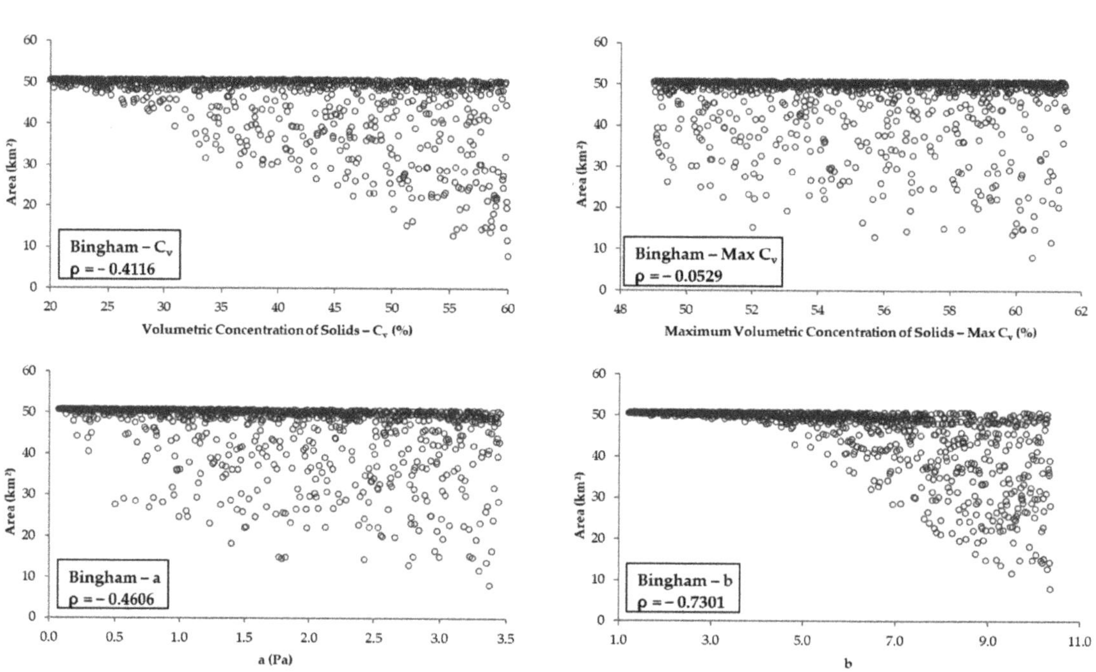

Figure 7. Relationship between flooded areas and C_V, Max C_V, *a*, and *b* in 1000 simulations with the Bingham model.

In the H-B model (Figure 8), the flooded areas ranged from 20.1 km^2 to 50.4 km^2, with a coefficient of variation of 25.7%. A very strong correlation ($\rho = -0.9738$) is observed between the increase in yield stress (τ_y) and the reduction in flooded area (Figure 8). The variation of the parameter C_V showed a weak correlation with the flood-simulated areas ($\rho = 0.2371$), and the parameters *K* and *n* showed no correlation (respectively, $\rho = 0.0113$ and $\rho = 0.0241$).

Both models reached the maximum flooded area threshold of approximately 50 km^2. It is observed that this maximum flooded area threshold refers to physical limitations related to the terrain, roughness, runoff volume, or different sets of rheological parameters, which were defined by the modeler and considered for the different scenarios. The results showed that, in these conditions, both models were able to represent the worst-case scenario with different sets of rheologically assigned variables.

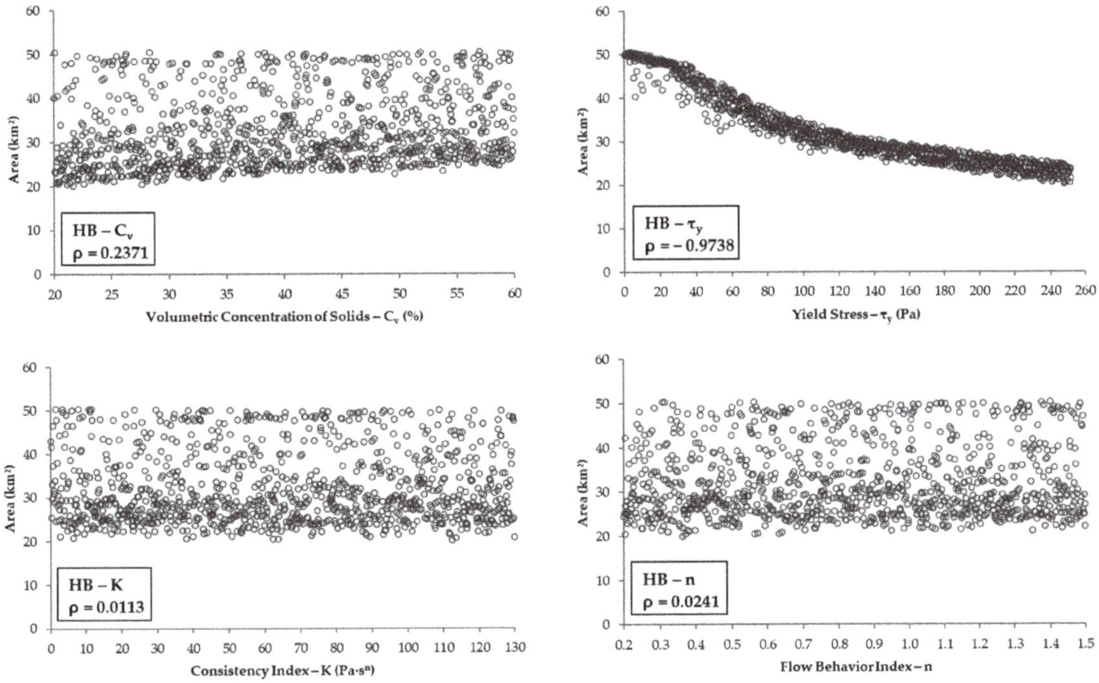

Figure 8. Relationship between inundated areas and C_v, τ_y, K, and n in 1000 simulations with the H-B model.

3.4. Rheological Parameters vs. H_{mean}

For a more general sensitivity assessment, the study area was discretized into two areas for evaluating the maximum depth variations in function of the rheological parameter determination: A1 and A2 (Figure 9).

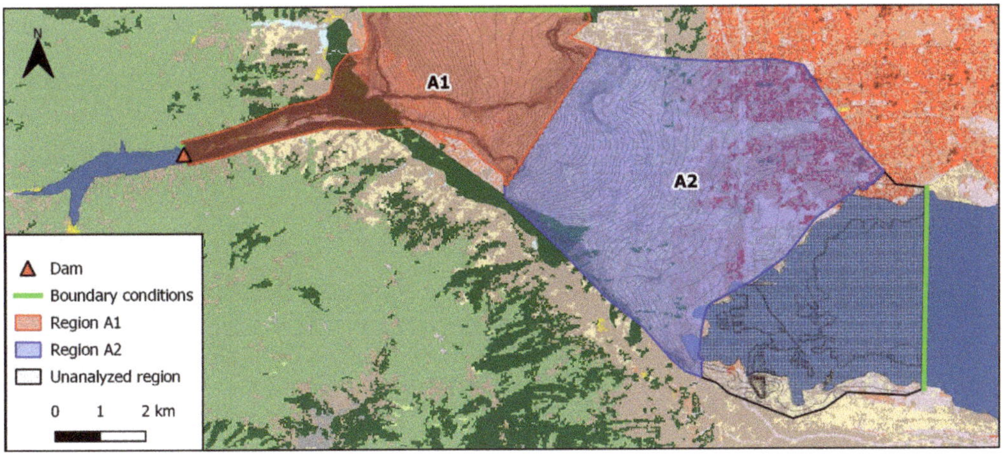

Figure 9. Discretized areas for sensitivity evaluation.

The results of maximum depths (H_{max}) in the computational mesh cells for the areas of interest (A1 and A2) (Figure 9) were evaluated according to Equation (6). This equation calculates the maximum average runoff height (H_{mean}) for each area by summing up the maximum flow depth (H_{max}) in each flooded cell ($H > 0$) within each area and dividing the result by the total number of flooded cells (N_{WC}) for each area [24].

$$H_{mean} = \sum_{j=0}^{N_{WC}} H_{max}(j) / N_{WC} \tag{6}$$

where j is the cell index; N_{WC} is the total number of flooded cells ($h > 0$); and H_{max} is the maximum flow depth.

The relationships between H_{mean} and each parameter were obtained for each area of interest for the Bingham model (Figure 10) and for the H-B model (Figure 11).

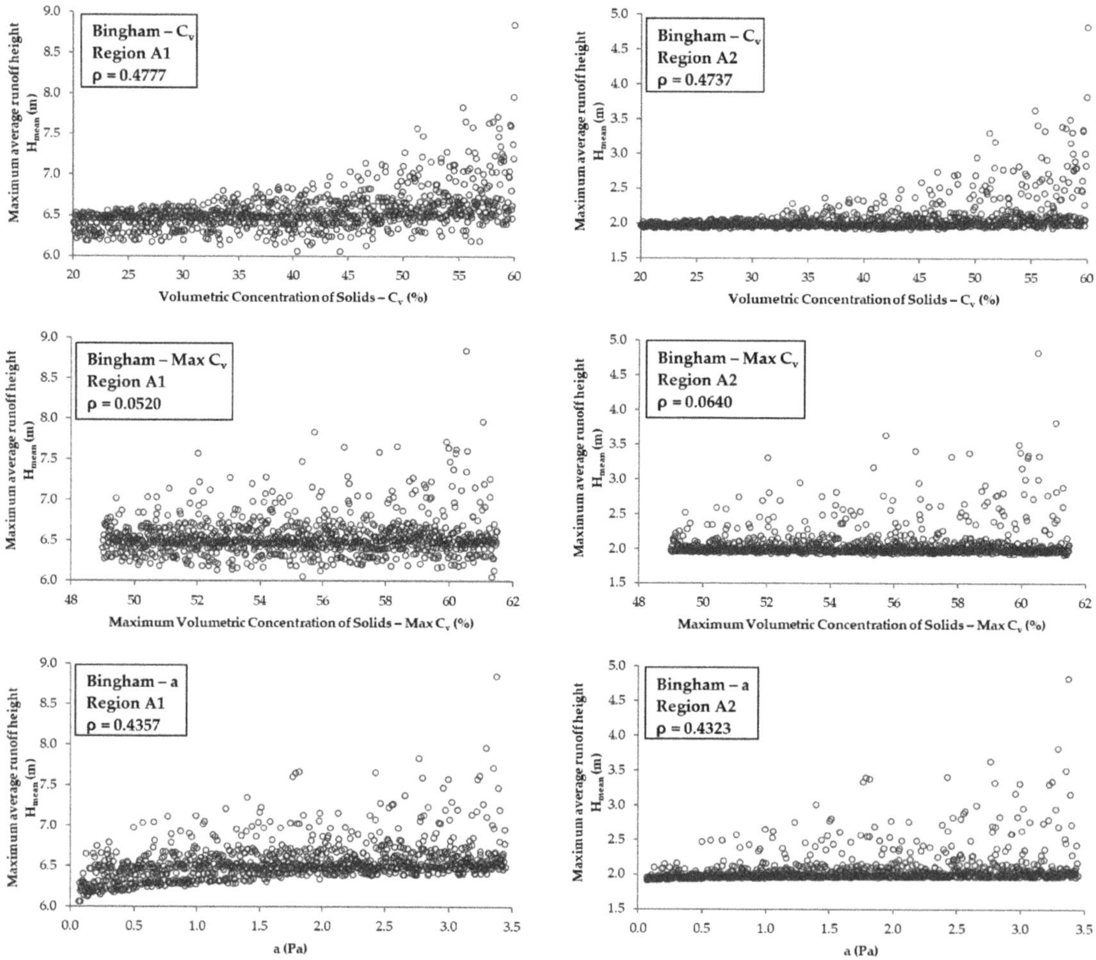

Figure 10. *Cont.*

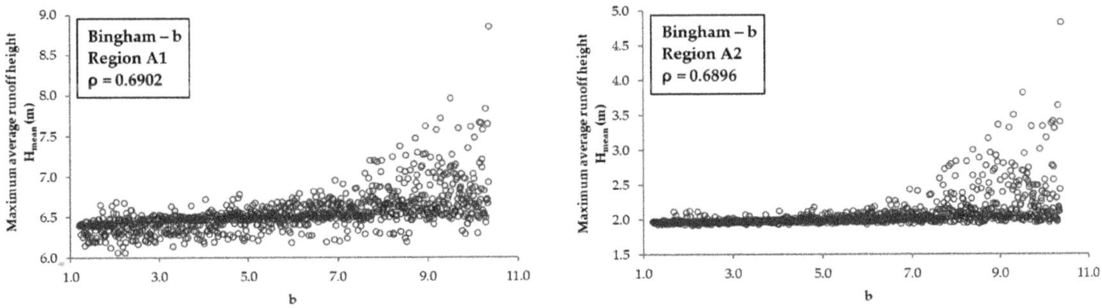

Figure 10. Relationships between maximum average runoff height (H_{mean}) and the parameters C_V, Max C_V, *a*, and *b* in the areas of interest, A1 and A2—Bingham Model results.

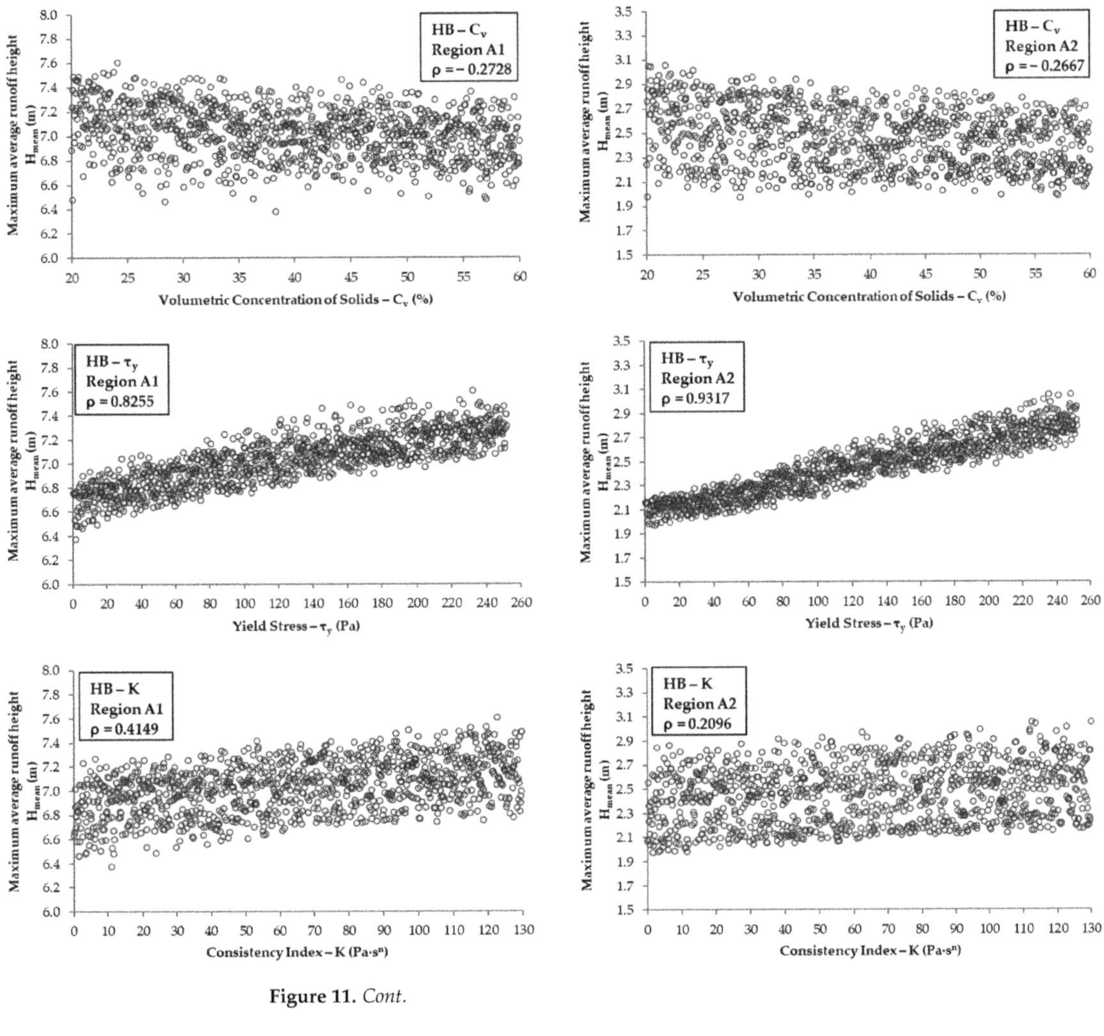

Figure 11. *Cont.*

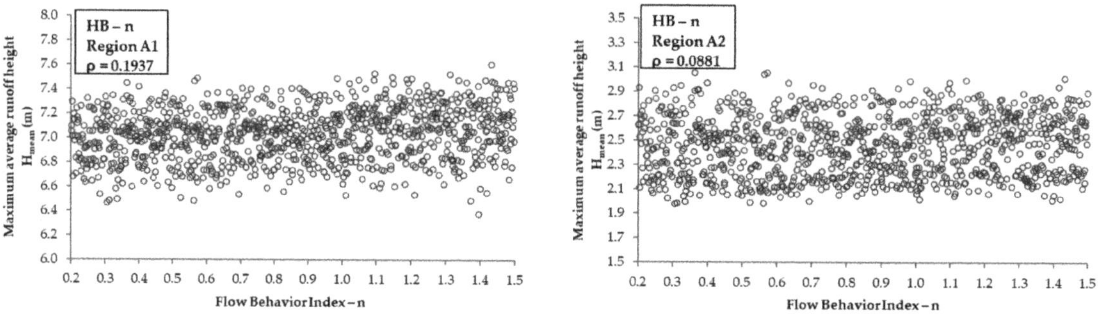

Figure 11. Relationships between the maximum average runoff height (H_{mean}) and the parameters C_v, τ_y, K, and n in the areas of interest, A1 and A2—H-B model results.

For the Bingham model, in region A1 (Figure 10), the simulated depths were greater due to the valley confinement condition immediately downstream of the dam, with average maximum depths in the cells ranging between 6.1 m and 8.8 m and a coefficient of variation equal to 3.8%. The depths reduced in the A2 region due to the spreading of the non-Newtonian fluid on the floored-shaped valley floodplain, ranging between 1.9 m and 4.8 m (COV equal to 11.8%). These results highlighted that the variation of rheological parameters has more influence on the simulated depths over the flatter region, further downstream of the dam (A2).

In region A1 (Figure 11), the resulting maximum average runoff height ranged from 6.4 m to 7.6 m and indicated a coefficient of variation of 3.0%. In the A2 region (Figure 11), the depth was reduced from upstream to downstream, ranging from 2.0 m to 3.1 m, with a COV of 9.8%. Thus, like the Bingham model, changes in the rheological parameters of the H-B model had a greater impact on the depths in the section further downstream of the simulated model, in flatter areas (A2).

For the Bingham model, in Table 3, it is noted that Spearman's correlation coefficient is approximately 0, i.e., an almost null correlation between maximum average runoff height and Max C_v in regions A1 and A2 (0.0520 and 0.0640, respectively). Between H_{mean} and b, it is verified that in both regions, A1 and A2, the larger values of b led to higher values of H_{mean}, with strong correlations in region A1 (0.6902) and in region A2 (0.6896).

Table 3. Spearman's rank correction coefficients between rheological parameters and H_{mean}.

Results/Parameters		Bingham				Herschel H-B			
		C_v	Max C_v	a	b	C_v	τ_y	K	n
H_{mean}	Region A1	0.4777	0.052	0.4357	0.6902	−0.2728	0.8255	0.4149	0.1937
	Region A2	0.4737	0.064	0.4323	0.6896	−0.2667	0.9317	0.2096	0.0881

For the H-B model (Table 3), the increase of τ_y led to the increase of H_{mean}, with Spearman correlation strong in region A1 (0.8255) and very strong in region A2 (0.9317). Spearman's correlation between parameter n and depths shows a weak correlation for both regions, equal to 0.1937 and 0.0881, respectively. By the formulation of the H-B model, a more evident effect of the parameters K and n on the results was expected; therefore, the parameter ranges may have evidenced the impact of the yield stress (τ_y) on the results. Therefore, in this scenario, the parameter τ_y was the one that caused the most noticeable effect on the flooded area and the maximum average runoff height results.

4. Discussion

The Bingham and Herschel-Bulkley (H-B) models presented the same minimum arrival time simulated values along the valley (among the 1000 simulations of each model). However, when evaluating the frequency distribution of arrival times (Appendix D), the

Bingham model proved to be more conservative once it presented lower values for arrival times (critical risk conditions) compared to the H-B model. With respect to the maximum arrival time values, the Bingham model indicated an extreme value in the furthest downstream section of the model. The Bingham model showed a larger coefficient of variation (variability of the data in relation to the mean) than H-B. Considering both models, the arrival times experienced greater variations further downstream of the dam, highlighting that the further downstream, the more uncertain the predictions related to this variable. Furthermore, the frequency distribution of the results in the last section shows that the Bingham model returned in most simulations arrival times close to the minimum simulated value (more conservative results), while the H-B model returned more distributed arrival times, presenting the Gamma distribution as the best fit.

Regarding the results of maximum depths along the simulated reach, the Bingham model also exhibited larger coefficients of variation in the downstream sections compared to the H-B model. The Bingham model resulted in greater depths than H-B, considering the maximum values of all simulations carried out. Nevertheless, most of the simulations with the Bingham model returned shallower depth values compared to H-B. The maximum depth results indicated the best fit to the exponential and normal distributions, in sections CS-00 and CS-11, respectively, in the Bingham model. And in the H-B model, the best fit occurred in the normal and lognormal distributions to the results of maximum depths, in sections CS-00 and CS-11. This differs from what was presented in [26], where the Gamma distribution was the best fit to the results of maximum depths and velocities in the sensitivity analysis of the MassaMov2D dynamic model to Bingham and Voellmy rheological parameters.

As for the flooded areas, the Bingham model presented a smaller coefficient of variation than the H-B model, even though the Bingham model presented a larger amplitude between the maximum and minimum values of the simulated area. The Bingham model resulted, in most simulations, in values close to the maximum value of the simulated area, once again more conservative without presenting a good fit to the probability distributions. The H-B model, on the other hand, presented more distributed flooded area values in the set of results obtained, with the best fit to the exponential distribution. For comparison purposes, the eight studies [40] that applied this same case study considering a water dam evaluated sensitivities and uncertainties related to erodibility coefficients, breach parameters, and propagation methods and models. In these studies, the simulated areas ranged from 27 km to 47 km, with a COV of 18.6%. This demonstrates how the rheological parameters have significant relevance in flood modeling since the flooded areas of the Bingham model showed COV equal to 17.9% and H-B equal to 25.7%.

The Bingham model resulted in a greater maximum average runoff height than the H-B model and showed greater coefficients of variation. The maximum average runoff height results suggest a normal and exponential distribution, respectively, in regions A1 (confined valley close to the dam) and A2 (flatter areas further from the dam) for the Bingham model. With the H-B model, in region A1, the H_{mean} frequency distribution showed a normal distribution of this result, and in region A2, the results were uniformly distributed.

In the Bingham model, the parameter b of the yield stress curve showed greater correlation with the results of flooded area and depths, and the parameter Max C_v was the one that showed less correlation. In the H-B model, the yield stress (τ_y) was the parameter that showed greater correlation with the variation of results, and the behavior index (n) expressed less correlation. In the results obtained by [24], it was observed that the FLO-2D model (Quadratic rheological model) is mainly sensitive to the parameter β_1 that composes the rheological curve of the dynamic viscosity, affecting mainly the results of the simulated flow depth. The study presented in [66] indicates that the results of maximum debris flow depths were more affected by the variation of the turbulent coefficient and the friction coefficient of the Voellmy rheological model in the 2D dynamic modeling with the RAMMS v.1.3.16 (Rapid Mass Movements) software. It is noticed that different hydrodynamic models have different sensitivities, which could justify future comparative tests using different approaches from this perspective.

Finally, the H-B model proved to be more sensitive to rheological parameter changes since the results varied considerably in the 1000 simulations performed. This sensitivity was expected by the H-B model formulation due to the variation of the flow behavior index (n), as pointed out in [37], although it was not the most sensitive parameter in the actual study. However, the changes in rheological parameters caused more significant impacts on the results with the Bingham model in terms of the amplitude (extreme values) of the results obtained. In this perspective, these results may indicate agreement with the conclusions presented in [8], considering that the Bingham model may be applied as a first approximation for modeling non-Newtonian flow as it presents conservative results. For a better understanding of the probabilistic variability of the simulations, it is indeed suitable to perform simulations considering the H-B model. The present study extended this understanding to dam-break analysis, leading to the same conclusions related to arrival time simulations.

5. Conclusions and Recommendations

Following the methodology developed, it was possible to bring considerable elements of answers to the main question raised within this article: "How does the rheological variability of the stored materials impact the results of the modeling of flooding caused by mining tailings dam ruptures?".

This article demonstrated the importance of determining rheological parameters since their variability brings relevant discrepancies in terms of simulated areas, depths, and arrival times. The Bingham model was shown to be more sensitive than the H-B model in terms of maximum and minimum values of simulated areas. Furthermore, in probabilistic terms, the Bingham model was shown to be much less sensitive, tending to represent higher simulated area values, while Herschel-Bulkley showed great variability of areas as a function of the probabilities associated with the rheology parameters. Bingham was considered, on the one hand, more uncertain since it can lead to considerably lower flooded area values and, on the other hand, more conservative since, in probabilistic terms, it tends to higher area results. Therefore, the use of this model within a deterministic scheme could bring significant uncertain results. In this context, the Herschel-Bulkley model proved to be more efficient for probabilistic analysis, presenting a smaller range of uncertainties with a better distributed probability distribution, which enables a better appreciation for risk management purposes.

This article also provided contributions related to the broad bibliographic survey carried out on the ranges of values for mining tailings rheological parameters. This literature review demonstrates the great variability of these parameters and may be valuable for establishing boundaries for sensitivity analysis in other case studies. The proposed automated methodology proved robust for the evaluation of uncertainties concerning the rheological parameters in the HEC-RAS. The evaluation considers, in probabilistic terms, the flooded areas, the maximum depths reached, and the arrival times of the flood wave. Finally, the algorithm developed in Python allows for automated estimation of uncertainty in the simulation results in probabilistic terms. It is available and can be adapted for other applications in further studies.

As suggestions and recommendations for future work, it is important to evaluate the impact of the sample space of the parameters in the evaluation of the sensitivity of hydrodynamic models. The application of global sensitivity analysis [24] is a relevant alternative, as it allows a complete and complex analysis, indicating the sensitivity as a function of different indicators. In addition, since there are several sources of uncertainty in these dam failure models, the inclusion of the variation of failure breach parameters, mobilized volumes, and Manning coefficient, together with rheological parameters, would be of great importance to understanding the significance of each parameter in the results of these models in a global analysis. This type of global uncertainty analysis may provide more robustness in the study of floods and associated risks.

Furthermore, it was stated that probabilistic methods can help in defining the rheological parameters due to the high heterogeneity of the material as well as quantifying the uncertainty in the risk assessment results, which is critical to assisting planning and decision-making in more comprehensive risk analyses. Consequently, the use of this probabilistic approach in real case studies could be an interesting direction for future studies searching for a better understanding and validation of different methods.

Author Contributions: Conceptualization, M.M. and J.E.; methodology, M.M. and J.E.; validation, M.M. and J.E.; formal analysis, M.M. and J.E.; investigation, M.M. and J.E.; resources, M.M. and J.E.; writing—original draft preparation, M.M.; writing—review and editing, M.M. and J.E.; visualization, M.M.; supervision, J.E.; project administration, J.E.; funding acquisition, J.E. All authors have read and agreed to the published version of the manuscript.

Funding: This research was funded by the Brazilian research institutions, foundations and agencies Coordenação de Aperfeiçoamento de Pessoal de Nível Superior (CAPES), Conselho Nacional de Desenvolvimento Científico e Tecnológico (CNPq), Fundação de Amparo à Pesquisa do Estado de Minas Gerais (FAPEMIG) and Pró-Reitoria de Pesquisa, Universidade Federal de Minas Gerais (PRPq/UFMG).

Data Availability Statement: The data that support the findings of this study are available in the supporting information of this article.

Conflicts of Interest: The authors declare no conflict of interest. The funders had no role in the design of the study; in the collection, analyses, or interpretation of data; in the writing of the manuscript; or in the decision to publish the results.

Appendix A

Table A1. Literature review with values of the rheological parameters of interest for the models.

Parameter [Unit]	Parameter Range	Reference	Tailings	Test/Method	Rheological Model/ Fitted Equation
Volumetric concentration of solids (C_V) [%]	21.21–35.78	[47]	Coal	n.a.	n.a.
	27.98–46.65	[47]	Copper		
	36.50–38.62	[47]	Gold		
	36.34–44.62	[47]	Lithium		
	11.09–19.66	[47]	Nickel		
	25.48–31.25	[47]	Uranium		
	11.82–50.11	[47]	Zinc		
	18.0–52.5	[7]	Base metal		
	21.8–55.3	[48]	Iron		
	32.5–65.0	[48]	Iron		
	9.3–48.4	[9]	Gold		
	34.75–57.14	[49]	Iron		
	58	[2]	Iron		
	47	[37]	Iron		
	23	[52]	Iron		
Maximum volumetric concentration of solids (Max C_V) [%]	61.5	[53]	-	-	n.a.
	53	[7]	Base metal	-	
	49.0–62.0 *	[50]	Copper	[67] **	
	59.9–76.5 *	[51]	Iron	[67] **	
	57.3–68.0 *	[51]	Gold	[67] **	
Dynamic viscosity (μ) [Pa·s]	0.03–0.49	[7]	Base metal	Viscometer	Bingham
	0.15–2.69	[48]	Iron	Viscometer	Bingham
	0.002–0.311	[9]	Gold	Viscometer	Bingham
	0.0071–0.4457	[49]	Iron	Rheometer	Quadrática
	1.09–1.46	[8]	Pyrophyllite	Rheometer	Bingham
	50	[37]	Iron	Calibration	Bingham
	30.0–100.0	[38]	Iron	Calibration	Full Bingham

Table A1. *Cont.*

Parameter [Unit]	Parameter Range	Reference	Tailings	Test/Method	Rheological Model/ Fitted Equation
Yield stress (τ_y) [Pa]	2.122–48.535	[47]	Coal	Viscometer	Herschel-Bulkley
	0.641–93.50	[47]	Copper	Viscometer	Herschel-Bulkley
	0.6–1.5	[47]	Gold	Viscometer	Herschel-Bulkley
	1.048–11.165	[47]	Lithium	Viscometer	Herschel-Bulkley
	1.564–37.110	[47]	Nickel	Viscometer	Herschel-Bulkley
	2.769–9.411	[47]	Uranium	Viscometer	Herschel-Bulkley
	0.652–100.260	[47]	Zinc	Viscometer	Herschel-Bulkley
	26.0–638.0	[7]	Base metal	Viscometer	Bingham
	19.36–602.82	[48]	Iron	Viscometer	Bingham
	59.59–2396.53 ***	[48]	Iron	Slump test	Bingham
	0.5–181.0	[9]	Gold	Viscometer	Bingham
	0.085–118.0	[49]	Iron	Rheometer	Quadrática
	12.0–23.0	[8]	Pyrophyllite	Rheometer	Herschel-Bulkley
	9.7–251.9	[27]	Synthetic	Rheometer	Herschel-Bulkley
	100.0–1000.0	[37]	Iron	Calibration	Bingham
	750.0–1000.0	[38]	Iron	Calibration	Full Bingham
a [Pa]	1	[7]	Base metal	Viscometer	Exponential
	21.381	[48]	Iron	Viscometer	
	0.0065	[48]	Iron	Slump test	
	0.08	[52]	Iron	Calibration	
	1.00×10^{-7}	[49]	Iron	Rheometer	
	~0.04–3.40	[10]	Copper	Rheometer	
	~0.40–3.45	[10]	Iron	Rheometer	
b [-]	12.2	[7]	Base metal	Viscometer	Exponential
	90.874	[48]	Iron	Viscometer	
	20.47	[48]	Iron	Slump test	
	40	[52]	Iron	Calibration	
	39.278	[49]	Iron	Rheometer	
	~1.2–5.0	[10]	Copper	Rheometer	
	~1.2–5.5	[10]	Iron	Rheometer	
Consistency index (K) [Pa·s^n]	0.034–6.409	[47]	Coal	Viscometer	Herschel-Bulkley
	0.008–130.0	[47]	Copper	Viscometer	
	0.108–0.221	[47]	Gold	Viscometer	
	0.222–1.515	[47]	Lithium	Viscometer	
	0.154–2.001	[47]	Nickel	Viscometer	
	0.065–0.097	[47]	Uranium	Viscometer	
	0.428–14.720	[47]	Zinc	Viscometer	
	0.69–1.96	[8]	Pyrophyllite	Rheometer	
Flow behavior index (n) [-]	0.4–1.0	[47]	Coal	Viscometer	Herschel-Bulkley
	0.192–1.347	[47]	Copper	Viscometer	
	0.705–0.744	[47]	Gold	Viscometer	
	0.766–1.020	[47]	Lithium	Viscometer	
	0.450–0.602	[47]	Nickel	Viscometer	
	0.742–0.913	[47]	Uranium	Viscometer	
	0.306–0.577	[47]	Zinc	Viscometer	
	0.84–1.14	[8]	Pyrophyllite	Rheometer	
	0.50–1.50	[37]	Teórico	Calibration	

* Values obtained based on tailings porosity; ** Proposes the equation: Max C_v = 1-pm′, to estimate the maximum concentration based on bed porosity (pm′); *** C_v of 65% in Slump Test; n.a.: not applicable.

Appendix B

Python language codes for automatic simulations with the Bingham and Herschel-Bulkley rheological models. The complete algorithm is also available at https://github.com/MCGI-UFMG/rheologyrasprob.git.

```
# -*- coding: utf-8 -*-
"""
@author: Malena Melo
Bingham Code
"""

# Importing Libraries

import win32com.client
import pandas as pd
import re as re
import numpy as np
import h5py
from osgeo import ogr
from osgeo import osr

RC = win32com.client.Dispatch("RAS610.HECRASCONTROLLER")
parametros = pd.read_csv('C:\param_bingham.txt', sep=" ")

# Parameter variation
for index, linha in parametros.iterrows():
    sim = index + 1
    cv = linha['Cv']
    cvmax = linha['Cvmax']
    a = linha['a']
    b = linha['b']

    # Reading the lines from the file unsteady
    with open('C:\ICOLD_Z_G.u01', 'r') as f:
        dados = f.readlines()

        # Finding the parameter line of interest and setting the line to be the old parameter
        for line in dados:
            if re.search(r'Non-Newtonian Constant Vol Conc', line):
                break
        antp_cv = line

        for line in dados:
            if re.search(r'Non-Newtonian Max Cv', line):
                break
        antp_cvmax = line
```

```

        for line in dados:
            if re.search(r'Non-Newtonian Yield Coef', line):
                break
        antp_a_b = line

    # Opening and saving a file to change the previous parameter line by the new value
    with open('C:\ICOLD_Z_G.u01', 'r') as f:
        dados = f.read()
    novop_cv = 'Non-Newtonian Constant Vol Conc=' + str(cv) + '\n'
    novop_cvmax = 'Non-Newtonian Max Cv=' + str(cvmax) + '\n'
    novop_a_b = 'Non-Newtonian Yield Coef=' + str(a) + ', ' + str(b) + '\n'

    print('Simulação %d - ' % sim, novop_cv, novop_cvmax, novop_a_b)
    dados = dados.replace(antp_cv, novop_cv)
    dados = dados.replace(antp_cvmax, novop_cvmax)
    dados = dados.replace(antp_a_b, novop_a_b)

    with open('C:\ICOLD_Z_G.u01', 'w') as f:
        f.write(dados)

    # Opening the project in RAS and simulating with changed parameters
    RC.ShowRAS()
    RC.Project_Open(r"C:\ICOLD_Z_G.prj")
    RC.Compute_CurrentPlan(None, None, True)
    RC.QuitRAS()

    # Saving results of maximum depths in each cell
    file = "C:\ICOLD_Z_G.p01.hdf"
    with h5py.File(file, "r") as hdf:
        celldepth = hdf.get(
            '/Results/Unsteady/Output/Output Blocks/Base Output/Unsteady Time Series/2D Flow Areas/jusante/Cell Hydraulic Depth')
        depthmax = np.amax(celldepth, axis=0)
        depthmax = h5py.File('depth_max.hdf', 'a')
        depthmax.create_dataset('sim_%d' % sim, data=depthmax)
        depthmax.close()

    # Saving arrival times to 0.30 (1 foot) per simulation performed
        for depth in tempos:
            cell = depth.tolist()
```

```
        for x in cell:
            if x > 0.3:
                tempo_cheg_t1.append(cell.index(x))
                break
        else:
            tempo_cheg_t1.append(0)
            continue

    tempo30 = np.array(tempo_cheg_t1)
    tempo_chegada30 = h5py.File('tempo_chegada30.hdf', 'a')
    tempo_chegada30.create_dataset('sim_%d' % sim, data=tempo30)
    tempo_chegada30.close()

# Accessing and saving flooded area results
driver = ogr.GetDriverByName('ESRI Shapefile')
hinputfile = driver.Open(r'C:\Inundation Boundary (Max Value_0).shp', 0)
in_layer = hinputfile.GetLayer(0)

# Transformation between projections
src_srs = in_layer.GetSpatialRef()
tgt_srs = osr.SpatialReference()
tgt_srs.ImportFromEPSG(3395)
transform = osr.CoordinateTransformation(src_srs, tgt_srs)

for feature in in_layer:
    geom = feature.GetGeometryRef()
    geom2 = geom.Clone()
    geom2.Transform(transform)
    area_m2 = geom2.GetArea()
    area_km2 = area_m2 / 1000000
    print('Area em km²: ', area_km2)
    area = h5py.File('area.hdf', 'a')
    area.create_dataset('sim_%d' % sim, data=area_km2)
    area.close()

# Storing the volume error in 1000 m3 and in percent per simulation
with open('C:\ICOLD_Z_G.p01.computeMsgs.txt', 'r') as f:
    dados = f.readlines()
    erro_vol = []
    erro_volperc = []

    for line in dados:
```

```
            if re.search(r'Overall Volume Accounting Error in 1000', line):
                erro_vol.append(line)
                with open('erro_vol.txt', 'a') as txt:
                    txt.write(str(erro_vol))
                break

        for line in dados:
            if re.search(r'Overall Volume Accounting Error as percentage', line):
                erro_volperc.append(line)
                with open('erro_volperc.txt', 'a') as txt:
                    txt.write(str(erro_volperc))
                break

    print('End simulation', sim)
    hinputfile.Destroy()
else:
    print("Simulations with LHS finished.")
```

```
# -*- coding: utf-8 -*-
"""
@author: Malena Melo
Herschel-Bulkley Code
"""

# Importing Libraries

import win32com.client
import pandas as pd
import re as re
import numpy as np
import h5py
from osgeo import ogr
from osgeo import osr

RC = win32com.client.Dispatch("RAS610.HECRASCONTROLLER")
parametros = pd.read_csv('C:\param_HB_icold.txt', sep=" ")

# Parameter variation
for index, linha in parametros.iterrows():
    sim = index + 1
    cv = linha['Cv']
```

```
    ty = linha['Ty']
    k = linha['K']
    n = linha['n']

    # Reading the lines from the file unsteady
    with open('C:\ICOLD_Z_G.u01', 'r') as f:
        dados = f.readlines()

        # Finding the parameter line of interest and setting the line to be the
old parameter
        for line in dados:
            if re.search(r'Non-Newtonian Constant Vol Conc', line):
                break
        antp_cv = line

        for line in dados:
            if re.search(r'User Yeild', line):
                break
        antp_ty = line

        for line in dados:
            if re.search(r'Herschel-Bulkley Coef', line):
                break
        antp_k_n = line

    # Opening and saving a file to change the previous parameter line by the new
value
    with open('C:\ICOLD_Z_G.u01', 'r') as f:
        dados = f.read()
    novop_cv = 'Non-Newtonian Constant Vol Conc=' + str(cv) + '\n'
    novop_ty = 'User Yeild=' + str(ty) + '\n'
    novop_k_n = 'Herschel-Bulkley Coef=' + str(k) + ', ' + str(n) + '\n'

    print('Simulação %d - ' % sim, novop_cv, novop_ty, novop_k_n)
    dados = dados.replace(antp_cv, novop_cv)
    dados = dados.replace(antp_ty, novop_ty)
    dados = dados.replace(antp_k_n, novop_k_n)

    with open('C:\ICOLD_Z_G.u01', 'w') as f:
        f.write(dados)

    # Opening the project in RAS and simulating with changed parameters
```

```
    RC.ShowRAS()
    RC.Project_Open(r"C:\ICOLD_Z_G.prj")
    RC.Compute_CurrentPlan(None, None, True)
    RC.QuitRAS()

    # Saving results of maximum depths in each cell
    file = "C:\ICOLD_Z_G.p01.hdf"
    with h5py.File(file, "r") as hdf:
        celldepth = hdf.get(
            '/Results/Unsteady/Output/Output Blocks/Base Output/Unsteady Time Se-
ries/2D Flow Areas/jusante/Cell Hydraulic Depth')
        depthmax = np.amax(celldepth, axis=0)
        depthmax = h5py.File('depth_max.hdf', 'a')
        depthmax.create_dataset('sim_%d' % sim, data=depthmax)
        depthmax.close()

    # Saving arrival times to 0.30 (1 foot) per simulation performed
        for depth in tempos:
            cell = depth.tolist()
            for x in cell:
                if x > 0.3:
                    tempo_cheg_t1.append(cell.index(x))
                    break
            else:
                tempo_cheg_t1.append(0)
                continue

        tempo30 = np.array(tempo_cheg_t1)
        tempo_chegada30 = h5py.File('tempo_chegada30.hdf', 'a')
        tempo_chegada30.create_dataset('sim_%d' % sim, data=tempo30)
        tempo_chegada30.close()

    # Accessing and saving flooded area results
    driver = ogr.GetDriverByName('ESRI Shapefile')
    hinputfile = driver.Open(r'C:\Inundation Boundary (Max Value_0).shp', 0)
    in_layer = hinputfile.GetLayer(0)

    # Transformation between projections
    src_srs = in_layer.GetSpatialRef()
    tgt_srs = osr.SpatialReference()
    tgt_srs.ImportFromEPSG(3395)
    transform = osr.CoordinateTransformation(src_srs, tgt_srs)
```

```
    for feature in in_layer:
        geom = feature.GetGeometryRef()
        geom2 = geom.Clone()
        geom2.Transform(transform)
        area_m2 = geom2.GetArea()
        area_km2 = area_m2 / 1000000
        print('Area em km²: ', area_km2)
        area = h5py.File('area.hdf', 'a')
        area.create_dataset('sim_%d' % sim, data=area_km2)
        area.close()

    # Storing the volume error in 1000 m3 and in percent per simulation
    with open('C:\ICOLD_Z_G.p01.computeMsgs.txt', 'r') as f:
        dados = f.readlines()
        erro_vol = []
        erro_volperc = []

        for line in dados:
            if re.search(r'Overall Volume Accounting Error in 1000', line):
                erro_vol.append(line)
                with open('erro_vol.txt', 'a') as txt:
                    txt.write(str(erro_vol))
                break

        for line in dados:
            if re.search(r'Overall Volume Accounting Error as percentage', line):
                erro_volperc.append(line)
                with open('erro_volperc.txt', 'a') as txt:
                    txt.write(str(erro_volperc))
                break

    print('End simulation ', sim)
    hinputfile.Destroy()
else:
    print("Simulations with LHS finished.")
```

Appendix C

Table A2. Specifications of the models and machines used in the probabilistic simulations.

Specification	Bingham	Herschel-Bulkley H-B
Model area	69.8 km^2	
2D mesh resolution	40 m on the slope centerline in the embedded valley 100 m in the downstream lake 50 m for the rest of the model	
Number of template cells	21,148	
Equation	Shallow Water Equations	
Simulation time frame	20 h	
Maximum number of computational iterations	20	
Computational Interval	Adjustable based on Courant: maximum = 1.0; minimum = 0.45 1.0–16.0 s	
Machine used	Processor AMD Ryzen 7 3700X. 8-Core. 3.6 GHz processing speed (4.4 GHz Turbo). 16 GB DDR4 RAM and 4 GB/s M,2 SSD	AMD Ryzen 3 3200G. 4-Core. 3.6 GHz processing speed. 8 GB DDR4 RAM and 6 GB/s Sata SSD
Average time per simulation	3.233 min/simulation	5.284 min/simulation

Appendix D

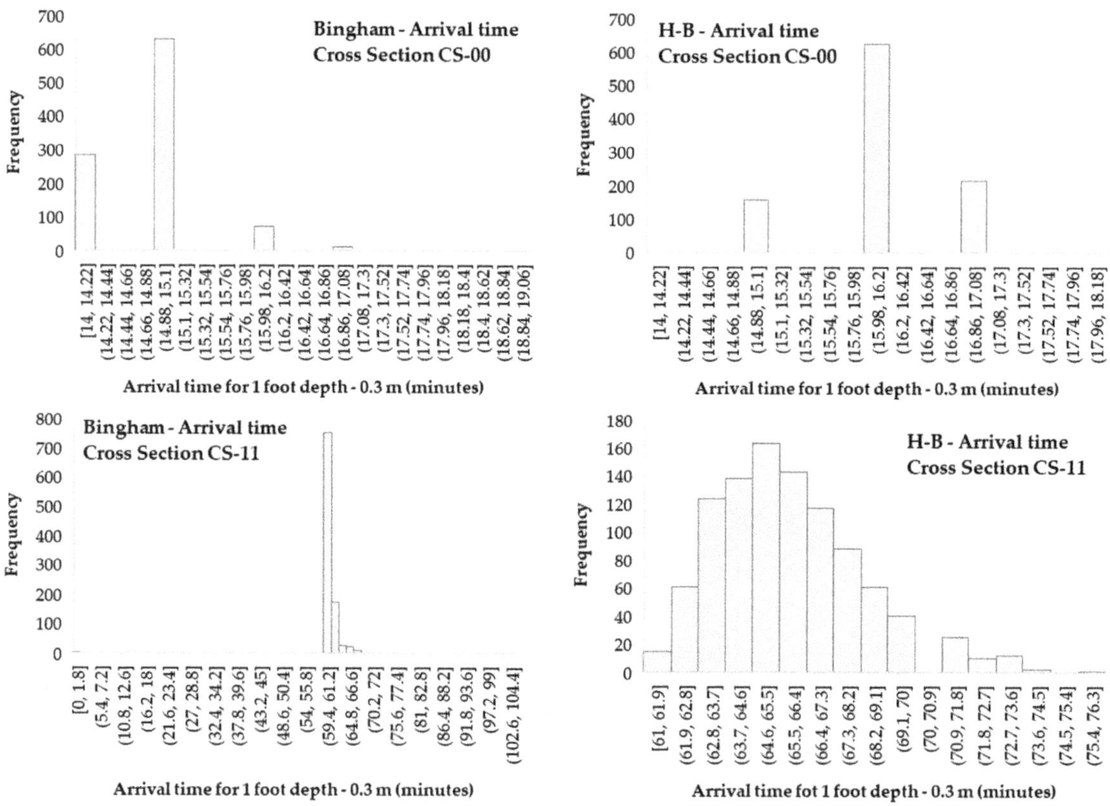

Figure A1. *Cont.*

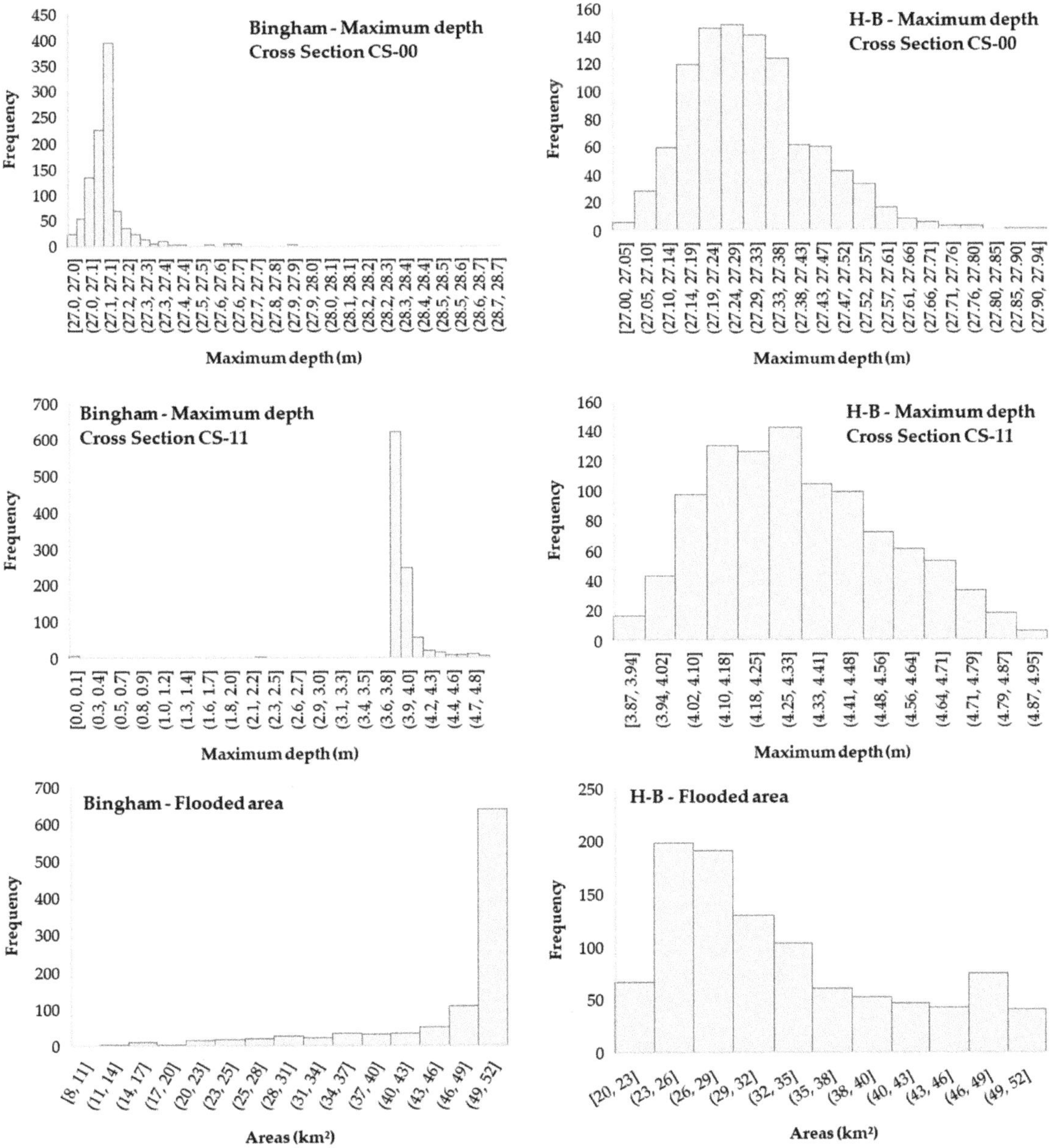

Figure A1. *Cont.*

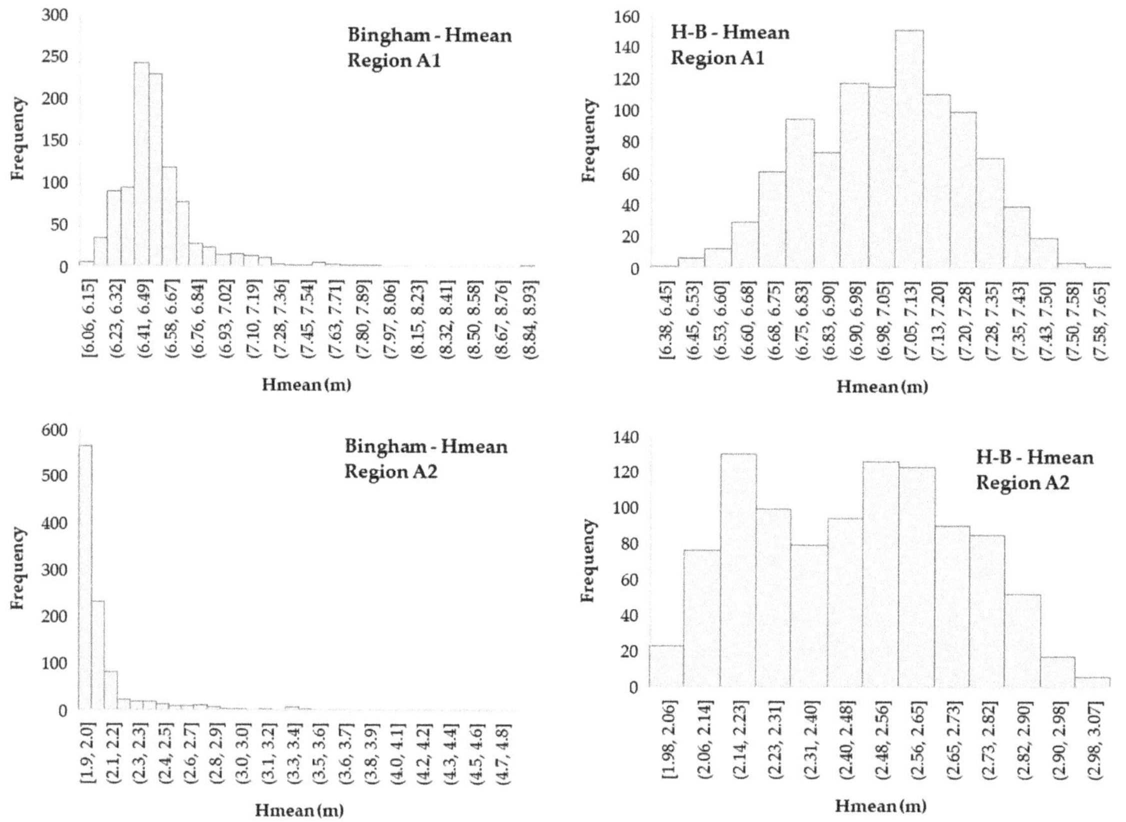

Figure A1. Probability distribution of Bingham and H-B model results.

References

1. Davies, M.P. Tailings Impoundment Failures: Are Geotechnical Engineers Listening? *Geotech. News* **2002**, *20*, 31–36.
2. Robertson, P.K.; de Melo, L.; Williams, D.J.; Wilson, G.W. Report of the Expert Panel on the Technical Causes of the Failure of Feijão Dam I. Available online: http://www.b1technicalinvestigation.com/ (accessed on 19 February 2022).
3. Morgenstern, N.R. et al Fundão Tailings Dam Review Panel. Available online: https://pedlowski.files.wordpress.com/2016/08/fundao-finalreport.pdf (accessed on 5 February 2022).
4. Julien, P.Y. *Erosion and Sedimentation*; Cambridge University Press: Cambridge, UK; New York, NY, USA, 1995; ISBN 978-0-521-44237-4.
5. Mezger, T. The Rheology Handbook: For Users of Rotational and Oscillatory Rheometers. In *The Rheology Handbook*; Vincentz Network: Hannover, Germany, 2020; ISBN 978-3-7486-0370-2.
6. O'Brien, J.S.; Julien, P.Y. Laboratory Analysis of Mudflow Properties. *J. Hydraul. Eng.* **1988**, *114*, 877–887. [CrossRef]
7. Faitli, J.; Gombkötő, I. Some Technical Aspects of the Rheological Properties of High Concentration Fine Suspensions to Avoid Environmental Disasters. *J. Environ. Eng. Landsc. Manag.* **2015**, *23*, 129–137. [CrossRef]
8. Jeong, S.-W. Shear Rate-Dependent Rheological Properties of Mine Tailings: Determination of Dynamic and Static Yield Stresses. *Appl. Sci.* **2019**, *9*, 4744. [CrossRef]
9. Zengeni, B.T. Bingham Yield Stress and Bingham Plastic Viscosity of Homogeneous Non-Newtonian Slurries. Ph.D. Thesis, Cape Peninsula University of Technology, Cape Town, South Africa, 2016.
10. Wang, X.; Wei, Z.; Li, Q.; Chen, Y. Experimental Research on the Rheological Properties of Tailings and Its Effect Factors. *Environ. Sci. Pollut. Res.* **2018**, *25*, 35738–35747. [CrossRef]
11. Jeyapalan, J.K.; Duncan, J.M.; Seed, H.B. Investigation of Flow Failures of Tailings Dams. *J. Geotech. Eng.* **1983**, *109*, 172–189. [CrossRef]
12. Yu, D.; Tang, L.; Chen, C. Three-Dimensional Numerical Simulation of Mud Flow from a Tailing Dam Failure across Complex Terrain. *Nat. Hazards Earth Syst. Sci.* **2020**, *20*, 727–741. [CrossRef]

13. Yang, S.-H.; Pan, Y.-W.; Dong, J.-J.; Yeh, K.-C.; Liao, J.-J. A Systematic Approach for the Assessment of Flooding Hazard and Risk Associated with a Landslide Dam. *Nat Hazards* **2013**, *65*, 41–62. [CrossRef]
14. O' Brien, J.S.; Julien, P.Y. *Physical Properties and Mechanics of Hyperconcentrated Sediment Flows*; Bowles, D.D., Ed.; Delineation of Landslide, Flash Flood, and Debris-Flow Hazards in Utah; Utah Water Research Laboratory: Logan, UT, USA, 1985; pp. 260–279.
15. D'Agostino, V.; Tecca, P.R. Some Considerations On The Application Of The FLO-2D Model For Debris Flow Hazard Assessment. In *Monitoring, Simulation, Prevention and Remediation of Dense and Debris Flows*; WIT Press: The New Forest, UK, 2006; Volume 90, pp. 159–170.
16. Naef, D.; Rickenmann, D.; Rutschmann, P.; McArdell, B.W. Comparison of Flow Resistance Relations for Debris Flows Using a One-Dimensional Finite Element Simulation Model. *Nat. Hazards Earth Syst. Sci.* **2006**, *6*, 155–165. [CrossRef]
17. Sosio, R.; Crosta, G.; Frattini, P. Field Observations, Rheologucal Testing and Numerical Modeling of a Debris-Flow Event. *Earth Surf. Process. Landf.* **2007**, *32*, 290–306. [CrossRef]
18. Rickenmann, D.; Laigle, D.; McArdell, B.W.; Hübl, J. Comparison of 2D Debris-Flow Simulation Models with Field Events. *Comput Geosci* **2006**, *10*, 241–264. [CrossRef]
19. Cesca, M.; D'Agostino, V. Comparison between FLO-2D and RAMMS in Debris-Flow Modelling: A Case Study in the Dolomites. In *Monitoring, Simulation, Prevention and Remediation of Dense Debris Flows II*; WIT Press: The New Forest, UK, 2008; Volume 60, pp. 197–206, ISBN 978-1-84564-118-4.
20. Lin, J.-Y.; Yang, M.-D.; Lin, B.-R.; Lin, P.-S. Risk Assessment of Debris Flows in Songhe Stream, Taiwan. *Eng. Geol.* **2011**, *123*, 100–112. [CrossRef]
21. Hungr, O. A Model for the Runout Analysis of Rapid Flow Slides, Debris Flows, and Avalanches. *Can. Geotech. J.* **1995**, *32*, 610–623. [CrossRef]
22. Arattano, M.; Franzi, L.; Marchi, L. Influence of Rheology on Debris-Flow Simulation. *Nat. Hazards Earth Syst. Sci.* **2006**, *6*, 519–528. [CrossRef]
23. Coussot, P.; Meunier, M. Recognition, Classification and Mechanical Description of Debris Flows. *Earth-Sci. Rev.* **1996**, *40*, 209–227. [CrossRef]
24. Zegers, G.; Mendoza, P.A.; Garces, A.; Montserrat, S. Sensitivity and Identifiability of Rheological Parameters in Debris Flow Modeling. *Nat. Hazards Earth Syst. Sci.* **2020**, *20*, 1919–1930. [CrossRef]
25. Iaccarino, G. Quantification of Uncertainty in Flow Simulations Using Probabilistic Methods. In Proceedings of the Non-Equilibrium Gas Dynamics from Physical Models to Hypersonic Flights, Rhode St. Genèse, Belgium, 8 September 2008.
26. Cepeda, J.; Quan Luna, B.; Nadim, F. Assessment of Landslide Run-out by Monte Carlo Simulations. In Proceedings of the Landslides, Risk & Reliability, Paris, France, 2 September 2013; pp. 2157–2160.
27. Contreras, S.; Castillo, C.; Olivera-Nappa, Á.; Townley, B.; Ihle, C.F. A New Statistically-Based Methodology for Variability Assessment of Rheological Parameters in Mineral Processing. *Miner. Eng.* **2020**, *156*, 106494. [CrossRef]
28. Kameda, J.; Okamoto, A. 1-D Inversion Analysis of a Shallow Landslide Triggered by the 2018 Eastern Iburi Earthquake in Hokkaido, Japan. *Earth Planets Space* **2021**, *73*, 116. [CrossRef]
29. Stowe, J.; Farrell, I.; Wingeard, E. Tailings Transport System Design Using Probabilistic Methods. *Min. Metall. Explor.* **2021**, *38*, 1289–1296. [CrossRef]
30. De La Rosa, Á.; Ruiz, G.; Castillo, E.; Moreno, R. Calculation of Dynamic Viscosity in Concentrated Cementitious Suspensions: Probabilistic Approximation and Bayesian Analysis. *Materials* **2021**, *14*, 1971. [CrossRef]
31. Quan Luna, B.; Blahut, J.; van Westen, C.J.; Sterlacchini, S.; van Asch, T.W.J.; Akbas, S.O. The Application of Numerical Debris Flow Modelling for the Generation of Physical Vulnerability Curves. *Nat. Hazards Earth Syst. Sci.* **2011**, *11*, 2047–2060. [CrossRef]
32. Wu, Y.-H.; Liu, K.-F.; Chen, Y.-C. Comparison between FLO-2D and Debris-2D on the Application of Assessment of Granular Debris Flow Hazards with Case Study. *J. Mt. Sci.* **2013**, *10*, 293–304. [CrossRef]
33. Pasculli, A.; Cinosi, J.; Turconi, L.; Sciarra, N. Learning Case Study of a Shallow-Water Model to Assess an Early-Warning System for Fast Alpine Muddy-Debris-Flow. *Water* **2021**, *13*, 750. [CrossRef]
34. Fallas Salazar, S.; Rojas González, A.M. Evaluation of Debris Flows for Flood Plain Estimation in a Small Ungauged Tropical Watershed for Hurricane Otto. *Hydrology* **2021**, *8*, 122. [CrossRef]
35. Bingham, E.C. *Fluidity and Plasticity.*; McGraw-Hill: New York, NY, USA, 1922.
36. Herschel, W.H.; Bulkley, R. Konsistenzmessungen von Gummi-Benzollösungen. *Kolloid-Z.* **1926**, *39*, 291–300. [CrossRef]
37. Ligier, P.-L. Implementation of Non-Newtonian Rheological Models in TELEMAC-2D. In Proceedings of the 2020 TELEMAC-MASCARET, Online, 15 October 2020; pp. 14–25.
38. Lumbroso, D.; Davison, M.; Body, R.; Petkovšek, G. Modelling the Brumadinho Tailings Dam Failure, the Subsequent Loss of Life and How It Could Have Been Reduced. *Nat. Hazards Earth Syst. Sci.* **2021**, *21*, 21–37. [CrossRef]
39. Da Silva, A.; Eleutério, J. Effectiveness of Dam-Breach Flood Alert in Mitigating Life-Losses—A Spatiotemporal Sectorisation Analysis in a High-Density Urban Area in Brazil. *Water* 2023, *under review*.
40. Zenz, G.; Goldgruber, M. ICOLD Proceeding, 12th International Benchmark Workshop on Numerical Analysis of Dams. In Proceedings of the ICOLD Proceedings, Gerald Zenz and Markus Goldgruber, Graz, Austria, 2 October 2013; Volume 12, p. 210.
41. Waele, A. *Viscometry and Plastometry*; Oil and Colour Chemists' Association: Manchester, UK, 1923.
42. Ostwald, W. Ueber die Geschwindigkeitsfunktion der Viskosität disperser Systeme. II. *Kolloid-Z.* **1925**, *36*, 157–167. [CrossRef]

43. Qian, N.; Wan, Z. *A Critical Review of the Research on the Hyperconcentrated Flow in China*; International Research and Training Centre on Erosion and Sedimentation: Beijing, China, 1986.
44. de Ferreira, F.O. Abordagem Matemática de Roll Waves em Escoamentos Hiperconcentrados Com Superfície Livre. Mater's Thesis, Universidade Estadual Paulista, Faculdade de Engenharia de Ilha Solteira, Ilha Solteira, Brazil, 2007.
45. Tarcha, B.A. Desafios na Medição da Tensão Limite de Escoamento de óleos Parafínicos. Mater's Thesis, Universidade Federal do Espírito Santo, Centro Tecnológico, Vitória, Espírito Santo, Brazil, 2014.
46. USACE (U. S. Army Corps of Engineers) HEC-RAS Mud and Debris Flow: Non-Newtonian User's Manual. Available online: https://www.hec.usace.army.mil/confluence/rasdocs/rasmuddebris (accessed on 5 June 2022).
47. Fitton, T.G.; Seddon, K.D. Relating Atterberg Limits to Rheology. In Proceedings of the Paste 2012, Australian Centre for Geomechanics, Perth, Australia, 16 April 2012; pp. 273–284.
48. Ribeiro, V.Q.F. Proposta de Metodologia para Avaliação do Efeito de Rupturas de Estruturas de Contenção de Rejeitos. Master's Thesis, Universidade Federal de Minas Gerais, Belo Horizonte, Brazil, 2015.
49. Machado, N.C. Retroanálise da propagação decorrente da ruptura da barragem do fundão com diferentes modelos numéricos e hipóteses de simulação. Master's Thesis, Universidade Federal de Minas Gerais, Belo Horizonte, Brazil, 2017.
50. Gitari, W.M.; Thobakgale, R.; Akinyemi, S.A. Mobility and Attenuation Dynamics of Potentially Toxic Chemical Species at an Abandoned Copper Mine Tailings Dump. *Minerals* **2018**, *8*, 64. [CrossRef]
51. Mahmood, A.A.; Elektorowicz, M. An Investigation of the Porosity Dependent Strength and Leachability of Mine Tailings Matrices Containing Heavy Metals. *Cogent Environ. Sci.* **2020**, *6*, 1743626. [CrossRef]
52. Gibson, S.; Moura, L.Z.; Ackerman, C.; Ortman, N.; Amorim, R.; Floyd, I.; Eom, M.; Creech, C.; Sánchez, A. Prototype Scale Evaluation of Non-Newtonian Algorithms in HEC-RAS: Mud and Debris Flow Case Studies of Santa Barbara and Brumadinho. *Geosciences* **2022**, *12*, 134. [CrossRef]
53. Bagnold, R.A. Experiments on a Gravity-Free Dispersion of Large Solid Spheres in a Newtonian Fluid under Shear. *Proc. R. Soc. Lond. A* **1954**, *225*, 49–63. [CrossRef]
54. Rickenmann, D. Hyperconcentrated Flow and Sediment Transport at Steep Slopes. *J. Hydraul. Eng.* **1991**, *117*, 1419–1439. [CrossRef]
55. Olsson, A.; Sandberg, G.; Dahlblom, O. On Latin Hypercube Sampling for Structural Reliability Analysis. *Struct. Saf.* **2003**, *25*, 47–68. [CrossRef]
56. McKay, M.D.; Beckman, R.J.; Conover, W.J. A Comparison of Three Methods for Selecting Values of Input Variables in the Analysis of Output from a Computer Code. *Technometrics* **1979**, *21*, 239–245. [CrossRef]
57. Pirouz, B.; Javadi, S.; Seddon, K. Thickener Performance Variability: Underflow Solids Concentration and Flowrate. In Proceedings of the Paste 2017, University of Science and Technology Beijing, Beijing, China, 16 June 2017; pp. 29–40.
58. Bezerra, R.; Eleutério, J. A Novel Full 2D Probabilistic Hydrodynamic Approach to Model and Map Floods from Dam Failures Considering Breach Parameters Uncertainty. *Water*, 2023; *under review*.
59. Goodell, M.C.R. *Breaking the HEC-RAS Code: A User's Guide to Automating HEC-RAS*, 1st ed.; H2ls: Portland, OR, USA, 2014; ISBN 978-0-9908918-0-2.
60. Dysarz, T. Application of Python Scripting Techniques for Control and Automation of HEC-RAS Simulations. *Water* **2018**, *10*, 1382. [CrossRef]
61. Papaioannou, G.; Vasiliades, L.; Loukas, A.; Aronica, G.T. Probabilistic Flood Inundation Mapping at Ungauged Streams Due to Roughness Coefficient Uncertainty in Hydraulic Modelling. *Adv. Geosci.* **2017**, *44*, 23–34. [CrossRef]
62. Hamouda, T. Impact of Micro-Topography and Bathymetry Modification on Inundation Modelling with Different Magnitudes Based on SRTM Data. Master's Thesis, UNESCO-IHE Institute for Water Education, Delft, The Netherlands, 2018.
63. Da Silva, A.Â.C.L.; Eleutério, J.C. Identifying and Testing the Probability Distribution of Earthfill Dam Breach Parameters for Probabilistic Dam Breach Modeling. *J. Flood Risk Manag.* **2023**, e12900. [CrossRef]
64. Froehlich, D. Empirical Model of Embankment Dam Breaching. In Proceedings of the River Flow 2016, St. Louis, MO, USA, 11–14 July 2016; CRC Press: Boca Raton, FL, USA, 2016; pp. 1821–1826.
65. FEMA (Federal Emergency Management Agency). *Federal Guidelines for Inundation Mapping of Flood Risks Associated with Dam Incidents and Failures*; FEMA: Washington, DC, USA, 2013; p. 145.
66. Hussin, H.Y. Probabilistic Run-out Modeling of a Debris Flow in Barcelonnette, France. Thesis, ITC: Faculty of Geo-Information Science and Earth Observation, Enschede, The Netherlands, 2011. Available online: http://essay.utwente.nl/84877/ (accessed on 30 July 2023).
67. Wooster, J.K.; Dusterhoff, S.R.; Cui, Y.; Sklar, L.S.; Dietrich, W.E.; Malko, M. Sediment Supply and Relative Size Distribution Effects on Fine Sediment Infiltration into Immobile Gravels. *Water Resour. Res.* **2008**, *44*. [CrossRef]

water

Article

Accurate Numerical Modeling for 1D Open-Channel Flow with Varying Topography

Zijian Xue, Ling Zhou * and Deyou Liu

College of Water Conservancy and Hydropower Engineering, Hohai University, 1 Xikang Road, Nanjing 210098, China; hhuxzj@163.com (Z.X.); liudyhhuc@163.com (D.L.)
* Correspondence: zlhhu@163.com

Abstract: In this paper, an accurate numerical model is presented for one-dimensional open-channel flows with varying topographies; the model is specifically applied to rectangular channels with variable widths. A pressure term is introduced in the shallow water momentum equation to construct a new conservation term, and the resulting non-conservation term is included in the source term to characterize the actual topography changes in the open-channel flow. Based on a Harten Lax and van Leer (HLL) Riemann solver for a homogeneous system, an upwind scheme is introduced into the model in which the flux is determined via randomly selecting a local Riemann solution state. This two-step random choice method enables the scheme to reach second-order accuracy in space. A Runge–Kutta scheme is introduced into the discretization of the source term of the system to achieve second-order accuracy. The proposed model is validated via a selection of steady and transient hydraulic problems with reference solutions. When compared with published experimental results, the predictions of the proposed model show a high degree of accuracy.

Keywords: open-channel flow; random choice method; Riemann problem; unsteady flow; varying topography

check for updates

Citation: Xue, Z.; Zhou, L.; Liu, D. Accurate Numerical Modeling for 1D Open-Channel Flow with Varying Topography. *Water* **2023**, *15*, 2893. https://doi.org/10.3390/w15162893

Academic Editors: Andrea Maranzoni, Francesca Aureli and Gabriella Petaccia

Received: 19 June 2023
Revised: 3 August 2023
Accepted: 4 August 2023
Published: 10 August 2023

1. Introduction

In the past, unsteady gravity flows occurring in sewers were simulated via numerical solutions for one-dimensional equations of momentum and continuity. To describe an incompressible flow on a nearly horizontal free surface under the effect of gravity, shallow water equations (SWEs) are used, assuming that the vertical acceleration component can be neglected. Compared with the Navier–Stokes equations, the computational solutions of SWEs require much less computational work while maintaining highly accurate flow physical characteristics. This hyperbolic, quasilinear system has been studied extensively in fluid dynamics, and its solutions have wide applications, such as in simulating tsunamis, river floods, and storm surges.

The numerical solutions of SWEs have been of interest to engineers and researchers for a long time. As a result, many efforts have been made to improve the accuracy and efficiency of SWE numerical solutions, such as developing high-order numerical schemes or coupling them with other models to account for complex physical processes. Due to the development of numerical schemes for the Euler equations in aerodynamics over the past 30 years, many numerical schemes used to solve one-dimensional gas-phase problems have been applied to the numerical simulations of hydraulic open-channel flows. For example, Fennema and Chaudhry [1] successfully applied the MacCormack scheme [2] to the numerical simulation of one-dimensional dam-break problems. Glaister [3] simulated shallow water flow using an approximate Riemann solver. Garcia-Navarro et al. [4] applied an explicit finite-difference TVD-MacCormack scheme to the numerical simulation of a one-dimensional open-channel flow. Savic and Holly [5] simulated a critical flow using the Godunov-type scheme. Hu, Yinying, et al. [6] simulated a free-surface transient flow based on a Godunov-type scheme with unsteady friction.

The presence of fluid discontinuities leads to Riemann problems in shallow water models. To deal with these shocks, there are well-established Riemann solvers used in numerical simulations, such as exact or approximate solutions and the linearized Roe method [7]. Among the most widely used methods for shallow water flows are the flux difference splitting (FDS) and flux vector splitting (FVS) methods [8], which are highly efficient and simple to implement. These methods have the advantage of providing sharp resolutions for discontinuities, but they can sometimes produce unwanted numerical oscillations. Researchers have proposed different techniques, such as limiters, slope limiters, and flux limiters, to mitigate these unwanted oscillations and attain more accurate solutions.

In 1965, Glimm [9] introduced a particular class of schemes as a critical component of a constructive proof of the existence of solutions for a set of nonlinear systems known as hyperbolic conservation laws. The scheme called the random choice method (RCM) is particularly noteworthy for its ability to effectively manage intricate wave interactions that include various forms of discontinuities, such as shock waves and material interfaces [10]. Because of these advantages, the RCM has been improved and applied in various fields. Su, Ying-Chin, et al. [11] developed a stable, extended RCM for conservation laws, and they successfully applied it to solve the compressible Euler equations in a transonic flow. Z. Wang and Q. Zhang [12] investigated a more general result in p-system, using an adapted RCM approach in the BV space. Recently, Olivier Hurisse [13] applied the RCM to solve the specific issue of advection in an indicator function and effectively expanded its application into two-dimensional domains. Based on the concepts utilized in the RCM, Gallout, T., and Hurisse, O., [14] proposed a two-step approach involving convection and projection for simulating front advection. They successfully extended this method to a multidimensional problem. P. Yun and Y.Q. Zhang [15] developed the random choice method to investigate the issue of the shape of a wedge in a planar, steady, supersonic potential flow. Xue et al. [16] applied the RCM technique to the field of transient mixed flows and found that this scheme effectively suppresses oscillation issues at the interface of flow state transitions.

In this work, an accurate numerical model is proposed for one-dimensional open-channel flows with changing widths and bed profiles. To discrete the hyperbolic part of the system, a two-step upwind scheme utilizing the random choice method is presented. The second-order scheme is used to handle the source term of the system. The proposed model is subjected to thorough examination and validation via numerical and experimental tests. The structure of this paper is as follows: (1) the governing equations of one-dimensional open-channel flows with varying topographies are presented; (2) an HLL Riemann solution for a conservational term and a two-step upwind scheme with the RCM are presented; (3) the stability constraints for the source term, discretized using the second-order Runge–Kutta method, are formulated; and (4) the model undergoes validation through a series of numerical simulations and practical experiments. Ultimately, the findings are consolidated and presented in a concise manner.

2. Mathematical Formulation

The geometry of the open-channel flow model is shown in Figure 1. Considering the assumptions that vertical movements have minimal impact, the water is incompressible, and the river channel features a gentle slope [17], one-dimensional unsteady flows in open channels can be expressed in a vector conservational form. The formulation is as follows:

$$\frac{\partial \mathbf{U}}{\partial t} + \frac{\partial \mathbf{F}}{\partial x} = \mathbf{S} \tag{1}$$

in which

$$\begin{aligned} \mathbf{U} &= \begin{bmatrix} A \\ Q \end{bmatrix}, \ \mathbf{F} = \begin{bmatrix} Q \\ \frac{Q^2}{A} + gI_1 \end{bmatrix}, \\ \mathbf{S} &= \begin{bmatrix} 0 \\ gA\left(S_0 - S_f\right) + gI_2 \end{bmatrix} \end{aligned} \tag{2}$$

$$\text{with } I_1(h) = \int_{\eta_b}^{\eta} (\eta - \tau)B(x,\tau)\mathrm{d}\tau \text{ and} \\ I_2(h) = \int_{\eta_b}^{\eta} (\eta - \tau)\frac{\partial B(x,\tau)}{\partial x}\mathrm{d}\tau \tag{3}$$

where **U** represents the conserved variables in the form of a vector; $A(x,\eta)$ is a function of the space and water surface elevation, which are predetermined and arbitrary, representing the cross-sectional area of the channel; Q is the flow discharge; **F** denotes the flux in the x direction; g is the gravitational acceleration; I_1 and I_2 are integral terms that are related to the hydrostatic pressure forces caused by the varying channel breadth; η is the water surface elevation; η_b denotes the bottom elevation; $B(x,\tau)$ is the channel width; **S** represents the source terms; S_0 represents the inclination or gradient of the channel bed; and S_f denotes the inclination or gradient of the energy line within the channel.

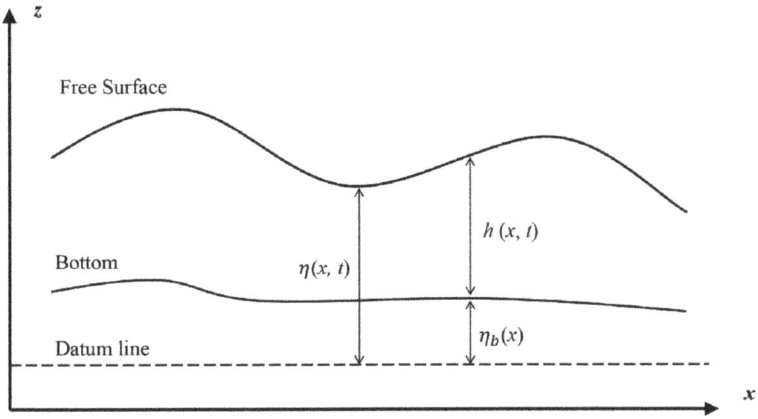

Figure 1. The geometry of the open-channel flow.

The terms in the momentum equation can be written as follows:

$$\frac{\partial}{\partial x}gI_1 - gI_2 = \frac{\partial}{\partial x}g\int_{\eta_b}^{\eta} (\eta - \tau)B(x,\tau)\mathrm{d}\tau - \int_{\eta_b}^{\eta} (\eta - \tau)\frac{\partial B(x,\tau)}{\partial x}\mathrm{d}\tau \\ = Ag\frac{\partial(\eta-\eta_b)}{\partial x} = Ag\frac{\partial h}{\partial x} \tag{4}$$

where the approximation comes from the simplified assumption of a rectangular cross-section of varying width $B(x)$, and the water depth gives $h = \eta - \eta_b$.

In source terms, the bottom slope gives $S_0 = -\partial_x\eta_b$ and the friction term S_f may be estimated using an empirical formula such as Manning's equation, namely:

$$S_f = \frac{{n_m}^2 Q|Q|}{A^2 R^{4/3}} \tag{5}$$

where R is the hydraulic radius and n_m is the Manning coefficient.

Based on this, the momentum equation can be rewritten by introducing a pressure term $P(x,A) = g\int_0^h A(x,\eta)\mathrm{d}\eta$ [18].

$$\frac{\partial P(x,A)}{\partial x} = Ag\frac{\partial \eta}{\partial x} + g\int_0^h \left(\frac{\partial A(x,\eta)}{\partial x}\right)_\eta \mathrm{d}\eta - Ag\frac{\partial \eta_b}{\partial x} \tag{6}$$

Compared with Equations (1)–(3), the formulation reads as follows:

$$\frac{\partial A}{\partial t} + \frac{\partial Q}{\partial x} = 0 \\ \frac{\partial Q}{\partial t} + \frac{\partial}{\partial x}\left[\frac{Q^2}{A} + P(x,A)\right] = g\int_0^h \left(\frac{\partial A(x,\eta)}{\partial x}\right)_\eta \mathrm{d}\eta - gA\frac{\partial \eta_b(x)}{\partial x} - gAS_f \tag{7}$$

For a rectangular cross-section of a varying width $B(x)$, the unsteady shallow water Equation (7) may be rewritten as

$$\begin{gathered} \frac{\partial \mathbf{W}}{\partial t}+\frac{\partial \mathbf{G}}{\partial x}=\mathbf{J} \\ \mathbf{W}=\begin{bmatrix} h \\ hu \end{bmatrix}, \mathbf{G}=\begin{bmatrix} hu \\ hu^2+\frac{1}{2}gh^2 \end{bmatrix}, \\ \mathbf{J}=\begin{bmatrix} -\frac{hu}{B}\frac{\partial B}{\partial x} \\ -\frac{hu^2}{B}\frac{\partial B}{\partial x}+gh\frac{\partial \eta_b}{\partial x}-ghS_f \end{bmatrix} \end{gathered} \tag{8}$$

3. Riemann Solver

The Jacobian Matrix of the flux vector $\mathbf{G}(\mathbf{W})$ in Equation (8) is expressed as

$$\mathbf{A}=\frac{\partial \mathbf{G}}{\partial \mathbf{W}}=\begin{bmatrix} 0 & 1 \\ c^2-u^2 & 2u \end{bmatrix} \tag{9}$$

set $c=\sqrt{gh}$ and u is the velocity components in x direction. The eigenstructure $\mathbf{A}=\mathbf{R\Lambda L}$ provides

$$\mathbf{R}=\begin{bmatrix} 1 & 1 \\ u-c & u+c \end{bmatrix}, \mathbf{\Lambda}=\begin{bmatrix} u-c & 0 \\ 0 & u+c \end{bmatrix}, \mathbf{L}=\begin{bmatrix} \frac{u+c}{2c} & -\frac{1}{2c} \\ -\frac{u-c}{2c} & \frac{1}{2c} \end{bmatrix} \tag{10}$$

The Riemann problem is the specific initial value problem of a non-linear hyperbolic system. The homogeneous conservation system yield the following:

$$\begin{gathered} \mathbf{W}_t+\mathbf{G}(\mathbf{W})_x=0 \\ \mathbf{W}(x,t=0)=\begin{cases} \mathbf{W}_L & \text{if } x<0 \\ \mathbf{W}_R & \text{if } x>0 \end{cases} \end{gathered} \tag{11}$$

The HLL approximate Riemann Solver was adopted in this paper, providing an approximation solution of the intercell numerical variables $\mathbf{W}_{i+1/2}$ according to following equations:

$$\mathbf{W}_{i+1/2}=\begin{cases} \mathbf{W}_L & \text{if } S_L>0 \\ \frac{S_R\mathbf{W}_R-S_L\mathbf{W}_L+\mathbf{G}_L-\mathbf{G}_R}{S_R-S_L} & \text{if } S_L\le 0\le S_R \\ \mathbf{W}_R & \text{if } S_R<0 \end{cases} \tag{12}$$

and the intercell numerical flux $\mathbf{G}_{i+1/2}$ is as follows:

$$\mathbf{G}_{i+1/2}=\begin{cases} \mathbf{G}_L & \text{if } S_L>0 \\ \frac{S_R\mathbf{G}_L-S_L\mathbf{G}_R+S_RS_L(\mathbf{W}_R-\mathbf{W}_L)}{S_R-S_L} & \text{if } S_L\le 0\le S_R \\ \mathbf{G}_R & \text{if } S_R<0 \end{cases} \tag{13}$$

where $(\mathbf{W}_L,\mathbf{W}_R)$ and $(\mathbf{G}_L,\mathbf{G}_R)$ are the solution and the flux vector at $t=t^n$, respectively. The wave speeds S_L and S_R represent the speed of the discontinuity at the two sides of the Riemann problem. For an open-channel flow, they are written as

$$\begin{gathered} S_L=\min(u_L-c_L,\hat{u}-\hat{c}_L) \\ S_R=\max(u_R+c_R,\hat{u}+\hat{c}_R) \end{gathered} \tag{14}$$

where the velocity (u_L,u_R) and the wave celerity (c_L,c_R) are the flow variables of the left and right states, $\mathbf{W}_L$, and $\mathbf{W}_R$, respectively. Using the Roe averaged method [19], $\hat{u}$ and $\hat{c}$ are given by the following:

$$\hat{c}=\sqrt{\frac{P_L-P_R}{A_L-A_R}} \text{ and } \hat{u}=\frac{u_L\sqrt{A_L}+u_R\sqrt{A_R}}{\sqrt{A_L}+\sqrt{A_R}} \tag{15}$$

The HLL approach provides a straightforward method for handling dry fronts, effectively addressing the challenges associated with regions of zero or near-zero flow velocities in which one of the data states has no water. If a dry bed exists, there is only one depression wave present in the Riemann system, which causes a loss of strict hyperbolicity. For the depression wave, namely, a single rarefaction wave, the head or toe speeds of the two characteristic waves are significant. The right dry bed state Riemann solution is given as

$$\mathbf{W}_{Right} = \begin{cases} \mathbf{W}_L & \text{if } x/t \le u_L - c_L \\ \mathbf{W}_{rare} & \text{if } u_L - c_L < x/t \le u_L + 2c_L \\ \mathbf{W}_R & \text{if } x/t > u_L + 2c_L \end{cases} \tag{16}$$

where the rarefaction wave solution at a point (x, t) is

$$\mathbf{W}_{rare} = \begin{cases} c = (u_L + 2c_L - x/t)/3 \\ u = (u_L + 2c_L + 2x/t)/3 \end{cases} \tag{17}$$

Also, for the case of left dry bed one has

$$\mathbf{W}_{Left} = \begin{cases} \mathbf{W}_L & \text{if } x/t \le u_R - 2c_R \\ \mathbf{W}_{rare} & \text{if } u_R - 2c_R < x/t \le u_R + c_R \\ \mathbf{W}_R & \text{if } x/t > u_R + c_R \end{cases} \tag{18}$$

and

$$\mathbf{W}_{rare} = \begin{cases} c = (-u_R + 2c_R + x/t)/3 \\ u = (u_R - 2c_R + 2x/t)/3 \end{cases} \tag{19}$$

4. Numerical Methodology

Equation (8) is discretized on a spatial grid over the interval [0; L]. The discretization grid divides the x-axis into space interval segments Δx and the t-axis into intervals of duration Δt. The ith control volume is positioned at the center and spans from $i{-}1/2$ to $i{+}1/2$.

The RCM scheme is based on an existence proof by Glimm [9] and was first successfully implemented by Chorin [20]. This method selects an x-position of random sampling and obtains Riemann solutions of the next time level, illustrated in Figure 2. To represent the value of W_i^{n+1} at the next time level, given the constant variables W_i^n of a cell i at a time step n, we can employ the following two steps:

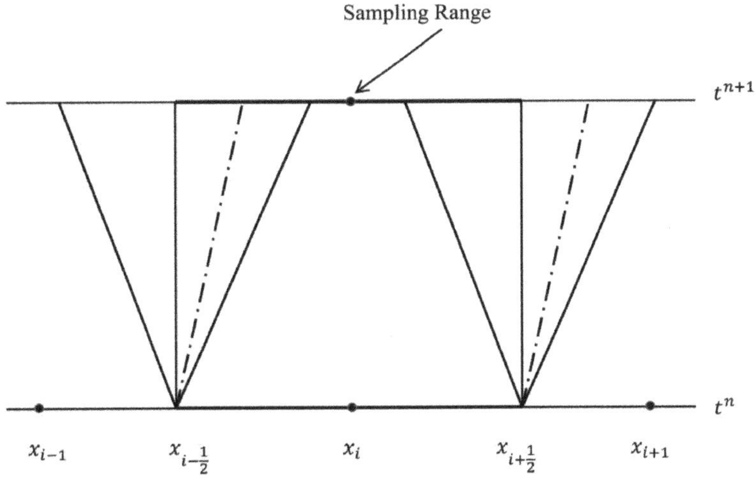

Figure 2. The RCM on a non-staggered grid.

Step **I**. Obtain the Riemann solutions $\mathbf{W}_{i-\frac{1}{2}}(x/t)$ and $\mathbf{W}_{i+\frac{1}{2}}(x/t)$ of the Riemann problems RP ($\mathbf{W}_{i-1}^n, \mathbf{W}_i^n$) and RP ($\mathbf{W}_i^n, \mathbf{W}_{i+1}^n$) at a general point ($x$, t).

Step **II**. Perform a random sampling of solutions in the cell i to obtain a state at time level t_{n+1}. The selected state is determined based on a random or quasi-random number R^n, which falls within the range of [0, 1].

$$\mathbf{W}_i^{n+1} = \begin{cases} \mathbf{W}_{i-1/2}\left(R^n \frac{\Delta x}{\Delta t}\right) & 0 \leq R^n \leq \frac{1}{2} \\ \mathbf{W}_{i+1/2}\left(R^n \frac{\Delta x}{\Delta t}\right) & \frac{1}{2} < R^n \leq 1 \end{cases} \tag{20}$$

Here, the van der Corput sequences are used to sample the solution [21], and were found to provide the best results [22].

One variation of the RCM involves a staggered grid and implements a two-step procedure. The non-linear systems' solutions are obtained through the following two steps:

Step **I**. Solve the local Riemann problems $\left(\mathbf{W}_{i-1}^n, \mathbf{W}_i^n\right)$ and $\left(\mathbf{W}_i^n, \mathbf{W}_{i+1}^n\right)$, which are defined by the constant states $\mathbf{W}_{i-1}^n$, $\mathbf{W}_i^n$ and $\mathbf{W}_{i+1}^n$. These Riemann problems are solved to determine the intermediate states for each cell interface.

$$\hat{\mathbf{W}}_{i-\frac{1}{2}}^{n+\frac{1}{2}}(x,t),\ \hat{\mathbf{W}}_{i+\frac{1}{2}}^{n+\frac{1}{2}}(x,t) \tag{21}$$

Randomly select local Riemann states; the solutions at time $t^{n+1/2}$ are calculated and mapped onto a secondary grid. This secondary grid serves as a mid-point grid or an intermediate grid where the numerical calculations take place as follows:

$$\mathbf{W}_{i-\frac{1}{2}}^{n+\frac{1}{2}} = \hat{\mathbf{W}}_{i-\frac{1}{2}}^{n+\frac{1}{2}}\left(\frac{R^n \Delta x}{\Delta t^{n+\frac{1}{2}}}\right),\ \mathbf{W}_{i+\frac{1}{2}}^{n+\frac{1}{2}} = \hat{\mathbf{W}}_{i+\frac{1}{2}}^{n+\frac{1}{2}}\left(\frac{R^n \Delta x}{\Delta t^{n+\frac{1}{2}}}\right) \tag{22}$$

Step II. Random-sample the Riemann problem $\hat{\mathbf{W}}_i^{n+1}(x,t)$, which is defined by the data $\mathbf{W}_{i-\frac{1}{2}}^{n+\frac{1}{2}}$ and $\mathbf{W}_{i+\frac{1}{2}}^{n+\frac{1}{2}}$, on the staggered grid. Subsequently, the solution is allocated to a new position at the time level t^{n+1}, aligning with the position it occupied at time t^n.

The final solution is written as

$$\mathbf{W}_i^{n+1} = \hat{\mathbf{W}}_i^n\left(\frac{R^n \Delta x}{\Delta t^{n+1}}\right) \tag{23}$$

Based on the RCM mentioned before, a two-step scheme is introduced which enables the scheme to reach second-order accuracy in space. Using the framework of the finite difference method, in the first step, calculate the Riemann problems RP $(\mathbf{W}_{i-1}^n, \mathbf{W}_i^n)$ and RP $(\mathbf{W}_i^n, \mathbf{W}_{i+1}^n)$ to find the respective solutions, $\hat{\mathbf{W}}_{i-\frac{1}{2}}^{n+\frac{1}{2}}(x,t)$, $\hat{\mathbf{W}}_{i+\frac{1}{2}}^{n+\frac{1}{2}}(x,t)$. After that, random-sample the solutions at time $t_{n+\frac{1}{2}}$ in a similar manner as in the RCM with a modified random number R^n, where R^n is the sequence of pseudo-random numbers in $[-1/2, 1/2]$. Next, with the solutions of $t = t_{n+\frac{1}{2}}$, the boundary fluxes of the cell i can be obtained, and $\mathbf{W}_i^n$ is advanced to $\mathbf{W}_i^{n+1}$ via the upwind scheme.

$$\mathbf{W}_i^{n+1} = \mathbf{W}_i^n - \frac{\Delta t}{\Delta x}\left[\mathbf{G}\left(\mathrm{W}_{i+\frac{1}{2}}^{n+\frac{1}{2}}\right) - \mathbf{G}\left(\mathrm{W}_{i-\frac{1}{2}}^{n+\frac{1}{2}}\right)\right] \tag{24}$$

Therefore, this higher-order random choice method can be considered a random generalization of the two-step Richtmeyer form of the Lax–Wendroff scheme. Here, the Lax–Friedrichs scheme is replaced by the classical random choice method, and the second leap-frog type step is identical to the Lax–Wendroff scheme.

5. Discretization of Source Terms and Stability Constraints

The source term **J** associated with the geometry in this study is incorporated into the solution via a time-splitting process, employing a second-order Runge–Kutta discretization which culminates in the ensuing explicit method.

1. Pure advection:

$$\mathbf{W}_i^{n+1} = \mathbf{W}_i^n - \frac{\Delta t}{\Delta x}\left(\mathbf{G}_{i+1/2}^{n+1/2} - \mathbf{G}_{i-1/2}^{n+1/2}\right) \tag{25}$$

2. Update with the source term by $\Delta t/2$:

$$\overline{\mathbf{W}}_i^{n+1} = \mathbf{W}_i^{n+1} + \frac{\Delta t}{2}\mathbf{J}\left(\mathbf{W}_i^{n+1}\right) \tag{26}$$

3. Re-update with the source term by Δt:

$$\overline{\overline{\mathbf{W}}}_i^{n+1} = \mathbf{W}_i^{n+1} + \Delta t\mathbf{J}\left(\overline{\mathbf{W}}_i^{n+1}\right) \tag{27}$$

where the evaluation of the source term $\mathbf{J}$, which appears in Equation (8), requires the definition of the variable width and bottom, namely,

$$\begin{aligned} \left(\frac{\partial B(x)}{\partial x}\right)_i &= \frac{B_{i+1/2}-B_{i-1/2}}{x_{i+1/2}-x_{i-1/2}} = \frac{\Delta B_i}{\Delta x_i} \\ \left(\frac{\partial \eta_b(x)}{\partial x}\right)_i &= \frac{\eta_{bi+1/2}-\eta_{bi-1/2}}{x_{i+1/2}-x_{i-1/2}} = \frac{\Delta \eta_{bi}}{\Delta x_i} \end{aligned} \tag{28}$$

and the grid energy line slope $\left(S_f\right)_i$, which may be obtained from Manning's equation, is as follows:

$$\left(S_f\right)_i = \frac{{n_m}^2 u_i|u_i|}{R_i^{4/3}} \tag{29}$$

where R is the hydraulic radius and n_m is the Manning roughness coefficient.

The stability constraint of the model includes two parts: the Courant–Friedrichs–Lewy (CFL) criterion and the source term constraint. The CFL constraint criterion is provided here.

$$\mathrm{Cr}_{\max} = \frac{\Delta t}{\Delta x_i}\mathrm{Max}_{i=0,1,2,\ldots Nx,Nx+1}\left(|u_i^n| + |c_i^n|\right) \leq 1 \tag{30}$$

where $\mathrm{Cr}_{\max}$ is he maximum Courant number at the time level n. From Equation (30), the permissible time step for the convective part is given by

$$\Delta t_{\max,\mathrm{CFL}} = \mathrm{Min}_{i=0,1,2,\ldots Nx,Nx+1}\left[\frac{\Delta x_i}{|u_i^n| + |c_i^n|}\right] \tag{31}$$

It can also be shown that the permissible time step for the source term is given by the following [23]:

$$\Delta t_{\max,\mathbf{J}} = \mathrm{Min}_{i=0,1,2,\ldots Nx,Nx+1}\left[-4\frac{\mathbf{W}_i^{n+1}}{\mathbf{J}\left(\mathbf{W}_i^{n+1}\right)}, -2\frac{\mathbf{W}_i^{n+1}}{\mathbf{J}\left(\overline{\mathbf{W}}_i^{n+1}\right)}\right] \tag{32}$$

where the ratio of vectors $\mathbf{W}/\mathbf{J}(\mathbf{W})$ is defined as $\left[\frac{A}{-\frac{hu}{B}B_x} \quad \frac{Q}{-\frac{hu^2}{B}B_x + gh(\eta_b)_x - ghS_f}\right]^T$. Since the same time step Δt must be used for the convective part and the source term, $\mathbf{W}_i^n$ must be used instead of $\mathbf{W}_i^{n+1}$. Finally, the maximum permissible time step including the convective part and the source term will be given by

$$\Delta t_{\max} = \mathrm{Min}_{i=0,1,2,\ldots Nx,Nx+1}\left[\Delta t_{\max,\mathbf{J}}, \Delta t_{\max,\mathrm{CFL}}\right] \tag{33}$$

6. Numerical Validations and Discussion

This section describes the validation of the present model against various open-channel flow tests with different widths/bottom topographies. The gravitational acceleration used in all tests was 9.81 $\mathrm{m/s^2}$, and the Courant number value was 0.8. The model was tested against a range of scenarios to evaluate its ability to accurately simulate open-channel flows and to ensure its suitability for future applications.

6.1. Dam Break Tests

In this test, the accuracy of the presented model was evaluated via dam break (dry and wet beds) simulations without friction. The first test scenario involved simulating the sudden opening of a sluice gate separating two pools of still water. The two pools are located at different depths with a 1000 m long frictionless, horizontal channel. Two ends of the pools are closed borders which can be calculated via a reflection boundary in the entire numerical simulation. The initial conditions provide the following:

$$\begin{cases} y = 10.0 \text{ m and } u = 0.0 \text{ m/s for } x \leq \ 500 \text{ m} \\ y = 3.0 \text{ m and } u = 0.0 \text{ m/s for } x > 500 \text{ m} \end{cases} \tag{34}$$

Figure 3 illustrates a comparison of the simulation results of the present and Godunov-type scheme (GTS) models, with the exact solution corresponding to the specified Courant number and grid size. The results of the simulation indicate that the present model has a great ability to handle discontinuities, particularly in case of the shock wave, where the resolution is more distinct compared to the Godunov-type scheme (GTS). For the rarefaction wave, the present model also shows a high level of accuracy, but wave oscillation appeared at the tail of the rarefaction wave.

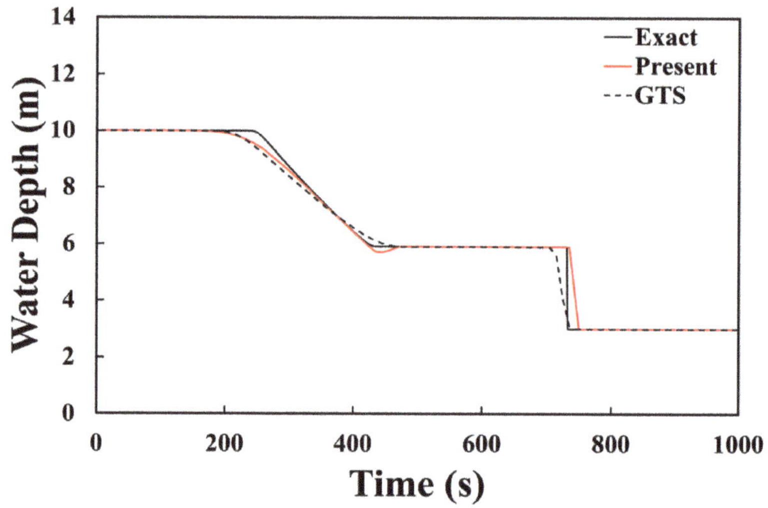

Figure 3. Water depth profile simulations compared with the exact solution at time = 36 s ($\Delta x = 10$ m).

To evaluate the accuracy of the model for dam break flows in dry bed scenarios, we considered a specific test case. In this analytical case, there is a frictionless horizontal channel measuring 10 m in length. The computational domain is divided into two halves: the right half has an initial water depth (h) of 0.3 m, while the left half represents a dry bed. At the beginning of the simulation, the water in the entire domain is in its initial state, with no motion. A Riemann solution of the dry bed in a shallow water flow is given in Equations (16)–(19). The water depth and velocity profiles at t = 1 s with a given Courant number and grid size are shown in Figure 4, respectively. As can be seen from the figures, the velocity profiles generated by the present model demonstrate good simulation capabilities at the dry–wet interface compared to the GTS. Actually, this demonstrates the advantages of the numerical scheme with the RCM currently employed. It allows for more accurate simulation results, especially when dealing with dry–wet interfaces.

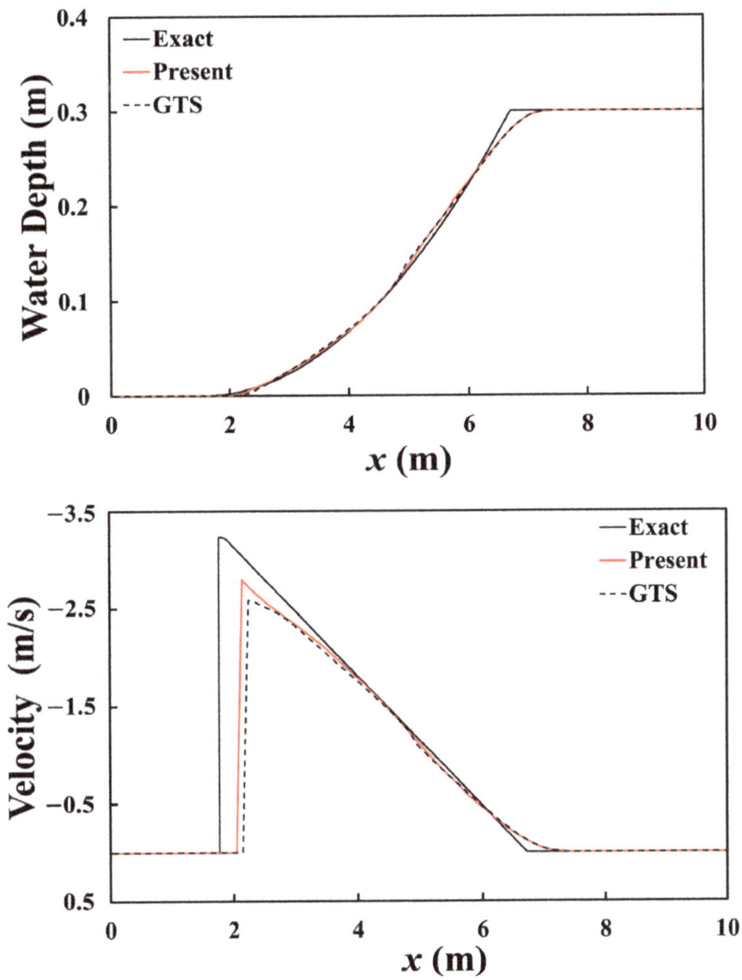

Figure 4. Simulation results for a dry bed problem at t = 1 s ($\Delta x = 0.1$ m).

6.2. Lobovsky et al.'s Dam Break ExperimentA

The second case in this section is about an experimental test, and a schematic of the experiment by Lobovsky et al. [24] is presented in Figure 5. The gray square area in Figure 5 represents still water at the initial moment, and the left side of the region is a dry bed. The points labeled H1, H2, H3, and H4 indicate the specific locations at which the water level was measured. These points serve as reference points for monitoring and analyzing the behavior of the water level at different positions along the tank. More details on the experimental configuration and the apparatus utilized can be found in Lobovsky et al. [24]. A comparison of the simulation results of the present model and the measurement data is shown in Figure 6. The water depths were standardized by referencing the initial water depth h = 0.3 m, and the time measurement on the x-axis was adjusted by scaling it according to $(g/h)^{1/2}$.

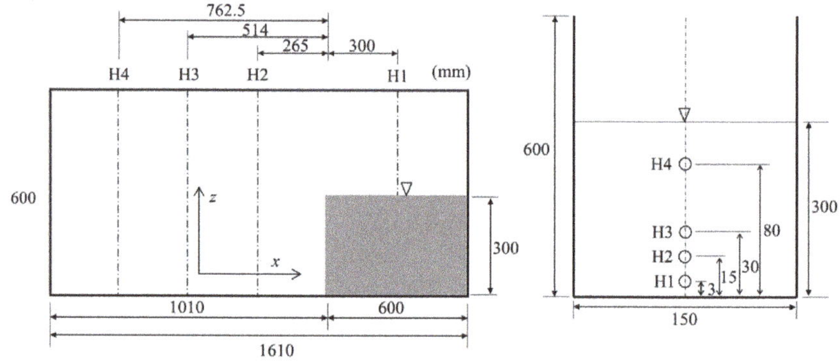

Figure 5. A sketch of the experiment by Lobovsky et al.

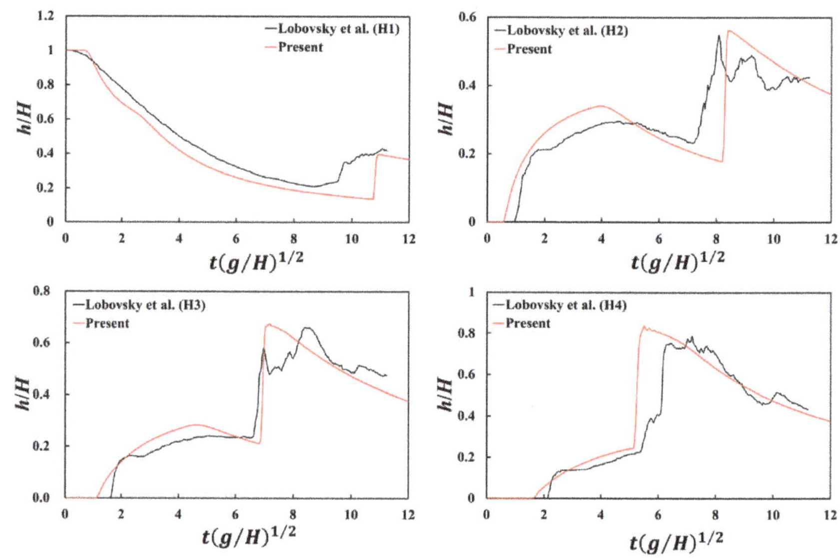

Figure 6. The simulated water levels at the designated points in relation to time are compared with the experimental data ($\Delta x = 0.061$ m).

Compared with the scale of the experiment, the grid size selected here is relatively dense ($Nx = 100$). Based on the observations from Figure 6, the simulation results are quite aligned with the measured values for the current model. The performance of the proposed model in this test is satisfactory, especially at the second peak of the four locations where very steep simulation results were achieved.

6.3. Laboratory Experiment of a Dam Break Flow over a Triangular Obstacle

A varying bed elevation is considered in this case. This test comes from the Concerted Action on Dam-break Modeling project (CADAM) [25], which presents effective benchmarks for verifying dam break flow simulations. As depicted in Figure 7, the experimental arrangement comprised a reservoir, a dam, and a rectangular channel measuring 38 m in length and 1.75 m in width. The dam separates the reservoir from a linear channel spanning 22.5 m in length. A symmetrical triangular elevation, measuring 0.4 m in height and extending 6 m in length, is located 28.5 m away from the upstream end. The channel is initially a dry bed with a constant Manning coefficient of $n_m = 0.0125$. In this test, the upstream end is assumed to be solid, and a chute is imposed downstream as a boundary condition.

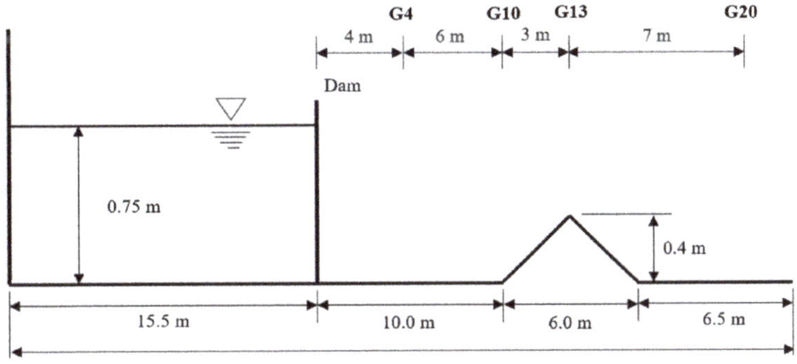

Figure 7. Layout of dam break flow over a hump.

At the instant of the initial time, the dam experiences a sudden breach, causing the initially stagnant water in the reservoir to rapidly flow onto the floodplain downstream. As shown in Figure 8, comparisons of the simulation results and experimental measurements were made at four gauges (G4, G10, G13, and G20). Several flow patterns can be observed in this test, like shock front propagation, a wet–dry front, partial reflections, and hydraulic jumps. The wet–dry front reaches location G4 about t = 1 s after the dam breaks. At t = 12 s, an increase in the incident water depth can be observed due to the shock wave. After that, a shock wave moving upstream forms because of the interaction between the incoming flow and the bump. As a result, a second increase in water depth can be observed at t = 35 s. At the same time, a rarefaction wave is generated and propagates downstream, leading to a reduction in the water level above the hump.

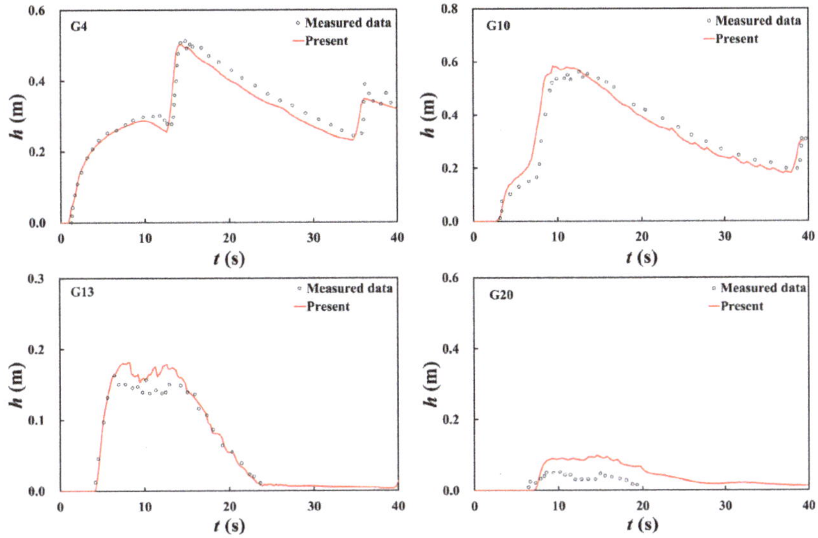

Figure 8. Dam break flow over an elevated section: chronological records of water depth at various measuring points: G4, G10, G13, and G20 ($\Delta x = 0.76$ m; $n_m = 0.0125$).

It is proved in this figure that the comparison between the numerical predictions of the present model and the measurements yields satisfactory results at the gauge locations, and the transitions between dry and wet conditions are simulated with high precision.

The current model produces oscillation issues; however, it accurately predicts the timing and maximum peak value of each of the two hydraulic surges. This confirms the effectiveness of the current method in dealing with a complex channel flow with non-uniform bed topography.

6.4. Dam Break Flow onto a Channel with a Local Constriction

Figure 9 depicts the layout of a channel featuring a gate positioned at a distance of 6.1 m from the initial end of the upstream. At a starting point of 7.9 m beyond the gate, the channel, which is 0.5 m in width, gradually narrows down to a width of 0.1 m over a stretch of 0.2 m. The channel then sustains this width of 0.1 m for an extent of 1 m before returning to its primary width over a span of 0.2 m. The initial static water depths located upstream and downstream of the gate are noted as 0.3 m and 0.003 m, respectively.

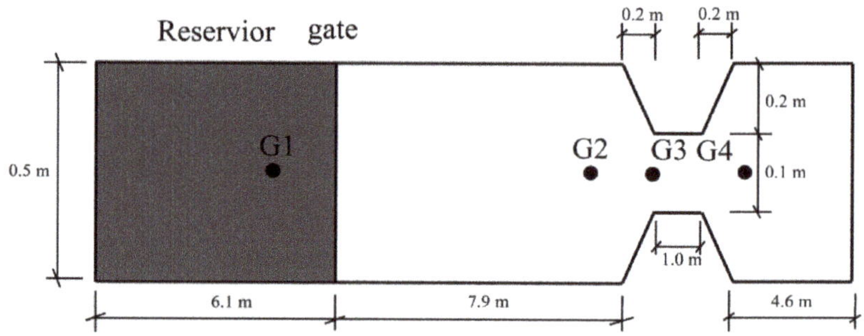

Figure 9. Geometry of the channel with the changing width.

In this test, the number of grid cells is set to 50, and a Manning coefficient of 0.01 is employed. The upstream boundary is assumed to be close, while the downstream boundary is set to be transmissive. The simulation was executed for a duration of 10 s following the collapse of the dam. Figure 10 illustrates a comparison between the forecasted water depth profiles at four designated measurement points with experimental data from the laboratory tests and the corresponding numerical outcomes generated by Goutal, N., and Maurel, F. [18].

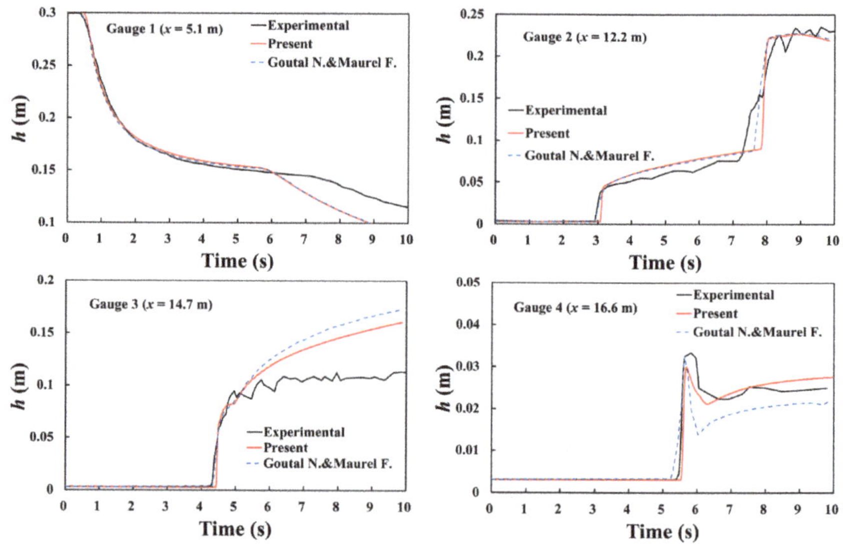

Figure 10. Dam break flow with a local constriction at four gauge points.

The measuring points are positioned along the central axis of the waterway, as indicated in Figure 9.

1. Gauge 1: 1.00 m upstream from the dam;
2. Gauge 2: 6.10 m downstream from the dam;
3. Gauge 3: 8.60 m downstream from the dam;
4. Gauge 4: 10.50 m downstream from the dam.

After the dam breaks, a surge of water is discharged downstream, propagating as a shock wave. Upstream of the dam, a rarefaction wave develops, resulting in a decrease in the water level, as depicted in Figure 10 at Gauge 1. Gauge 2 experiences a sudden surge in water depth when the shock front arrives after approximately 3 s. At t = 8 s, another abrupt change in depth is observed when the reflected shock from the narrow passage reaches the gauge. Inside the narrow passage at Gauge 3, only one sudden increase in water depth occurs. Downstream of the narrow passage at Gauge 4, the water level experiences a sudden increase caused by the incoming pressure surge, after which it drops to a relatively stable level.

The present model and the predictions made by Goutal, N., and Maurel, F., are quite similar. Especially at Gauge points 3 and 4, the predicted water depth ofthe present model after 7 s is more consistent with the experimental data than the results from Goutal, N., and Maurel, F.. Overall, this test validates the model in dealing with a complex channel flow with a varying width.

6.5. Channel Contraction Test Problem

To validate the numerical schemes, simulations were conducted to replicate a set of steady flows in domains with changing topographies. This specific quasi-equilibrium flow issue was introduced by MacDonald et al. [26] and is widely recognized as a benchmark for assessing numerical schemes. It involves all source terms described by Equation (8), which includes diverse channel widths, changing bed elevations, and frictional effects.

The flow occurs in a channel with a length of 1000 m and exhibits specific analytical characteristics. The depth variation features a hydraulic jump located at the mid-point of the channel and is expressed as a function of the horizontal coordinate x.

$$h(x) = \begin{cases} -\frac{1}{40} + \left[1 + 2\left(\frac{x}{1000} - \frac{1}{2}\right)^2\right]^{-1} & 0 \le x \le 500 \\ a_0 \exp\left(\frac{x}{4000} - \frac{1}{4}\right) + \sum_{k=1}^{3} a_k \exp\left(15k - 30k\frac{x}{1000}\right) & 500 \le x \le 1000 \end{cases} \tag{35}$$

where $a_0 = 1.5$, $a_1 = -0.230680$, $a_2 = 0.248267$, and $a_3 = -0.228271$.

As shown in Figure 11, variable channel width is given by

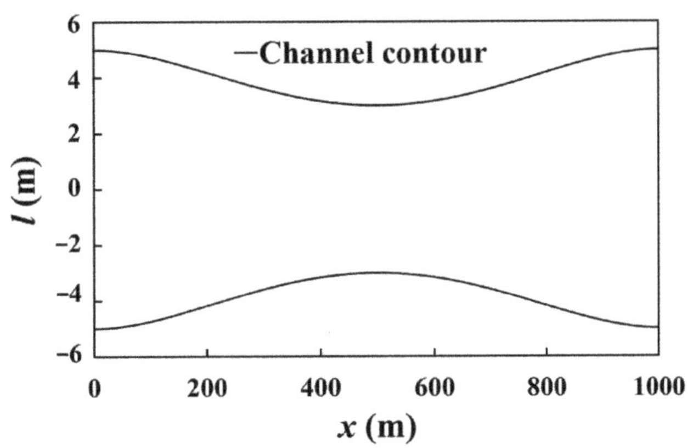

Figure 11. Width contour $B(x)$.

$$B(x) = 10 - 64\left[\left(\frac{x}{1000}\right)^2 - 2\left(\frac{x}{1000}\right)^3 + \left(\frac{x}{1000}\right)^4\right] \quad (36)$$

And the friction term S_f is

$$S_f(x,t) = {n_m}^2 u(x,t)|u(x,t)|\left(\frac{2}{l(x)} + \frac{1}{h(x,t)}\right)^{4/3} \quad (37)$$

The bed slope $S_0(x)$ is

$$S_0(x) = S_f(x) + \left(1 - \frac{u^2(x)}{gh(x)}\right)h'(x) - \frac{u^2(x)}{g}\frac{l'(x)}{l(x)} \quad (38)$$

The particular data for this test case include a Manning resistance coefficient of $n_m = 0.02$, and the physical boundary conditions for this simulation include an upstream discharge of 20 m^3/s and a downstream water depth of 1.5 m. The discretization of the channel is accomplished using a uniform grid spacing of Δx = 2.5 m. The simulation was initialized using these specific boundary and initial conditions, and the simulation ran until a steady state was reached. The exact bed elevation, water surface level, and velocity profiles obtained from the numerical simulation are plotted in Figure 12, and they show a good agreement with the expected analytical profiles. The evaluation of numerical and analytical profiles is an essential step in validating the precision of the numerical approach used in the simulation. The predicted free surface level matches the analytical profiles, indicating that the current model has achieved a satisfactory level of accuracy for simulating steady flows.

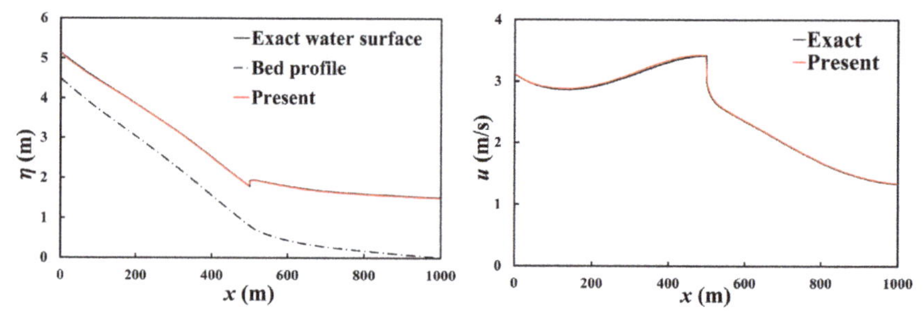

Figure 12. Numerical and analytical solutions for the steady flow test.

7. Conclusions

In this paper, a novel model was proposed for simulating a 1D open-channel flow with a changing width and bed profile. For the hyperbolic problems in the homogeneous system, a higher-order RCM in the framework of an upwind scheme was used by sampling a local Riemann solution state at random with an HLL Riemann solver. The Runge–Kutta scheme was introduced for the discretization of the source term of the system to achieve second-order accuracy and ensure accurate numerical simulations.

To test the proposed model, five specific tests were carried out. These tests encompassed dam break flow tests and different configurations involving variable widths, bottom elevations, and frictions. The first test focused on the simulation of an unsteady dam break flow in a flat bed channel under both dry and wet downstream conditions for which the presented model was evaluated compared to the Godunov-type scheme and the exact solutions. The second, third and fourth tests were unsteady dam break tests that considered the source term, and the calculation results of the presented model were compared with the experimental data. The last test was a steady flow test including a varying channel width, changing bed elevation, and friction, which was used to verify the stability of the model. The main findings of the proposed model can be summarized as follows:

(1) The present model has a great ability to handle discontinuities, especially in the shock wave or the dry–wet interface, which exhibits a higher level of resolution compared

to the Godunov-type scheme (GTS). For the rarefaction wave, the present model also shows a high level of accuracy, but wave oscillation appeared.

(2) By comparing the experimental results, the performance of the proposed model is found to be satisfactory, particularly in accurately simulating dry-to-wet transitions. The presented model produces oscillation issues but achieves accurate predictions for the timings and maximum peak values of the hydraulic surges. These confirm the effectiveness of the current method in presence of source term discretization.

(3) The simulated free surface level matches precisely with the analytical profiles while maintaining satisfactory accuracy for steady discharge solutions in the steady flow test. This demonstrates the ability of the presented model to handle source terms and to maintain equilibrium for a steady flow.

The proposed model is inherently accuracy and stable, and it deserves further research and application.

Author Contributions: Conceptualization, L.Z. and D.L.; methodology, Z.X.; formal analysis, Z.X.; Investigation, Z.X.; writing original draft preparation, Z.X.; writing-review and editing. All authors helped in reviewing and editing the manuscript; funding acquisition, D.L. All authors have read and agreed to the published version of the manuscript.

Funding: This research received no external funding.

Data Availability Statement: The study's data are available from the corresponding author by request.

Acknowledgments: The writers gratefully acknowledge the financial support for this research from the Fundamental Research Funds for the Central Universities (Grant Nos. B200203060 and 2018B43114), the Graduate Research and Innovation Project of Jiangsu Province (Grant No. KYCX20_0473), the National Natural Science Foundation of China (Grant Nos. 51839008 and 51679066), and the Fok Ying Tong Education Foundation (Grant No. 161068).

Conflicts of Interest: The authors declare no conflict of interest.

References

1. Fennema, R.J.; Hanif Chaudhry, M. Simulation of one-dimensional dam-break flows. *J. Hydraul. Res.* **1987**, *25*, 41–51. [CrossRef]
2. MacCormack, R.W. The effect of viscosity in hypervelocity impact cratering. *Front. Comput. Fluid Dyn.* **2002**, *2022*, 27–43. [CrossRef]
3. Glaister, P. Approximate Riemann solutions of the shallow water equations. *J. Hydraul. Res.* **1988**, *26*, 293–306. [CrossRef]
4. Garcia-Navarro, P.; Alcrudo, F.; Saviron, J.M. 1-D open-channel flow simulation using TVD-McCormack scheme. *J. Hydraul. Eng.* **1992**, *118*, 1359–1372. [CrossRef]
5. Savic, L.J.; Holly Jr, F.M. Dam break flood waves computed by modified Godunov method. *J. Hydraul. Res.* **1993**, *31*, 187–204. [CrossRef]
6. Hu, Y.; Zhou, L.; Pan, T.; Fang, H.; Li, Y.; Liu, D. Godunov-type solutions for free surface transient flow in pipeline incorporating unsteady friction. *AQUA-Water Infrastruct. Ecosyst. Soc.* **2022**, *71*, 546–562. [CrossRef]
7. Roe, P.L. Approximate Riemann solvers, parameter vectors, and difference schemes. *J. Comput. Phys.* **1981**, *43*, 357–372. [CrossRef]
8. Steger, J.L.; Warming, R. Flux vector splitting of the inviscid gasdynamic equations with application to finite-difference methods. *J. Comput. Phys.* **1981**, *40*, 263–293. [CrossRef]
9. Glimm, J. Solution in the Large for Nonlinear Hyperbolic Systems of Equations. *Comm. Pure. Appl. Math.* **1965**, *18*, 697–715. [CrossRef]
10. Zhang, Q.; Sheng, W. A random choice method based on the generalized Riemann problem for the Euler equations in gas dynamics. *J. Comput. Phys.* **2021**, *441*, 110431. [CrossRef]
11. Su, Y.-C.; Hong, J.M.; Chou, S.-W. An extension of Glimm's method to the gas dynamical model of transonic flows. *Nonlinearity* **2013**, *26*, 1581. [CrossRef]
12. Wang, Z.; Qi, Z. Periodic solutions to p-system constructed through Glimm scheme. *J. Math. Anal. Appl.* **2016**, *435*, 1088–1098. [CrossRef]
13. Hurisse, O. On the use of Glimm-like schemes for transport equations on multidimensional domain. *Int. J. Numer. Methods Fluids.* **2021**, *93*, 1235–1268. [CrossRef]
14. Gallout, T.; Hurisse, O. Convergence of a multidimensional Glimm-like scheme for the transport of fronts. *IMA J. Numer. Anal.* **2021**, *42*, 2924–2958. [CrossRef]
15. Yun, P.; Zhang, Y. An Inverse Problem for Determining the Shape of the Wedge in Steady Supersonic Potential Flow. *J. Math. Fluid Mech.* **2023**, *25*, 25. [CrossRef]

16. Xue, Z.; Zhou, L.; Liu, D.; Che, T.C. A Random Choice Scheme for Transient Mixed Flows. *J. Hydraul. Eng.* **2023**, *149*, 04023026. [CrossRef]
17. Colella, P. Glimm's method for gas dynamics. *SIAM J. Sci. Comput.* **1982**, *3*, 76–110. [CrossRef]
18. Goutal, N.; Maurel, F. A finite volume solver for 1D shallow-water equations applied to an actual river. *Int. J. Numer. Meth. Fluids* **2002**, *38*, 1–19. [CrossRef]
19. Roe, P.L.; Pike, J. Efficient Construction and Utilisation of Approximate Riemann Solutions. In *Computing Methods in Applied Science and Engineering*; Elsevier: Amsterdam, The Netherlands, 1984. [CrossRef]
20. Chorin, A.J. Random Choice solution of Hyperbolic Systems. *J. Comput. Phys.* **1976**, *22*, 517. [CrossRef]
21. Toro Eleuterio, F. *Reimann Solvers and Numerical Methods for Fluid Dynamics*, 3rd ed.; Springer: Berlin/Heidelberg, Germany, 2009. [CrossRef]
22. Fiedler, F.R.; Ramirez, J.A. A numerical method for simulating discontinuous shallow flow over an infiltrating surface. *Int. J. Numer. Methods Fluids* **2000**, *32*, 219–239. [CrossRef]
23. León, A.S.; Ghidaoui, M.S.; Schmidt, A.R.; Garcia, M.H. Efficient second-order accurate shock-capturing scheme for modeling one-and two-phase water hammer flows. *J. Hydraul. Eng.* **2008**, *134*, 970–983. [CrossRef]
24. Lobovsky, L.; Botia-Vera, E.; Castellana, F.; Mas-Soler, J.; Souto-Iglesias, A. Experimental investigation of dynamic pressure loads during dam break. *J. Fluids Struct.* **2014**, *48*, 407–434. [CrossRef]
25. Morris, M. CADAM: Concerted Action on Dambreak Modeling—Final Report. Rep. SR 571. HR Wallingford. 2000. Available online: http://eprints.hrwallingford.com/id/eprint/447 (accessed on 27 June 2023).
26. MacDonald, I.; Baines, M.J.; Nichols, N.K.; Samuels, P.G. Analytic benchmark solution of open-channel flows. *J. Hydraul. Eng.* **1997**, *123*, 1041–1045. [CrossRef]

water

MDPI

Article

Non-Equilibrium Bedload Transport Model Applied to Erosive Overtopping Dambreach

Sergio Martínez-Aranda [1,*], Javier Fernández-Pato [2] and Pilar García-Navarro [1]

[1] Fluid Dynamic Technologies I3A, University of Zaragoza, C. Maria de Luna 3, 50018 Zaragoza, Spain
[2] Estación Experimental Aula Dei EEAD, CSIC, Avda. Montañana 1005, 50059 Zaragoza, Spain
* Correspondence: sermar@unizar.es

Abstract: Bedload sediment transport is an ubiquitous process in natural surface water flows (rivers, dams, coast, etc), but it also plays a key role in catastrophic events such as dyke erosion or dam breach collapse. The bedload transport mechanism can be under equilibrium state, where solid rate and flow carry capacity are balanced, or under non-equilibrium (non-capacity) conditions. Extremely transient surface flows, such as dam/dyke erosive collapses, are systems which always change in space and time, hence absolute equilibrium states in the coupled fluid/solid transport rarely exist. Intuitively, assuming non-equilibrium conditions in transient flows should allow to estimate correctly the bedload transport rates and the bed level evolution. To get insight into this topic, a 2D Finite Volume model for bedload transport based on the non-capacity approach is proposed in this work. This non-equilibrium model considers that the actual bedload sediment discharge can be delayed, spatial and temporally, from the instantaneous solid carry capacity of the flow. Furthermore, the actual solid rate and the adaptation length/time is governed by the temporal evolution of the bedload transport layer and the vertical exchange solid flux. The model is tested for the simulation of overtopping dyke erosion and dambreach opening cases. Numerical results seems to support that considering non-equilibrium conditions for the bedload transport improves the general agreement between the computed results and measured data in both benchmarking cases.

Keywords: dam breaching; overtopping erosion; bedload transport; non-equilibrium transport; finite volume methods; breach side stability

Citation: Martínez-Aranda, S.; Fernández-Pato, J.; García-Navarro, P. Non-Equilibrium Bedload Transport Model Applied to Erosive Overtopping Dambreach. *Water* **2023**, *15*, 3094. https://doi.org/10.3390/w15173094

Academic Editors: Francesca Aureli, Andrea Maranzoni and Gabriella Petaccia

Received: 8 August 2023
Revised: 23 August 2023
Accepted: 26 August 2023
Published: 29 August 2023

1. Introduction

In surface water flows, such as river or coasts, the bed material particles are transported by two main mechanisms: suspended load transport accounts for the sediment particles moving within the flow column, whereas bedload transport involves the particles moving in a relatively thin layer over the static bed stratum. In this bedload transport layer, the solid particles are in contact with the underlying static stratum and hence the particle velocity is much lower than that of the the surrounding fluid [1]. The bedload transport plays a key role in the riverbed morphology since it can lead to marked and rapid variations in the bed level. Bedload transport can occur under a steady equilibrium state or, contrarily, under transient non-equilibrium conditions. The classical equilibrium approach considers that the actual bedload rate is the capacity of the flow to carry solid weight and hence only depending on instantaneous local flow features. The capacity bedload rate can be estimated using different empirical closure relations found in literature [2] and is the most widespread approach in current bedload transport models [3–10].

Contrarily, in bedload models based on the non-equilibrium assumption, the actual transport rates must be computed through the advection of the solid particles in the bedload layer and the mass exchange with the underlying static riverbed [11–15], and usually it does not agree with the local load carry capacity of the flow. Intuitively, assuming non-equilibrium conditions in transient flows should allow to estimate correctly the solid transport rates and the bed evolution, but this topic remains uncertain nowadays. Van-Rijn [16] suggested that the bedload rate can be assumed equal to the solid carry capacity event in moderate unsteady flows, since the adjustment of the transported solid load to new local flow conditions proceeds enough rapid close to the riverbed surface. However, this assumption was only conceptual and not supported by theoretical or numerical results. Later, analysing numerically the time scale of the processes involved in bedload transport, Cao et al. [13,17] found that the bedload rate was able adapt rapidly to the local carry capacity in flood cases. These works seem to justify the widespread application of the equilibrium/capacity bedload models [17]. However, this fact is not supported by any results in case of highly transient erosive dambreach flows.

Assuming non-equilibrium conditions for the bedload transport requires to compute locally both the thickness of the bedload transport layer and the bulk velocity of the solid particles. Furthermore, the net exchange flux of solid particles between static riverbed and moving bedload layer also has to be estimated, since it works as a source/sink of solid load. Therefore, non-equilibrium bedload models for transient flows involve a high uncertainty in the estimation of the parameters associated to entrainment/deposition vertical flux and the transport layer thickness and velocity. In the specialized literature, different strategies have been reported to overcome this uncertainty. El Kadi Abderrezzak et al. [18] assumed non-storage conditions for the transport layer and approximated the net solid flux through the interface between the static bed and the moving transport layer by the divergence between the actual and capacity transport rates. This theoretical model also introduced the adaptation length concept, fixed in time and space, leading to the quasi-steady relaxation flux approach for bedload models. Based on the same adaptation length concept, refs. [11,19,20] proposed more complex approaches to determine the net exchange flux through the interface but accounting now for the mass storage in the bedload transport layer. These models assume the actual transport rate as one of the unknown conservative variables, but they estimate the bulk bedload layer velocity using the closure relations found in literature [21–23] developed in laboratory under solid carry capacity conditions.

Recently, novel complex developments applying the non-capacity approach in bedload transport models have been derived, most of them based on the pioneering work Charru [24]. This work studies, at a grain scale, the mechanical interaction between moving fluid (clear water) and sediment particles at the interface between the static riverbed and the moving bedload layer. Based on this particle-scale momentum balance, Zech et al. [25] modelled the net exchange flux through the interface as a function of the shear stress exerted by the flow in the bedload transport layer and the shear strength for the incipient motion at the upper boundary of the static underlying stratum. Based also on [24], Fernández-Nieto et al. [14,26] derived closure relationships for the erosion and deposition rates at the static-moving interface, leading to a novel expression for the actual solid transport rate under non-equilibrium conditions. Under equilibrium conditions, this new expression for the solid transport reduces to the classical formula for bedload rate. Some interesting works have been recently published modelling bedload transport rates in cohesive clay–silt–gravel mixtures [27,28]. Finally, Bohorquez and Ancey [15] performed a grain-scale analysis of the particle activity at the bedload transport layer and proposed expressions for the entrainment/deposition rates including an additional diffusion term in the non-capacity bedload transport equation.

In this work, the novel generalized non-equilibrium bedload transport formulation developed by Martínez-Aranda et al. [29,30] for 2D bedload transport models is used to simulate dyke overtopping erosion and dam breach collapse processes. The equilibrium and non-equilibrium approaches are compared for different synthetic tests and experimental benchmarking cases. The main goal of this work is to test the suitability of the non-equilibrium approach for these highly unsteady erosive flows. This manuscript organized as follows: in Section 2 the system of conservation equations for 2D bedload transport is introduced; the numerical method is detailed in Section 3; the model results for one synthetic test and two experimental benchmarking cases are analysed in Section 4; and finally the main conclusions are drawn in Section 5.

2. Mathematical Modelling of 2D Generalized Bedload Transport

The 2D governing equations for bedload transport assuming non-equilibrium approach can be written as a compact conservation laws system [30], which includes the conservation laws for the water mass and the depth-averaged momentum of the flow, together with the transport equation for the erodible bed material:

$$\frac{\partial \mathbf{U}}{\partial t} + \frac{\partial \mathbf{F}(\mathbf{U})}{\partial x} + \frac{\partial \mathbf{G}(\mathbf{U})}{\partial y} = \mathbf{S_b}(\mathbf{U}) + \mathbf{S_\tau}(\mathbf{U}) \tag{1}$$

where $\mathbf{U} = (h,\ hu,\ hv,\ z_b)^T$ are the conservative variables, $\mathbf{F}(\mathbf{U})$ and $\mathbf{G}(\mathbf{U})$ are the conservative fluxes along the x- and y-coordinates respectively:

$$\mathbf{U} = \begin{pmatrix} h \\ hu \\ hv \\ z_b \end{pmatrix} \quad \mathbf{F}(\mathbf{U}) = \begin{pmatrix} hu \\ hu^2 + \frac{1}{2}gh^2 \\ huv \\ q_{bx} \end{pmatrix} \quad \mathbf{G}(\mathbf{U}) = \begin{pmatrix} hv \\ huv \\ hv^2 + \frac{1}{2}gh^2 \\ q_{by} \end{pmatrix} \tag{2}$$

being h the flow depth, $(u,\ v)$ the components of the depth-averaged velocity $\mathbf{u}$ along the x- and y-coordinates respectively, z_b is the erodible bed elevation, $(q_{bx},\ q_{by})$ are the components of the non-equilibrium bedload discharge $\mathbf{q_b}$ along the x- and y-coordinates respectively and g is the gravity acceleration. On the right hand side, the terms $\mathbf{S_b}(\mathbf{U})$ includes the bed-pressure momentum source and $\mathbf{S_\tau}(\mathbf{U})$ accounts for the frictional momentum dissipation along the x- and y-coordinates:

$$\mathbf{S_b}(\mathbf{U}) = \begin{pmatrix} 0 \\ -gh\frac{\partial z_b}{\partial x} \\ -gh\frac{\partial z_b}{\partial y} \\ 0 \end{pmatrix} \quad \mathbf{S_\tau}(\mathbf{U}) = \begin{pmatrix} 0 \\ -\tau_{bx}/\rho_w \\ -\tau_{by}/\rho_w \\ 0 \end{pmatrix} = \begin{pmatrix} 0 \\ -gh\,S_{fx} \\ -gh\,S_{fy} \\ 0 \end{pmatrix} \tag{3}$$

being $(\tau_{bx},\ \tau_{by})$ the components of the boundary shear stress vector $\boldsymbol{\tau_b}$ between the flow and the movable bed along the x- and y-coordinates, respectively, and (S_{fx}, S_{fy}) the corresponding frictional energy loss. For the sake of simplicity, in this work the turbulent Manning friction formulation is adopted:

$$S_{fx} = \frac{n^2\,|\mathbf{u}|u}{h^{4/3}} \qquad S_{fy} = \frac{n^2\,|\mathbf{u}|v}{h^{4/3}} \tag{4}$$

where $|\mathbf{u}| = \sqrt{u^2+v^2}$ is the depth-averaged flow velocity modulus and n is the bulk Manning roughness parameter depending on the bed material.

The bedload solid discharge in classical morpho-dynamical models is considered under equilibrium (capacity) conditions and usually modelled using the Grass law [31], which scales exponentially the solid transport rate with the flow velocity as $\mathbf{q_b} = A_g|\mathbf{u}|^{m-1}\mathbf{u}$, being A_g a flow-bed interaction constant factor and m an behaviour exponent (usually $m = 3$). More complex equilibrium models applied to realistic bedload transport tests estimate the instantaneous local flow-bed interaction factor from the hydro-dynamical flow features, hence $A_g \equiv G(h,\theta)$ [2,10,32], being θ the dimensionless Shields stress at the bed surface. According to Martínez-Aranda et al. [29] and Martínez-Aranda et al. [30], the bedload transport rate assuming a non-equilibrium approach can be expressed following a modified Grass-type law as:

$$q_{bx} = \frac{1}{1-\xi}G(h,\theta,\eta)\ |\mathbf{u}|^2 u \qquad q_{by} = \frac{1}{1-\xi}G(h,\theta,\eta)\ |\mathbf{u}|^2 v \tag{5}$$

being ξ the bed layer porosity and G a interaction factor between the flow and erodible riverbed. This interaction factor depends not only on the flow depth h and the dimensionless Shields stress θ, but also on the bedload transport layer thickness η. Figure 1 shows a schematic representation of the bedload generalized bedload mechanism at grain-scale.

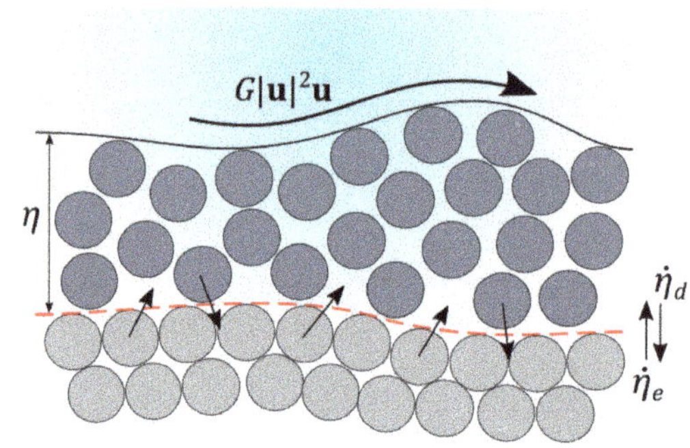

Figure 1. Sketch of the generalized bedload transport model at grain-scale.

It is useful to split G in a product of three factors $\Gamma_1(h)$, $\Gamma_2(\theta)$ and $\Gamma_3(\eta)$ as:

$$G = \Gamma_1(h)\,\Gamma_2(\theta)\,\Gamma_3(\eta) \tag{6}$$

Table 1 provides expressions for functions Γ_1, Γ_2 and Γ_3 using empirical relations found in literature, where r_s is the relative density of the sediment particles $r_s = \rho_s/\rho_w$, d_s is the characteristic diameter of the bed sediment, k_d and k_e are the two constants associated to the deposition and erosion exchange rates, respectively, between the underlying static bed stratum and the bedload transport layer (moving layer). The term $\Delta\theta$ accounts for the excess of Shields stress over the critical threshold θ_c for the incipient motion threshold of the solid particles, calculated as:

$$\begin{aligned} \Delta\theta &= \begin{cases} \theta - \theta_c & \text{if } \theta > \theta_c \\ 0 & \text{Otherwise} \end{cases} \\ \text{with:} \quad \theta &= \frac{|\boldsymbol{\tau}_b|}{(\rho_s - \rho_w)g d_s} = \frac{n^2\,|\mathbf{u}|^2}{(r_s - 1)\,d_s\,h^{1/3}} \end{aligned} \tag{7}$$

This generalized Grass-type model (6) allows to consider capacity (equilibrium) or non-equilibrium approaches for the bedload transport only by selecting the proper estimation procedure for the bedload layer thickness η. On one hand, assuming the equilibrium approach leads to the instantaneous adaptation of the bedload transport rate $\mathbf{q_b}$ to its capacity value $\mathbf{q_b^*}$ and the equilibrium transport layer thickness η^* can be estimated as:

$$\eta \equiv \eta^* = \frac{k_e}{r_s\,k_d}\,\Delta\theta\,d_s \tag{8}$$

On the other hand, the non-equilibrium assumption forces the necessity of estimating the temporal evolution of the bedload transport layer thickness η using the conservation equation:

$$\frac{\partial \eta}{\partial t} + \frac{\partial\,q_{bx}}{\partial x} + \frac{\partial\,q_{by}}{\partial y} = (\dot{\eta}_e - \dot{\eta}_d) \tag{9}$$

where $\dot{\eta}_e$ and $\dot{\eta}_d$ are the entrainment and deposition vertical rates, respectively, between the the underlying static stratum and the bedload transport layer. Following [24,29], the volumetric entrainment and deposition rates can be expressed as:

$$\dot{\eta}_e = k_e \frac{\Delta\theta}{r_s} \sqrt{(r_s - 1) g d_s} \qquad \dot{\eta}_d = k_d \frac{\eta}{d_s} \sqrt{(r_s - 1) g d_s} \tag{10}$$

Table 1. Grass-type interaction factor G for transport rate formulations.

Formulation	$\Gamma_1(h)$	$\Gamma_2(\theta)$	$\Gamma_3(\eta)$	θ_c
MPM	$\frac{n^3\sqrt{g}}{(r_s-1)\sqrt{h}}$	$8\sqrt{\Delta\theta}\,\theta^{-3/2}$	$\frac{r_s k_d}{k_e}\frac{\eta}{d_s}$	0.047
Nielsen	$\frac{n^3\sqrt{g}}{(r_s-1)\sqrt{h}}$	$12\,\theta^{-1}$	$\frac{r_s k_d}{k_e}\frac{\eta}{d_s}$	0.047
Fernandez-Luque	$\frac{n^3\sqrt{g}}{(r_s-1)\sqrt{h}}$	$5.7\sqrt{\Delta\theta}\,\theta^{-3/2}$	$\frac{r_s k_d}{k_e}\frac{\eta}{d_s}$	0.037
Wong	$\frac{n^3\sqrt{g}}{(r_s-1)\sqrt{h}}$	$3.97\sqrt{\Delta\theta}\,\theta^{-3/2}$	$\frac{\rho_s/\rho_f\, k_d}{k_e}\frac{\eta}{d_s}$	0.0495
Smart	$\frac{n^2}{(r_s-1)h^{1/3}}$	$4.2\, S_b^{0.6}\theta^{-1}$	$\frac{r_s k_d}{k_e}\frac{\eta}{d_s}$	0.047
Wu	$\frac{n^3\sqrt{g}}{(r_s-1)\sqrt{h}}$	$0.0053\,\theta_c^{-2.2}\sqrt{\Delta\theta}\,\theta^{-3/2}$	$\frac{r_s k_d}{k_e}\frac{\eta}{d_s}$	0.030

In the Smart formula, the parameter S_b accounts for the bed slope along the bedload transport direction.

3. Numerical Scheme for the 2D Bedload Transport Model

In order to obtain the numerical solution applying Finite Volume (FV) methods, the spatial domain is discretized into a fixed-in-time mesh of triangular cells and system (1) is integrated in each computational cell using the Gauss theorem:

$$\frac{d}{dt}\int_{\Omega_i} \mathbf{U}\, d\Omega + \sum_{k=1}^{NE} \left[\mathbf{F}(\mathbf{U})\, n_x + \mathbf{G}(\mathbf{U})\, n_y\right]_k l_k = \int_{\Omega_i} \mathbf{S_b}(\mathbf{U})\, d\Omega + \int_{\Omega_i} \mathbf{S_\tau}(\mathbf{U})\, d\Omega \tag{11}$$

being Ω_i the inner domain and NE the number of edges for the i cell, $\mathbf{n} = (n_x, n_y)$ the outward unit normal vector and l_k the length of the kth edge. Assuming a piecewise representation of the conservative variables $\mathbf{U}$ at each i cell, $\mathbf{U}_i$, the updating formulation between from time $t = t^n$ to the next step $t = t^{n+1}$ is expressed as:

$$\mathbf{U}_i^{n+1} = \mathbf{U}_i^n - \frac{\Delta t}{A_i}\sum_{k=1}^{\mathrm{NE}} \mathbf{R}_k^{-1}\, \boldsymbol{\mathcal{F}}_k^{\downarrow -}\, l_k \tag{12}$$

where A_i is the i cell area, $\Delta t = t^{n+1} - t^n$ is the time step, $\mathbf{R}_k^{-1}$ is the inverse of the rotation matrix $\mathbf{R}_k$ at kth edge [33] and $\boldsymbol{\mathcal{F}}_k^{\downarrow -}$ is the numerical normal flux throughout the kth edge. This numerical normal flux includes the convective flux augmented with the non-conservative bed-pressure contribution and the frictional momentum dissipation integrated thought the edge.

In this work, the upwind computation of $\boldsymbol{\mathcal{F}}_k^{\downarrow -}$ is achieved using a fully-coupled Roe-type Riemann solver for bedload transport (*FCM* solver) [34]. This augmented 5-waves solver is based on computing the approximated solution of the linear Riemann problem:

$$\frac{\partial \mathbf{R}_k \mathbf{U}}{\partial t} + \widetilde{\mathbf{M}}_k \frac{\partial \mathbf{R}_k \mathbf{U}}{\partial \hat{n}} = \mathbf{R}_k \widetilde{\mathbf{T}}_k \tag{13}$$

where $\hat{n}$ denotes the normal axis to the cell edge and superscripts $\widetilde{\ }$ indicates edge-averaged quantities. The pseudo-Jacobian matrix $\widetilde{\mathbf{M}}_k = (\widetilde{\mathbf{J}} - \widetilde{\mathbf{H}})_k$ of the coupled system includes the approximated Jacobian matrix of the conservative convective flux $\widetilde{\mathbf{J}}_k$ and the non-conservative matrix $\widetilde{\mathbf{H}}_k$ of the bed-pressure momentum contribution at the edge. On the right hand side, the term $\widetilde{\mathbf{T}}_k$ denotes the momentum dissipation due to flow-bed frictional stress, integrated through the kth edge. Details on the flux computation have been extensively reported in Martínez-Aranda et al. [34] but, for the sake of brevity, $\boldsymbol{\mathcal{F}}_k^{\downarrow -}$ at the kth edge is computed as:

$$\boldsymbol{\mathcal{F}}_k^{\downarrow -} = \mathbf{F}(\mathbf{R}_k \mathbf{U}_i^n) + \sum_{m-} \left[(\widetilde{\lambda}_m \widetilde{\alpha}_m - \widetilde{\sigma}_m)\, \widetilde{\mathbf{e}}_m\right]_k^n \tag{14}$$

where $\widetilde{\lambda}_{m,k}$ are the wave celerities at the edge, i.e., the eigenvalues of the pseudo-Jacobian matrix $\widetilde{\mathbf{M}}_k$ whereas $(\widetilde{\mathbf{e}}_m)_k$ are the corresponding eigenvectors. The coefficients $\widetilde{\alpha}_{m,k}$ denotes the wave strengths which accounts for the discontinuity on the conservative variables, whereas the coefficients $\widetilde{\sigma}_{m,k}$ are the source strengths involving the frictional momentum contribution at the edge. The subscript $m-$ under the sums indicate that only the waves travelling inward the i cell are considered, leading to the upwind computation of the flux at the edge.

Furthermore, the stability of the explicitly computed numerical solution is addressed by the dynamical limitation of the global time step using a CFL condition. The local time step allowed for the kth edge is estimated assuming that the absolute maximum of the eigenvalues corresponds to the fastest wave celerity. Hence the global time step Δt is limited by minimum of the local time steps, i.e., the more restrictive instantaneous local flow features (see Martínez-Aranda et al. [34]).

3.1. Transport Layer Updating with Capacity and Non-Capacity Approaches

Once the conservative variables have been updated using (12), the bedload transport rate $\mathbf{q_b}$ is computed at the next time level t^{n+1} as:

$$\begin{aligned} (\mathbf{q_b})_i^{n+1} &= \frac{1}{1-\xi} G_i^{n+1} \left[(u_i^{n+1})^2 + (v_i^{n+1})^2 \right] \mathbf{u}_i^{n+1} \\ \text{with:} \quad G_i^{n+1} &= \Gamma_1(h_i^{n+1})\, \Gamma_2(\theta_i^{n+1})\, \Gamma_3(\eta_i^{n+1}) \end{aligned} \tag{15}$$

where the hydrodynamic quantities h_i^{n+1}, u_i^{n+1}, v_i^{n+1} and θ_i^{n+1} are directly extracted from $\mathbf{U}_i^{n+1}$. Nevertheless, equation (15) also requires to compute the updated bedload transport layer thickness η_i^{n+1} at the next time level. The updating of the transport layer thickness depends on the assumption made for the bedload transport:

- Equilibrium hypothesis: The new transport layer thickness is directly computed as:

$$\eta_i^{n+1} = \frac{k_e}{r_s\, k_d}\, (\Delta\theta)_i^{n+1}\, d_s \tag{16}$$

  where $(\Delta\theta)_i^{n+1}$ is the Shields excess (7) at the i cell computed from the updated conservative variables $\mathbf{U}_i^{n+1}$.
- Non-capacity approach: This leads to the necessity of solving equation (9) each time step. The updating formula for the transport layer thickness η is expressed as:

$$\eta_i^{n+1} = \eta_i^n - \frac{\Delta t}{A_i} \sum_{k=1}^{NE} F_k^{\eta\downarrow}\, l_k + \Delta t\, (\dot{\eta}_e - \dot{\eta}_d)_i^n \tag{17}$$

  being $F_k^{\eta\downarrow}$ the numerical normal flux at the kth intercell edge for the transport layer equation and $\dot{\eta}_{e,i}^n$ and $\dot{\eta}_{d,i}^n$ the values of the entrainment and deposition rates for the i cell at time n, respectively. To compute the numerical flux $F_k^{\eta\downarrow}$, the bedload transport layer Equation (9) is projected along the normal direction to the edge $\hat{n}$ and approximated by the local scalar Riemann problem:

$$\frac{\partial \eta}{\partial t} + \widetilde{\lambda}_{\eta,k} \frac{\partial \eta}{\partial \hat{n}} = 0 \tag{18}$$

$$\text{with:}\ \eta(\hat{n},0) = \begin{cases} \eta_i^n & \text{if} \quad \hat{n} < 0 \\ \eta_j^n & \text{if} \quad \hat{n} > 0 \end{cases} \tag{19}$$

where the virtual bedload wave celerity $\widetilde{\lambda}_{\eta,k}$ is defined as:

$$\widetilde{\lambda}_{\eta,k} = \left(\frac{\delta q_{bn}}{\delta \eta}\right)_k \tag{20}$$

being $\delta q_{bn} = (q_{bx}n_x + q_{by}n_y)_j^n - (q_{bx}n_x + q_{by}n_y)_i^n$ and $\delta\eta = \eta_j^n - \eta_i^n$, and the bedload numerical flux $F_k^{\eta\downarrow}$ is computed as:

$$F_k^{\eta\downarrow} = \begin{cases} (q_{bx}n_x + q_{by}n_y)_i^n & \text{if} \quad \widetilde{\lambda}_{\eta,k} > 0 \\ (q_{bx}n_x + q_{by}n_y)_j^n & \text{if} \quad \widetilde{\lambda}_{\eta,k} < 0 \end{cases} \tag{21}$$

The cell-centered exchange rates $\dot{\eta}_{e,i}^n$ and $\dot{\eta}_{d,i}^n$ between the underlying static stratum and the transport layer are computed as:

$$(\dot{\eta}_e)_i^n = k_e \frac{(\Delta\theta)_i^n}{r_s}\sqrt{(r_s - 1)gd_s} \qquad (\dot{\eta}_d)_i^n = k_d \frac{\eta_i^n}{d_s}\sqrt{(r_s - 1)gd_s} \tag{22}$$

3.2. Morphological Collapse Mechanism

In realistic applications, the maximum bed level slope is limited due to morphological collapse processes. In order to implement an efficient bed slope control mechanism, we propose the non-iterative limitation of the numerical beload flux provided for the augmented Riemann solver at each cell edge imposing geometrical restrictions. The bed level updating expressions for the left L and right R cells sharing the edge k_0 read:

$$\begin{aligned} z_{bL}^{n+1} &= z_{bL}^n - \frac{\Delta t}{A_L}\left[(q_b^{\downarrow} l)_{k_0} + \ldots + (q_b^{\downarrow} l)_{k_{NL}}\right] \\ z_{bR}^{n+1} &= z_{bR}^n - \frac{\Delta t}{A_R}\left[(q_b^{\downarrow} l)_{k_0} + \ldots + (q_b^{\downarrow} l)_{k_{NR}}\right] \end{aligned} \tag{23}$$

where k_{NL} and k_{NR} refer to the non-shared edges of the left and right cells respectively and $(q_b^{\downarrow})_{k_0}$ denotes the forth component of the numerical flux vector $\mathcal{F}^{\downarrow-}$ at the common edge k_0 between cells L and R. This value is provided by the Riemann solver, i.e., the numerical upwind bedload normal flux at the edge ($q_b^{\downarrow} \equiv \mathcal{F}_4^{\downarrow-}$). Therefore, the evolution of the local bed slope at edge k_0 can be expressed as:

$$(z_{bR} - z_{bL})^{n+1} = (z_{bR} - z_{bL})^n + \frac{A_L + A_R}{A_L\, A_R}(q_b^{\downarrow} l)_{k_0}\,\Delta t \tag{24}$$

Assuming a collapse bed slope angle δ_b along the edge-normal distance between cell centres d_n, two conditions arise for the morphological limiting of the numerical bedload flux depending on the sign of the local bed slope:

- Positive bed slope if $0 < (z_{bR} - z_{bL})^{n+1} < \tan\delta_b\, d_n$

$$q_b^{\downarrow} \leq \frac{\tan\delta_b\, d_n - (z_{bR} - z_{bL})^n}{l\,\Delta t}\frac{A_L\, A_R}{A_L + A_R} \tag{25}$$

- Negative bed slope if $-(\tan\delta_b\, d_n) < (z_{bR} - z_{bL})^{n+1} < 0$

$$q_b^{\downarrow} \geq \frac{-\left[\tan\delta_b\, d_n + (z_{bR} - z_{bL})^n\right]}{l\,\Delta t}\frac{A_L\, A_R}{A_L + A_R} \tag{26}$$

Actually, this limitation on the numerical bedload flux $q_b^{\downarrow}$ acts as a bed form control mechanism, including artificial diffusion in the numerical bed level solution and avoiding the appearance of shock waves in the bed level with local slopes larger than the morphological collapse limit.

4. Numerical Results and Discussion

4.1. Adaptation of the Non-Equilibrium Bedload Rate to Equilibrium States

This new generalised model for the non-equilibrium bedload transport (5)–(6) requires to compute the thickness η of the transport layer using (9), but it is able to account for the spatial and temporal delay of the actual bedload rate with respect to the solid transport capacity in highly unsteady flows. Moreover, if the local flow features allow to develop steady states in the bedload transport process, the entrainment and deposition rates tend to balance ($\dot{\eta}_e = \dot{\eta}_d$) and the solid rate recovers the equilibrium condition, characterized by the transport layer thickness η^* (8). Therefore, the equilibrium state is a particular case of the generalised non-equilibrium model where $\eta = \eta^*$.

In order to illustrate the temporal-spatial behaviour of the generalised non-equilibrium bedload model, a simple idealised test is included here forcing the following flow conditions

$$h|\mathbf{u}|(t,x,y) = 2\,\mathrm{m}^2\mathrm{s}^{-1} \qquad \forall t,\ \forall x$$

$$|\mathbf{u}|(t,x,y) = \begin{cases} 2\,\mathrm{m/s} & \text{if} \quad t = 0\,\mathrm{s}, \quad \forall x \\ 4\,\mathrm{m/s} & \text{if} \quad t > 0\,\mathrm{s}, \quad 20\,\mathrm{m} \le x \le 70\,\mathrm{m} \\ 2\,\mathrm{m/s} & \text{if} \quad t > 0\,\mathrm{s}, \quad \text{otherwise} \end{cases}$$

with $\rho_s = 2650\,\mathrm{kg/m^3}$, $\xi = 0.35$, $d_s = 1\,\mathrm{mm}$, $n = 0.035$, $k_e/k_d = 20$ and the MPM relation (see Table 1). Using (15)–(16), these local flow features lead to a uniform bedload discharge at the initial time $t = 0\,\mathrm{s}$ and a stepped capacity transport rate $|\mathbf{q}_\mathbf{b}^*|$ for any $t > 0\,\mathrm{s}$, as depicted in Figure 2.

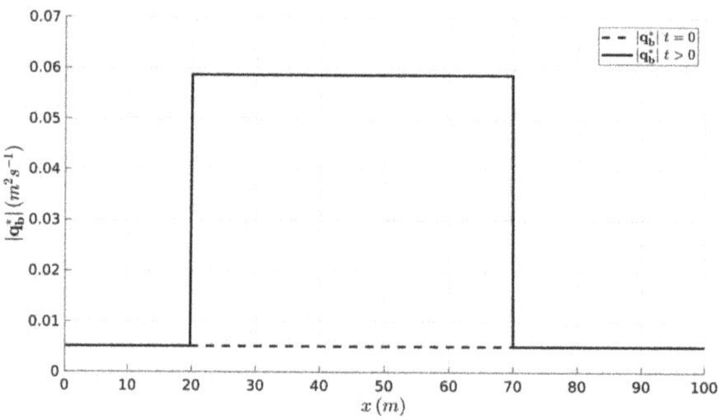

Figure 2. Bedload transport rate evolution for classical equilibrium formulations.

The generalised non-equilibrium model introduces a temporal and spatial delay of the actual transport rate with respect to the capacity value. Figure 3 shows the temporal evolution of the bedload rate $|\mathbf{q}_\mathbf{b}|$ in the spatial domain computed using the non-equilibrium model (15)–(17), with (top) $k_e = 0.06$ and (bottom) $k_e = 0.015$. As time increases, the actual solid rate adapts progressively to the capacity rate. Nevertheless, sudden spatial changes of the local flow features, as occurs at $x = 20\,\mathrm{m}$ and $x = 70\ \mathrm{m}$, need a length for the actual transport rate to adapt to the transport capacity value even with $t \to \infty$. Comparing the adaptation of the beload rate to the flow carry capacity with (top) $k_e = 0.06$ and (bottom) $k_e = 0.015$, it is worth noting that the temporal and spatial delay of the non-equilibrium solid discharge respect to the corresponding equilibrium state increases as the entrainment constant k_e decreases.

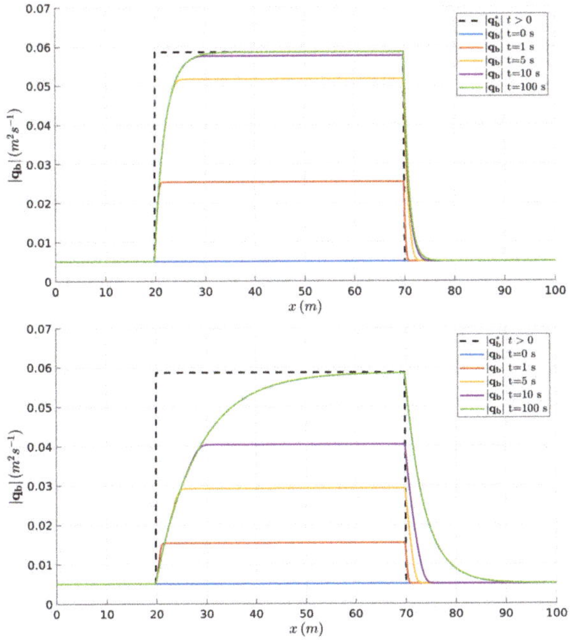

Figure 3. Bedload transport rate evolution for the generalised non-equilibrium model with (**top**) $k_e = 0.06$ and (**bottom**) $k_e = 0.015$.

Furthermore, the net exchange flux through the static-moving bed layers interface can be expressed as $N_{exch} = \dot{\eta}_d - \dot{\eta}_e$ (22). Previous non-equilibrium (non-capacity) bedload models [12,20,35,36] assumed a spatial length L_b for the adaptation of the actual bedload solid rate to its equilibrium state and estimate the net exchange vertical flux as

$$N_{exch} = \frac{1}{1-\xi} \frac{|\mathbf{q_b}| - |\mathbf{q_b^*}|}{L_b} \tag{27}$$

where L_b is a constant parameter which needs to be calibrated for each case. Comparison of method (27) to determine the entrainment/deposition flux N_{exch} with the method proposed in this work ($N_{exch} = \dot{\eta}_d - \dot{\eta}_e$) demonstrates that the current model leads to a dynamic adaptation length L_b which depends on the local flow features and scales following

$$L_b \propto \frac{\sqrt{\theta}}{k_e} d_s \tag{28}$$

According to (28), higher boundary Shields stresses θ are associated longer distances for the adaptation of the actual bedload discharge to its equilibrium state (see Figure 3). Contrarily, the lower the entrainment parameter k_e, the larger the dynamic adaptation length. Therefore, high Shields stresses θ and low entrainment factors k_e enhance non-equilibrium bedload transport states, i.e., highly erosive flow features lead to larger adaptation lengths. Instead, most of the previous models assume a global constant value based on the dominant bed form [36].

To analyse the influence of the Shields stress on the dynamic adaptation of actual non-equilibrium bedload rate to the equilibrium state, we use the above case but setting a entrainment constant $k_e = 0.025$ and increasing flow discharges $h|\mathbf{u}|(t,x,y) = [1,2,3,4]\,\mathrm{m^2s^{-1}}$. For $t = 0\,\mathrm{s}$ the flow is uniform, with $h = 1\,\mathrm{m}$ along the whole spatial domain regardless of the discharge, whereas for $t > 0\,\mathrm{s}$ a steady step in the flow features is set between 20 m $\leq x \leq 70\,\mathrm{m}$ so that $|\mathbf{q_b^*}| = 2|\mathbf{q_b^*}|_{t=0}$ regardless of the flow discharge. The shear stress at the bed surface, hence the Shields stress θ, increases progressively with the flow discharge as $\theta = [1.18, 4.71, 10.6, 18.9]$ respectively. Figure 4 shows the non-equilibrium bedload rate $|\mathbf{q_b}|$ for $t = 100$ s, normalised by the initial uniform equilibrium value $|\mathbf{q_b^*}|_{t=0}$, along the whole spatial domain. As the Shields stress at the bed surface increases, the adaptation length increases and enhances the non-capacity state in the bedload transport.

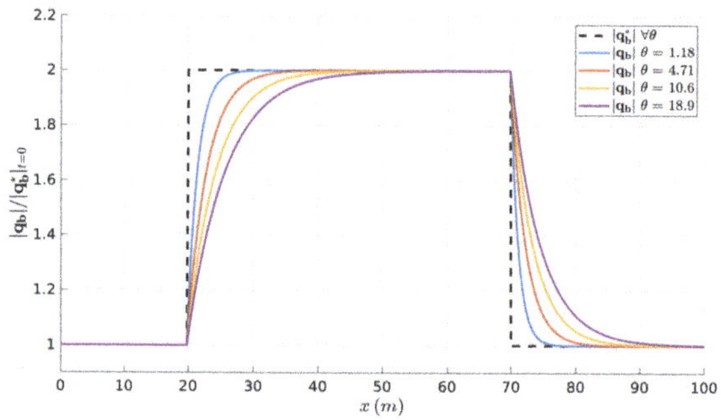

Figure 4. Normalised bedload transport rate for the generalised non-equilibrium model with increasing Shields stresses θ.

4.2. Dike Breaking by Overtopping Flow Erosion

This experimental test case was carried out by Tingsanchali and Chinnarasri [3] in a 35 m long 1 m wide straight rectangular cross-section flume. In the middle of the flume, a trapezoidal dyke was constructed with a non-cohesive sand of medium diameter $d_s = 0.86$ mm, solid particle density $\rho_s = 2650\,\mathrm{kg/m^3}$ and associated bulk porosity $\xi = 0.34$. The dyke height over the flume floor was 0.80 m and the crest width was 0.30 m, whereas the upward slope S_u and downward slope S_d depends on the laboratory case. The initial water surface elevation was 0.83 m over the flume floor upstream the dyke. During each experimental case, a constant inlet discharge q_{inlet} was set at the flume upstream section. In the original work [3], the temporal evolution of the flow discharge at the dyke crest section and the upstream reservoir level were reported during the whole experiment. Temporal bed elevation data were also provided at probes $P1 = 15$ cm, $P2 = 65$ cm and $P3 = 115$ cm downstream the upward vertex of the dyke crest.

For this case, simulations have been performed using a single-row mesh of 3500 rectangular cells with $\Delta x = 0.01$ m and $\Delta y = 1$ m. The Manning's roughness coefficient used has been $n = 0.010$ for the flume floor and $n = 0.015$ for the sediment material. The CFL = 0.50 was set for a full simulated period of 140 s. The Smart formula (Table 1) has been used to compute the bedload transport rate whereas the collapse mechanism has been neglected in this case. Two different benchmarking experiments, C1 and C2, have been used to study the capabilities of both the equilibrium and the non-equilibrium approaches (see Table 2).

Table 2. Setup for benchmarking cases C1 and C2.

Case	S_u	S_d	q_{inlet} (L/s)	k_e	k_d
C1	1V:3H	1V:5H	1.05	0.24	0.012
C2	1V:3H	1V:3H	1.23	0.34	0.017

The left column in Figure 5 shows the computed dam profile evolution for both benchmarking tests C1 and C2, considering the equilibrium and non-equilibrium bedload transport assumption, whereas the right column depicts the temporal evolution of the dam level at probes P1, P2 and P3. The black lines show the available experimental data for each test. Analysing the results obtained in both cases C1 and C2, the assumption of non-equilibrium bedload transport leads to a delay on the erosive process in the dam region and hence the change on the bed level is reduced for early stages after the flow starts. However, as the dam erosion progresses, the non-equilibrium bedload transport tends to approach to the equilibrium state and the dam profile level predicted by the non-equilibrium model converges to that obtained by the equilibrium model. It is worth noting that, in case C2, a marked scouring is observed in the experiments just downstream of the dam crest at time $t = 60\,\mathrm{s}$. This scouring was caused likely by 3D behaviour of the flow at the upstream region of the eroded dam. The proposed depth-averaged bedload model is not able to capture the appearance and development of this feature in the bed profile, regardless of the equilibrium or the non-equilibrium approach is selected, due to its 3D nature.

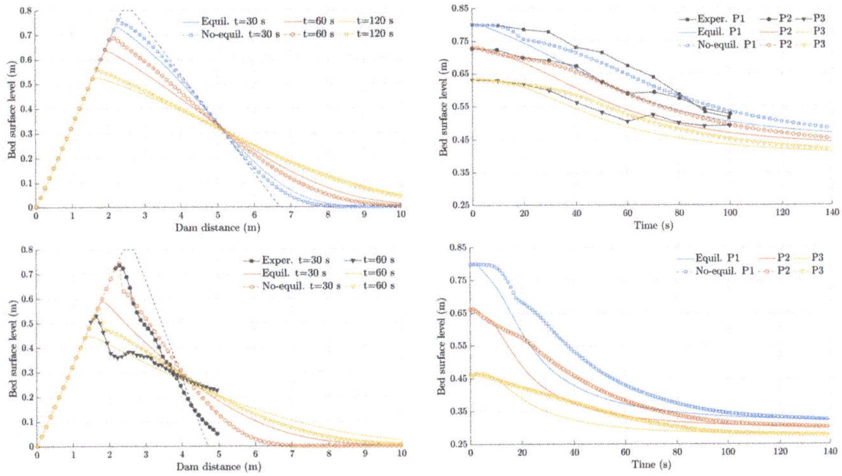

Figure 5. Dam profile evolution for (**top row**) case C1 and (**bottom row**) case C2 with equilibrium and non-equilibrium approaches: (**left**) dam profile at different times and (**right**) bed level evolution at P1, P2 and P3.

Table 3 shows the Root Mean Square Error (RMSE) of the computed results with both the equilibrium and non-equilibrium models respect to the measured bed level evolution at P1, P2 and P3 in case C1 (available experimental data). Also, for case C2, the RMSE of the computed bed profiles at times $t = 30\,\mathrm{s}$ and $t = 60\,\mathrm{s}$ respect to the dam profiles measured in the laboratory for the same times are shown in Table 3. Generally, the computed dam level changes with the non-equilibrium approach show a much better agreement with the measured data than those obtained assuming the equilibrium restriction.

Table 3. RMSE for dam level evolution data in benchmarking cases C1 and C2.

Data Series	Equil.	No-Equil.
C1—Bed level in P1	0.057 m	0.023 m
C1—Bed level in P2	0.053 m	0.022 m
C1—Bed level in P3	0.033 m	0.020 m
C2—Bed level profile at t = 30 s	0.112 m	0.061 m
C2—Bed level profile at t = 60 s	0.029 m	0.039 m

In order to get insight into the non-equilibrium bedload behaviour, the temporal evolution of the thickness of the transport layer η at probes P1, P2 and P3, normalised by the corresponding equilibrium value η^* computed from the instantaneous local hydrodynamic features using (8), is depicted in Figure 6 for cases C1 and C2. During the early stages after the overtopping flow starts, non-equilibrium states in the bedload transport layer occur all along the downward dam slope. The closer the dyke crest, the more enhanced is the non-equilibrium state of bedload transport and hence the normalised value η/η^* generally shows lower values at P1 location than that at probes P2 and P3 during the first flow stages. Then, as the dam erosion progresses, the thickness of the transport layer tends progressively to the equilibrium value ($\eta/\eta^* \approx 1$) all along the dyke downward side.

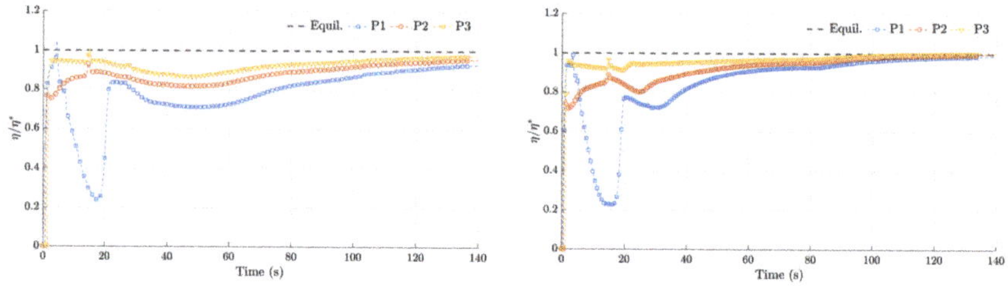

Figure 6. Temporal evolution of the transport layer thickness at P1, P2 and P3: (**left**) case C1 and (**right**) case C2.

The temporal evolution of the computed and measured discharge at the dyke crest and the water surface level at the the upstream reservoir have been plotted in Figure 7 for cases C1 and case C2. Also the Root Mean Square Error (RMSE) of the computed data with respect to the available measured data have been reported in Table 4 considering both equilibrium and non-equilibrium approaches. Considering non-equilibrium conditions for the bedload transport improves clearly the general agreement between the computed and measured evolution of the crest discharge and reservoir level in both benchmarking cases.

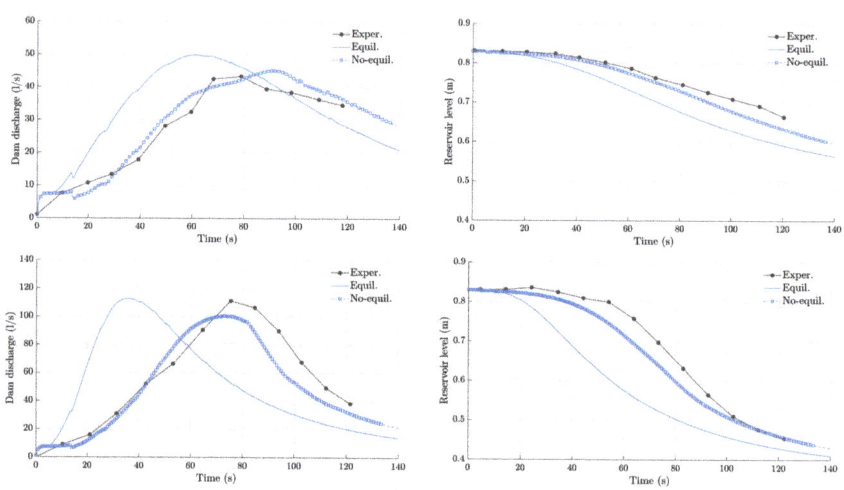

Figure 7. Computed and measured temporal evolution of (**left column**) the discharge at the dyke crest and (**right column**) the upstream reservoir water surface level: (**top row**) case C1 and (**bottom row**) case C2.

Table 4. RMSE for the crest discharge and reservoir level data in benchmarking cases C1 and C2.

Data Series	Equil.	No-Equil.
C1—Dam discharge	10.94 L/s	3.10 L/s
C1—Reservoir level	0.055 m	0.019 m
C2—Dam discharge	45.02 L/s	11.76 L/s
C2—Reservoir level	0.116 m	0.034 m

4.3. Breach Opening in Homogeneous Dam

This benchmark case aims to explore qualitatively the differences on the numerical modelling of the dambreach opening process using both bedload equilibrium and non-equilibrium approaches. Here we consider one of the experimental dambreach tests carried out at the Laboratório Nacional de Engenharia Civil (LNEC), Lisbon, Portugal, as part of the 6th Workshop on River and Sedimentation Hydrodynamics and Morphodynamics (Online, 24–26 November 2021), supported by Experimental Methods and Instrumentation Committee of the IAHR. The main results obtained during this laboratory campaign are unpublished yet. The experimental data available for this work were provided to the numerical modellers during the first blind stage and hence they are limited. Figure 8 shows a 3D sketch of the initial experimental setup. A homogeneous trapezoidal dam was constructed at the end of a 6 m length 1.2 m width rectangular cross-section channel. The dam was 0.45 m height, 1V:2H upward slope, 1V:2.5H downward slope, 2.20 m base length and 0.17 m crest width. Downstream the dam, there exists a 3.6 m × 2.0 m × 0.55 m settling basin, where the elevation of the free surface is zero relatively to the bottom of the upstream reservoir (dam base).

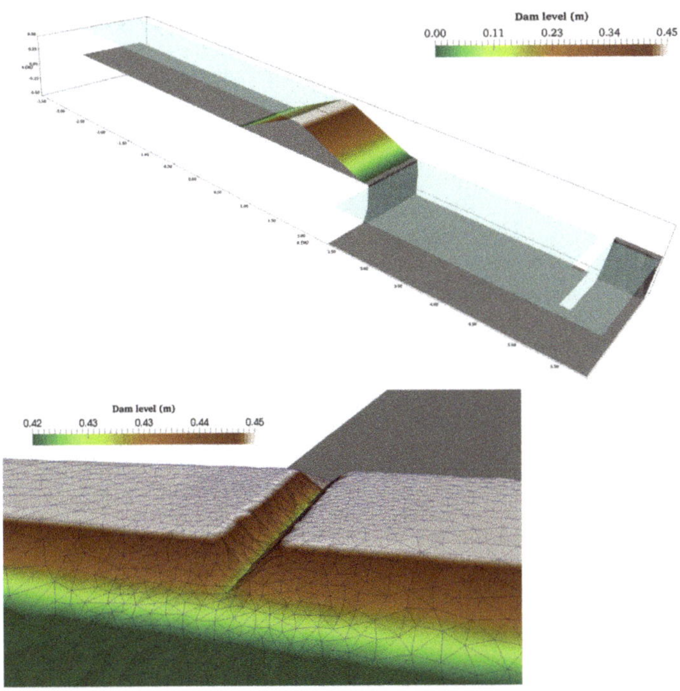

Figure 8. 3D sketch of the experimental setup (initial condition) at the Laboratório Nacional de Engenharia Civil (LNEC, Lisbon). Detail of the initial triangular notch at the dam crest.

The dam was composed of a non-cohesive silty sand with 25% of fines, solid density 2650 kg/m^3, median diameter $d_s = 0.320$ mm, relative compaction of 90% of the Standard Proctor reference value, 10% optimal water content for compaction. At the centerline of the dam crest, a triangular cross-section notch with 1V:1H side slopes and 1.5 cm deep (see Figure 8) was carved to guide the dambreach process during the early flow stages. Initially, the water level at the upstream reservoir was equal to the height of the model dam (0.45 m) and the free surface at the downstream basin is at the dam toe elevation. Note that the water at the downstream basin is spilt if the elevation is higher than the dam toe. Once the flow started at the notch region and the breach opening process began, the water level at the upstream reservoir was maintained approximately constant by changing the upstream inlet discharge. The maximum inlet discharge allowed in the upstream reservoir is $0.030\,m^3/s$. The test was conducted until the maximum discharge at the upstream reservoir was attained. During the test, the temporal evolution of the inlet discharge at the upstream reservoir was recorded using an electromagnetic flowmeter whereas the water level was obtained from an acoustic probe following the free-surface.

This test has been simulated using an unstructured mesh of almost 16,500 triangular cells, with $50\,cm^2$ of base cell area and a progressively refined region at the dam and notch zones of $0.5\,cm^2$ minimum cell area. At the inlet section of the upstream reservoir, the time series of inlet discharge recorded in the experiment has been imposed during the simulation, whereas free outflow has been allowed at the outlet section at the downstream basin. The Manning's roughness coefficient $n = 0.0125$ has been used for the sediment material in the Manning-Strickler relation $n = 1/21 d_s^{1/6}$. The porosity of the dam layer has been estimated as $\xi = 0.335$ from the bulk weight of the compacted dam material (90% of the Standard Proctor) and the entrainment and deposition constants have been set to $k_e = 0.06$ and $k_d = 0.003$ respectively. The Smart formula (Table 1) has been used to compute the bedload transport rate with the critical Shields stress for the incipient motion $\theta_c = 0.047$. The morphological mechanism plays a key role in this case to model the breach opening due to the collapse of the lateral sides. Based on experimental qualitative observations, a collapse bed slope angle $\delta_b = 75°$ has been set (see Section 3.2). CFL = 0.95 for a full simulated period of 4500 s.

Figures 9–12 show a 3D view of the dam topography and free water surface computed with the non-equilibrium model at $t = 1500$ s, $t = 2650$ s, $t = 3566$ s and $t = 4317$ s respectively. Also the thickness of the bedload transport layer and the flow velocity field at the breach region are shown. The breach opens progressively, increasing its cross-wise width due to the side wall collapse thanks to the collapse mechanism. A couple of coherent vortices appear downstream the dam, as the flow through the breach goes into the downward basin. The thickness of the transport layer increases progressively along the breach, allowing that the mobilized material incorporated to the flow from the breach sides can be transported downstream and reaches the settling basin.

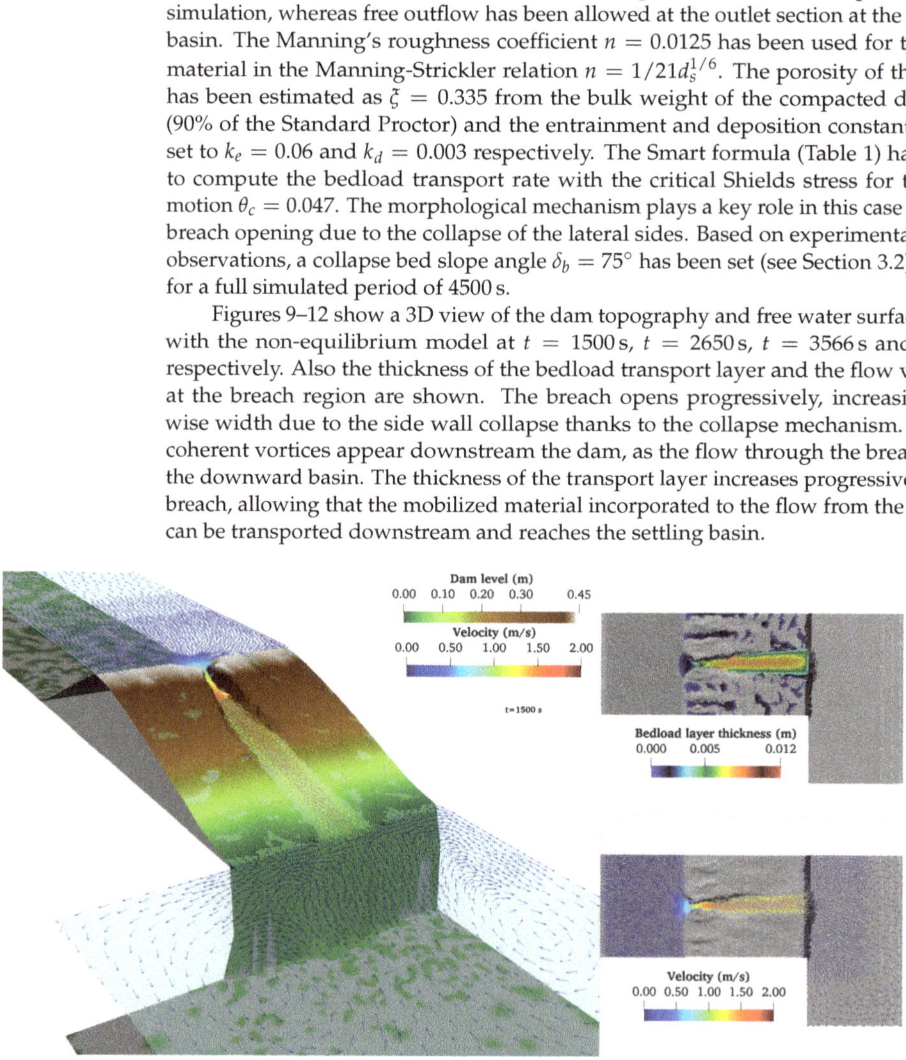

Figure 9. Dambreach opening at $t = 1500$ s with non-equilibrium model: (**left**) 3D view of the dam level and water surface; (**right top**) thickness of the bedload transport layer and (**right bottom**) depth-averaged velocity field.

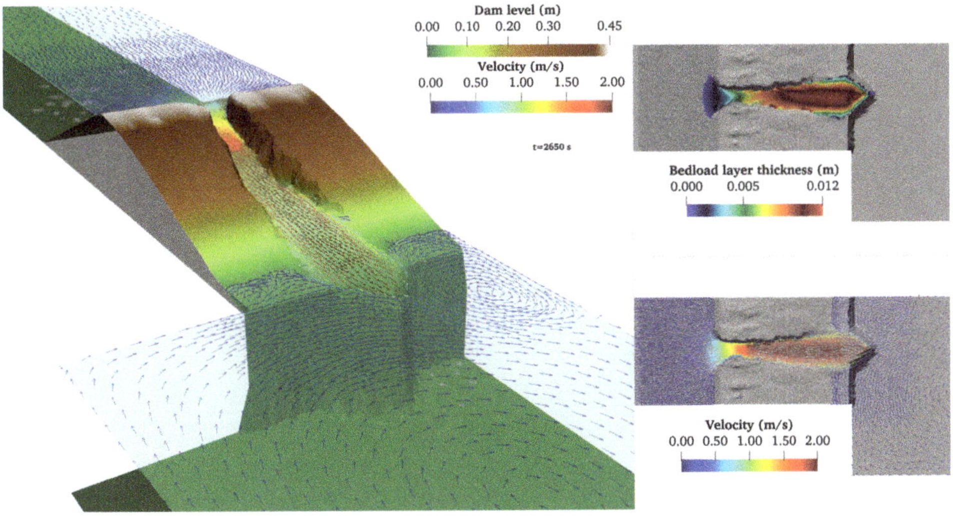

Figure 10. Dambreach opening at $t = 2650\,\text{s}$ with non-equilibrium model: (**left**) 3D view of the dam level and water surface; (**right top**) thickness of the bedload transport layer and (**right bottom**) depth-averaged velocity field.

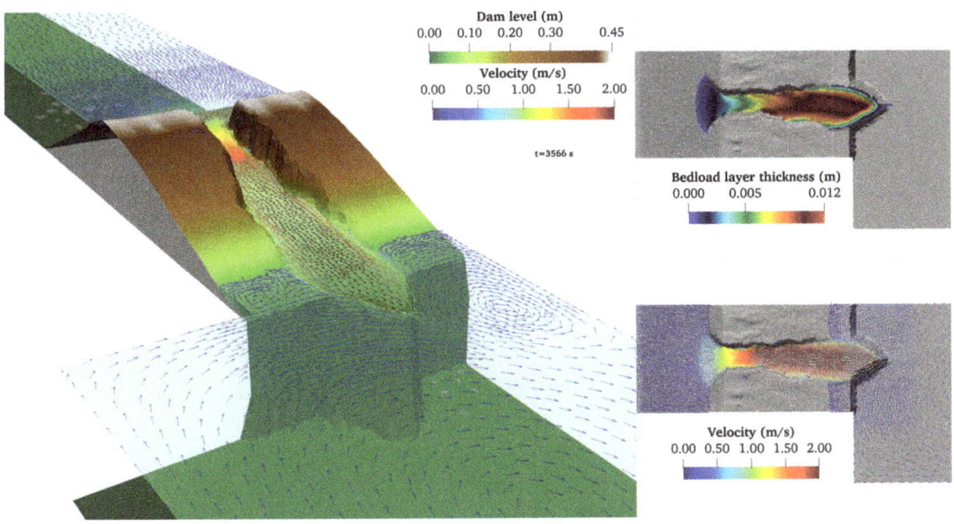

Figure 11. Dambreach opening at $t = 3566\,\text{s}$ with non-equilibrium model: (**left**) 3D view of the dam level and water surface; (**right top**) thickness of the bedload transport layer and (**right bottom**) depth-averaged velocity field.

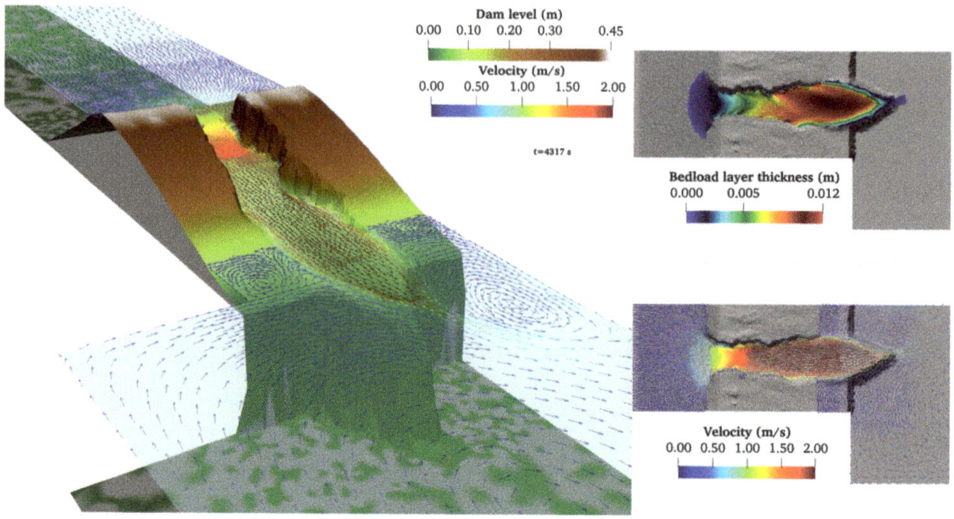

Figure 12. Dambreach opening at $t = 4317$ s with non-equilibrium model: (**left**) 3D view of the dam level and water surface; (**right top**) thickness of the bedload transport layer and (**right bottom**) depth-averaged velocity field.

Differences appears in the breach opening process depending on the equilibrium or non-equilibrium approaches selected. Figure 13 depicts (left) the total breach discharge and (right) the upstream reservoir level during the test computed with both equilibrium and non-equilibrium models. The total water discharge at the breach crest cross-section is almost similar using both approaches. Nevertheless, as the breach opens, the reservoir level decreases more markedly with the equilibrium assumption than that using the non-equilibrium model. Note that the experimental test was carried out maintaining an almost-constant water level at the upstream reservoir (solid black line in Figure 13 right). Therefore, the non-equilibrium model seems to approximate better the breach opening process than the equilibrium approach.

Figure 13. Dambreach evolution with equilibrium and non-equilibrium approaches: (**left**) dam discharge and (**right**) upstream reservoir level.

Finally, the left column of Figure 14 shows cross-section wetted-area at the dam crest section whereas the right column depicts the averaged flow velocity at the same section. Assuming non-equilibrium bedload transport leads to a slower flow velocity during the early stages of the breach opening and a higher wetted area along the breach region. This hydrodynamic combination reduces the erosive capability of the flow at this region and delays the breach opening process.

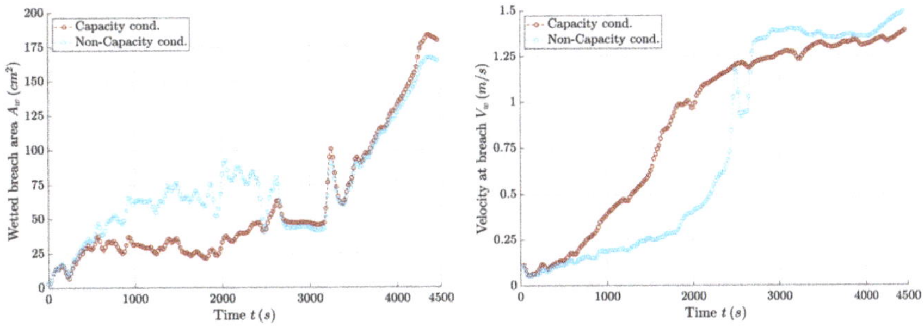

Figure 14. Dambreach evolution with equilibrium and non-equilibrium approaches: (**left**) cross-section wetted-area of the breach and (**right**) averaged flow velocity at the dam crest section.

5. Conclusions

In this work, a recently developed 2D generalized bedload transport model [30], able to account for equilibrium and non-equilibrium transport conditions, is applied to study the dam breach and dyke erosion processes. This generalized model is based on the grain-scale analysis of the net exchange flux through the static-moving bed layers interface and provide new expressions for the erosion and deposition rates $\dot{\eta}_e$ and $\dot{\eta}_d$, respectively. The proposed bedload transport model is supported by a generalized Grass-type formula for the non-equilibrium 2D bedload rate $\mathbf{q_b}$ as a function of the transport layer thickness η, which must be solved depending on equilibrium or non-equilibrium approach is considered. The 2D morpho-dynamical system of flow-bed conservation equations is solved using and fully-coupled augmented Roe solver [34] whereas the bedload transport layer equation is solved using an upwind scheme. Furthermore, the maximum bed slope is limited using an efficient non-iterative morphological collapse mechanism, which imposes geometrical restrictions on each cell edge.

The proposed model has been benchmarked against dyke erosion and dambreach opening laboratory experiments, involving highly unsteady flows. The non-equilibrium model is able to predict the measured data in all the cases tested, improving the results obtained with the classical equilibrium approach. Furthermore, the effects of the entrainment and deposition constants, k_e and k_d respectively, have also been analysed. When non-equilibrium model is applied, the adaptation of the actual bedload solid rate to the flow carry capacity becomes non-instantaneous and the appearance of bed changes is delayed in time and space. The temporal and spatial delay of the non-equilibrium bedload rate respect to the corresponding flow carry capacity increases as the entrainment constant k_e decreases. Moreover, as the local Shield stress θ exerted by the flow in the bedload transport layer grows, the equivalent dynamic adaptation length L_b is increased and hence the non-equilibrium states are promoted (see Section 4.1).

In general, considering the unsteady dambreach flow configuration, better predictions of the breach opening mechanism can be obtained using the non-equilibrium approach for the bedload transport rate with moderate values of the exchange entrainment and deposition parameters respect those obtained using the equilibrium model. Numerical results seems to support that considering non-equilibrium conditions for the bedload transport improves the general agreement between the computed results and measured data for overtopping dyke erosion and dambreach opening cases, delaying the breach opening process. However, the non-equilibrium approach requires a careful calibration of the exchange parameters in regions with complex transient processes.

Author Contributions: Conceptualization, S.M.-A. and P.G.-N.; methodology, S.M.-A.; software, S.M.-A. and J.F.-P.; validation, S.M.-A.; resources, P.G.-N.; data curation, S.M.-A.; writing and original draft preparation, S.M.-A.; review and editing, J.F.-P. and P.G.-N.; supervision, P.G.-N.; funding acquisition, S.M.-A. and P.G.-N. All authors have read and agreed to the published version of the manuscript.

Funding: This work has been carried out under the project JIUZ2022-IAR-03: High Performance Computing Tools for Smart Agriculture and Forestry, funded by the University of Zaragoza. This work has been partially funded by the Aragón Government (DGA) through the Fondo Europeo de Desarrollo Regional (FEDER).

Institutional Review Board Statement: Not applicable.

Informed Consent Statement: Not applicable.

Data Availability Statement: Not applicable.

Acknowledgments: Thanks are due to Rui M.L. Ferreira (Instituto Superior Técnico, Universidade de Lisboa, Lisbon, Portugal) for the invitation to participate in the 6th Workshop on River and Sedimentation Hydrodynamics and Morphodynamics (Online, 24–26 November 2021). The resulting experimental data provided for validation of these numerical models is gratefully appreciated.

Conflicts of Interest: The authors declare no conflict of interest.

Abbreviations

The following abbreviations are used in this manuscript:

FV	Finite Volume
RMSE	Root Mean Square Error

References

1. Yang, C. *Sediment Transport: Theory and Practice*; McGraw-Hill Inc.: New York, NY, USA, 1996.
2. Juez, C.; Murillo, J.; Garcia-Navarro, P. Numerical assessment of bed-load discharge formulations for transient flow in 1D and 2D situations. *J. Hydroinform.* **2013**, *15*, 1234–1257. [CrossRef]
3. Tingsanchali, T.; Chinnarasri, C. Numerical modelling of dam failure due to flow overtopping. *Hydrol. Sci. J.* **2001**, *46*, 113–130. [CrossRef]
4. Cao, Z.; Day, R.; Egashira, S. Coupled and Decoupled Numerical Modeling of Flow and Morphological Evolution in Alluvial Rivers. *J. Hydraul. Eng.* **2002**, *128*, 306–321. [CrossRef]
5. Hudson, J.; Sweby, P.K. Formulations for Numerically Approximating Hyperbolic Systems Governing Sediment Transport. *J. Sci. Comput.* **2003**, *19*, 225–252. [CrossRef]
6. Goutière, L.; Soares-Frazão, S.; Savary, C.; Laraichi, T.; Zech, Y. One-Dimensional Model for Transient Flows Involving Bed-Load Sediment Transport and Changes in Flow Regimes. *J. Hydraul. Eng.* **2008**, *134*, 726–735. [CrossRef]
7. Castro-Díaz, M.; Fernández-Nieto, E.; Ferreiro, A. Sediment transport models in Shallow Water equations and numerical approach by high order finite volume methods. *Comput. Fluids* **2008**, *37*, 299–316. [CrossRef]
8. Murillo, J.; García-Navarro, P. Weak solutions for partial differential equations with source terms: Application to the shallow water equations. *J. Comput. Phys.* **2010**, *229*, 4327–4368. [CrossRef]
9. Juez, C.; Murillo, J.; García-Navarro, P. A 2D weakly-coupled and efficient numerical model for transient shallow flow and movable bed. *Adv. Water Resour.* **2014**, *71*, 93–109. [CrossRef]
10. Martínez-Aranda, S.; Murillo, J.; García-Navarro, P. A 1D numerical model for the simulation of unsteady and highly erosive flows in rivers. *Comput. Fluids* **2019**, *181*, 8–34. [CrossRef]
11. Wu, W.; Wang, S. One-Dimensional Modeling of Dam-Break Flow over Movable Beds. *J. Hydraul. Eng.* **2007**, *133*, 48–58. [CrossRef]
12. El Kadi Abderrezzak, K.; Paquier, A. One-dimensional numerical modeling of sediment transport and bed deformation in open channels. *Water Resour. Res.* **2009**, *45*. [CrossRef]
13. Cao, Z.; Li, Z.; Pender, G.; Hu, P. Non-capacity or capacity model for fluvial sediment transport. *Proc. Inst. Civil Eng. Water Manag.* **2012**, *165*, 193–211. [CrossRef]
14. Fernández-Nieto, E.; Lucas, C.; Morales-de Luna, T.; Cordier, S. On the influence of the thickness of the sediment moving layer in the definition of the bedload transport formula in Exner systems. *Comput. Fluids* **2014**, *91*, 87–106. [CrossRef]
15. Bohorquez, P.; Ancey, C. Particle diffusion in non-equilibrium bedload transport simulations. *Appl. Math. Model.* **2016**, *40*, 7474–7492. [CrossRef]
16. Van-Rijn, L. *Principles of Sediment Transport in Rivers, Estuaries and Coastal Seas*; Aqua Publications: Amsterdam, The Netherlands, 1993; Chapter 10, Bed Material Transport, Erosion and Deposition in Non-Steady and Non-Uniform Flow.
17. Cao, Z.; Hu, P.; Pender, G. Multiple Time Scales of Fluvial Processes with Bed Load Sediment and Implications for Mathematical Modeling. *J. Hydraul. Eng.* **2011**, *137*, 267–276. [CrossRef]
18. El Kadi Abderrezzak, K.; Paquier, A.; Gay, B. One-dimensional numerical modelling of dam-break waves over movable beds: application to experimental and field cases. *Environ. Fluid Mech.* **2008**, *8*, 169–198. [CrossRef]
19. Zhang, S.; Duan, J. 1D finite volume model of unsteady flow over mobile bed. *J. Hydrol.* **2011**, *405*, 57–68. [CrossRef]

20. Soliman, M.R.; Ushijima, S. Equilibrium and Non-equilibrium Sediment Transport Modeling Based On Parallel MACS Algorithm. *J. Jpn. Soc. Civil Eng.* **2013**, *69*, 79–86. [CrossRef]
21. van Rijn, L. Sediment Transport, Part I: Bed Load Transport. *J. Hydraul. Eng.* **1984**, *110*, 1431–1456. [CrossRef]
22. Lee, H.; Hsu, I. Investigation of Saltating Particle Motions. *J. Hydraul. Eng.* **1994**, *120*, 831–845. [CrossRef]
23. Ferreira, R.M.L.; Franca, M.J.; Leal, J.G.A.B.; Cardoso, A.H. Mathematical modelling of shallow flows: Closure models drawn from grain-scale mechanics of sediment transport and flow hydrodynamics. *Can. J. Civil Eng.* **2009**, *36*, 1605–1621. [CrossRef]
24. Charru, F. Selection of the ripple length on a granular bed sheared by a liquid flow. *Phys. Fluids* **2006**, *18*, 121508. [CrossRef]
25. Zech, Y.; Soares-Frazão, S.; Spinewine, B.; Grelle, N.L. Dam-break induced sediment movement: Experimental approaches and numerical modelling. *J. Hydraul. Res.* **2008**, *46*, 176–190. [CrossRef]
26. Fernández-Nieto, E.D.; Luna, T.M.d.; Narbona-Reina, G.; Zabsonré, J.d.D. Formal deduction of the Saint-Venant-Exner model including arbitrarily sloping sediment beds and associated energy. *Math. Model. Numer. Anal.* **2017**, *51*, 115–145. [CrossRef]
27. Singh, U.; Ahmad, Z. Transport rate and bed profile computations for clay–silt–gravel mixture. *Environ. Earth Sci.* **2019**, *78*, 432. [CrossRef]
28. Singh, U.K.; Jamei, M.; Karbasi, M.; Malik, A.; Pandey, M. Application of a modern multi-level ensemble approach for the estimation of critical shear stress in cohesive sediment mixture. *J. Hydrol.* **2022**, *607*, 127549. [CrossRef]
29. Martínez-Aranda, S.; Murillo, J.; García-Navarro, P. A comparative analysis of capacity and non-capacity formulations for the simulation of unsteady flows over finite-depth erodible beds. *Adv. Water Resour.* **2019**, *in print*. [CrossRef]
30. Martínez-Aranda, S.; Meurice, R.; Soares-Frazão, S.; García-Navarro, P. Comparative analysis of HLLC- and Roe-based Models for the simulation of a dam-break flow in an erodible channel with a 90° bend. *Water* **2021**, *13*, 1840. [CrossRef]
31. Grass, A. *Sediments Transport by Waves and Currents*; Department of Civil Engineering, University College: London, UK, 1981.
32. Murillo, J.; García-Navarro, P. An Exner-based coupled model for two-dimensional transient flow over erodible bed. *J. Comput. Phys.* **2010**, *229*, 8704–8732. [CrossRef]
33. Godlewski, E.; Raviart, P.A. *Numerical Approximation of Hyperbolic Systems of Conservation Laws*; Springer: New York, NY, USA, 1996.
34. Martínez-Aranda, S.; Murillo, J.; García-Navarro, P. Comparison of new efficient 2D models for the simulation of bedload transport using the augmented Roe approach. *Adv. Water Resour.* **2021**, *153*, 103931. [CrossRef]
35. Wu, W.; Vieira, D.; Wang, S. One-Dimensional Numerical Model for Nonuniform Sediment Transport under Unsteady Flows in Channel Networks. *J. Hydraul. Eng.* **2004**, *130*, 914–923. [CrossRef]
36. Wu, W. *Computational River Dynamics*; CRC Press: London, UK, 2007.

water

MDPI

Article

Effectiveness of a Dam-Breach Flood Alert in Mitigating Life Losses: A Spatiotemporal Sectorisation Analysis in a High-Density Urban Area in Brazil

André Felipe Rocha Silva [1] and Julian Cardoso Eleutério [2,*]

[1] Postgraduate Program in Sanitation, Environment and Water Resources, Federal University of Minas Gerais, Belo Horizonte 31270-901, Minas Gerais, Brazil; andrefelipe94@ufmg.br

[2] Department of Hydraulic and Water Resources Engineering, Federal University of Minas Gerais, Belo Horizonte 31270-901, Minas Gerais, Brazil

* Correspondence: julian.eleuterio@ehr.ufmg.br

Abstract: The integration of early warning and evacuation systems (EWES) with estimations for mitigating the loss of life in flood risk assessments marks an advancement towards developing robust emergency action plans for dam breaks. Through the simulation of diverse EWES scenarios, the impact of these systems, coupled with community preparedness, on minimising the potential for loss of life could be calculated. This study was conducted in the theoretical context of a dam break located upstream from a densely populated urban region in Brazil. Hydrodynamic and agent-based models were utilised to estimate potential loss of life across various scenarios and simulations. The Monte Carlo approach, in combination with the LifeSim model, was applied to assess how factors such as warning issuance timing, evacuation strategies and community responses impact the model's outcomes. Sensitivity analysis was performed considering the overall exposed area and specific areas at risk for different spatiotemporal EWES strategies. The results of simulations highlighted the EWES' great potential for risk mitigation and displayed optimal times for warning issuance. The warning diffusion and the protective action initiation parameters proved crucial for improving EWES. The spatiotemporal sectorisation of the alert and evacuation was also an effective strategy to optimise EWES. This methodology should allow for further similar tests and incite EWES improvements based on consistent loss of life alleviation simulations.

Keywords: agent-based modelling; dam breach; emergency action plans; flood risk; LifeSim

check for updates

Citation: Silva, A.F.R.; Eleutério, J.C. Effectiveness of a Dam-Breach Flood Alert in Mitigating Life Losses: A Spatiotemporal Sectorisation Analysis in a High-Density Urban Area in Brazil. *Water* **2023**, *15*, 3433. https://doi.org/10.3390/w15193433

Academic Editors: Francesca Aureli, Andrea Maranzoni and Gabriella Petaccia

Received: 29 July 2023
Revised: 28 August 2023
Accepted: 26 September 2023
Published: 29 September 2023

1. Introduction

Dams located upstream of urban areas present a substantial risk, primarily attributed to the large volume and velocities of water involved in the possibility of a flood event triggered by their failure [1,2]. The high level of vulnerability and exposure in urbanised downstream valleys intensifies this risk [3–5]. Loss of life (LOL) is considered the most severe consequence of floods. This consequence carries significant weight due to its objective quantifiability and crucial role in shaping the public perception of disasters [6]. From 1998 to 2017, of 7000 recorded disasters worldwide 43% were attributed to flood events. These floods resulted in damages exceeding USD 650 billion, affecting over 2 billion people and causing over 142,000 fatalities [7].

The frequency of floods is increasing on a global scale, along with the number of people affected and the incurred fatalities [8]. However, the number of people impacted and fatalities per event show a decreasing trend [9,10]. This indicates advancements in flood risk management and assessment, which is crucial for decision-making in risk mitigation, including resource allocation for structural or non-structural measures [11,12].

Early warning and evacuation systems (EWES) are essential in minimising LOL during disasters. By providing anticipated alerts and facilitating organised evacuations, these systems play a vital role in mitigating the risk of casualties [13,14]. Different emergency management interventions during the 1953 flood in Canvey Island demonstrated the potential to mitigate LOL [15]. Studies conducted on the 2019 Brumadinho tailings dam failure in Brazil, which led to over 260 fatalities and extensive damages [16], highlight the potential benefits of early warning systems and population preparedness. According to Lumbroso et al. [17], loss of life could have been entirely avoided if an EWES was efficiently implemented with a projected alert time of 15 min and a well-prepared population. Similarly, the floods experienced in Germany in July 2021, resulting in more than 180 fatalities, underscore the importance of implementing operational forecasting models to enhance warning systems and facilitate an adequate response from the affected population [14,18]. In contrast, a study conducted over another tailings dam breach that occurred in Brazil in 2007 revealed that if no warning and evacuation had been implemented, fatalities would probably have been another consequence of the accident [19].

Hence, LOL and EWES analyses effectively evaluate flood risks, ensure the integrity of safety structures and facilitate comprehensive emergency planning [20,21]. Numerous models for estimating life losses have been developed since the 1970s [20,22–30]. Generally, according to Jonkman et al. [28], the estimation of life losses in these models is determined by three main factors: the number of people potentially at risk, the effectiveness of evacuation and shelter strategies to determine the number of people exposed to the event and the estimation of the exposure/fatality rate (Equation (1)).

$$N = F_d(1 - F_E)N_{par} \quad (1)$$

where N is the estimate LOL; F_d is the fatality rate; F_E is the evacuation rate; and N_{par} is the number of people at risk.

Models for estimating life losses are classified as empirical, which relates population mortality solely to event characteristics, and mechanistic (agent-based), which provides a more detailed simulation of the flood, allowing for modelling individual behaviour and causes of fatality [31]. Empirical models are mainly suitable for preliminary and rapid estimation. Agent-based models are recommended when the simulation is conducted to support decision-making processes and assist in selecting measures to reduce the potential impacts of flooding [32].

Among the models currently available, only the Life Safety Model [27] and LifeSim [26] are built on agent-based principles, which allow for dynamic analyses. HEC-LifeSim 1.0.1 software fully incorporates the agent-based LifeSim model [33]. This model stands out as one of the most up-to-date and comprehensive for the proposed purposes, widely used in North American consulting and insurance companies [34]. It has also been applied in various studies [19,35–42], and its application is growing in different countries (Australia, Brazil, Morocco and Switzerland). However, it still presents some obstacles due to the need for a large quantity and quality of input data [43]. The model is structured through a modular modelling system, where modules exchange information through a database with multiple layers and tables of geographic information systems. The model has a probabilistic approach via Monte Carlo analysis and consists of four main modules: the flood-routine module (FRM), which includes a network representing flood characteristics throughout the area and time; the shelter loss module (SLM), which simulates the exposure of individuals in buildings during each event as a result of structural damages and submersion criteria; the warning and evacuation module (WEM), which simulates the distribution and dynamics of the population at risk after the issuance of the alert; and the loss of life module (LLM), which estimates the losses through probability distributions obtained from a set of historical flood events [44,45].

A significant advantage of applying these models, which has been little explored in the literature, is the ability to measure, in advance, the efficiency of alert and evacuation systems with different characteristics, such as the decision-making process of stakeholders involved in hazard identification and warning issuance [46], the mechanisms of warning dissemination and their influence [47,48], as well as the preparedness and prior knowledge of the population regarding mobilisation and evacuation to safe places [49,50]. The spatial and temporal sectorisation of alerts and evacuations adapting the system to the specific characteristics of each sector can bring significant benefits in terms of system efficiency [51–53].

Although some studies developed a LOL analysis related to different EWES scenarios, the sectorisation of flood warning and evacuation throughout LOL simulation is still unexplored in the literature. Additionally, uncertainty analyses related to different EWES scenarios are rare [15,17,19]. Especially for highly urbanised areas, if simultaneous alert and evacuation procedures are deployed for significant populations, traffic problems and specific population behaviour characteristics may compromise evacuation, which cannot be appropriately analysed without a sectorisation approach. Therefore, this article aims to explore the potential for enhancing the effectiveness of alert systems through the sectorisation of alerts and evacuations based on the spatial and temporal characteristics of land use and occupancy in the downstream valley. In order to achieve this objective, the hypothetical failure of the Pampulha Lake water dam (Brazil), located upstream of a high-density urban area, is analysed for different alert and evacuation scenarios.

2. Case Study

As a case study, the hypothetical failure of the Pampulha Lake Dam (PL.Dam) and its potential LOL were selected for different EWES approaches in its high-density downstream valley, situated in Belo Horizonte City, the State of Minas Gerais, Brazil (Figure 1). The area is highly populated, accounting for around 13,381 people potentially exposed to the dam-break flood risk.

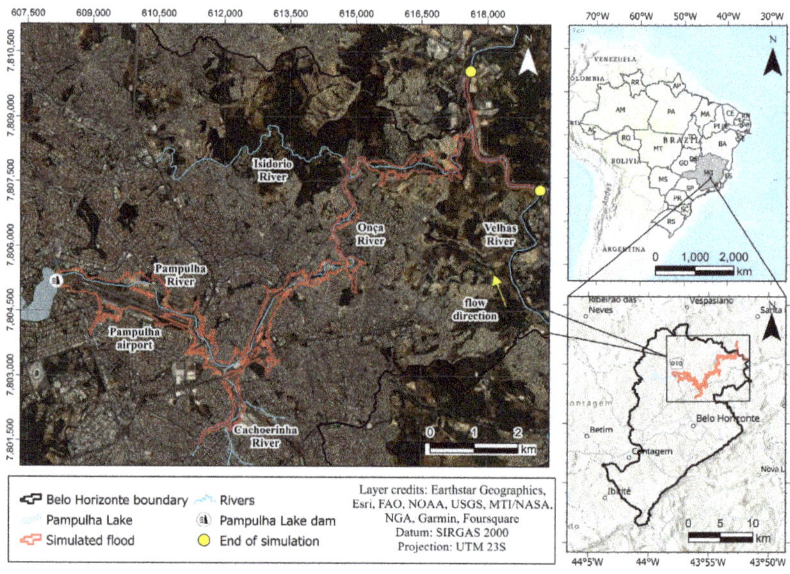

Figure 1. Pampulha Lake reservoir and its dam and downstream valley location.

The PL.Dam collapsed in 1954 because of internal erosion (piping) caused by inadequate positioning of its drains, resulting in the gradual water flow from the reservoir over the downstream floodplain (Figure 2). It led to limited economic losses, as the valley had few urban developments at that time. Four years after the accident, the dam was reconstructed with new configurations for the overflow system and currently has a capacity of 20 hm^3 [54].

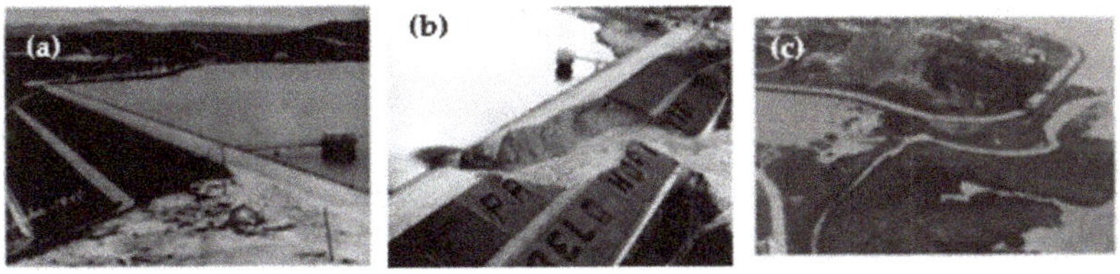

Figure 2. Photos (**a**) before, (**b**) during and (**c**) after the 1954 Pampulha Lake dam failure. Source: adapted from Vianini Neto [54].

A local study created hydrodynamic simulations aiming to represent the 1954 accident [54]. This study also gathered current dam information, which is more relevant for the purpose of this article. As the conditions are entirely different, we highlight that this article does not intend to match the real flood event or its consequences, considering only the current risk situation. Currently, the dam serves the functions of sediment containment, landscape enhancement and flood control, and the downstream valley has had a high urbanisation rate since then (Figure 3).

Figure 3. Image of the current downstream valley occupation of the Pampulha Lake dam.

The maximum height of the dam is 20 m, with a crest at an elevation of 805 m. It has a width of 20 m and a length of 450 m. The spillway system consists of a linear overflow of the side channel type. Additionally, there is an auxiliary tulip overflow [54]. The contributing basin upstream of its reservoir spans an area of 98.3 km^2 [54]. This basin exhibits a pronounced level of urbanisation, closely accompanied by a significant presence of sanitation-related issues. These challenges, in turn, contribute to water quality degradation within the Pampulha Lake [55].

3. Materials and Methods

In order to understand the spatiotemporal effect of alerts and mobilisation, the analyses developed have two main goals: to estimate uncertainty in flood alert and mobilisation times in relation to land use, event time occurrence and anticipation; and to elucidate if alert sectorisation could improve LOL alleviation from the perspective of organising mobilisation according to spatiotemporal event characteristics. From this perspective, the HEC-LifeSim model [33] is used for examining the potential effectiveness of different sectorised and non-sectorised alert and evacuation systems adapted to dam-failure floods.

The probabilistic and agent-based simulations in HEC-LifeSim, encompassing its four main modules (flood routine module—FRM; shelter module—SLM; warning and evacuation module—WEM; and loss of life module—LLM), adhere to the timeline for individuals exposed to the flood event (Figure 4). Commencing when the imminent threat is identified, the timeline encompasses the first delay associated with the necessary time to issue the warning. Subsequently, the timeline extends until the agent reaches a safe location.

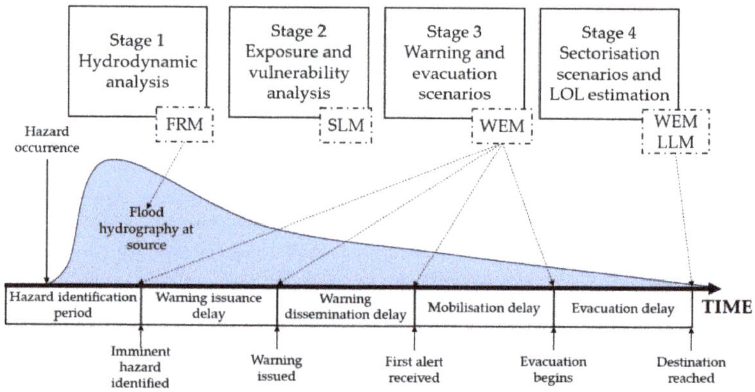

Figure 4. LifeSim modules and their relationship with the methodological approach.

In order to achieve the goals of this article, the development of the proposed analyses is based on the implementation of the different modules of the HEC-LifeSim model, which is organised in this article into four stages (Figure 4): (1) evaluation of potential flood dynamics; (2) analysis of exposure and vulnerability in the affected area; (3) development of alert and mobilisation scenarios in order to estimate uncertainty related to EWES; and (4) development of specific alert and evacuation spatiotemporal scenarios based on the sectorisation of the EWES. Hence, the analysis of the effectiveness of the warning system in terms of LOL alleviation potential is achieved for the different tested scenarios.

The application of the presented methodological stages (Figure 4) in the PL.Dam is detailed in the following sections of this article.

3.1. *Dynamic Flood Evaluation*

The purpose of simulating the flood is solely to determine one specific hypothetical hazardous scenario for enabling LOL analysis and the evaluation of uncertainty and EWES efficiency under different scenarios. Therefore, the analysis of flood dynamics followed the basic premises for a deterministic dam-breach flood analysis [56].

Based on the dam description (Table 1), a piping breach simulation was performed to generate a flood hydrograph, according to Froehlich [57]. A sunny day scenario was considered, taking into account a trigger in the elevation corresponding to the normal operation of the reservoir. Consequently, the influence of the dam spillway system was not considered in this particular case.

Table 1. Dam characteristics by [54] to obtain the breach hydrograph, applying the proposed equations of Froehlich [57].

Parameter	Value
Crest length (m)	450
Height (m)	20
Crest elevation (m)	805
Maximum operation elevation (m)	801
Volume at maximum operation elevation (hm^3)	10

The flood wave propagation through the valley was simulated through solving shallow water equations via the Eulerian-Lagrangian method implemented in the HEC-RAS 6.4 model [58], considering two-dimensional flow. The digital elevation model, with a resolution of one metre and derived from publicly accessible data (https://bhmap.pbh.gov.br, accessed on 25 July 2023), was used in the study, incorporating necessary modifications to address potential flow limitations caused by singularities. The Manning coefficient of 0.2 was applied to building areas, locations with steep slopes, and immediately downstream of the dam. In contrast, a coefficient of 0.1 was employed for the remaining areas [58]. Upstream conditions accounted for the dam breach outflow hydrograph, while downstream conditions considered the normal depth and average flow during the wet season in the Velhas River (Figure 1).

The downstream section of the PL.Dam covered approximately 16.7 kilometres, following the course of the Pampulha River until its confluence with the Onça River, which flows into the Velhas River, marking the municipal boundary of Belo Horizonte (Figure 1). The maximum depths and velocities and the flood wave arrival times are shown in Figure 5 and summarised in Table 2.

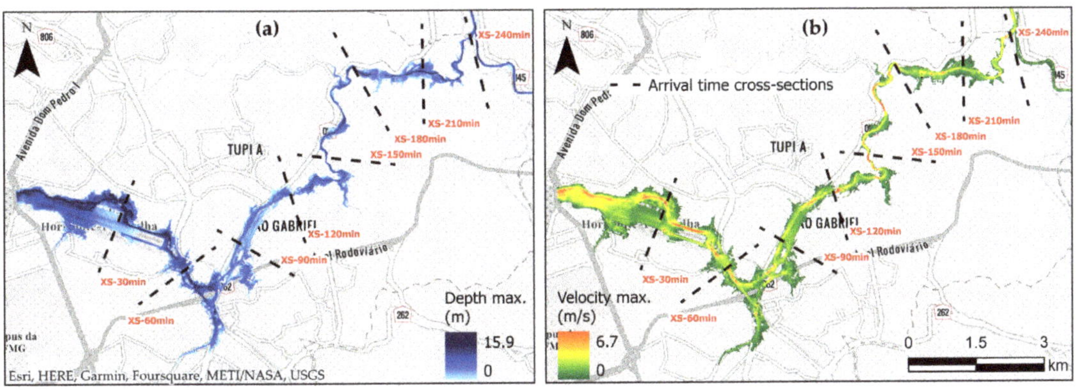

Figure 5. Maximum depths (**a**) and velocities (**b**) with arrival time cross-sections for each 30 min.

Table 2. Summarised results of the dam-breach simulation in each arrival time cross-section.

Cross-Section	Average Maximum Depth (m)	Average Maximum Velocity (m/s)
XS-30 min	4.8	1.0
XS-60 min	3.9	0.9
XS-90 min	2.4	0.9
XS-120 min	2.8	1.2
XS-150 min	4.6	2.9
XS-180 min	4.1	1.3
XS-210 min	4.0	0.5
XS-240 min	4.8	0.9

This flood scenario was the input for the HEC-LieSim flood-routine module (FRM), according to (Figure 4), including all the required details concerning the flood hydrodynamics through the downstream valley, allowing for the computation of flood depths, velocities and flood wave arrival times over the floodplain and the different elements at risk.

3.2. Exposure and Vulnerability Analysis of the Valley

Population exposure mapping is essential for characterising elements at risk in areas potentially impacted by flood events [59]. Microscale analysis is required to optimise investments by evaluating the cost–benefit ratio of structural and non-structural measures to reduce the risk [60], analysing loss of life, the efficiency of warning systems and population evacuation [61]. The LifeSim simulations are based on microscale data.

Exposure and vulnerability analysis was conducted in order to fulfil the LifeSim Shelter Loss Model (SLM) data input requirements. This analysis involves the representation of buildings as points, wherein information about the structures, such as the number of storeys, occupancy type and construction material, is considered. Additionally, the model requires data regarding the population in these buildings during daytime and nighttime and their mobility patterns. However, a significant challenge lies in acquiring the requisite high-resolution data [62] to represent the SLM accurately.

Based on the more recent available census data, public municipality layers and a buffer of 100 metres in the inundation boundary, 38,002 buildings with populations of 101,865 and 106,520 individuals were estimated during daytime and nighttime, respectively. It was observed that approximately 7.5% of the total population in both periods consisted of individuals aged 65 years or older, which was considered as a factor when representing the population's mobility characteristics. Besides residential places, the authors considered public places such as schools, hospitals, and companies and population allocation based on census data about work and education characteristics.

It is crucial to acknowledge that the primary focus of this study, as stated in the dynamic flood simulation section, is not to tackle or elucidate the uncertainties related to the analysis of exposure and vulnerability. Instead, the purpose of conducting exposure and vulnerability analysis is solely to parameterise the SLM in order to facilitate LOL analysis and evaluate uncertainty and efficiency within the context of EWES in various scenarios. The estimated population was aggregated within a statistical grid, a geographic unit derived from Brazilian census data with a spatial resolution of 200 × 200 metres, corresponding to the daytime period used in this study (Figure 6).

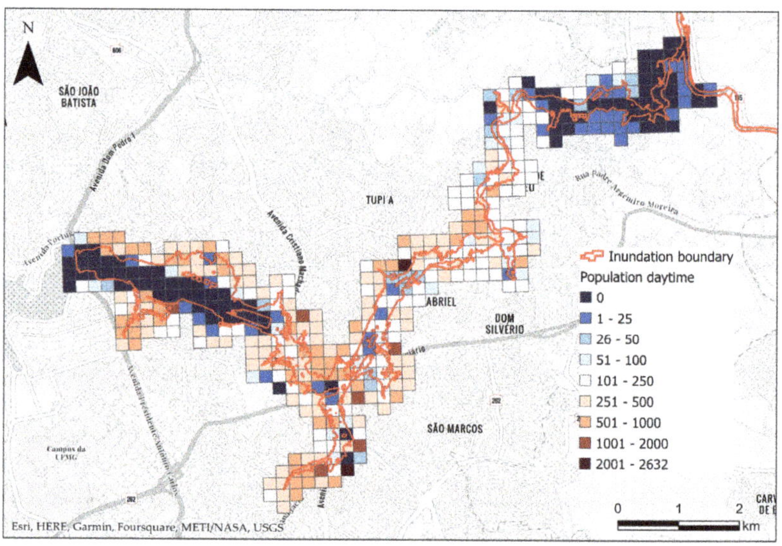

Figure 6. Downstream exposure and vulnerability analysis and the inundation boundary.

Based on the direct exposure of the population revealed by overlaying the generated exposure maps and the inundation boundary, it is estimated that a daytime rupture event can directly impact a population of 13,381 people. The total number of exposed buildings reaches 4417, with 3820 buildings classified as residential, 587 as commercial and 10 as public places.

3.3. Warning and Evacuation Scenarios

The timeline depicting the warning and evacuation process is represented in LifeSim under the flood hydrograph in Figure 4. The timeline commences with hazard identification and encompasses subsequent stages, including warning issuance, warning dissemination, mobilisation and evacuation delays, ending when the agent reaches the designated destination.

For the delay in issuing the warning, which is defined as the time from when the managers receive the notification of the imminent hazard to the issuance of an evacuation order to the public, Sorensen and Mileti [63] highlight the scarcity of studies on the time it takes to decide to issue an alert. Most of the research conducted on this topic is related to chemical emergencies. However, since many chemical accidents involve fast-moving events that require quick decisions and have a potential for fatalities, the results are relevant to other dangerous situations, such as dam failures. Based on the analysis of these historical cases, Sorensen and Mileti [63] present Equation (2) to represent warning issuing delay.

$$p = 1 - e^{(-at^{b})} \tag{2}$$

where p is the probability of delay in the decision; t is the time [min]; and a and b are coefficients specific for adjusting empirical data.

For the delay in warning dissemination, which is the period between the issuance of the warning and when it is received by the population at risk, Sorensen and Mileti [64] consider two main factors: the sending of the first alert, which covers the use of technologies, dissemination channels and repetition of the alert; and the receipt of the first alert, which involves population and event characteristics that influence the alert diffusion process. The authors discuss a specific model of first alert diffusion. Variables that influence it are highlighted, such as direct alert, which is the alert directly received through a communication channel, and indirect alert, which is the alert received through informal messages from friends, neighbours, colleagues and relatives. Thus, based on the analysis of historical cases of floods, volcanic eruptions and chemical accidents with a high variability of alert diffusion time, Sorensen and Mileti [64] present the following Equation (3).

$$\Delta W/\Delta t = Pu_t \ * \ (B_t + C_t - B_t \ * \ C_t) \tag{3}$$

where W is the alerted population; $\frac{\Delta W}{\Delta t}$ is the rate of the alerted population at the time t; Pu_t is the population that has not been warned at the time t; B_t is the coefficient related to the effectiveness of the direct alert system at time t; and C_t is the coefficient associated with the effectiveness of the indirect alert at time t.

For the delay in the initiation of mobilisation, which is the time it takes for the population to take a protective measure after receiving the alert, Sorensen and Mileti [65], through the analysis of mobilisation in historical cases of hurricanes, floods and material accidents, developed a model relating the population's response to the mobilisation speed and the average time for the initiation of mobilisation, as presented below (Equation (4)).

$$P_{t_h} = 1 - e^{(-({t_h}^2)/a_m {b_m}^2)} \tag{4}$$

where P_{t_h} is the cumulative probability of being mobilised at time t_h; t_h is the time [h]; a_m is the coefficient related to the mobilisation speed; and b_m is the coefficient related to the average time for mobilisation.

Each equation presented above is accompanied by coefficients that depend on population characteristics, the issued alert and the event under consideration. HEC-LifeSim v 1.0.1 already incorporates the recommended values for these coefficients with an uncertainty range into the warning and evacuation module (WEM). However, it is also possible to define specific coefficients for a particular case and construct customised curves based on them [66].

Regarding evacuation delay, the Greenshields et al. [67] transportation model is used to represent the effects of traffic density and road capacity on vehicle speed. These aspects are quite relevant in the valley downstream of the PL.Dam, where the evacuation process encompasses over 13,000 people. The interaction of these aspects can significantly decrease the speed at which vehicles can reach designated safe areas.

The number of vehicles is obtained by the occupancy rate determined by the user. This rate is only considered for people evacuating in vehicles, a parameter also defined by the user. In order to determine evacuation routes, the Dijkstra's optimisation algorithm [68] is employed to identify the path (imported by the OpenStreetMap database) with the shortest time between the building and the corresponding destination.

In this work, Monte Carlo simulations were conducted to analyse the warning and evacuation processes in two parts. The first part involved a sufficiently long interval in representing the time between the hazard identification and the warning issuance, ranging from 0 to 24 h, starting at the onset of the rupture and characterised by uniform uncertainty. This interval was used to evaluate the optimal response time for the emergency managers. The second part of the analysis focused on the influence of uncertainty in the recommended values of warning dissemination delay and mobilisation time proposed by Sorensen and Mileti in [64,65]. We examined two types of parameterisation—the first type involved fast warning dissemination and a population with high perception and preparedness for mobilisation. The second type consisted of the same warning dissemination rate, but with a population with low levels of perception and preparedness (Figure 7). Only the mobilisation pattern was varied because this aspect has more impact on time and the evacuated percentage of the population achieved than warning dissemination times.

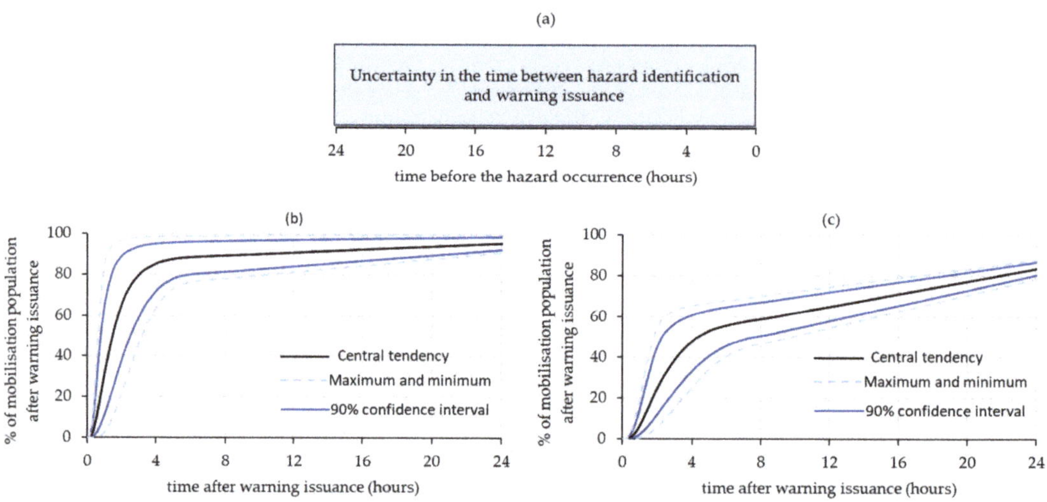

Figure 7. Two parts of analysis in the warning and evacuation process and their values: (**a**) between the hazard identification and the warning issuance; and uncertainties in the recommended values of warning dissemination delay and mobilisation time proposed by Sorensen and Mileti [61,62], considering (**b**) fast warning dissemination and high perception and preparedness of the population and fast warning dissemination and (**c**) low perception and preparedness of the population.

Before these analyses, tests were conducted to assess the evacuation mode (50% on foot and 50% using vehicles, as well as 100% on foot) to identify routes to avoid and the safest evacuation points. The goal was to optimise the evacuation process, whether on foot or using vehicles. However, the results were not highly sensitive to changes in this parameter. Since the region of interest had a significant population, evacuating by vehicles, although it could promote fast displacements, could be limited due to traffic limitations. In contrast, congestion does not affect pedestrian evacuation but may lead to slower speeds than vehicle evacuation in certain situations. Based on these preliminary tests, the 50%/50% on foot and by vehicle evacuation mode was adopted for all the simulations.

3.4. Sectorisation Scenarios and LOL Estimation

Partitioning the downstream valley into distinct sectors and assessing their outcomes using EWES and LOL simulations allows for identifying unique characteristics and proposing specific actions for each sector. In this study, we choose to evaluate sectors based on local conditions and using the distance of approximately 2 kilometres along the flow path of the flood wave to determine the sector's limits. This discretisation aligns spatially with the arrival time cross-sections, delineated every 30 min. A total of eight sectors were analysed. Among these sectors, the first three exhibited the highest concentration of population and buildings directly impacted by flooding (Figure 8).

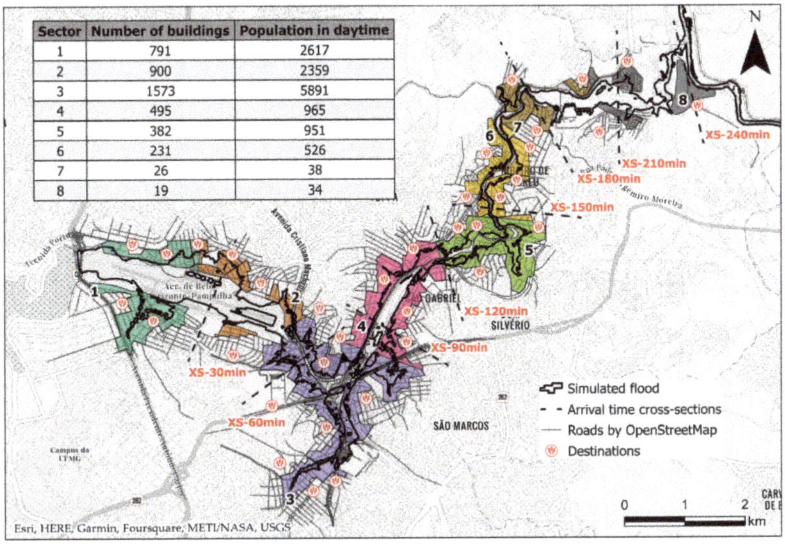

Sector	Number of buildings	Population in daytime
1	791	2617
2	900	2359
3	1573	5891
4	495	965
5	382	951
6	231	526
7	26	38
8	19	34

Figure 8. Sectors, their number of buildings and population directly impacted by the flood, and the roads and destinations used in EWES and LOL evaluation.

In HEC-LifeSim, the loss of life module (LLM) involves the application of probability distributions for fatality rates derived from the research conducted by McClelland and Bowles [44] and subsequently updated by Aboelata et al. [45]. The updates were based on the analysis of 54 historical cases, further subdivided into approximately 250 homogeneous regions based on event characteristics. The majority of these cases pertained to floods resulting from dam failures.

The LLM involves three distinct curves corresponding to different flood risk zones: the chance zone, which encompasses an area with a high probability of LOL; the compromised zone, representing an area with a moderate probability of LOL; and the safe zone, defining an area with a very low probability of LOL. To determine the risk zone allocation for an agent or groups of agents affected by the flood while inside a building, a submergence threshold is employed, considering the mobility of each agent. If the floodwater surpasses this threshold, the affected agents are allocated to the respective risk zone based on this threshold. When the flood reaches vehicles or agents during the evacuation process, the stability criteria indicated via USACE [33] are applied. If the floodwater exceeds these criteria, the affected agents are allocated to the chance zone. Conversely, if the stability criteria are not exceeded, the affected agents are assigned to the safe zone.

A recent version of the model, LifeSim v.2.0, has been implemented [69]. The Loss of Life Module (LLM) in this new version considers two flood zones: high hazard and low hazard. During the development process, the developed team encountered challenges in identifying and characterising an intermediate flood zone. To address this, they conducted a comprehensive review and expansion of the existing database compiled from the studies conducted by McClelland and Bowles [49] and Aboelata et al. [50]. Nevertheless, specific issues identified in the 2.0 version were reported to the Risk Management Center of the U.S. Army Corps of Engineers. Therefore, the more reliable HEC-LifeSim v.1.0.1 was used for the tests. With some modifications implemented in the LLM in order to bring it closer to the latest version, the probability distribution curves of fatality rates for high-hazard and low-hazard zones were considered in the model, in which the high-hazard zone combines the chance and the compromised zones, while the low-hazard zone comprises the safe zone.

Ten thousand simulations were conducted using HEC-LifeSim v.1.0.1 to account for uncertainties. These uncertainties encompassed a range of values defined in the Warning and Evacuation Module (WEM) and the application of fatality rate distributions derived from the LLM. The evacuation routes were established utilising road data obtained from OpenStreetMap. Safe destinations were determined based on local plans (Figure 8).

4. Results and Discussion

An apparent effect can be observed in the potential reduction of fatalities by increasing the alert systems and on-site mobilisation capacity, facilitating a more efficient evacuation in an emergency (Figures 9 and 10).

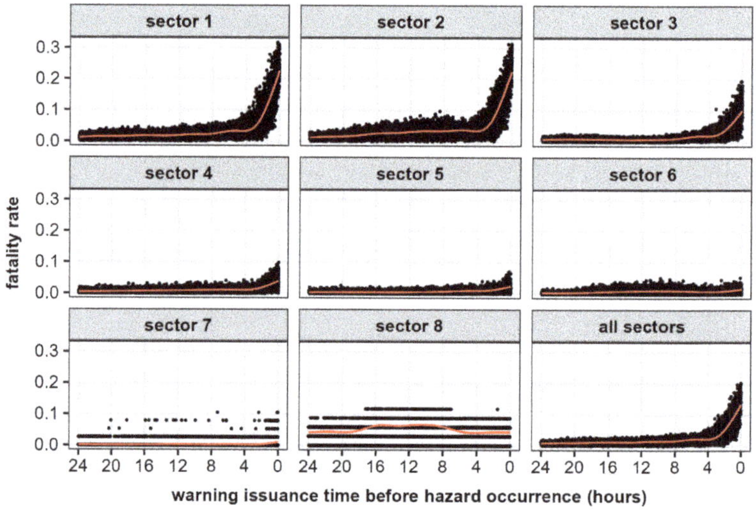

Figure 9. Fatality rates for the high mobilisation scenario summarised by sector and general estimation.

The first three sectors exhibit the highest sensitivity in terms of fatality rates. In cases of high mobilisation, there is an observed trend towards optimising the desired scenario of issuing warnings four hours before the occurrence of the hazard (Figure 9). However, in cases of low mobilisation, such a favourable scenario is not observed, as the likelihood of LOL rates exceeding 0.1 remains prevalent even when alerts are issued up to 12 h before the dam-breach event, specifically for sectors 1 and 2 (Figure 10).

The influence of the warning and evacuation module (WEM) on LOL in the other sectors is comparatively less pronounced compared to the three initial sectors. This observation can be attributed to two primary factors. First, the relatively lower hazard posed by the flood in this area might contribute to a diminished impact on the other sectors. Second, the arrival time of the floodwater, which is around 90 min in sector 4, could also permit a longer time for the population to reach safe places. In the last two sectors, the LOL does not demonstrate any apparent relationship to both warning issuance and mobilisation times.

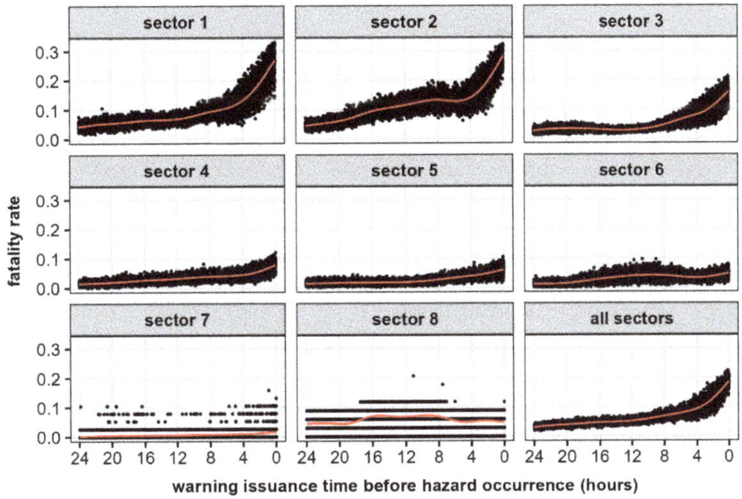

Figure 10. Fatality rates for the low mobilisation scenario summarised by sector and general estimation.

The exclusive impact of mobilisation can be assessed by examining the range of LOL rates estimates for the same time of warning issuance. This range is observed to be wider when the warning is issued closer to the hazard occurrence. It implies that even in an optimal mobilisation scenario, a delay in issuing the warning has a more pronounced impact on the estimation of LOL.

Upon comparing the quantiles of high preparedness and mobilisation (Table 3), the significant impact of this delay on the warning and evacuation timeline becomes evident. The median of the low mobilisation scenario is observed to exceed the 0.9 quantile of the high mobilisation scenario. Thus, determining the appropriate curve or range of curves of a specific case study to employ in the LifeSim model becomes an essential step to LOL evaluation. Moreover, the improvement of mobilisation could significantly impact the mitigating of LOL.

Table 3. Quantiles of LOL for each scenario of mobilisation by sector and general simulation.

Sector	High Mobilisation							Low Mobilisation						
	0.05	0.25	0.50	0.75	0.90	0.95	0.99	0.05	0.25	0.50	0.75	0.90	0.95	0.99
1	13	33	53	83	176	345	638	106	150	189	276	450	578	738
2	16	38	64	102	165	304	574	108	175	258	327	433	542	692
3	16	38	61	98	200	341	700	128	171	214	347	609	774	1024
4	0	4	7	13	19	26	49	11	21	30	42	54	65	86
5	0	2	5	9	14	18	32	9	15	21	31	44	53	68
6	0	1	3	6	10	12	17	4	9	15	21	26	29	35
7	0	0	0	0	0	1	1	0	0	0	0	1	1	3
8	0	1	2	3	3	3	4	0	1	2	3	3	4	4
All	59	129	197	304	570	1032	1950	416	563	713	1037	1613	2023	2499

Depending on the specific case study, the pattern curves in the WEM of LifeSim may not accurately represent the local context. Therefore, adjusting the definition of the curve locally using Equations (2)–(4) is recommended, along with additional studies on the definition of coefficient equations [70,71]. These adjustments can ensure the appropriate application of the curves to the local conditions.

In order to examine a more specific influence of mobilisation on estimating LOL, two predetermined instances of warning issuance time were established and simulated.

These simulations were designed to evaluate the effects of mobilisation in isolation. In this analysis, the focus was to evaluate the potential gains related to high levels of mobilisation. Values closer to one indicate a more optimised mobilisation, as depicted in the sampled curve of Figure 7b. Conversely, values closer to zero signify a less favourable scenario.

It is worth noting that when the warning is issued at the moment of the dam-breach event, the impact of mobilisation does not exhibit a clear trend (Figure 11), mainly in the first two sectors, compared to when the warning is issued four hours before the occurrence of the hazard (Figure 12).

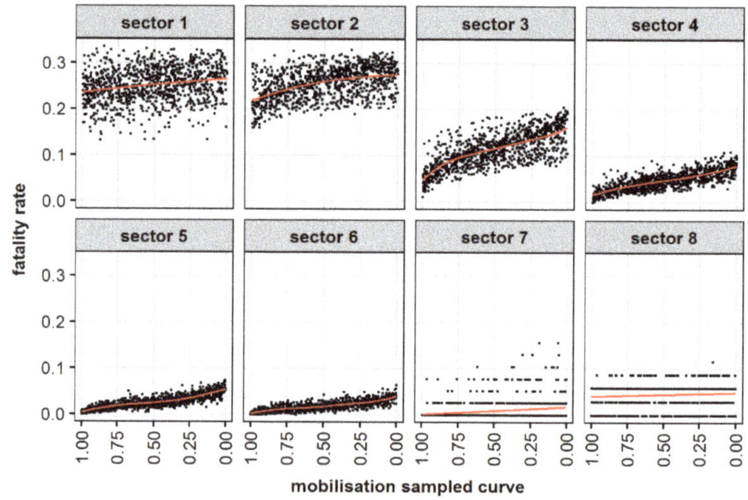

Figure 11. Fatality rates for the high mobilisation scenario summarised by sector when the warning is issued at the moment of dam breach.

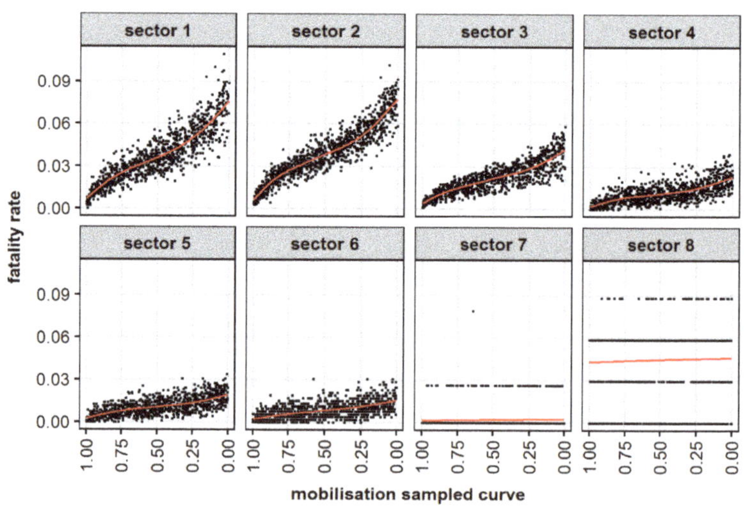

Figure 12. Fatality rates for the high mobilisation scenario summarised by sector when the warning is issued four hours before dam breach.

Even with optimal preparedness and awareness of the population, there is insufficient time to observe a substantial impact on the evaluation of LOL for the first two sectors.

However, in the simulation focused on the critical time of four hours, the mobilisation primarily affects these regions.

In high mobilisation scenarios, it was concluded that a minimum of four hours was required for anticipated warning issuance for significant alleviation of LOL (Figure 9). Three scenarios of sectored downstream areas were analysed to assess if this four-hour timeframe represents an optimal event anticipation and to evaluate the impact of spatial and temporal sectorization (Figure 13). These scenarios involved varying the time of anticipated warning issuance, ranging from the moment of hazard occurrence to four hours, four to eight hours and eight to twelve hours. These sectored downstream scenarios were compared to a reference scenario of a four-hour fixed warning issuance in a non-sectored downstream valley, with the results summarised within the respective sectors. To isolate the specific impact of warning issuance, the dissemination of this issuance and the mobilisation of the population were set without uncertainty, considering the median curve presented in Figure 7b.

(**a**) sectored warning issued between the moment and 4 h before the dam breach.

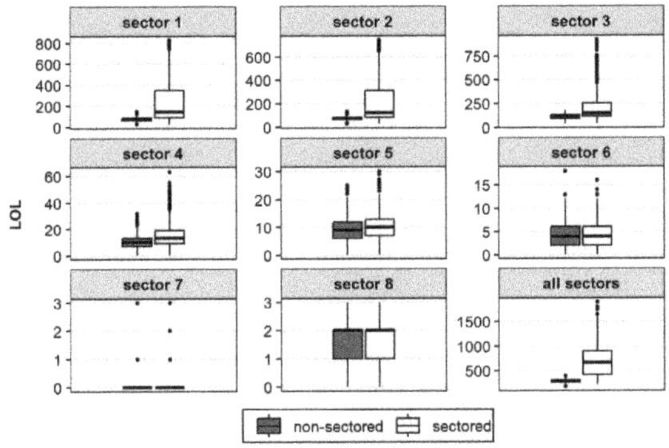

(**b**) sectored warning issued between 4 and 8 h before the dam breach.

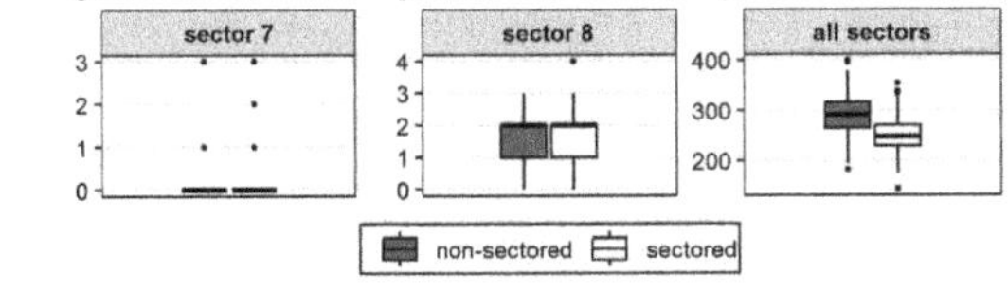

(**c**) sectored warning issued between 8 and 12 h before the dam breach.

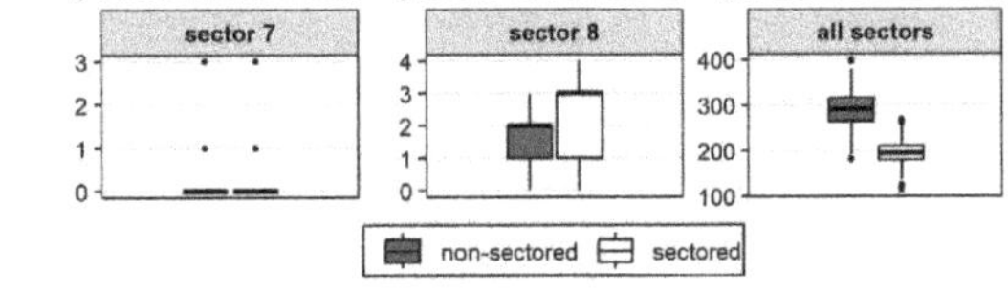

Figure 13. Comparison between a non-sectored simulation, where a fixed warning was issued four hours prior to the occurrence of the hazard, and sectored simulations with warnings issued within different time ranges: (**a**) between the moment of a dam breach and 4 h, (**b**) between 4 and 8 h and (**c**) between 8 and 12 h.

The first four hours (Figure 13a) reveal a notable sensitivity in estimating the LOL, particularly within the first three sectors. However, this sensitivity diminishes between four and eight hours (Figure 13b), with a minor difference of two fatalities simulated in the second sector. A slight decrease in the mean values of LOL from eight to twelve hours is also evident (Figure 13c), except for in the third sector, where the mean value experiences a reduction of over 40%.

It suggests that having prior knowledge limited to four hours may not be an optimal scenario for reducing the LOL in the third sector. While the other sectors exhibit stabilisation or only a slight difference in the absolute values of the median after the initial four hours, the LOL in the third sector continues to show significant changes, as a function of their specific vulnerability, exposure and hydrodynamic characteristics.

It is crucial to maintain awareness that although the design of an efficient EWES and the strategic implementation of sectorisation can unquestionably result in a notable decrease in the LOL, the perceptions and patterns exposed in this paper can be specific to the actual case study. In other cases, the results may vary. Several studies have unveiled a noticeable trend regarding mitigating LOL in EWES scenarios with different magnitudes. However, it was noticed that during critical junctures, mainly when an imminent hazard is near, the optimal course of action for the population might involve seeking shelter within buildings. This recommendation is based on observations that, during such moments, a significant influx of vehicles can reduce the road capacity, consequently reelevating the LOL [35,37,40].

The results and specificities of the study underline the importance of LOL simulations in obtaining a deeper understanding of the functioning of the EWES implemented or suggested. These findings and their patterns can help define alternative methods of disseminating warnings and allocating resources to increase awareness and preparedness among at-risk populations.

5. Conclusions

The estimation of loss of life (LOL) plays a vital role in flood emergency planning. This study assesses the impact of early warning and evacuation systems (EWES) on the potential LOL in the hypothetical scenario of a dam breach at Pampulha Lake. Specifically, the analysis focuses on examining the impact of this event on various sectors located in a downstream valley of the dam.

The LOL estimation revealed that a minimum desirable time for issuing warnings before the hazardous event was four hours, particularly in a scenario with high population mobilisation. Furthermore, it was revealed that the level of preparedness and perception of mobilisation, especially in the initial three sectors, significantly influenced the mitigation of LOL.

6. Further Research

The proposed sectorisation approach was crucial for understanding the effectiveness of dam-breach flood alerts in mitigating loss of life. It revealed that the effect of different warning and evacuation strategies might influence the areas at risk differently, according to their vulnerability, exposure and the hazard and EWES spatiotemporal characteristics. Therefore, it is recommended that similar approaches should be used in other case studies when EWES effectiveness is aimed for, especially for high-density population areas.

Furthermore, the estimation inputs for LOL calculations involve several uncertainties. A dam-breach flood event exhibits significant variability due to uncertainties associated with dam-breach parameters [72,73]. Acquiring data for exposure and vulnerability analysis at the microscale, particularly in this case, poses challenges, necessitating the consideration of various hypotheses regarding the transfer of characteristics from another scale to the microscale [62]. Furthermore, relying solely on a sample of historical events to characterise the warning and evacuation processes may not adequately represent the specific circumstances of a region in different contexts from those used in this analysis. In this study, our primary objective was to analyse the impact of LOL alleviation. Further research could consider and discuss these cited sources of uncertainties in the estimation model aiming to enhance the understanding of these uncertainties.

Author Contributions: Conceptualisation, A.F.R.S. and J.C.E.; methodology, A.F.R.S. and J.C.E.; validation, A.F.R.S. and J.C.E.; formal analysis, A.F.R.S. and J.C.E.; investigation, A.F.R.S. and J.C.E.; resources, A.F.R.S. and J.C.E.; writing—original draft preparation, A.F.R.S. and J.C.E.; writing—review and editing, A.F.R.S. and J.C.E.; visualisation, A.F.R.S.; supervision, J.C.E.; project administration, J.C.E.; funding acquisition, J.C.E. All authors have read and agreed to the published version of the manuscript.

Funding: This research was funded by the Brazilian research institutions, foundations and agencies Coordenação de Aperfeiçoamento de Pessoal de Nível Superior (CAPES), Conselho Nacional de Desenvolvimento Científico e Tecnológico (CNPq), Fundação de Amparo à Pesquisa do Estado de Minas Gerais (FAPEMIG) and Pró-Reitoria de Pesquisa, Universidade Federal de Minas Gerais (PRPq/UFMG).

Data Availability Statement: The data are available upon request.

Conflicts of Interest: The authors declare no conflict of interest. The funders had no role in the design of the study; in the collection, analyses, or interpretation of data; in the writing of the manuscript; or in the decision to publish the results.

References

1. Ge, W.; Sun, H.; Zhang, H.; Li, Z.; Guo, X.; Wang, X.; Qin, Y.; Gao, W.; van Gelder, P. Economic Risk Criteria for Dams Considering the Relative Level of Economy and Industrial Economic Contribution. *Sci. Total Environ.* **2020**, *725*, 138139. [CrossRef] [PubMed]
2. Sarchani, S.; Koutroulis, A.G. Probabilistic Dam Breach Flood Modeling: The Case of Valsamiotis Dam in Crete. *Nat. Hazards* **2022**, *114*, 1763–1814. [CrossRef]
3. Mishra, A.; Mukherjee, S.; Merz, B.; Singh, V.P.; Wright, D.B.; Villarini, G.; Paul, S.; Kumar, D.N.; Khedun, C.P.; Niyogi, D.; et al. An Overview of Flood Concepts, Challenges, and Future Directions. *J. Hydrol. Eng.* **2022**, *27*, 2164. [CrossRef]
4. Chai, J.; Wu, H.-Z. Prevention/Mitigation of Natural Disasters in Urban Areas. *Smart Constr. Sustain. Cities* **2023**, *1*, 4. [CrossRef]
5. Zheng, Q. Method for a New Risk Assessment of Urban Inundation: G-DEMATEL-AHP. *MethodsX* **2023**, *10*, e101997. [CrossRef]
6. Jonkman, S.N.; Van Gelder, P.H.A.J.M.; Vrijling, J.K. An Overview of Quantitative Risk Measures for Loss of Life and Economic Damage. *J. Hazard Mater* **2003**, *99*, 1–30. [CrossRef]
7. CRED; UNISDR. *Economic Losses, Poverty and Disasters 1998–2017*; UNISDR: Geneva, Switzerland, 2018.
8. Zhang, J.; Xu, W.; Liao, X.; Zong, S.; Liu, B. Global Mortality Risk Assessment from River Flooding under Climate Change. *Environ. Res. Lett.* **2021**, *16*, 064036. [CrossRef]
9. Hu, P.; Zhang, Q.; Shi, P.; Chen, B.; Fang, J. Flood-Induced Mortality across the Globe: Spatiotemporal Pattern and Influencing Factors. *Sci. Total Environ.* **2018**, *643*, 171–182. [CrossRef]
10. Merz, B.; Blöschl, G.; Vorogushyn, S.; Dottori, F.; Aerts, J.C.J.H.; Bates, P.; Bertola, M.; Kemter, M.; Kreibich, H.; Lall, U.; et al. Causes, Impacts and Patterns of Disastrous River Floods. *Nat. Rev. Earth Environ.* **2021**, *2*, 592–609. [CrossRef]
11. Hammond, M.J.; Chen, A.S.; Djordjević, S.; Butler, D.; Mark, O. Urban Flood Impact Assessment: A State-of-the-Art Review. *Urban Water J.* **2015**, *12*, 14–29. [CrossRef]
12. Kreibich, H.; Van Loon, A.F.; Schröter, K.; Ward, P.J.; Mazzoleni, M.; Sairam, N.; Abeshu, G.W.; Agafonova, S.; AghaKouchak, A.; Aksoy, H.; et al. The Challenge of Unprecedented Floods and Droughts in Risk Management. *Nature* **2022**, *608*, 80–86. [CrossRef]
13. Kreibich, H.; Hudson, P.; Merz, B. Knowing What to Do Substantially Improves the Effectiveness of Flood Early Warning. *Bull. Am. Meteorol. Soc.* **2021**, *102*, E1450–E1463. [CrossRef]
14. Thieken, A.H.; Bubeck, P.; Heidenreich, A.; von Keyserlingk, J.; Dillenardt, L.; Otto, A. Performance of the Flood Warning System in Germany in July 2021—Insights from Affected Residents. *Nat. Hazards Earth Syst. Sci.* **2023**, *23*, 973–990. [CrossRef]
15. Lumbroso, D.; Davison, M. Use of an Agent-Based Model and Monte Carlo Analysis to Estimate the Effectiveness of Emergency Management Interventions to Reduce Loss of Life during Extreme Floods. *J. Flood Risk Manag.* **2018**, *11*, S419–S433. [CrossRef]
16. Rotta, L.H.S.; Alcântara, E.; Park, E.; Negri, R.G.; Lin, Y.N.; Bernardo, N.; Mendes, T.S.G.; Souza Filho, C.R. The 2019 Brumadinho Tailings Dam Collapse: Possible Cause and Impacts of the Worst Human and Environmental Disaster in Brazil. *Int. J. Appl. Earth Obs. Geoinf.* **2020**, *90*, 102–119. [CrossRef]
17. Lumbroso, D.; Davison, M.; Body, R.; Petkovšek, G. Modelling the Brumadinho Tailings Dam Failure, the Subsequent Loss of Life and How It Could Have Been Reduced. *Nat. Hazards Earth Syst. Sci.* **2021**, *21*, 21–37. [CrossRef]
18. Apel, H.; Vorogushyn, S.; Merz, B. Brief Communication: Impact Forecasting Could Substantially Improve the Emergency Management of Deadly Floods: Case Study July 2021 Floods in Germany. *Nat. Hazards Earth Syst. Sci.* **2022**, *22*, 3005–3014. [CrossRef]
19. Silva, A.F.R.; Eleutério, J.C. Analysis of Flood Warning and Evacuation Efficiency by Comparing Damage and Life-Loss Estimates with Real Consequences Related to the São Francisco Tailings Dam Failure in Brazil. *Nat. Hazards Earth Syst. Sci.* **2023**, *23*, 3095–3110. [CrossRef]
20. Ge, W.; Jiao, Y.; Wu, M.; Li, Z.; Wang, T.; Li, W.; Zhang, Y.; Gao, W.; van Gelder, P. Estimating Loss of Life Caused by Dam Breaches Based on the Simulation of Floods Routing and Evacuation Potential of Population at Risk. *J. Hydrol.* **2022**, *612*, 128059. [CrossRef]

21. Jiao, H.; Li, W.; Ma, D. Assessment of Life Loss Due to Dam Breach Using Improved Variable Fuzzy Method. *Sci. Rep.* **2022**, *12*, 3237. [CrossRef]
22. Friedman, D.G. *Computer Simulation in Natural Hazard Assessment*; Institute of Behavioral Science, University of Colorado: Boulder, CO, USA, 1975.
23. Brown, C.A.; Graham, W.J. Assessing the Threat to Life from Dam Failure. *J. Am. Water Resour. Assoc.* **1988**, *24*, 1303–1309. [CrossRef]
24. DeKay, M.L.; McClelland, G.H. Predicting Loss of Life in Cases of Dam Failure and Flash Flood. *Risk Anal.* **1993**, *13*, 193–205. [CrossRef]
25. Graham, W. *A Procedure for Estimating Loss of Life Caused by Dam Failure. Sediment*; U.S. Department of the Interior Bureau of Reclamation: Denver, CO, USA, 1999.
26. Aboelata, M.; Bowles, D.S. *LIFESim: A Model for Estimating Dam Failure Life Loss*; Utah State University: Logan, UT, USA, 2005.
27. Johnstone, W.M.; Sakamoto, D.; Assaf, H.; Bourban, S. Architecture, Modelling Framework and Validation of BC Hydro's Virtual Reality Life Safety Model. In Proceedings of the International Symposium on Stochastic Hydraulics, Nijmegen, The Netherlands, 23–24 May 2005; Volume 2005.
28. Jonkman, S.N.; Vrijling, J.K.; Vrouwender, A.C.W.M. Methods for the Estimation of Loss of Life Due to Floods: A Literature Review and a Proposal for a New Method. *Nat. Hazards* **2008**, *46*, 353–389. [CrossRef]
29. U.S. Department of the Interior Bureau of Reclamation. *RCEM—Reclamation Consequence Estimating Methodology—Guidelines for Estimating Life Loss for Dam Safety Risk Analysis*; U.S. Department of the Interior Bureau of Reclamation: Washington, DC, USA, 2015.
30. Huang, D.; Yu, Z.; Li, Y. Calculation Method and Application of Loss of Life Caused by Dam Break in China. *Nat. Hazards* **2017**, *85*, 39–57. [CrossRef]
31. Jonkman, S.N.; Maaskant, B.B.; Kolen, B.B.; Needham, J.T.J. Loss of Life Estimation—Review, Developments and Challenges. In Proceedings of the III European Conference on Flood Risk Management, Lyon, France, 17–21 October 2016; Volume 7.
32. Zhuo, L.; Han, D. Agent-Based Modelling and Flood Risk Management: A Compendious Literature Review. *J. Hydrol.* **2020**, *591*, 125600. [CrossRef]
33. USACE HEC-LifeSim. *Life Loss Estimation. User's Manual—Version 1.0.1*; USACE: Davis, CA, USA, 2018.
34. Needham, J.; Fields, W.; Lehman, W. The US Army Corps of Engineers Scalable Approach to Estimating Loss of Life from Flooding. In Proceedings of the III European Conference on Flood Risk Management, Lyon, France, 17–21 October 2016; p. 6.
35. Hill, P.; Kavanagh, C.; Lang, S. Applications of Simulation Model to Estimate Potential Loss of Life. In Proceedings of the XXVI International Congress on Large Dams, Viena, Austria, 1–7 July 2018.
36. Kalinina, A.; Spada, M.; Burgherr, P. Alternative Life-Loss Rates for Failures of Large Concrete and Masonry Dams in Mountain Regions of OECD Countries. In Proceedings of the XXVIII International European Safety and Reliability Conference, ESREL 2018, Trondheim, Norway, 17–21 June 2018; p. 9.
37. Leong-Cuzack, T.; Nielsen, C.; Kavanagh, C.; Watt, S. Quantitative Assessment of Dam Safety Emergency Management Using HEC-LifeSim: Is It Feasible? In Proceedings of the Australian Committee on Large Dams Conference, Auckland, New Zealand, 9–12 October 2019; p. 11.
38. Wang, H. Lessons Learnt From Evacuation Modelling for Dam Failure Consequence Assessments. In Proceedings of the Australian Committee on Large Dams Conference, Auckland, New Zealand, 9–12 October 2019; p. 9.
39. Tomura, S.; Chiba, M.; Yamamoto, T.; Uemura, F.; Masuya, S.; Omura, N.; Yoshida, T.; Takeda, A.; Hoshino, T.; Yamada, T.; et al. Fatality Estimation by Life Loss Evaluation Model for the Large-Scale Floods under Future Climate. In Proceedings of the 22nd Congress of the International Association for Hydro-Environment Engineering and Research-Asia Pacific Division, IAHR-APD 2020: "Creating Resilience to Water-Related Challenges", Sapporo, Japan, 14–17 September 2020; pp. 1–8.
40. El Bilali, A.; Taleb, A.; Boutahri, I. Application of HEC-RAS and HEC-LifeSim Models for Flood Risk Assessment. *J. Appl. Water Eng. Res.* **2021**, *9*, 336–351. [CrossRef]
41. Kalinina, A.; Spada, M.; Burgherr, P. Quantitative Assessment of Uncertainties and Sensitivities in the Estimation of Life Loss Due to the Instantaneous Break of a Hypothetical Dam in Switzerland. *Water* **2021**, *13*, 22. [CrossRef]
42. El Bilali, A.; Taleb, I.; Nafii, A.; Taleb, A. A Practical Probabilistic Approach for Simulating Life Loss in an Urban Area Associated with a Dam-Break Flood. *Int. J. Disaster Risk Reduct.* **2022**, *2022*, 103011. [CrossRef]
43. Lang, S. Current Australian and International Practices for Dam Failure Consequence Assessments. In Proceedings of the Australian Committee on Large Dams Conference, Melbourne, Australia, 10 November–10 December 2018; Australian National Committee on Large Dams: Melbourne, VIC, Austrália, 2018; pp. 1–6.
44. McClelland, D.M.; Bowles, D.S. *Estimating Life Loss for Dam Safety Risk Assessment—A Review and New Approach*; Institute for Water Resources, U.S. Army Corps of Engineers: Alexandria, VA, USA, 2002.
45. Aboelata, M.; Bowles, D.S.; McClelland, D.M. A Model for Estimating Dam Failure Life Loss. In Proceedings of the Australian Committee on Large Dams Risk Workshop, Launceston, TAS, Australia, 10 October 2003; p. 18.
46. Sorensen, J.H.; Lindell, M.K.; Baker, E.J.; Lehman, W.P. Community Response to Hurricane Threat: Estimates of Warning Issuance Time Distributions. *Weather Clim. Soc.* **2020**, *12*, 837–846. [CrossRef]
47. Lindell, M.K.; Sorensen, J.H.; Baker, E.J.; Lehman, W.P. Community Response to Hurricane Threat: Estimates of Warning Diffusion Time Distributions. *Weather Clim. Soc.* **2021**, *22*, 04021007. [CrossRef]

48. Koll, C.; Lindell, M.K.; Chen, C.; Wang, H. Emergency Warning Dissemination in a Multiplex Social Network. *JASSS* **2023**, *26*, 4946. [CrossRef]
49. Lindell, M.K.; Sorensen, J.H.; Baker, E.J.; Lehman, W.P. Community Response to Hurricane Threat: Estimates of Household Evacuation Preparation Time Distributions. *Transp. Res. D Transp. Environ.* **2020**, *85*, 102457. [CrossRef]
50. Chen, C.; Koll, C.; Wang, H.; Lindell, M.K. An Interdisciplinary Agent-Based Evacuation Model: Integrating the Natural Environment, Built Environment, and Social System for Community Preparedness and Resilience. *Nat. Hazards Earth Syst. Sci.* **2023**, *23*, 733–749. [CrossRef]
51. Alaeddine, H.; Serrhini, K.; Maizia, M.; Néron, E. A Spatiotemporal Optimization Model for the Evacuation of the Population Exposed to Flood Hazard. *Nat. Hazards Earth Syst. Sci.* **2015**, *15*, 687–701. [CrossRef]
52. Lee, Y.H.; Kim, H.I.; Han, K.Y.; Hong, W.H. Flood Evacuation Routes Based on Spatiotemporal Inundation Risk Assessment. *Water* **2020**, *12*, 2271. [CrossRef]
53. Bopp, E.; Douvinet, J. Alerting People Prioritising Territories over Technologies. A Design Framework for Local Decision Makers in France. *Appl. Geogr.* **2022**, *146*, 102769. [CrossRef]
54. Vianini Neto, L. Estudo de Ruptura Da Barragem Da Pampulha, Em Belo Horizonte: Retroanálise Da Brecha Do Acidente de 1954 e Ruptura Hipotética Nas Condições Atuais. Master's Thesis, Federal University of Minas Gerias, Belo Horizonte, Brazil, 2016.
55. Silva, T.F.G.; Vinçon-Leite, B.; Giani, A.; Figueredo, C.C.; Petrucci, G.; Lemaire, B.; Von Sperling, E.; Tassin, B.; Seidl, M.; Khac, V.T.; et al. Modelagem Da Lagoa Da Pampulha: Uma Ferramenta Para Avaliar o Impacto Da Bacia Hidrográfica Na Dinâmica Do Fitoplâncton. *Eng. Sanit. Ambient.* **2016**, *21*, 95–108. [CrossRef]
56. Psomiadis, E.; Tomanis, L.; Kavvadias, A.; Soulis, K.X.; Charizopoulos, N.; Michas, S. Potential Dam Breach Analysis and Flood Wave Risk Assessment Using HEC-RAS and Remote Sensing Data: A Multicriteria Approach. *Water* **2021**, *13*, 364. [CrossRef]
57. Froehlich, D.C. Embankment Dam Breach Parameters and Their Uncertainties. *J. Hydraul. Eng.* **2008**, *134*, 1708–1721. [CrossRef]
58. Brunner, G.W. *HEC-RAS. River Analysis System. Hydraulic Reference Manual—Version 6.0*; U.S. Army Corps of Engineers: Davis, CA, USA, 2021.
59. Merz, B.; Kreibich, H.; Schwarze, R.; Thieken, A. Review Article "Assessment of Economic Flood Damage". *Nat. Hazards Earth Syst. Sci.* **2010**, *10*, 1697–1724. [CrossRef]
60. De Moel, H.; Jongman, B.; Kreibich, H.; Merz, B. Flood Risk Assessments at Different Spatial Scales. *Mitig. Adapt. Strat. Glob. Chang.* **2015**, *20*, 865–890. [CrossRef] [PubMed]
61. Mauro, M.D.; De Bruijn, K.M.; Meloni, M. Quantitative Methods for Estimating Flood Fatalities: Towards the Introduction of Loss-of-Life Estimation in the Assessment of Flood Risk. *Nat. Hazards* **2012**, *63*, 1083–1113. [CrossRef]
62. Paprotny, D.; Mengel, M. Population, Land Use and Economic Exposure Estimates for Europe at 100 m Resolution from 1870 to 2020. *Sci. Data* **2023**, *10*, 372. [CrossRef] [PubMed]
63. Sorensen, J.; Mileti, D. *First Alert and/or Warning Issuance Time Estimation for Dam Breaches, Controlled Dam Releases, and Levee Breaches or Overtopping*; U.S. Army Corps of Engineers: Davis, CA, USA, 2015.
64. Sorensen, J.; Mileti, D. *First Alert or Warning Diffusion Time Estimation for Dam Breaches, Controlled Dam Releases and Levee Breaches or Overtopping*; U.S. Army Corps of Engineers: Davis, CA, USA, 2015.
65. Sorensen, J.; Mileti, D. *Protective Action Initiation Time Estimation for Dam Breaches, Controlled Dam Releases, and Levee Breaches or Overtopping*; U.S. Army Corps of Engineers: Davis, CA, USA, 2015.
66. Needham, J.; Sorensen, J.; Mileti, D.; Lang, S. Warning and Mobilization of Populations at Risk of Dam Failure. In Proceedings of the ANCOLD Conference, Adelaide, Australia, 17–19 October 2016; pp. 1–10.
67. Greenshields, B.D.; Channing, W.; Miller, H. A Study of Traffic Capacity. In Proceedings of the XIV Annual Meeting of the Highway Research Board, Washington, DC, USA, 18–20 November 1935.
68. Dijkstra, E.W. A Note on Two Problems in Connexion with Graphs. *Numer. Math.* **1959**, *1*, 269–271. [CrossRef]
69. USACE LifeSim. *Version 2.0. Hydrologic Engineering Center*; U.S. Army Corps of Engineers: Davis, CA, USA, 2021.
70. Sorensen, J.; Mileti, D. *Influence Weights and Measures for the Factors Shaping First Alert/Warning Delay, Diffusion and Protective Action Initiation Curves for Dam Breaches, Controlled Dam Releases, and Levee Breaches or Overtopping*; U.S. Army Corps of Engineers: Davis, CA, USA, 2015.
71. Sorensen, J.; Mileti, D. *Interview Schedule. Community Warning Issuance, Diffusion, and Protective Action Initiation Estimation*; U.S. Army Corps of Engineers: Davis, CA, USA, 2015.
72. Bello, D.; Alcayaga, H.; Caamaño, D.; Pizarro, A. Influence of Dam Breach Parameter Statistical Definition on Resulting Rupture Maximum Discharge. *Water* **2022**, *14*, 1776. [CrossRef]
73. Rizzo, C.; Maranzoni, A.; D'Oria, M. Probabilistic Mapping and Sensitivity Assessment of Dam-Break Flood Hazard. *Hydrol. Sci. J.* **2023**, *68*, 700–718. [CrossRef]

water

Article

2D and 3D Numerical Simulation of Dam-Break Flooding: A Case Study of the Tuzluca Dam, Turkey

Cagri Akgun [1,*], Salim Serkan Nas [2] and Akin Uslu [2]

1 Igdir Vocational School of Higher Education, Igdir University, 76000 Igdir, Turkey
2 Civil Engineering Department, Gumushane University, 29100 Gumushane, Turkey; serkannas@gmail.com (S.S.N.); usluakin@hotmail.com (A.U.)
* Correspondence: cagriakgun2063@gmail.com

Abstract: Unlike river floods, floods caused by dam breaks occur much more abruptly and at higher water velocities. The failure of a dam due to any reason can pose a significant threat to settlements in the downstream area due to the high volume of water in its reservoir. In this case study, 2D and 3D numerical failure simulations of Tuzluca Dam, to be built in the Igdir province of Turkey, was performed using the Flow3D Hydro software. The Shallow Water and Reynolds-Averaged Navier-Stokes (RANS) equations were used to simulate flood propagation. In the numerical models, RNG k-ε was chosen as the turbulence model. In the analyses carried out under different scenarios, it was observed that in both methods (2D and 3D), significant flood depths would occur in two downstream village settlements. Within the scope of the study, the mortality rates in different dam-break scenarios of the Tuzluca Dam were also investigated. It has been observed that the dam failure time and breach geometry have a significant impact on the flood propagation and the loss of life in settlements. Additionally, a dam-break experiment in literature dam-break experiment was numerically solved using the Flow3D Hydro software to validate the numerical model, and a satisfactory harmony was observed between the experimental results and the numerical models. Although the 3D analysis provided a more accurate representation of the experimental results, it took approximately 20 times longer to complete the dam failure simulation in the case study compared to the 2D model. As a result, it has been determined that for comprehensive studies such as those of dam failures, 2D analyses may be more practical, while for smaller-scale scenarios or situations requiring greater precision, 3D modelling is appropriate.

Keywords: dam-break modelling; dam overtopping; flood propagation; Flow3D; validation of numerical models

Citation: Akgun, C.; Nas, S.S.; Uslu, A. 2D and 3D Numerical Simulation of Dam-Break Flooding: A Case Study of the Tuzluca Dam, Turkey. *Water* **2023**, *15*, 3622. https://doi.org/10.3390/w15203622

Academic Editor: Paolo Mignosa

Received: 8 September 2023
Revised: 9 October 2023
Accepted: 12 October 2023
Published: 16 October 2023

1. Introduction

Dams, constructed to retain flowing water in river valleys, have been utilized for centuries to meet humanity's water needs. The collapse of a dam for any reason results in floods occurring at much more abrupt and rapid water velocities compared to river floods due to the high water mass in its reservoir. Therefore, dam collapses represent low-probability but high-risk disaster events [1–3].

Historical dam breaks have shown a high occurrence of casualties and property losses. In 1979, the collapse of the Machuchu-2 Dam in India resulted in 5000 deaths, while the collapse of the Val De Stava Dam in Italy in 1985 caused the loss of 286 lives. Similarly, the collapse of the Patel Dam in Kenya in 2018 resulted in 47 deaths, and the collapse of the Brumadinho Dam in Brazil in 2019 claimed the lives of 270 people, providing recent concrete examples in this regard [4,5].

Dam breaks can occur due to various factors, both internal and external. Research has shown that factors such as overtopping, piping (internal erosion), earthquakes, seepage, and structural issues can lead to dam breaks [6,7]. According to the data from the

International Commission on Large Dams (ICOLD), 44% of the past dam breaks occurred due to overtopping, while 34% occurred due to piping [8]. There has been no dam failure in Turkey, the study region, and only in 2012, a flood occurred in the construction area as a result of the rupture of the derivation tunnel cover during the construction of a dam in Adana province. Regardless of the cause, all dam breaks begin with the formation of a breach, which grows and leads to the break of the dam [2]. In the case of dam failure caused by overtopping, breach initiation typically occurs in the dam crest and tends to grow downward, often assuming a trapezoidal shape. On the other hand, in the case of failure caused by piping, commonly observed in embankment dams, breaches start at any location in the dam body and grow in both the transverse and longitudinal directions, forming a trapezoidal or rectangular shape [9–11].

Studies on dam failure provide valuable information for the development of early warning systems and evacuation plans in downstream areas. Conducting possible dam-break simulations and establishing early warning systems contribute to preventing or mitigating disasters for existing and planned dams [3,12]. In recent years, many studies have been carried out on flood control and prediction methods using remote sensing and GIS technology. Li et al. [13], in their study, analyzed the public's awareness of the debris flow using 3D remote sensing methods and made suggestions to improve people's behavior concerning this issue. Liu et al. [14] stated that monitoring urban water resources using remote sensing and geostatistical technologies will benefit flood control. Oborie and Rowland [15] developed an impact flood prediction method based on hydrological and morphological data using remote sensing and GIS techniques for river basins.

The problem of dam breaks has been the subject of numerous studies focusing on statistical, experimental, and numerical solutions. Researchers such as MacDonald and Langridge-Monopolis [9], Von Thun and Gillette [16], Froehlich [6,17], and Xu and Zhang [18] have developed empirical formulas based on statistical analysis of past dam breaks with reliable data to estimate parameters such as the average breach width, peak discharge, and failure time for potential dam breaks.

Experimental studies on dam breaks have been used not only to focus on hydraulic parameters such as the breach geometry, time-varying outflow hydrograph, flow velocity, and flow depths, but also to validate numerical simulations related to dam breaks [19,20]. Bella et al. [21] experimentally investigated the case of a sudden dam break in straight and curved horizontal channels. They monitored the pointwise variations of the water levels at five different points in the downstream region for conditions of curved and smooth channel beds. Lauber and Hager [22] conducted an experimental dam-break study in a rectangular smooth channel and compared the results with the analytical solutions developed by Ritter [23] for dam breaks. Soares-Frazao [24] performed an experimental dam-break test in a rectangular horizontal channel, with a triangular bottom sill placed in the downstream region, to examine the behavior of the flood wave during sudden changes in the slope using video cameras and level gauges. Hui et al. [25] experimentally investigated sudden dam breaks in a horizontal rectangular channel under different reservoir depths and wet and dry downstream conditions. They compared the obtained results with the analytical methods developed by Stoker [26], Lin et al. [27], and Ritter [23].

Many studies have used the shallow water equations, also known as the Saint-Venant equations, for the numerical solution of the dam-break problem [28]. Albano et al. [29] conducted 1D and 2D dam failure analyses for the San Giuliano and Gleno Dams using the Saint-Venant equations and developed a Geographic Information System-based flood mapping method for masonry arch and gravity dams. Riha et al. [30] examined the flood propagation effect of multiple dam failures by performing 2D dam failure analysis of three dams on the Cižina River (Czech Republic) using the HECRAS v. 5.0.7 software. Psomiadis et al. [31] performed 2D failure analysis of the Bramianos dam using two different surface models; the dam failure analyses were performed using a digital elevation model (DEM) obtained from reliable satellite photographs and a digital surface model (DSM) obtained with an unmanned aerial vehicle (UAV), and the results were compared. Bello et al. [32]

used the HEC-RAS and McBreach v 5.0.7 software to conduct dam failure analysis of the Chacrillas dam. Sensitivity analyses performed in the study revealed that the most significant factor influencing the maximum discharge after failure is the failure time.

Although the Saint-Venant equations provide significant advantages in terms of the solution time and ease, they have important limitations, such as neglecting vertical acceleration and the dominance of hydrostatic pressure conditions in the flow; that is, the water depth is considered to have a small ratio compared to the length in the horizontal direction. In these cases, the flow is represented in 1 or 2 dimensions. However, a real dam break is a fluid dynamics problem where the 3-dimensional components of velocity and acceleration are active, sudden topographic changes can occur, and hydrostatic pressure conditions do not dominate [6,33,34]. The complexity of the hydraulic components and the wide extent of flood impact in a real dam break require long computation times and computers with a high processing power for numerical solutions. Therefore, a significant portion of the 3D dam-break studies in the literature have been conducted by creating and solving a flood wave in small-scale horizontal channels using numerical models. Larocque et al. [35] solved a previously published dam failure experiment numerically in 3D using the LES and k-e turbulence models. Robb and Vasquez [36] numerically solved two previously published experimental studies. Within the scope of the study, the results of the experiments solved in 2D using the Telemac-2D software and in 3D using the Flow3D and Openfoam software were compared. Tahershamsi et al. [37] investigated the effect of the breach ratio and reservoir geometry on the peak discharge in dam failure using both experimental and numerical methods. Through 3D analyses conducted using the Ansys Fluent v. 6.3.26 software, they developed empirical formulas based on the reservoir shape factor and dam breach ratio. Franco et al. [38] conducted numerical analysis of the 1958 Lituya Bay tsunami event. In the 3D analyses performed using the Flow3D software, they examined the propagation of the flood wave caused by the tsunami and compared it with historical data. Kocaman et al. [39] numerically simulated their laboratory-developed dam-break experiment using the Flow3D software. The results of the numerical simulations conducted using the Shallow Water Equations (SWE) and Reynolds-Averaged Navier-Stokes (RANS) equations were compared. Additionally, within the scope of their study, the experiment was repeated for different tailwater depths to investigate the effect of tailwater on flood propagation.

When reviewing the literature, it is observed that due to the long solution times associated with 3D dam-break analyses, dam-break analyses at the actual topographical scale are performed in 1D and 2D. Furthermore, for similar reasons, studies comparing 2D and 3D analyses in dam-break scenarios have been conducted in scaled-down experimental setups.

In this case study, dam-break analyses were conducted at a real dam topography scale, considering the gaps in the literature, using 2D Shallow Water Equations (SWE) and 3D Navier-Stokes Equations (RANS). The study area was selected as the planned Tuzluca Dam in the eastern-most province of Turkey; namely, Iğdır. The suitability of 3D modeling in large-scale topographies was explored by comparing the results of the analyses conducted using the two different methods. Furthermore, the impact of different breach geometries, failure modes, and failure times on flood propagation was investigated by implementing three different scenarios. The risk situations and possible loss of life in settlements in the downstream area were evaluated in the conducted analyses. Another significance of the study is that the selected Tuzluca Dam is still in the planning phase, so the findings and risks obtained in the research can provide some revision opportunities to decision-makers regarding the project.

2. Materials and Methods

2.1. Study Area and Dam Characteristics

Tuzluca Dam, which has been included in the investment program by the public, and the necessary plans and projects have been prepared. It is planned to be built on the 40°07′52″ N and 43°38′12″ E coordinate lines on the Aras River, which forms the Turkey–Armenia border in Igdir, Turkey. The location map is given in Figure 1, below.

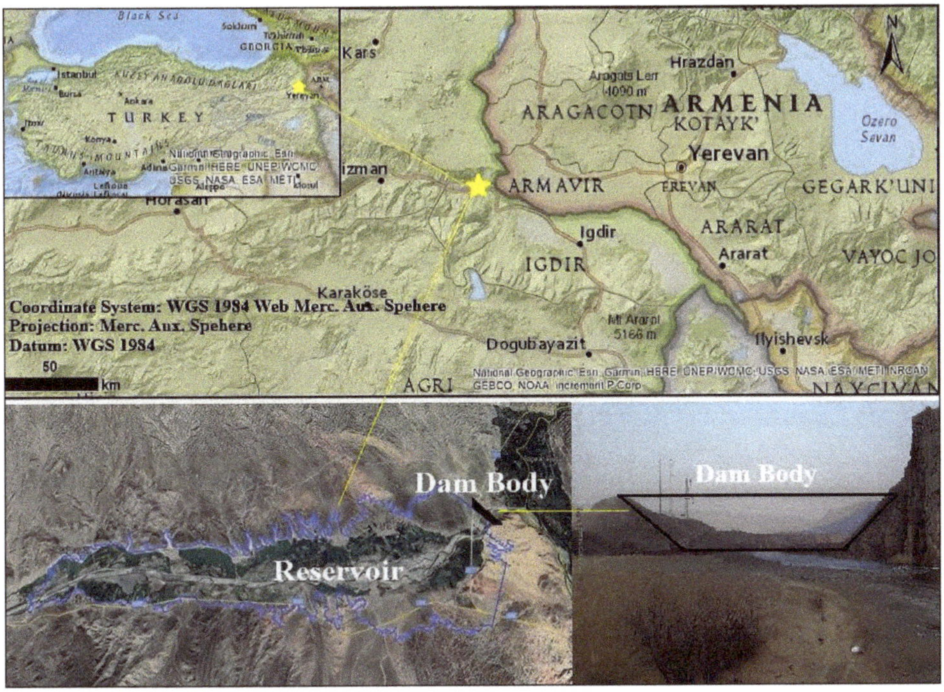

Figure 1. Location map of the Tuzluca Dam.

The study area covers a large area of 101.5 km^2. The settlements of Halimcan, Halikisla, Turabi, Surmeli, and Calpala on the Turkish border, and the settlements of Bagaran and Yervandashat on the Armenian border are potential at-risk settlements in the downstream region of the dam.

The Tuzluca Dam is planned to be a rock-fill dam with a height of 45 m. The dam, with a reservoir capacity of 265 million m^3, is designed to generate hydroelectric power, with an installed capacity of 20 MW, while also providing irrigation benefits. Three villages will be submerged by the reservoir upon the construction of the dam. The cross-section of the dam body is shown in Figure 2, below.

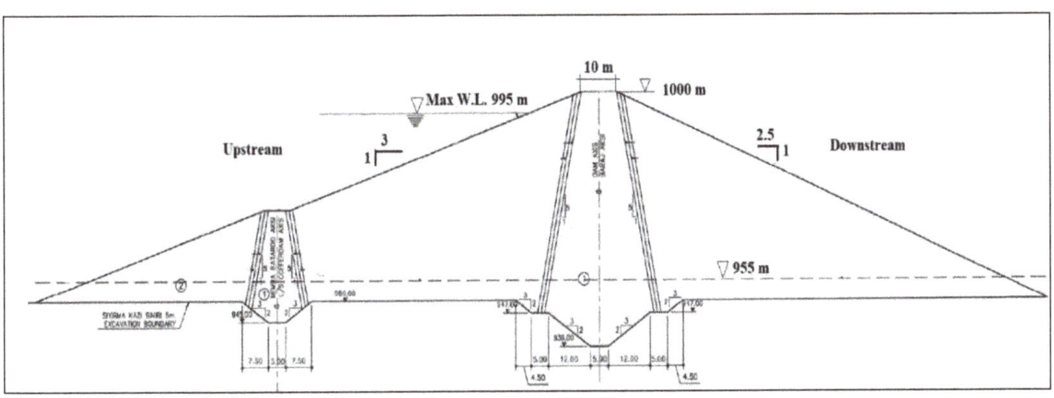

Figure 2. Section of Dam Body [40].

2.2. Geometric Data

In hydraulic modelling, accurately representing the geometry of the study area is an essential requirement; particularly in extensive studies such as dam breaches, obtaining topographic data through remote sensing methods provides significant convenience to users [41,42].

In this study, a digital elevation model (DEM) with a resolution of 12.5 m from the ALOS satellite was used to represent the topographic conditions. The ALOS satellite, launched by the Japan Aerospace Exploration Agency (JAXA) in 2006, was developed to contribute to various studies such as cartography, resource exploration, disaster monitoring, and environmental change analysis [43]. The reliability of the DEM data from the ALOS satellite has been tested for flood risk analysis in various studies and is considered to be successful [44,45]. The DEM data for the study area were converted to the Stereolithography (STL) format using the Global Mapper v. 24.0 software and transferred to the Flow3D software.

2.3. Estimation of Dam Breach Parameters

One of the most significant uncertainties in dam failure analyses is the estimation of the breach parameters. In this context, calculations were made using the methods developed by MacDonald and Langridge-Monopolis [9], Bureau of Reclamation [46], Von Thun and Gillette [16], Froehlich [6,17], and Xu and Zhang [18] to calculate the breach dimensions and failure time. The breach parameters calculated according to the relevant methods are presented in Table 1.

Table 1. Calculated breach parameters.

Method	B_t (m)		B_{ave} (m)		B_b (m)		T_f (h)	
	O	P	O	P	O	P	O	P
MacDonald and Langridge-Monopolis [9]	341	224			161	224	3	
Bureu of Reclamation [46]			135				1.5	
Von Thun and Gillette [16]			167				1	
Froehlich [17]			258	184			2.4	
Froehlich [6]			218	156			2	
Xu and Zhang [18]	380	222	287	168	217	87	2.4	2.3

Note(s): The values used in the simulations are underlined. B_t: breach top width, B_{ave}: breach average width, B_b: breach bottom width, T_f: failure time, O: overtopping, P: piping.

When comparing the methods in the literature, it is observed that the most unfavorable conditions in terms of the breach dimensions in overtopping scenarios are obtained using the method proposed by Xu and Zhang [18], while the methods proposed by Von Thun and Gillette [16] are found to be particularly unfavorable in the case of failure time. Consequently, the Xu and Zhang [18] method was employed for the overtopping scenarios. As the rectangular breach condition for dam failures caused by piping can only be applied using the MacDonald and Langridge-Monopolis [9] method, the MacDonald and Langridge-Monopolis [9] method was chosen to examine different breach geometries in the piping scenario. In the 3rd scenario, the effect of reducing the failure time to half an hour on flood propagation was investigated, considering breach dimensions calculated using the method proposed by Xu and Zhang [18]. Information about the scenarios is visualized in Figure 3.

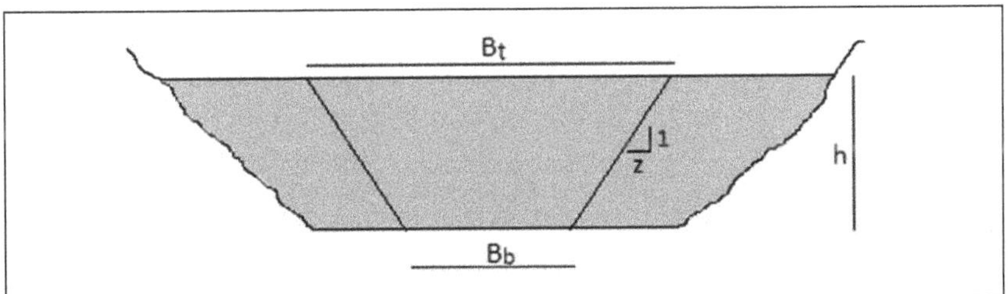

	Failure Type	Bt (m)	Bb (m)	h (m)	z	t (h)	Method
Scenario 1	Overtopping	380	217	45	1.8	1	Xu and Zhang
Scenario 2	Overtopping	380	217	45	1.8	0.5	Xu and Zhang
Scenario 3	Piping	224	224	45	0	1	MacDonald and Langridge-Monopolis

Figure 3. Information regarding dam failure scenarios in Flow3D.

2.4. Flow3D Numeric Model

In this study, the possible failure simulation of the Tuzluca Dam was numerically modelled as 2D and 3D using the Flow3D software to investigate the impact of flooding on downstream settlements under different scenarios. Flow3D is a computational fluid dynamics (CFD) software specifically developed to solve fluid flow equations [47].

The software utilizes the Volume of Fluid (VOF) method to track the free surface of the fluid in the calculation grids and employs the Fractional Area-Volume Obstacle Representation (FAVOR) algorithm to define solid objects [33,48]. In this study, the Renormalized Group (RNG) k-ε model, widely used in the dam-break analyses in the literature and validated in many studies, was selected as the turbulence model. This model can accurately model turbulent flow conditions with both high and low Reynolds numbers [49–55].

Flow3D offers two different approaches to deal with surface roughness. In the first approach, the surface roughness is obtained by pre-processing the solid object model (STL file) of the topography using the FAVOR technique. Additionally, due to the fact that the mesh blocks created do not perfectly conform to the terrain surface, there is also roughness depending on the angle formed. As a second option, Flow3D provides the possibility to define custom surface roughness. This feature can be used in cases where additional roughness is introduced in the topography, such as the presence of vegetation [38]. In the simulations conducted within the scope of the study, a roughness value of 1 m was used in areas with vegetation. In other areas, this coefficient was set to 0.

The Flow3D software uses artificial probes to detect the physical properties of water waves at any given point. In this study, probes were assigned to 7 settlements at risk of flooding to examine the water depths in post-failure settlements (Figure 4).

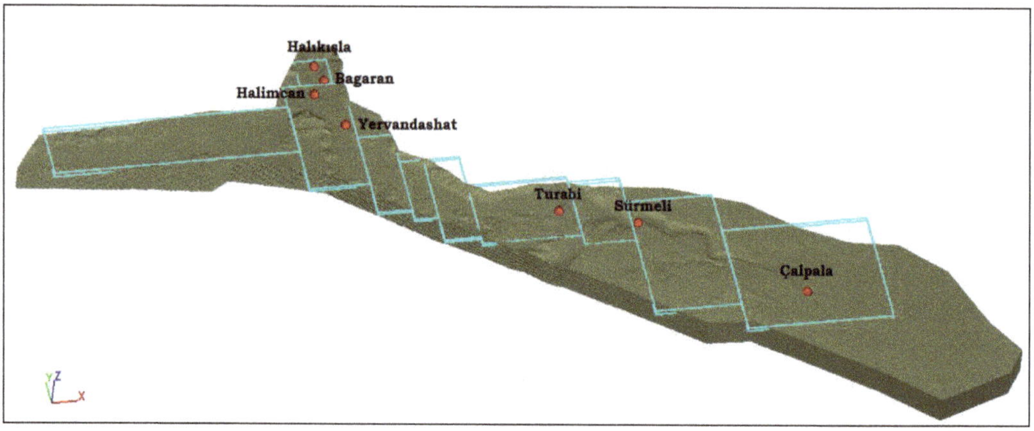

Figure 4. Display of topography, mesh blocks and probes in Flow3D software.

The numerical model was solved using two different methods—the 2D Shallow Water equation and 3D Navier Stokes equations—and the results were compared.

2.4.1. Two-Dimensional Shallow Water Equations

2D dam failure analyses can be performed by activating the Shallow Water model in the Flow3D software. The solution equation of the Shallow Water model is given below.

$$\frac{\partial V_f F}{\partial t} + \frac{\partial u A_x F}{\partial x} + \frac{\partial v A_y F}{\partial y} = 0 \tag{1}$$

In these equations, u represents the velocity of the fluid in the x direction, A_x represents the area occupied by the fluid in the calculation cell in the x direction, V_f represents the volume occupied by the fluid in the calculation cell, v represents the velocity of the fluid in the y direction, A_y represents the area occupied by the fluid in the calculation cell in the y direction, and F represents the fluid depth.

In the 2D analysis, the mesh size was determined as 10 m × 10 m in the areas where settlements with the risk of casualties are located, and 20 m × 20 m in the other regions. A total of 606,248 rectangular mesh cells were created in 10 different mesh blocks.

2.4.2. Three-Dimensional Reynolds-Averaged Navier-Stokes Equations

Flow3D can solve the three-dimensional motion equations (Reynolds-averaged Navier-Stokes equations) and continuity equations simultaneously using the finite volume method.

$$\frac{\partial}{\partial x_i}(u_i A_i) = 0 \tag{2}$$

$$\frac{\partial u_i}{\partial t} + \frac{1}{V_f}\left(u_j A_j \frac{\partial u_i}{\partial x_j}\right) = -\frac{1}{\rho}\frac{\partial p}{\partial x_i} + G_i + f_i \tag{3}$$

In these equations, u_i represents the velocity of the fluid in the i direction, A_i represents the area occupied by the fluid in the calculation cell in the i direction, V_f represents the volume occupied by the fluid in the calculation cell, p represents the pressure, G_i represents the gravitational acceleration in the i direction, and f_i represents the turbulence stresses in the i direction [49].

In the 3D analysis, the mesh size was determined as 10 m × 10 m × 1 m in the areas where settlements with the risk of casualties are located, and 20 m × 20 m × 1 m in the other regions. A total of 14,823,728 rectangular mesh cells were created in 10 different mesh blocks. The general characteristics of the 2D and 3D numerical model are summarized in Table 2. All scenarios were modelled with a water level of 1000 m (dam crest elevation) along the dam axis at the moment of break. The visualization of the numerical model with the reservoir is shown in Figure 5.

Table 2. Characteristics of the numerical model in Flow3D.

General	Finish time	1. Scenario 7200 s 2. Scenario 4700 s 3. Scenario 6500 s
	Simulation units	SI
	Flow Mode	Incompressible
Physics	Gravity z component	$-9.81\ \mathrm{m/sn^2}$
	Turbulence model	RNG k-ε
Fluids	Material	Water at 20 °C
	Density	$1000\ \mathrm{kg/m^3}$
Meshing and Geometry	Size of cells	10 m × 10 m; 20 m × 20 m
	Total number of real cells *	606,248; 14,823,728
	Mesh type	Cartesian
	Boundary conditions **	X_{min} = Wall, Symmetry X_{max} = Symmetry, Outflow Y_{min} * = Wall Y_{max} * = Wall Z_{min} = Wall Z_{max} = Outflow

Note(s): * The total number of meshes in 2D analysis is 606,248. Because 1 m mesh size is used in the z direction in 3D analysis, the total number of meshes is 14,823,728. ** In the first mesh block, the boundary condition X_{min} = Wall was applied, while in the other mesh blocks, the boundary condition X_{min} = Symmetry was applied to ensure continuity. In the last mesh block, the boundary condition X_{max} = Outflow was applied, while in the other mesh blocks, the boundary condition X_{max} = Symmetry was applied.

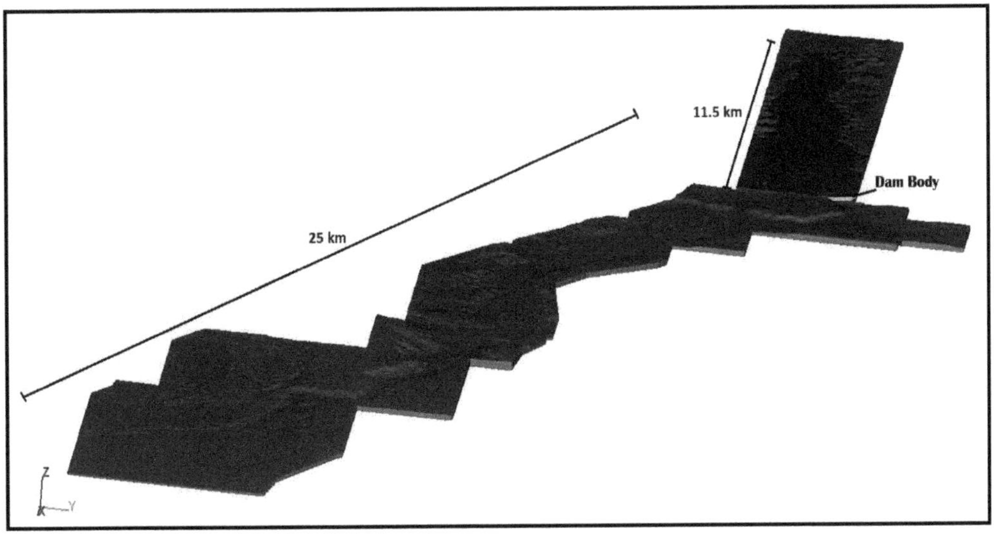

Figure 5. 3D solid model of the Tuzluca Dam and reservoir in Flow3D.

The failure of an actual embankment dam occurs gradually due to the time-dependent growth of a breach [11,19]. In the analysis conducted in this study, the breach formation was modelled by introducing moving parts into the system. The Flow3D software allow the definition of moving body elements along the dam axis to create the dam body. These parts were assigned various speeds in the vertical and horizontal directions to achieve the desired final breach width and opening time for each scenario. When the simulations were performed, visualizations of the dam body and water flow were obtained for both trapezoidal and rectangular breach geometries, as shown in Figure 6.

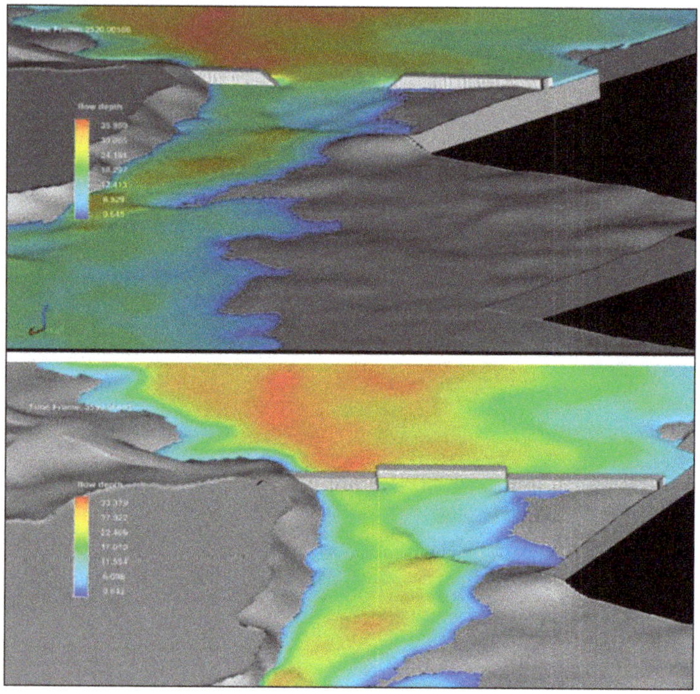

Figure 6. Trapezoidal and rectangular breach formations in simulations in Flow3D.

2.5. Model Validation

Laboratory experiments are often used to validate numerical models of dam breaks [20]. In this study, the dam-break experiment conducted by Soares-Frazao [24] was numerically solved in the Flow3D software to validate the numerical model, and the results were compared. The experiment consisted of a rectangular channel with closed ends, measuring 0.5 m in width and 5.6 m in length. By placing a vertical gate at a distance of 2.39 m from the starting point, a reservoir area with a height of 0.111 m and a length of 2.39 m was created. In addition, there is a triangular bottom sill, with a crest height of 0.065 m and a side slope of 0.14, located 1.61 m downstream of the dam (gate). There is water at a depth of 0.02 m in the section after the sill (Figure 7). A dam failure experiment was performed by removing the gate, and the water depths were measured using a three level-gauge and a digital camera. The locations of the level gauge are given in Table 3.

Table 3. Position of level gauges of Soares-Frazao (2007).

Level Gauge	x (m)	y (m)
G1	5.575	0.25
G2	4.925	0.25
G3	3.935	0.25

Note(s): The dam gate is located at $x = 0$.

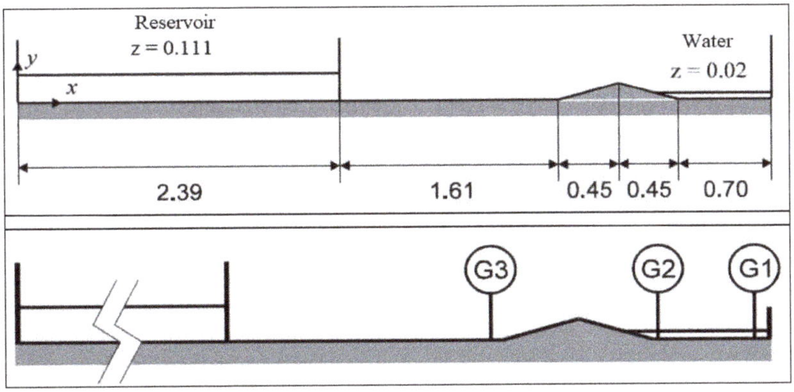

Figure 7. Experimental model of Soares-Frazao (2007) and level gauges.

The experimental model was set up in the Flow3D software using the same measurements (Figure 8.) Probe points were assigned to the positions given in Table 3 to compare the numerical model results with the experimental data.

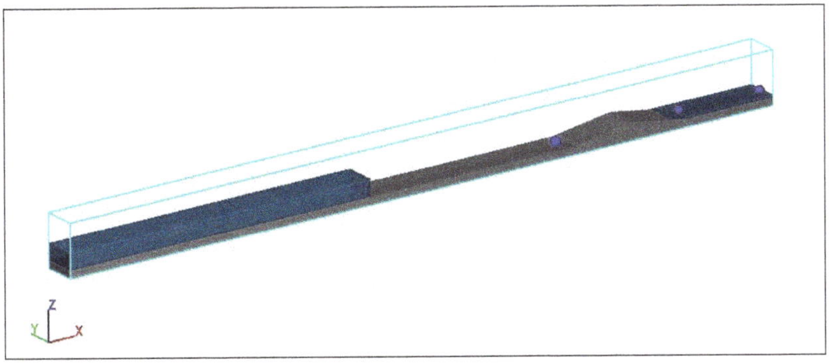

Figure 8. Soares-Frazao (2007) experimental model created in Flow3D software.

In the numerical simulation, the grid cells were selected with a resolution of 0.02 m in the x and y directions for the 2D analysis. It was selected to be 0.02 m $\times$ 0.02 m $\times$ 0.005 m for the 3D analysis. A total of 7000 and 254,800 mesh cells were used in the 2D and 3D analyses, respectively. The RNG k-ε turbulence model was selected for the simulations. The 45-s experimental data, which measured the changes in the water level over time at the G1, G2, and G3 points, were compared with the measurement results at the G1, G2, and G3 probe points in the numerical simulations.

3. Results and Discussion

Three different simulations were carried out for three scenarios determined over two different types of failure (overtopping, piping), different breach sizes, and different failure times (Figure 3). The analyses were conducted using a computer equipped with an Intel i7, 6-core processor, and 16 GB of RAM. While the average solution time for each scenario was 18 h in the 2D analysis, this time was 15 days in the 3D analysis.

The orthogonal form of the study area for the 2D analyses in the Flow3D software provides more precise results. For this reason, the topography is represented by orthogonal mesh blocks in the 2D analysis.

The changes in the flood propagation and flow depths over time for each scenario in the 2D analyses are shown in the Figures 9–11 below.

When examining the flood propagation simulations for each scenario, it can be observed that in Scenario 2, where the dam-break time is shorter, as expected, the flood propagation velocity is higher. In Scenario 3, where the failure occurs due to piping and has a smaller breach geometry, the flood propagation velocity is the lowest.

To further analyze the arrival times of a flood and the effects of flood propagation on the at-risk settlements, probe points were assigned to seven settlements. According to the simulation results, it was observed that only the Halimcan and Surmeli villages were affected by the flood wave. As the other villages were located in higher areas, only agricultural lands were affected. The flood depths in the village settlements for each scenario are shown in Figure 12.

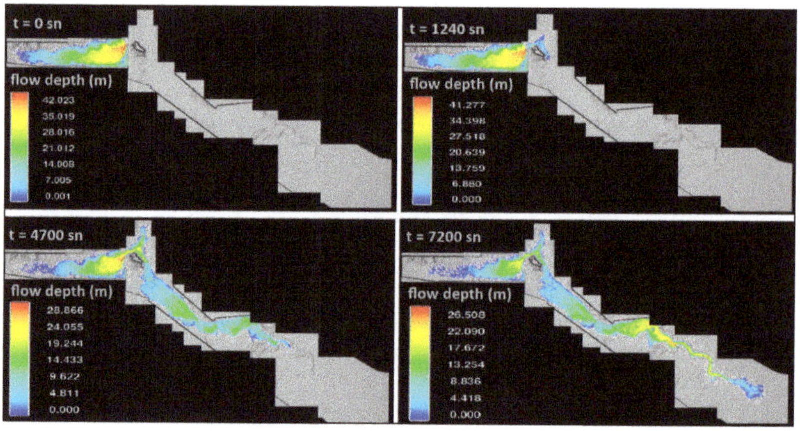

Figure 9. Flood depths and propagation over time for Scenario 1 in 2D analyses in Flow3D.

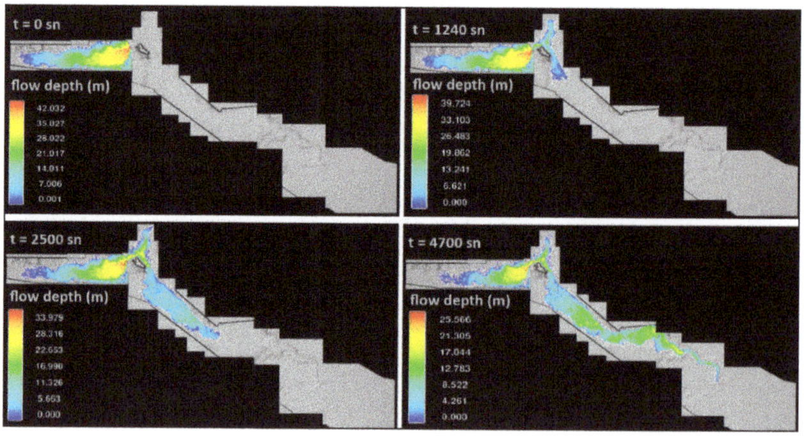

Figure 10. Flood depths and propagation over time for Scenario 2 in 2D analyses in Flow3D.

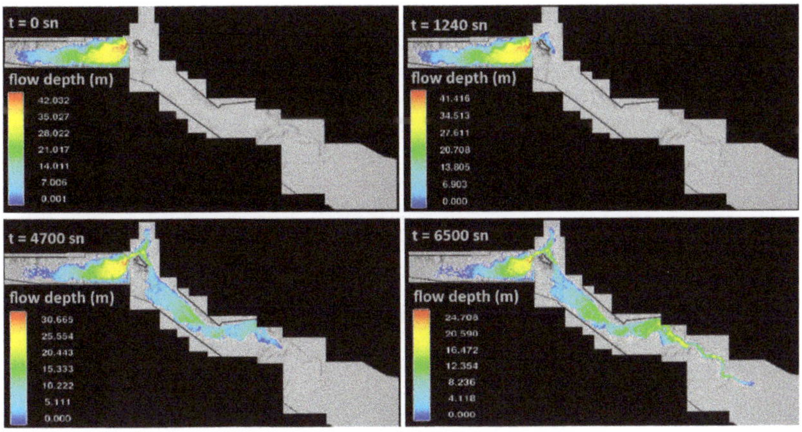

Figure 11. Flood depths and propagation over time for Scenario 3 in 2D analyses in Flow3D.

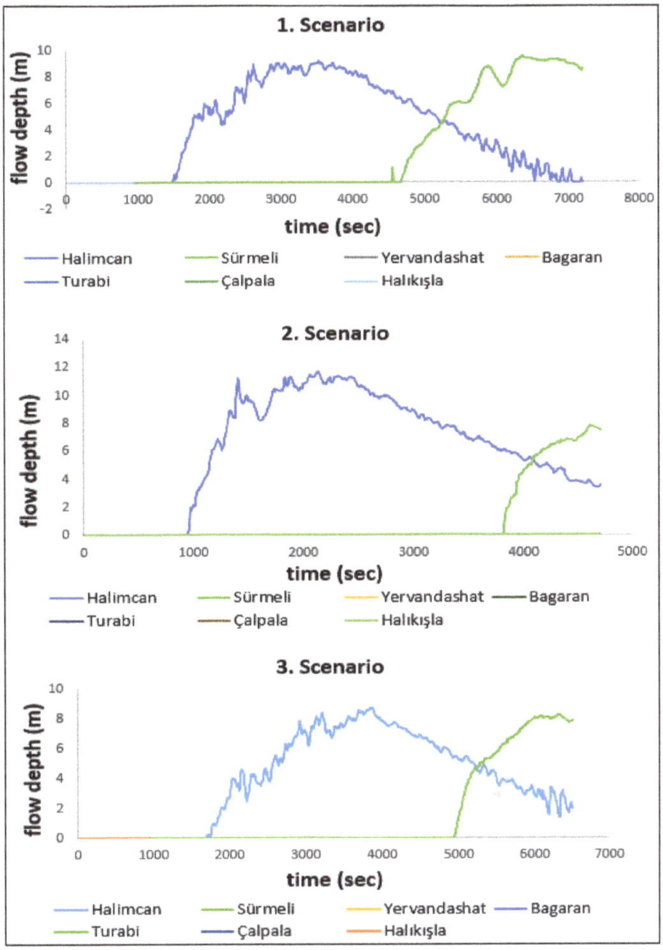

Figure 12. Flow depths in settlements for Scenarios 1, 2, and 3 in 2D analyses in Flow3D.

When the 2D dam failure simulation results were examined, it was seen that Halimcan village, located approximately 1.6 km from the dam axis, experienced maximum water depths of 9.12 m, 11.26 m, and 8.68 m for Scenarios 1, 2, and 3, respectively. In Surmeli village, located approximately 15 km from the dam axis, maximum water depths of 9.61 m, 8.15 m, and 8.20 m were observed for Scenarios 1, 2, and 3, respectively. Additionally, the arrival times of floodwaters at Halimcan village were observed to be 24 min, 16 min, and 28 min for Scenarios 1, 2, and 3, respectively, while for Surmeli village, the corresponding times were 75 min, 62 min, and 82 min (Figure 12).

The changes in the flood propagation and flow depths over time for each scenario in the 3D analyses are shown in the Figures 13–15, below.

Similar to the 2D simulation results, only the Halimcan and Surmeli villages were affected by the flood waves. The flood depths in the village settlements for each scenario in the 3D analysis are shown in Figure 16.

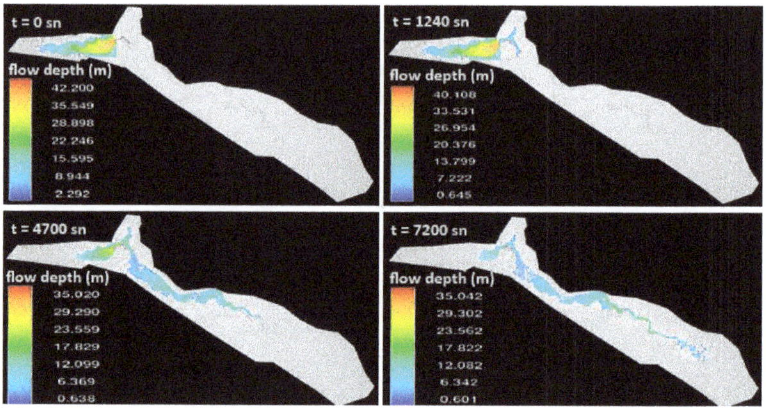

Figure 13. Flood depths and propagation over time for Scenario 1 in 3D analyses in Flow3D.

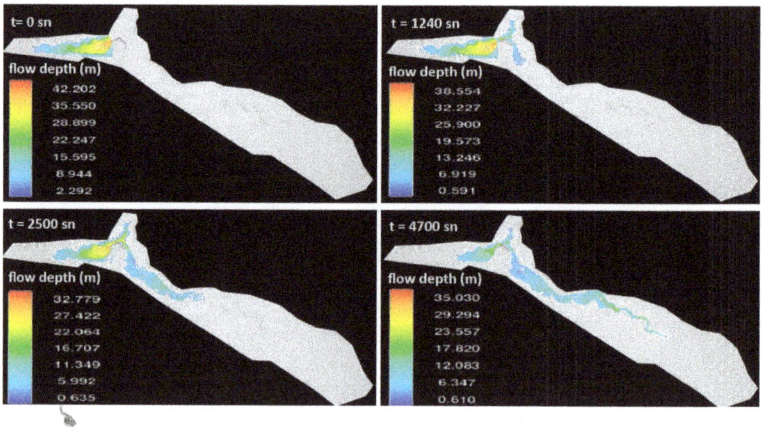

Figure 14. Flood depths and propagation over time for Scenario 2 in 3D analyses in Flow3D.

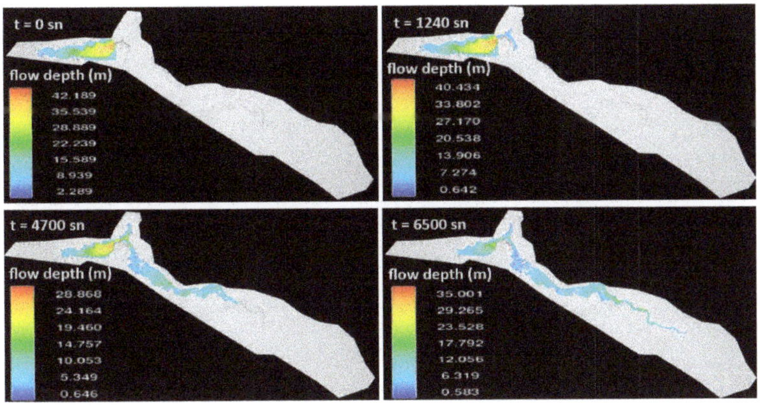

Figure 15. Flood depths and propagation over time for Scenario 3 in 3D analyses in Flow3D.

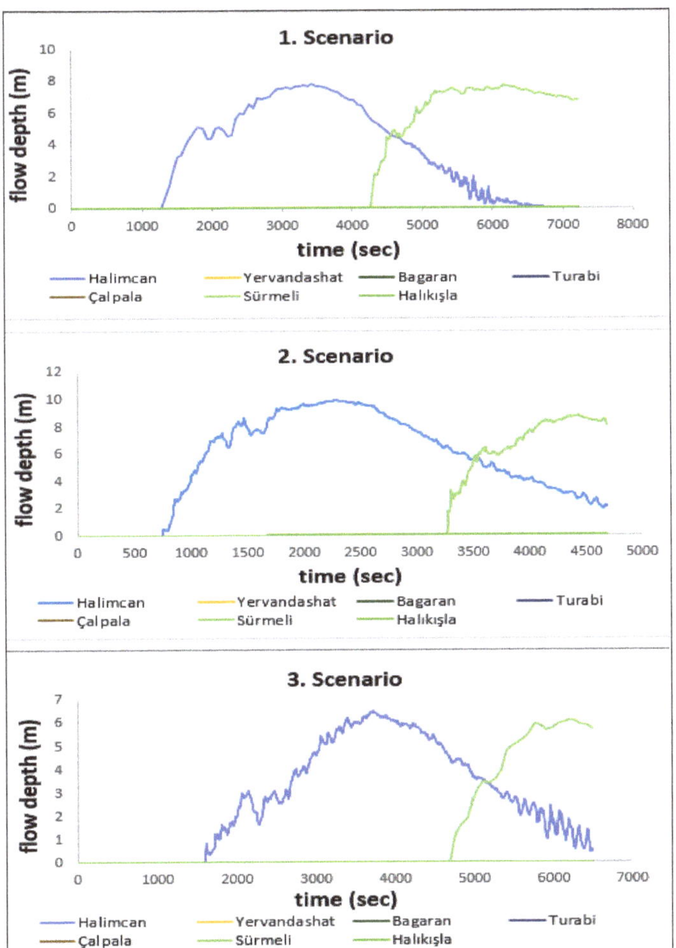

Figure 16. Flow depths in settlements for Scenarios 1, 2, and 3 in 3D analyses in Flow3D.

When the 3D dam failure simulation results were examined, it was seen that Halimcan village experienced maximum water depths of 7.8 m, 9.9 m, and 6.45 m for Scenarios 1, 2, and 3, respectively. In Surmeli village, maximum water depths of 7.6 m, 8.9 m, and 6.1 m were observed for Scenarios 1, 2, and 3, respectively. Additionally, the arrival times of floodwaters at Halimcan village were observed to be 22 min, 13 min, and 26 min for Scenarios 1, 2, and 3, respectively, while for Surmeli village, the corresponding times were 71 min, 54 min, and 79 min.

In order to obtain the flood hydrographs that would occur at the dam axis, a Flux Surface was placed at the dam axis in Flow3D. The flood hydrographs obtained in the 2D and 3D analyses for all scenarios are shown in Figure 17.

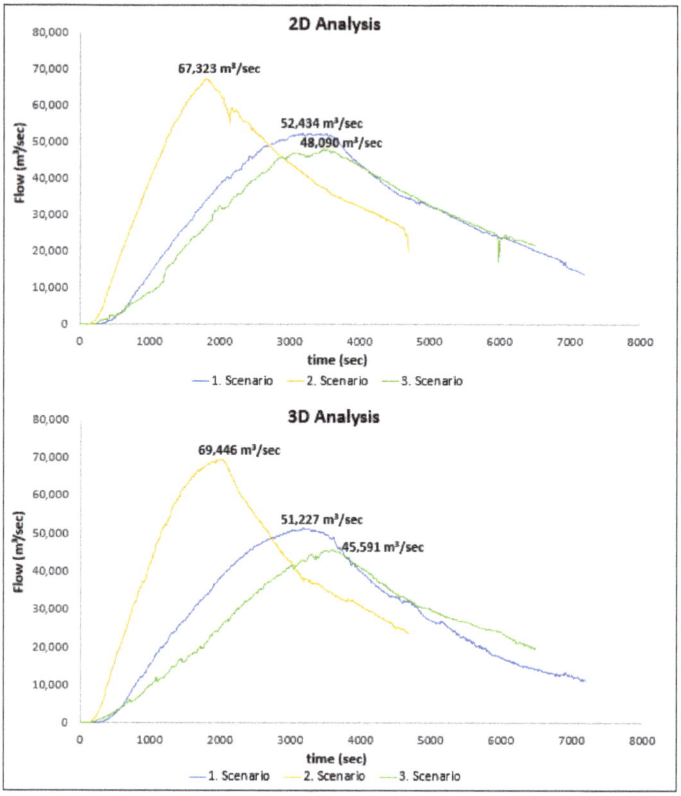

Figure 17. The flood hydrographs on the dam axis for Scenarios 1, 2, 3 in 2D and 3D Analyses.

In the conducted 2D and 3D simulations, it was observed that the highest peak discharge occurred in Scenario 2, which had a shorter failure time, while the lowest peak discharge occurred in Scenario 3, which had the smallest breach cross-sectional area.

The peak discharges for Scenarios 1, 2, and 3 in the 2D simulation were 52,434 m^3/s, 67,323 m^3/s, and 48,090 m^3/s, respectively. The times taken to reach the peak discharge were observed as 3175 s, 1805 s, and 3529 s, respectively. The peak discharges for Scenarios 1, 2, and 3 in the 3D simulation were 51,227 m^3/s, 69,446 m^3/s, and 45,591 m^3/s, respectively. The times taken to reach the peak discharge were observed as 3262 s, 2025 s, and 3578 s, respectively (Figure 17).

For the results of the 2D analysis, when the effect of the failure time on the propagation of the flood was examined in Scenarios 1 and 2 with the same breach geometry, in Scenario 2 (30 min), where the failure time is shorter, the peak flow value was 28% higher. This situation is also compatible with the study conducted by Bello et al. [32], in which the failure time was stated as the most important parameter affecting flood propagation in dam failures. When Scenarios 1 and 3 with the same failure time and different breach geometries are compared, it is seen that the peak flow value is 9% lower in the 3rd scenario, where the breach cross sectional area is approximately 30% smaller. It was observed that the results were compatible with similar studies in the literature [10,41,56].

When the peak flows were compared according to the 3D analysis results, it was observed that the peak flow value in the 2nd scenario was 35% larger than the 1st scenario. When Scenarios 1 and 3 are compared, it is seen that the peak flow value is 12% lower in the 3rd scenario. It was observed that the results were compatible with similar studies in the literature [10,41,56].

In the validation study, the dam failure experiment performed by Soares-Frazao [24] was solved numerically through 2D and 3D analyses using the Flow3D software, and the results were compared. The simulations performed on a computer with an Intel i7 processor, 6 cores, and 16 GB of RAM took 3 min for the 2D analysis and 30 min for the 3D analysis. The comparison of the water level changes at the G1, G2, and G3 points with the experiment results is shown in Figure 18.

The numerical model demonstrated satisfactory compatibility with the experimental results obtained using both the 2D and 3D methods. The 2D model, which solves the shallow water equations, showed successful results at the G3 point, where there are no obstacles, but it showed a relatively poorer performance at the G2 and G1 points, located downstream of the sill. It has been observed that 3D analysis, which solves using Navier-Stokes equations, is more successful in general, and especially in cases of sudden topographic changes, hydraulic jumps, and the timing of water movement. This observation is consistent with the studies that involve similar comparisons in the literature [33,39,57–60].

To determine the amount of deviation between the experimentally determined water depths and those obtained from the numerical simulations, the Root Mean Square Error (*RMSE*) has been calculated for three probe points. The results are presented in Table 4.

$$RMSE = \sqrt{\frac{1}{n}\sum_{i=1}^{n} (x - x_i)^2} \tag{4}$$

Table 4. RMSE Values.

Probe	*RMSE* (m)	
	3D	**2D**
G3	0.0047	0.0048
G2	0.0050	0.0051
G1	0.0048	0.0050

To summarize, Figures 9–11 and Figures 13–15, respectively, depict the flood propagation after the dam failure for the 2D and 3D models. Upon examination of the figures, it has been observed that there is no significant difference between the results of the two different solutions. When the risk assessment of the settlements in the downstream area is considered, as seen in Figures 12 and 16, only the Halimcan and Surmeli settlements are affected by the floodwaters in both models. Although there is a slight difference in the water depth of a few meters between the 2D and 3D model results in the settlements, there is a general similarity in the study area. Figure 17 shows the hydrographs obtained at the dam axis for both methods. It can be observed that the peak discharge and time to reach the peak discharge of the hydrographs obtained using the 2D and 3D models are very close to each other, with a difference of approximately 3–5%.

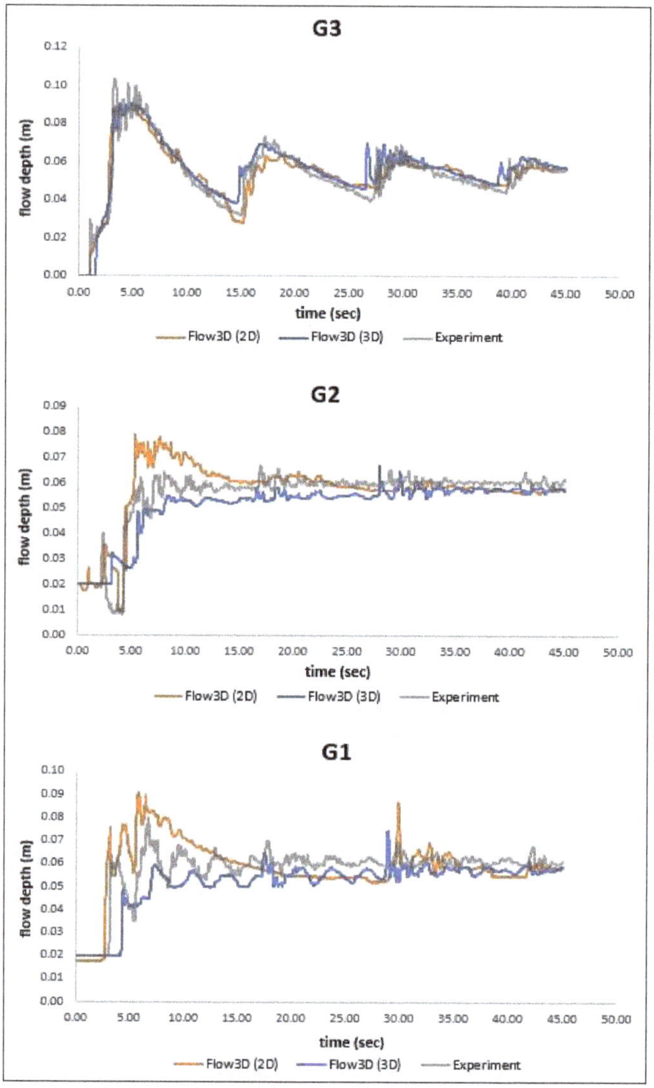

Figure 18. Comparison of experimental and numerical models.

Although 3D models are more successful than 2D models when compared to experimental studies, they have a significant disadvantage at the time point, so much so that in the analyses conducted within the scope of this study, the solution time of the 3D model was about 15 days (357 h), while the solution of the 2D model took 18 h. In this regard, it is thought that 2D modelling may be more useful for large-scale work areas. This situation was stated in studies comparing 2D and 3D numerical solutions of the dam failure experimental setup [36,39].

Within the scope of this study, the mortality rates in the different dam breach scenarios of the Tuzluca Dam were also investigated. In the literature, several methods have been developed to estimate the potential casualties resulting from dam failures (Dekay and McClelland [61], Graham [62], Jonkman [63]). These methods are empirical formulas developed through regression analyses based on data from past dam failures.

In this study, a modified Graham method, which is based on flood simulations, contains more parameters compared to the other methods used. This method is widely used in China, where many dam failures have occurred in the past [64,65].

According to this method, the number of casualties is calculated as follows:

$$L_{OL} = Par.\ f.\ a \tag{5}$$

In these equations, L_{OL} represents loss of life, *Par* represents people at risk, *f* represents the fatality rate, and *a* represents the flood severity coefficient.

$$a = m_1 + bm_2 \tag{6}$$

$$m_1 = \sum_{i=1}^{4} s_i\theta_i \tag{7}$$

$$m_2 = \sum_{i=1}^{9} n_i t_i \tag{8}$$

In these equations, the b and f coefficients are suggested to be 0.25 and 0.75, respectively [62,65–67]. S_i and n_i, used in calculating the m_1 and m_2 values, indicate the value of the factors affecting the loss of life; the expressions θ_i and t_i represent the degree of influence of the factors in question.

The s_1, s_2, s_3, and s_4 values used in calculating the m_1 value are values depending on the population at risk, flood severity, warning time, and local people's awareness about dam failures, respectively. The information regarding these values is given in Table 5. The values of θ_1, θ_2, θ_3, and θ_4 show the degree of effect of the values of s_1, s_2, s_3, and s_4, respectively. The values in question were determined as $\theta_1 = \theta_2 = 0.2$ and $\theta_3 = \theta_4 = 0.3$ according to the Analytical Hierarchy Process (AHP), which is one of the multi-criteria decision-making methods widely used in the literature [65].

Table 5. Suggested ranges of s_i values [65].

s_i	**Population at Risk (Person)**	**Flood Severity**	**Warning Time (h)**	**Local People's Awareness about Dam Failures**
0.80–1.00	$>10^5$	Extremely high	$W_t < 0.25$	Extremely unclear
0.60–0.80	10^4–10^5	High	$0.25 < W_t < 0.50$	Unclear
0.40–0.60	10^3–10^4	Middle	$0.50 < W_t < 0.75$	Generally clear
0.20–0.40	10^2–10^3	Low	$0.75 < W_t < 1.00$	Clear
0.01–0.20	1–10^2	Extremely low	$W_t > 1.00$	Extremely clear

The n_1, n_2, n_3, n_4, n_5, n_6, n_7, n_8, and n_9 values used in calculating the m_2 value represent, respectively, the young population rate, weather conditions, the time of dam failure, distance to the dam, evacuation and rescue capability, dam height, storage capacity, downstream slope, and building abrasion resistance. The information regarding these values is given in Table 6. Similarly, t_1, t_2, t_3, t_4, t_5, t_6, t_7, t_8, and t_9 indicate the degree of influence of the values n_1, n_2, n_3, n_4, n_5, n_6, n_7, n_8, and n_9, respectively. According to the Analytical Hierarchy Process (AHP) method, these values were determined as $t_1 = t_3 = t_4 = 0.2$, $t_2 = t_5 = 0.1$, and $t_6 = t_7 = t_8 = t_9 = 0.05$ [65].

Table 6. Suggested ranges of n_i values [65].

N_i	0.80–1.00	0.60–0.80	0.40–0.60	0.20–0.40	0.01–0.20
Young population rate (%)	0–20	20–40	40–60	60–80	80–100
Weather conditions	Heavy storm	Rainstorm	Moderate rain	Sprinkle	Sunny days
Time of dam failure	Holiday in the morning	Working day in the morning	Holiday at night	Working days at night	Daytime
Distance to the dam (km)	0–5	5–10	10–20	20–50	>50
Evacuation and rescue capability	Very unsuccessful	Unsuccessful	General	Successful	Very successful
Dam height (m)	>70	30–70	15–30	10–15	<10
Reservoir capacity	Very Large	Large	Medium	Small	Very Small
Downstream slope	Valley	Mountain	Hill	Plain	Beach
Building abrasion resistance	Extremely weak	Weak	Middle	Strong	Extremely strong

The s_i and n_i values were determined by taking into account the 3D analysis results on the basis of the location and scenario. The results are presented in Tables 7 and 8.

Table 7. s_i values selected according to settlement and simulation scenarios.

s_i	Halimcan			Sürmeli		
	1. Scenario	**2. Scenario**	**3. Scenario**	**1. Scenario**	**2. Scenario**	**3. Scenario**
s_1	0.10	0.10	0.10	0.30	0.30	0.30
s_2	0.60	0.80	0.50	0.60	0.80	0.50
s_3	0.70	0.90	0.65	0.10	0.20	0.05
s_4	0.70	0.70	0.70	0.70	0.70	0.70

Table 8. n_i values selected according to settlement and simulation scenarios.

n_i	Halimcan	Sürmeli
	1., 2. and 3. Scenarios	**1., 2. and 3. Scenarios**
n_1	0.50	0.50
n_2	0.40	0.40
n_3	0.50	0.50
n_4	0.90	0.50
n_5	0.40	0.40
n_6	0.65	0.65
n_7	0.60	0.60
n_8	0.80	0.90
n_9	0.80	0.80

Note(s): n_i values are the same on a scenario-by-scenario basis as they include social and environmental factors.

Using the data in Tables 7 and 8 and Equations (5)–(8), the possible casualties were calculated according to the settlements and scenarios in the event of the failure of the Tuzluca Dam.

The estimated casualties for the settlements of Halimcan, with a population of 41, and Surmeli, with a population of 434, are presented in Table 9.

Table 9. Projected loss of life in the event of Tuzluca Dam failure.

	Halimcan L_{OL} (Person)	**Sürmeli L_{OL} (Person)**	**Total L_{OL} (Person)**
1. Scenario	22	179	201
2. Scenario	25	202	227
3. Scenario	21	170	191

When Table 9 is examined, it is seen that there will be serious losses of life in the case of the possible collapse of the Tuzluca Dam. In Halimcan village, which is close to the dam axis (1.6 km), it is expected that 54%, 61%, and 51% of the population will lose their lives in the 1st, 2nd, and 3rd scenarios, respectively. In Surmeli village, which is approximately 15 km away from the dam axis, it is expected that 41%, 47%, and 39% of the population will lose their lives in the 1st, 2nd, and 3rd scenarios, respectively. The total losses of life was 13% higher in Scenario 2 (30 min), where the failure time was shorter, than in Scenario 1 with the same breach geometry. When Scenarios 1 and 3, with the same failure time and different breach geometries, are compared, it is seen that total loss of life is 5% lower in the 3rd scenario, where the breach cross sectional area is approximately 30% smaller. This shows that the dam failure time and breach geometry have an impact on the loss of life that may occur.

4. Conclusions

In this study, 2D and 3D dam-break simulations of the Tuzluca Dam, planned for the Igdir province in Turkey, were conducted using the Flow3D software. The same computer was used in all simulations. The propagation of flood waves was investigated under three different scenarios based on the breach geometry, failure type, and failure times. To validate the numerical model, the dam-break experiment conducted by Soares-Frazao [24] was numerically simulated using the Flow3D software. While both numerical models (2D and 3D) exhibited sufficient compatibility with the experimental results, the 3D analysis was more successful in this regard. In the dam failure simulations of the case study, although the 2D and 3D models provided similar results, the 3D analysis required approximately 20 times more computation time compared to the 2D analysis. Consequently, it has been concluded that for extensive studies like dam failures, 2D analyses may be more practical. For smaller-scale scenarios or situations necessitating higher precision, 3D modelling is deemed suitable. Additionally, if high-performance computers with 32 or 64 cores are accessible, the computation time for dam failure simulations employing 3D analysis can be mitigated.

In addition, the possible loss of life was estimated within the scope of the study, and it was seen that serious loss of life could occur in the Halimcan and Sürmeli villages. The analyses revealed that the breach geometry, especially the dam failure time, has a significant impact on the flood propagation and potential loss of life. The information obtained regarding the flood hydrographs, flood depths in the settlements, and the time it takes for the flood to reach the settlements offers valuable insights into the failure scenario of the Tuzluca Dam. It is believed that the obtained information will guide the preparation of disaster and emergency action plans for the region. For example, the settlement of Halimcan, located approximately 1.6 km downstream of the dam axis and experiencing significant flood depths in all scenarios, has not been included in the resettlement plan prepared by the government. Due to the low population in Halimcan (41), taking the action to include it in a resettlement plan at a reasonable cost could be a measure to reduce the risk in the region. Another example would be the village of Surmeli, with a population of 434, located approximately 15 km downstream of the dam axis, where serious flood depths occur in all scenarios. Various measures are proposed to mitigate the potential damage to settlements in the event of a dam failure. These measures include increasing the existing vegetation in the study area, installing audible and visual warning systems to provide rapid and effective information to the public in the event of a dam failure, educating the local population about the risk of dam failure, and appropriate response procedures. It is also recommended to construct cost-effective small-scale water retention facilities in certain areas and apply soil conservation techniques to increase water absorption.

In the literature, dam-break simulations have been conducted for either existing or collapsed dams. Considering past dam breaks and their adverse effects, dam-break analyses should be conducted for planning-stage dams, similar to this study, to identify the necessary risk conditions before dam construction and integrate them into resettlement and emergency action plans. Additionally, similar studies can be carried out using different

CFD software (Fluent, Mike3D, Openfoam, etc.) as an alternative to Flow3D, and the results can be compared.

Author Contributions: Conceptualization, C.A. and S.S.N.; software, C.A. and A.U.; analysis, C.A.; writing—original draft preparation, review, and editing, C.A., S.S.N. and A.U. All authors have read and agreed to the published version of the manuscript.

Funding: This study was supported by the Scientific Research Projects Coordination Unit of Gumushane University (Grant number: 21.F5110.02.01).

Data Availability Statement: Data sharing is not applicable to this article.

Conflicts of Interest: The authors declare no conflict of interest.

References

1. Viseu, T.; de Almeida, A.B. Dam-break risk management and hazard mitigation. *State Art Sci. Eng.* **2009**, *36*, 211–239.
2. Xiong, Y. Dam break analysis using HEC-RAS. *J. Water Resour. Prot.* **2011**, *3*, 370–379. [CrossRef]
3. Le, T.T.H.; Nguyen, V.C. Numerical study of partial dam–break flow with arbitrary dam gate location using VOF method. *Appl. Sci.* **2022**, *12*, 3884. [CrossRef]
4. Adamo, N.; Al-Ansari, N.; Sissakian, V.; Laue, J. Dam safety and overtopping. *J. Earth Sci. Geotech. Eng.* **2020**, *10*, 41–78.
5. Lumbroso, D.; Davison, M.; Body, R.; Petkovšek, G. Modelling the Brumadinho tailings dam failure, the subsequent loss of life and how it could have been reduced. *Nat. Hazards Earth Syst. Sci.* **2021**, *21*, 21–37. [CrossRef]
6. Froehlich, D.C. Embankment Dam breach parameters and their uncertainties. *J. Hydraul. Eng.* **2008**, *134*, 1708–1721. [CrossRef]
7. Najar, M.; Gül, A. Investigating the influence of dam-breach parameters on dam-break connected flood hydrograph. *Tech. J.* **2022**, *33*, 12501–12524. [CrossRef]
8. International Commission onLarge Dams. *Dam Failures—Statistical Analysis*; Bulletin No. 99; International Commission on Large Dams: Paris, France, 1995.
9. MacDonald, T.C.; Langridge-Monopolis, J. Breaching characteristics of dam failures. *J. Hydraul. Eng.* **1984**, *110*, 567–586. [CrossRef]
10. Karakaya, K. Numerical Simulation of the Kirazliköprü Dam Failure on the Gökırmak River. Master's Thesis, Middle East Technical University, Ankara, Turkey, 2005.
11. Wu, W. *Introduction to DLBreach—A Simplified Physically-Based Dam/Levee Breach Model*; Technical Report; Clarkson University: Potsdam, NY, USA, 2016.
12. Ackerman, C.T.; Brunner, G.W. Dam failure analysis using HEC-RAS and HEC-GEORAS. In Proceedings of the 2008 World Environmental and Water Resources Congress, Ahupua'a, Honolulu, HI, USA, 12–16 May 2008.
13. Li, W.; Zhu, J.; Fu, L.; Zhu, Q.; Xie, Y.; Hu, Y. An augmented representation method of debris flow scenes to improve public perception. *Int. J. Geogr. Inf. Sci.* **2021**, *35*, 1521–1544. [CrossRef]
14. Liu, Z.; Xu, J.; Liu, M.; Yin, Z.; Liu, X.; Yin, L.; Zheng, W. Remote sensing and geostatistics in urban water-resource monitoring: A review. *Mar. Freshw. Res.* **2023**, *74*, 747–765. [CrossRef]
15. Oborie, E.; Rowland, D.R. Flood influence using GIS and remote sensing based morphometric parameters: A case study in Niger delta region. *J. Asian Sci. Res.* **2023**, *13*, 1–15. [CrossRef]
16. Von Thun, J.L.; Gillette, D.R. *Guidance on Breach Parameters*; Internal Memorandum Rep.; U.S. Department of the Interior, Bureau of Reclamation: Denver, CO, USA, 1990.
17. Froehlich, D.C. Embankment dam breach parameters revisited. In Proceedings of the First International Conference on Water Resources Engineering, San Antonio, TX, USA, 14–18 August 1995; American Society of Civil Engineers: New York, NY, USA, 1995; pp. 887–891.
18. Xu, Y.; Zhang, L.M. Breaching parameters for earth and rockfill dams. *J. Geotech. Geoenviron. Eng.* **2009**, *135*, 1957–1970. [CrossRef]
19. Khoshkonesh, A.; Nsom, B.; Bahmanpouri, F.; Dehrashid, F.A.; Adeli, A. Numerical study of the dynamics and structure of a partial dam-break flow using the VOF method. *Water Resour. Manag.* **2021**, *35*, 1513–1528. [CrossRef]
20. Aureli, F.; Maranzoni, A.; Petaccia, G.; Soares-Frazão, S. Review of experimental investigations of dam-break flows over fixed bottom. *Water* **2023**, *15*, 1229. [CrossRef]
21. Bella, S.W.; Elliot, R.C.; Chaudry, M.H. Experimental results of two dimensional dam-break flows. *J. Hydraul. Res.* **1992**, *30*, 225–252. [CrossRef]
22. Lauber, G.; Hager, W.H. Ritter's dambreak wave wevisited. In Proceedings of the 27 IAHR Congress, San Francisco, CA, USA, 10–15 August 1997; pp. 258–262.
23. Ritter, A. Die fortpflanzung de wasserwellen. *Z. Ver. Dtsch. Ingenieure* **1892**, *36*, 947–954. (In German)
24. Soares-Frazão, S. Experiments of dam-break wave over a triangular bottom sill. *J. Hydraul. Res.* **2007**, *45*, 19–26. [CrossRef]
25. Hui, L.; Liu, H.; Guo, L.; Lu, S. Experimental study on the dam-break hydrographs at the gate location. *J. Ocean. Univ. China* **2017**, *16*, 697–702.
26. Stoker, J.J. *Water Waves*; Interscience Publishers Inc.: Hoboken, NJ, USA, 1957.

27. Lin, B.; Gong, Z.; Wang, L. Dam-site hydrographs due to sudden release. *Sci. Sin.* **1980**, *23*, 1570–1582.
28. Quecedo, M.; Pastor, M.; Herreros, M.I.; Merodo, J.A.F.; Zhang, Q. Comparison of two mathematical models for solving the dam break problem using the FEM method. *Comput. Methods Appl. Mech. Engrg.* **2005**, *194*, 3984–4005. [CrossRef]
29. Albano, R.; Mancusi, L.; Adamowski, J.; Cantisani, A.; Sole, A. A GIS tool for mapping dam-break flood hazards in Italy. *ISPRS Int. J. Geo-Inf.* **2019**, *8*, 250. [CrossRef]
30. Ríha, J.; Kotaška, S.; Petrula, L. Dam break modeling in a cascade of small earthen dams: Case study of the Cižina River in the Czech Republic. *Water* **2020**, *12*, 2309. [CrossRef]
31. Psomiadis, E.; Tomanis, L.; Kavvadias, A.; Soulis, K.X.; Charizopoulos, N.; Michas, S. Potential dam breach analysis and flood wave risk assessment using HEC-RAS and remote sensing data: A multicriteria approach. *Water* **2021**, *13*, 364. [CrossRef]
32. Bello, D.; Alcayaga, H.; Caamaño, D.; Pizarro, A. Influence of dam breach parameter statistical definition on resulting rupture maximum discharge. *Water* **2022**, *14*, 1776. [CrossRef]
33. Kocaman, S. Experimental and Theoretical İnvestigation of Dam Break Problem. Ph.D. Thesis, Department of Civil Engineering Institute of Natural and Applied Sciences University of Çukurova, Adana, Türkiye, 2007.
34. Vasquez, J.A. Testing River2D and Flow3D for sudden dam-break flow simulations. In Proceedings of the Conference on Canadian Dam Association, Whistler, BC, Canada, 3–8 October 2009.
35. Larocque, L.A.; Imran, J.; Chaudhry, M.H. 3D numerical simulation of partial breach dam break flow using the LES and k–ε turbulence models. *J. Hydraul. Res.* **2013**, *51*, 145–157. [CrossRef]
36. Robb, D.M.; Vasquez, J.A. Numerical simulation of dam–break flows using depth-averaged hydrodynamic and three-dimensional CFD models. In Proceedings of the 22nd Canadian Hydrotechnical Conference, Montreal, QC, Canada, 29 April–2 May 2015.
37. Tahershamsi, A.; Hooshyaripor, F.; Razib, S. Reservoir's geometry impact of three dimensions on peak discharge of dam-failure flash flood. *Sci. Iran.* **2017**, *25*, 1931–1942. [CrossRef]
38. Franco, A.; Moernaut, J.; Schneider-Muntau, B.; Strasser, M.; Gems, B. The 1958 Lituya Bay tsunami–prevent bathymetry reconstruction and 3D numerical modelling utilising the computational fluid dynamics software Flow-3D. *Nat. Hazards Earth Syst. Sci.* **2020**, *20*, 2255–2279. [CrossRef]
39. Kocaman, S.; Evangelista, S.; Guzel, H.; Dal, K.; Yilmaz, A.; Viccione, G. Experimental and numerical investigation of 3D dam-break wave propagation in an enclosed domain with dry and wet bottom. *Appl. Sci.* **2021**, *11*, 5638. [CrossRef]
40. Republic of Turkey General Directorate of Electricity Works and Studies. Feasibility Report of the Lower Aras Basin Tuzluca Dam and HEPP Project, 2004.
41. Bharath, A.; Shivapur, A.V.; Hiremath, C.G.; Maddamsetty, R. Dam break analysis using HEC-RAS and HEC-GeoRAS: A case study of Hidkal dam, Karnataka state, India. *Environ. Chall.* **2021**, *5*, 100401. [CrossRef]
42. Annis, A.; Nardi, F.; Petroselli, A.; Apollonio, C.; Arcangeletti, E.; Tauro, F.; Belli, C.; Bianconi, R.; Grimaldi, S. UAV-DEMs for small-scale flood hazard mapping. *Water* **2020**, *12*, 1717. [CrossRef]
43. Sena, N.C.; Veloso, G.V.; Fernandes-Filho, E.I.; Francelino, M.R.; Schaefer, C.E.G. Analysis of terrain attributes in different spatial resolutions for digital soil mapping application in southeastern Brazil. *Geoderma Reg.* **2020**, *21*, e00268. [CrossRef]
44. Khojeh, S.; Ataie-Ashtiani, B.; Hosseini, S.M. Effect of DEM resolution in flood modeling: A case study of Gorganrood River, Northeastern Iran. *Nat. Hazards* **2022**, *112*, 2673–2693. [CrossRef]
45. Uysal, G.; Tasci, E. Analysis of downstream flood risk in the failure of Batman Dam with two-dimensional hydraulic modeling and satellite data. *J. Nat. Hazards Environ.* **2023**, *9*, 39–57.
46. USBR. *Downstream Hazard Classification Guidelines*; ACER Tech. Memorandum Rep. No. 11; U.S. Department of the Interior, Bureau of Reclamation: Denver, CO, USA, 1988.
47. Yin, Y.P.; Huang, B.L.; Chen, X.T.; Liu, G.N.; Wang, S.C. Numerical analysis of wave generated by the Qianjiangping landslide in three Gorges Reservoir. *J. Int. Consort. Landslides* **2015**, *12*, 355–364. [CrossRef]
48. Flow Science Inc. *Flow-3D Version 10.0 User's Manual*; Flow Science Inc.: Santa Fe, New Mexico, 2007.
49. Tannehill, J.C.; Anderson, D.A.; Pletcher, R.H. *Computational Fluid Mechanics and Heat Transfer*, 2nd ed.; Taylor & Francis: New York, NY, USA, 1997; p. 792.
50. Kaheh, M.; Kashefipour, M.A.; Dehghani, A.A. Comparison of k-ε and RNG k-ε turbulent models for estimation of velocity profiles along the hydraulic jump on corrugated beds. In Proceedings of the 6th International Symposium on Environmental Hydraulics, Athens, Greece, 23–25 June 2010.
51. Dehdar-behbahani, S.; Parsaie, A. Numerical modeling of flow pattern in dam spillway's guide wall. Case study: Balaroud dam, Iran. *Alex. Eng. J.* **2016**, *55*, 467–473. [CrossRef]
52. Oliva, J.B.; Valentin, M.G. Comparative Evaluation of Two Different 3D OpenFOAM Modules in a Dam-Break Test. Bachelor's Thesis, Higher Technical School of Road, Canals Engineering, Barcelona, Spain, 2017.
53. Ersoy, H.; Karahan, M.; Gelişli, K.; Akgun, A.; Anılan, T.; Sunnetci, M.O.; Yahsi, B.K. Modelling of the landslide-induced impulse waves in the Artvin Dam reservoir by empirical approach and 3D numerical simulation. *Eng. Geol.* **2019**, *249*, 112–128. [CrossRef]
54. Karahan, M.; Ersoy, H.; Akgun, A. A 3D numerical simulation-based methodology for Assessment of landslide-generated impulse waves: A case study of the Tersun Dam reservoir (NE Turkey). *J. Int. Consort. Landslides* **2020**, *17*, 2777–2794. [CrossRef]
55. Aksoy, A.O.; Dogan, M.; Guven, S.O.; Tanır, G.; Guney, M.S. Experimental and numerical investigation of the flood waves due to partial dam break. *Iran. J. Sci. Technol. Trans. Civ. Eng.* **2022**, *46*, 4689–4704. [CrossRef]

56. Shahrim, M.F.; Ros, F.C. Estimation of breach outflow hydrograph using selected regression breach equations. In *IOP Conference Series: Earth and Environmental Science*; IOP Publishing: Bristol, UK, 2020; p. 476.
57. Hien, L.T.H.; Nguyen, V.C. Investigate Impact Force of Dam-Break Flow against Structures by Both 2D and 3D Numerical Simulations. *Water* **2021**, *13*, 344. [CrossRef]
58. Zhang, T.; Peng, L.; Feng, P. Evaluation of a 3D unstructured-mesh finite element model for dam-break flood. *Comput. Fluids* **2018**, *160*, 64–77. [CrossRef]
59. Dai, S.; He, Y.; Yang, J.; Ma, Y.; Jin, S.; Liang, C. Numerical study of cascading dam- break characteristics using SWEs and RANS. *Water Supply* **2020**, *20*, 348–360. [CrossRef]
60. Ozmen-Cagatay, H.; Kocaman, S. Dam-break flows during initial stage using SWE and RANS approaches. *J. Hydraul. Res.* **2010**, *48*, 603–611. [CrossRef]
61. DeKay, M.L.; McClelland, G.H. Predicting loss of life in cases of dam failure and flash flood. *Risk Anal.* **1993**, *13*, 193–205. [CrossRef]
62. Graham, W.J. A procedure for estimating loss of life caused by dam failure. *Sediment River Hydral.* **1999**, *6*, 1–43.
63. Jonkman, S.N. Loss of Life Estimation in Flood Risk Assessment. Ph.D. Thesis, Delft University, Delft, The Netherlands, 2007.
64. Xiaoling, W.; Xuefei, A.; Ruirui, S.; Weiping, G. The comprehensive risk analysis of dam-break consequences based on Numerical Simulation. *Appl. Mech. Mater.* **2012**, *229*, 1850–1853.
65. Sun, R.; Wang, X.; Zhou, Z.; Ao, X.; Sun, X.; Song, M. Study of the comprehensive risk analysis of dam-break flooding based on the numerical simulation of flood routing. Part I: Model development. *Nat. Hazards* **2014**, *73*, 1547–1568. [CrossRef]
66. Sun, R.; Wang, X.; Zhou, Z.; Ao, X.; Sun, X.; Song, M. Study of the comprehensive risk analysis of dam-break flooding based on the numerical simulation of flood routing. Part II: Model application and results. *Nat. Hazards* **2014**, *72*, 675–700.
67. Zhou, K.F. Study on Analysis Method for Loss of Life Due to Dam Breach. Master's Thesis, Nanjing Hydraulic Research Institute, Nanjing, China, 2006.

MDPI
St. Alban-Anlage 66
4052 Basel
Switzerland
www.mdpi.com

Water Editorial Office
E-mail: water@mdpi.com
www.mdpi.com/journal/water

www.ingramcontent.com/pod-product-compliance
Lightning Source LLC
Chambersburg PA
CBHW061609220926
51850CB00005B/459